Advances in Natural and Technological Hazards Research

Volume 40

For further volumes:
http://www.springer.com/series/6362

DanLing Tang · GuangJun Sui
Gad Lavy · Dmitry Pozdnyakov
Y. Tony Song · Adam D. Switzer
Editors

Typhoon Impact and Crisis Management

Springer

Editors
DanLing Tang
Chinese Academy of Sciences
South China Sea Institute of Oceanology
Guangzhou
People's Republic of China

GuangJun Sui
Guangdong University of Foreign Studies
Guangzhou
People's Republic of China

Gad Lavy
NorthWest Research Associates
Seattle, WA
USA

Dmitry Pozdnyakov
Nansen International Environmental
and Remote Sensing Centre
St. Petersburg
Russia

and

Nansen International Environmental
and Remote Sensing Centre
Bergen
Norway

Y. Tony Song
Jet Propulsion Laboratory NASA
Pasadena, CA
USA

Adam D. Switzer
Earth Observatory of Singapore
Nanyang Technological University
Singapore

ISSN 1878-9897 ISSN 2213-6959 (electronic)
ISBN 978-3-642-40694-2 ISBN 978-3-642-40695-9 (eBook)
DOI 10.1007/978-3-642-40695-9
Springer Heidelberg New York Dordrecht London

Library of Congress Control Number: 2013955554

Printed on acid-free paper

Springer is part of Springer Science+Business Media (www.springer.com)

Preface

Typhoons are tropical cyclones that are accompanied by strong winds, heavy rains, and large waves, sometimes coinciding with high tides;

Impacts of typhoons can be huge on marine ecosystems, oceanic environments, and coastal communities;

And typhoons are also frequent and influential natural hazards, causing enormous economic losses.

Crisis management is vital to mitigate typhoon disasters and save lives, and therefore,

Management and natural science issues are examined concomitantly in this book.

DanLing (lingzis) Tang
GuangJun Sui

Acknowledgments

It took much time and great efforts to successfully complete this book "Typhoon Impact and Crisis Management". It is the result of cooperation and confidence of many people.

We wish to express our sincere gratitude to our editorial board member Drs. Gad Levy, Y. Tony Song, Dmitry Pozdnyakov, and Adam D. Switzer, and all reviewers for their time, passion, and all efforts to improve this book. We would also like to thank Dr. Dounald Thomas for his contribution on the English editorial work for some chapters of this book.

We appreciate Dr. WangKun Chen, Dr. SuFen Wang, Dr. AiMing Yang, and Dr. Gang Pan, and all our research team members, including HaiJun Ye, Yi Su, Li Zhou, for their assistances in our work.

The following research programs and projects have made contributions to the related research: (1) National Natural Sciences Foundation of China (31061160190, 40976091, NSFC–RFBR Project-41211120181); (2) Guangdong Sciences Foundation (2010B031900041, 8351030101000002), China; (3) Innovation Group Program of State Key Laboratory of Tropical Oceanography (LTOZZ1201), China; (4) Dragon 3 of ESA and NRSCC (2012-ID 10705).

And finally, we must thank our son, Mr. Yi Sui and our mother Ms. MuLan Li, for their understanding and support for our research. We love them.

Guangzhou, China, Oct 22, 2012 GuangJun Sui

DanLing Tang

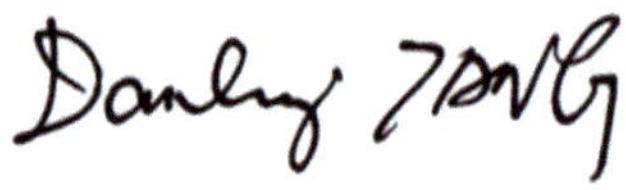

Contents

Part III Typhoon Impacts on Marine Ecosystems

Part IV Typhoon Impacts and Global Changes

Part V Typhoon Impacts on Economics

Part VI Crisis Management for Typhoon Disaster

Contributors

Y. D. Afanasyev Department of Physics and Physical Oceanography, Memorial University, Newfoundland, Canada; Nansen International Environmental and Remote Sensing Centre, St. Petersburg, Russia

Georges Allaert Institute for Sustainable Mobility, Ghent University, Vrijdagmarkt 10/301, Ghent 9000, Belgium, e-mail: georges.allaert@ugent.be

Peter Black SAIC, Inc. and Naval Research Laboratory, Monterey, CA, USA

L. P. Bobylev Nansen Centre, St. Petersburg, Russia; Nansen Centre, Bergen, Norway

Leonid Bobylev Nansen International Environmental and Remote Sensing Centre, St. Petersburg, Russia; Nansen Environmental and Remote Sensing Centre, Bergen, Norway

Lisa T.-C. Chang Department of Environmental Management, Tungnan University, New Taipei City, Taiwan; Risk Frontiers Natural Hazards Research Centre, Macquarie University, Macquarie, Australia

Yang Chaoyu South China Sea Branch, State Oceanic Administration, Guangzhou 510301, People's Republic of China; Départment de Biologie/Pavillon Alexandre-Vachon, Université Laval Québec (Québec), G1V 0A6 Québec, Canada

B. Chapron Department of Oceanography and Ecosystem Dynamics, IFREMER, Brest, France

He Chen Guangdong University of Foreign Studies, 510420 Guangzhou, People's Republic of China

Kuo-Shu Chen Institute of Marine Geology and Chemistry, and Asia–Pacific Ocean Research Center, National Sun Yat-sen University, Kaohsiung 80424, Taiwan

Meng-Hsien Chen Department of Marine Biotechnology and Resources, National Sun Yat-sen University, Kaohsiung, Taiwan

Wang-Kun Chen Department of Environment and Property Management, Jinwen University of Science and Technology, New Taipei 23154, Taiwan, People's Republic of China

Y.-H. Cheng Department of Environmental Informatics, National Taiwan Ocean University, Keelung, Taiwan

Kevin K. W. Cheung Department of Environment and Geography, Macquarie University, Sydney, NSW 2109, Australia, e-mail: kevin.cheung@mq.edu.au

Weiyu Ding Guangzhou Institute of Tropical Marine Meteorology, China Meteorological Administration, Guangzhou, People's Republic of China

Yang DingTian State Key Laboratory of Oceanography in the Tropics, South China Sea Institute of Oceanology, Chinese Academy of Sciences, Guangzhou 510301, People's Republic of China; State Key Laboratory of Lake Science and Environment, Nanjing 210008, People's Republic of China, e-mail: dtyang@scsio.ac.cn

Pavel Golubkin Nansen International Environmental and Remote Sensing Centre, St. Petersburg, Russia; Russian State Hydrometeorological University, St. Petersburg, Russia

Gwo-Ching Gong Taiwan Ocean Research Institute, National Applied Research Laboratories, Kaohsiung 85243, Taiwan; Institute of Marine Environmental Chemistry and Ecology, and Center of Excellence for Marine Bioenvironment and Biotechnology, National Taiwan Ocean University, Keelung 20424, Taiwan

C.-R. Ho Department of Environmental Informatics, National Taiwan Ocean University, Keelung, Taiwan

S.-J. Huang Department of Environmental Informatics, National Taiwan Ocean University, Keelung, Taiwan

Chin-Chang Hung Institute of Marine Geology and Chemistry, and Asia–Pacific Ocean Research Center, National Sun Yat-sen University, Kaohsiung 80424, Taiwan; Taiwan Ocean Research Institute, National Applied Research Laboratories, Kaohsiung 85243, Taiwan, e-mail: cchung@mail.nsysu.edu.tw

Peng Jia Horn Point Laboratory, University of Maryland Center for Environmental Science, P.O. Box 775, Cambridge, MD 21613, USA

Kristina B. Katsaros Northwest Research Associates, Inc., Bellevue, WA, USA, e-mail: katsaros@whidbey.net

N.-J. Kuo Department of Environmental Informatics, National Taiwan Ocean University, Keelung, Taiwan

Ming-An Lee Department of Environmental Biology and Fisheries Science, National Taiwan Ocean University, Keelung, Taiwan

Gad Levy NorthWest Research Associates, Seattle, WA, USA

ChongYin Li Institute of Meteorology, PLA University of Science and Technology, Nanjing, People's Republic of China

Ming Li Horn Point Laboratory, University of Maryland Center for Environmental Science, P.O. Box 775, Cambridge, MD 21613, USA, e-mail: mingli@umces.edu

Yubin Li Department of Environment and Geography, Macquarie University, Sydney, NSW 2109, Australia

Cheng-Hsin Liao Department of Environmental Biology and Fisheries Science, National Taiwan Ocean University, Keelung, Taiwan

Fangni Lin Guangzhou Central Meteorological Observatory, Guangdong Meteorological Administration, Guangzhou, People's Republic of China

Jianghao Lin Guangdong University of Foreign Studies, Guangzhou, China

A. K. Liu NASA Goddard Space Flight Center, Greenbelt, MD, USA

L. M. Mitnik V.I. Il'ichev Pacific Oceanological Institute (POI), Vladivostok, Russia

Leonid Mitnik V.I. Il'ichev Pacific Oceanological Institute, FEB RAS, Vladivostok, Russian Federation

Evgeny Morozov Nansen International Environmental and Remote Sensing Centre, St. Petersburg, Russia

Gang Pan State Key Laboratory of Tropical Oceanography, Research Center for Remote Sensing and Marine Ecology and Environment, South China Sea Institute of Oceanology, Chinese Academy of Sciences, Guangzhou, People's Republic of China

Dmitry Petrenko Nansen International Environmental and Remote Sensing Centre, St. Petersburg, Russia; Russian State Hydrometeorological University, St. Petersburg, Russia

Dmitry Pozdnyakov Nansen International Environmental and Remote Sensing Centre, St. Petersburg, Russia; Nansen Environmental and Remote Sensing Centre, Bergen, Norway

Huiying Pu Institute of Industrial Economy, Jinan University, Guangzhou 510632, People's Republic of China

V. Ramaswamy National Institute of Oceanography, Dona Paula, Goa 403004, India, e-mail: rams@nio.org; rams.goa@gmail.com

Hussain Aziz Saleh Higher Commission for Scientific Research, 30151, Damascus, Syria; Institute for Sustainable Mobility, Ghent University, Vrijdagmarkt 10/301, Ghent 9000, Belgium, e-mail: hussain.saleh@ugent.be

Peter H. Santschi Laboratory for Oceanography and Environmental Research, Texas A&M University at Galveston, 200 Seawolf Parkway, Galveston, TX 77553, USA

Yung-Yen Shih Institute of Marine Geology and Chemistry, and Asia–Pacific Ocean Research Center, National Sun Yat-sen University, Kaohsiung 80424, Taiwan

GuangJun Sui Guangdong Research Institute for International Strategies, Guangdong University of Foreign Studies, Guangzhou, People's Republic of China; College of Economics and Trade, Guangdong University of Foreign Studies, Guangzhou 510420, China

Yi. Sui Nansen International Environmental and Remote Sensing Centre, St. Petersburg, Russia; Department of Oceanography, University of Dalhousie, Halifax, Canada; Department of Physics and Physical Oceanography, Memorial University, St. John's, NL, Canada

Adam D. Switzer Earth Observatory of Singapore, Nanyang Technological University, 639798, Singapore; Division of Earth Sciences, School of Mathematical and Physical Sciences, Nanyang Technological University, 639798, Singapore, e-mail: aswitzer@ntu.edu.sg

DanLing Tang Research Center for Remote Sensing of Marine Ecology and Environment, State Key Laboratory of Tropical Oceanography, South China Sea Institute of Oceanography, Chinese Academy of Sciences, Xingang Xi Road 164, Guangzhou 510300, People's Republic of China; Graduate University of Chinese Academy of Sciences, Beijing 100049, People's Republic of China; Nansen International Environmental and Remote Sensing Centre, St. Petersburg, Russia, e-mail: lingzistdl@126.com

Qilin Wan Guangzhou Institute of Tropical Marine Meteorology, China Meteorological Administration, Guangzhou, People's Republic of China, e-mail: qlwan@grmc.gov.cn

Dong Xiao Wang State Key Laboratory of Tropical Oceanography, South China Sea Institute of Oceanology, Chinese Academy Science, Guangzhou, China, e-mail: dxwang@scsio.ac.cn

SuFen Wang Nansen International Environmental and Remote Sensing Centre, St. Petersburg, Russia; Research Center for Remote Sensing of Marine Ecology and Environment, State Key Laboratory of Tropical Oceanography, South China Sea Institute of Oceanography, Chinese Academy of Sciences, Xingang Xi Road 164, Guangzhou 510300, People's Republic of China

Xiaohong Wang Department of Mathematics and Computer Science, Salisbury University, 1101 Camden Avenue, Salisbury, MD 21801, USA

Xin Wang State Key Laboratory of Tropical Oceanography, South China Sea Institute of Oceanology, Chinese Academy Science, Guangzhou, People's Republic of China

Zhangjin Wei Information College in Guangdong University of Foreign Studies, Guangzhou, People's Republic of China

Chau-Ron Wu Department of Earth Sciences, National Taiwan Normal University, Taipei, Taiwan

Aimin Yang Guangdong University of Foreign Studies, Guangzhou, People's Republic of China

Hai Jun Ye Research Center for Remote Sensing and Marine Ecology and Environment, State Key Laboratory of Tropical Oceanography, South China Sea Institute of Oceanology, Chinese Academy of Sciences, Guangzhou, People's Republic of China; Graduate University of the Chinese Academy of Sciences, Beijing, People's Republic of China; Nansen International Environmental and Remote Sensing Centre, St. Petersburg, Russia

Chang Yi Institute of Ocean Technology and Marine Affairs, National Cheng Kung University, Tainan, Taiwan

Fengling Yu Earth Observatory of Singapore, Nanyang Technological University, 639798, Singapore

Jie Yu Research Center for Remote Sensing of Marine Ecology and Environment, State Key Laboratory of Tropical Oceanography, South China Sea Institute of Oceanography, Chinese Academy of Sciences, Xingang Xi Road 164, Guangzhou 510300, People's Republic of China; Graduate University of Chinese Academy of Sciences, Beijing 100049, People's Republic of China; Scientific Observing and Experimental Station of South China Sea Fishery Resources and Environments, Ministry of Agriculture, South China Sea Fisheries Research Institute, Chinese Academy of Fishery Sciences, Guangzhou 510300, People's Republic of China

Jinnan Yuan Guangzhou institute of tropical marine meteorology, China Meteorological Administration, Guangzhou, People's Republic of China

Elizaveta Zabolotskikh Nansen International Environmental and Remote Sensing Centre, St. Petersburg, Russia; Satellite Oceanography Laboratory, Russian State Hydrometeorological University and Nansen Centre, Malookhtinsky Prospect 98, Saint Petersburg, Russia, e-mail: liza@rshu.ru

Wen Zhou School of Energy and Environment, Guy Carpenter Asia-Pacific Climate Impact Centre, City University of Hong Kong, Hong Kong, People's Republic of China

Acronyms and Abbreviations

ACI	Absolute composite index
AGE	Applied general equilibrium
AGW	Anthropogenic global warming
AHP	Analytical hierarchy process
AMSR	Advanced microwave scanning radiometer
AMSU	Advanced microwave sounding unit
ANN	Artificial neural network
ATDI	A typhoon disaster index
AVL	Average body length
BCNMV	Bottom central nearby maximum velocity
CB	Capacity building
CBA	Cost-benefit analysis
CC	Climate change
CGE	Computable general equilibrium
Chl-a	Chlorophyll a
CMA	Central meteorological bureau
CMORPH	Climate prediction center morphed products
CMR	Central mountain range
CPC	Communist party of china
CRED	Centre for research on epidemiology of disasters
CSA	Canadian space agency
CSTARS	Center for southeastern tropical advanced remote sensing
CT	Cold tongue
CTD	Conductivity-temperature-depth
CTDSS	Coastal typhoon decision serving system
DB	Data base
DD	Disaster degree
DDI	Disaster degree index
DELI	Disaster economic loss index
DG	Disaster grade
DI	Disaster index
DLD	Disaster loss degree

DLI	Disaster loss rate
DM	Disaster management
DRR	Disaster risk reduction
DSFS	Dead sea fault system
DSS	Decision support system
DTBs	Dual trawl boats
EI	Environmental instability
EIA	Environment impact assessment
EM-DAT	The international disaster management
EMDSS	Evacuation management decision support system
EMS	East mediterranean sea
ENSO	El Niño–southern oscillation
ENVISAT	Environmental satellite
ERS	Emergency response system
ERS-2	Environmental remote sensing satellite-2
ETES	Evacuation time estimates
ETS	Equitable threat score
EWSs	Early warning systems
FA	Fractal analysis
FAO	Food and agriculture organisation
FEMA	Federal emergency management agency
FPHLM	Florida public hurricane loss model
FSEM	Fuzzy synthetic evaluation method
FSN	Fish species number
FT	Fuzzy theory
FYP	Five-year-plan
GDP	Gross domestic product
GEF	Global environmental facility
GIS	Geographic information system
GMS5	Geostationary meteorological satellite-5
GoS	Government of syria
H1N1	Hemagglutinin1 neuraminidase1
HAZUS-MH	Hazards U.S. multi-hazard
HDD	Hazard destruction degree
HDI	Hazard destruction degree
HDM	Harrod-Dorma model
HFA	Hyogo framework for action
HNI	Hainan island
HPI	Hazard potential indices
IBA	Important bird area
IBTrACS	International best track archive for climate stewardship
ICZM	Integrated coastal zone management

IO	Input–output analysis
IPCC	Intergovernmental panel on climate change
IPO	Interdecadal pacific oscillation
IPP	Integrated primary production
IR	Infrared channel
IRM	Integrated risk management
ISDR	International strategy for disaster reduction
JMA	Japan meteorological agency
JPL	Jet propulsion laboratory
JTWC	Joint typhoon warning center
LSD	Least significant discrepancy
MA	Multi-variable analysis
Med. Sea	Mediterranean sea
MLD	Mixed layer depth
MLIM	Multiple linear interdependent models
MODIS	Moderate resolution imaging spectroradiometer
MPAs	Marine protected areas
MSL	Mean sea level
MSW	Maximum sustained wind
MTSAT	Multi-functional transport satellite
NASA	National aeronautics and space administration
NCDR	National science and technology center for disaster reduction
NCEP/NCAR	National centers for environmental prediction/National center for atmospheric research
NEOC	National emergency operations center
NGOs	Non-governmental organizations
NN	Neural network
NOAA	National oceanic atmospheric administration
NSC	National research council
NTD	New Taiwan dollar
NTOU	National Taiwan ocean university
NWP	Numerical weather prediction
OFDA	Office of foreign disaster assistance
OIR	Office of insurance regulation
OSCAR	Ocean Surface Current Analyses Real-time
PHRaM	Parametric hurricane rainfall model
PLS	Partial least squares
PR	Pattern recognition
PRE	Pearl river estuary
PSC	Phytoplankton size classes
RA	Risk assessment
RCI	Relative composite index

REMS	Regional evacuation modeling system
RMB	Ren min bi
RMW	Radius of maximum wind
SAR	Synthetic aperture radar
SATIS	Safe Taiwan information system
SCR	Syrian coastal region
SCS	South china sea
SD	Sustainable development
SEA	Strategic environment assessment
SeaWiFs	Sea-viewing wide field-of-view sensor
SINC	Syria's initial national communication
SL	Sea level
SLA	Sea level anomaly
SLR	Sea level rise
SNA	System of national accounts
SON	Sept-oct-nov
SSC	Sea surface current
S.P.	Syrian pounds
SST	Sea surface temperature
STI	Shanghai typhoon institute
STI	Shanghai typhoon research institute
STS	Strong tropical storm
STY	Strong typhoon
Super TY	Super typhoon
SVM	Support vector machine
SWOT	Strength, weakness, opportunity, and threat
TBB	Blackbody or brightness temperature
TC	Tropical cyclone
TCDI	Typhoon comprehensive disaster index
TCEI	Typhoon comprehensive evaluation index
TCGMP	Typhoon central ground minimum pressure
TCII	Typhoon central ground minimum pressure
TCNMWV	Typhoon central nearby maximum wind velocity
TDCG	Typhoon disaster comprehensive grade
TDI	Typhoon disaster Index
TFDSS	Typhoon forecast and decision service system
TISS	Typhoon integrated service system
TMI	TRMM microwave imager
TN	Transect north
TRMM	Tropical rainfall measurement mission
TS	Tropical storm
TS	Transect south

TY	Typhoon
UNDP	United nation development program
UNFCCC	United nations framework convention on climate change
UNISDR	United nations international strategy for disaster reduction
UTC	Coordinated universal time
VGPM	Vertically generalized production model
VIS	Visible channel
VOA	Analysis of variance
WMS	West mediterranean Sea
WNP	West north pacific
WP	Warm pool
W_E	Ekman pumping velocity
f	The coriolis parameter
ρ_{a}	The density of air
C_D	The drag coefficient
U_t	Translation speed of the typhoon
$\vec{\tau}$	Wind stress
$\vec{\mathrm{U}}_{10}$	Wind vector
$\Delta\rho$	The density difference between the upper and lower layers
$\Delta\eta$	Thermocline displacement
ρ	The density of seawater

Chapter 1
Introduction

DanLing Tang and GuangJun Sui

Natural hazards have, throughout history, caused public concern worldwide. Typhoons are some of the most frequent and influential natural hazards in the world.

Typhoons are tropical cyclones that develop in the tropical seas. They are called typhoons in the Northwest Pacific Ocean and hurricanes in the North Atlantic Ocean and East Pacific Ocean. Typhoons are always accompanied by strong winds, heavy rains, and large waves.

Typhoons have a huge impact on marine ecosystems and the oceanic environment. They are one of the most frequent and serious natural hazards in many parts of coastal Asia, causing enormous economic losses. At the same time, they may have positive impacts on marine ecosystems and also have significant negative impacts on coastal environments.

Crisis management is an important element that is called for in preparation for typhoon disasters. The scientific and technical aspects of understanding, monitoring and forecasting tropical storms and the management of the impacts of landfall typhoons are the focus of this book. This book links natural sciences and social sciences via typhoon studies, it provides new information about typhoon impacts and crisis management by taking advantage of satellite remote sensing technology.

This book "Typhoon Impact and Crisis Management" has 27 chapters in 6 parts. They are: Remote sensing of typhoons; Typhoon impacts on various environments, in particular Marine ecosystems and the Relation to global changes. The last two parts cover typhoon impacts on economic and crisis management for typhoon disasters. The framework of this book is illustrated by the figure below.

D. L. Tang (✉)
South China Sea Institute of Oceanlogy, Chinese Academy of Sciences, Guangzhou, China
e-mail: lingzistdl@126.com

G. J. Sui
Guangdong University of Foreign Studies, Guangzhou, China
e-mail: sgj@mail.gdufs.edu.cn

D. L. Tang and G. J. Sui (eds.), *Typhoon Impact and Crisis Management*,
Advances in Natural and Technological Hazards Research 40,
DOI: 10.1007/978-3-642-40695-9_1,

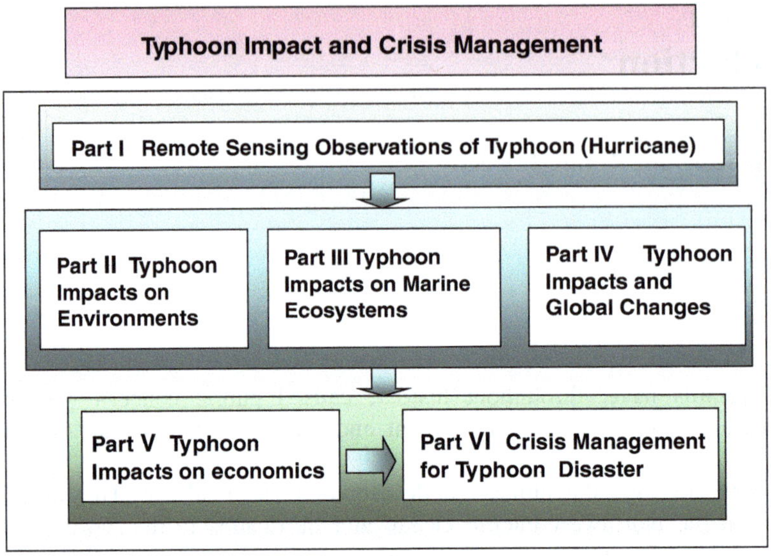
Typhoon Impact and Crisis Management
Part I Remote Sensing Observations of Typhoon (Hurricane)
Part II Typhoon Impacts on Environments
Part III Typhoon Impacts on Marine Ecosystems
Part IV Typhoon Impacts and Global Changes
Part V Typhoon Impacts on economics
Part VI Crisis Management for Typhoon Disaster

Part I
Remote Sensing Observations of Typhoon (Hurricane)

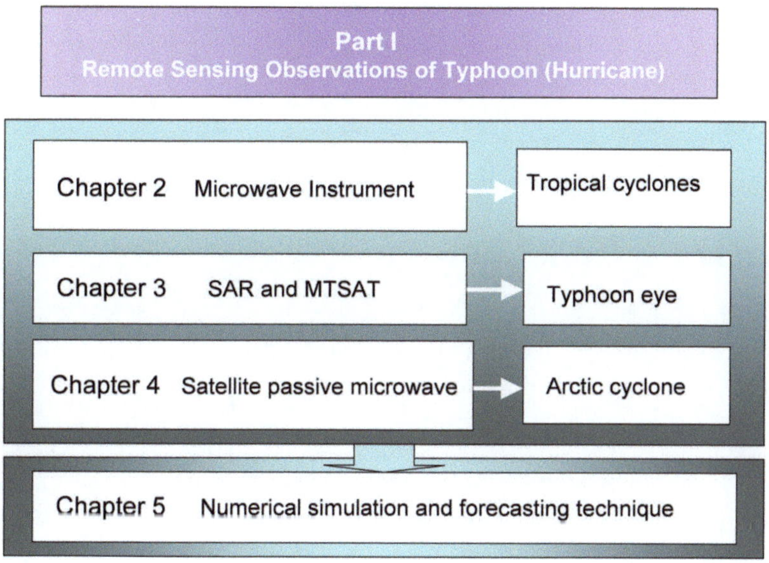

Chapter 2
Microwave Instruments for Observing Tropical Cyclones

Kristina B. Katsaros, Leonid Mitnik and Peter Black

Abstract Development of microwave technologies and our ability to penetrate into Tropical Cyclones (TCs) by instrumented aircraft and observe from satellites have contributed much of the knowledge and understanding that exist today. We can now follow the structure and development of a storm from inception through the many stages towards a dangerous typhoon, hurricane or cyclone, as they are variously called around the globe. The data from satellite microwave radiometers, scatterometers, altimeters, Synthetic Aperture Radars (SARs), microwave sounders, a rain radar and a cloud profiling radar as well as coastal radars, airborne radars and microwave radiometers have all contributed to changing the fields of both TC research and TC operational forecasting.

Keywords Tropical cyclones · Microwave remote sensing · Satellite and aircraft observations

Here we highlight some crucial steps towards current capabilities and describe a short history of these developments and current value of the technology to the field of TC study. We outline the long development of a Stepped-Frequency Microwave Radiometer (SFMR) and the most recent descendant of the same, a scanning microwave radiometer (HIRAD) operated on U.S. research and operational aircraft. These instruments measure surface wind and rain rate inside a storm remotely from safe altitudes. We show the amazing small-scale details of TC structure revealed by SAR's, which have the highest resolution of all satellite microwave instruments. The 30-year development of scatterometry for obtaining

K. B. Katsaros (✉)
Northwest Research Associates, Inc, Bellevue, WA 98052, USA
e-mail: katsaros@whidbey.net

L. Mitnik
V.I. Il'ichev Pacific Oceanological Institute, FEB RAS, Vladivostok, Russian Federation

P. Black
SAIC, Inc. and Naval Research Laboratory, Monterey, CA, USA

D. L. Tang and G. J. Sui (eds.), *Typhoon Impact and Crisis Management*,
Advances in Natural and Technological Hazards Research 40,
DOI: 10.1007/978-3-642-40695-9_2, © Springer-Verlag Berlin Heidelberg 2014

surface wind vectors and the multi-faceted evolution of microwave radiometry along with the associated imagers and sounders are reviewed. The use of altimetry to reveal the hidden heat content in subsurface eddies in the ocean, which provide fuel for intensification of TCs, is discussed. The first spaceborne weather radar on the Tropical Rainfall Measuring Mission, that lasted for 10 years, is shown to greatly enhance knowledge of TC structure through its large observational data base of the world's TCs.

This broad review has to be limited to a few salient examples rather than a complete review of each development's history. It concludes with a description of some promising ideas for future advances, such as dual frequency scatterometers, scanning altimeters, scanning SFMRs and a constellation of microwave radiometers with one or more unifying rain radar(s) for inter-calibration between sensors in the Global Precipitation Measuring (GPM) Mission. The latest sensors for salinity measurements may also contribute information about the effects on the upper ocean by the heavy rainfall associated with TC's.

2.1 Introduction

This chapter is focused on observations by microwave remote sensing for research on tropical cyclones, TCs, for use in diagnosing and thereby aid forecasting of tropical cyclones. We will discuss satellite and airborne microwave techniques with both passive and active microwave sensors.

Early work with microwave sensors in space began with try-outs on short space flights and Shuttle missions and many aircraft surveys in the late 1960s and through the 1970s. The possibilities of retrieving sea surface temperature, near-surface winds, total atmospheric water vapor content, total cloud liquid water content, rain intensity and sea ice parameters were documented by Basharinov et al. (1974) in Russia and similar observations were made by U.S. National Aeronautics and Space Administration, NASA (e.g. Nordberg et al. 1969). A certain amount of exchange of information and face to face meetings between the U.S. and the Soviet Union scientists took place in the field of remote sensing even during the cold war days. Figure 2.1 shows many of the pioneers in the field and by its dated look gives a sense of how much time has already passed with access to this technology.

The biggest boost to microwave remote sensing from space came with the dedicated mission, SEASAT, launched in June 1978 by the NASA and enduring for just 3 months (e.g. Jones et al. 1982; Njoku 1982; Katsaros and Brown 1991). The four major techniques that are discussed below were used on this satellite, microwave radiometry (the Scanning Multifrequency Microwave Radiometer, SMMR), scatterometery (the SEASAT scatterometers, SCATT), the SEASAT altimeter (ALT) and the SAR (the SEASAT SAR). The name SEASAT indicates that these sensors give meteorological and oceanographic data over the oceans, while over land the varying emissivity of the background results in signal complexity and often ambiguous interpretation.

Fig. 2.1 Participants of final symposium on the Bering Sea experiment at the A.I. Voeykov Main Geophysical Observatory (MGO), May 1973. *Back row* Dr. W. J. Webster, Leading Engineer M. A. Prokophiev (MGO), Mr. P. Tibido, Prof. W. J. Campbell, Dr. V. V. Melentyev, Head of scientific program onboard MGO Flying Laboratory IL-18, Mr. D. B. Ross, Dr. P. Gloersen, Dr. Yu. I. Rabinovich (MGO), Dr. Yu. I. Loshchilov (AARI), *Second row* Dr. R. O. Ramseier, Dr. E. M. Shul'gina (MGO), MrsGloersen, Dr. E. P. Dombkovskaya (Hydrometcenter USSR), Dr. G. P. Khokhlov (AARI), *Front row* N. Latter (NASA interpreter), Dr. T. T. Wilheit, Dr. L. M. Martsinkevich (NPO Planeta). Photo provided by Prof. Vladimir Melentyev of the NIERSC/ St. Petersburg

Radiometry has dominated microwave remote sensing from space with data from another SMMR instrument launched by NASA in 1978 on NIMBUS VII and with the Special Sensor Microwave Imager, SSM/I, in the US Defense Meteorological Satellite Program (DMSP). The first one in a long series of SSM/Is was launched in 1989 (Hollinger et al. 1990). The SEASAT experiment resulted in the original development of algorithms for deriving geophysical variables from the signals. It has jokingly been suggested that it was a blessing that the satellite did not last very long, so that the data collected were thoroughly evaluated for biases and errors. Interpretation in terms of geophysical properties of the atmosphere and the sea required detailed study of the comparison data sets from measurements on or over the ocean. Currently, the heritage instrument, the Special Sensor Microwave Imager Sounder, SSMIS, continues with similar wavelengths as the SSM/I providing images and in addition providing soundings.

Scatterometers only began to be routinely launched in 1991 with the first European Remote Sensing (ERS) satellite and many have followed. SAR's have not been launched routinely by either the European Space Agency (ESA) or the U.S. The best examples of SAR data for tropical cyclone work comes from the

European Envisat and Canadian RADARSAT 1 and 2 in wide swath SAR observations (e.g. Vashon and Katsaros 1999; Horstmann et al. 2005; Mitnik et al. 2005a; Reppucci et al. 2010). Envisat Advanced SAR (ASAR) was launched in 2002. RADARSAT 1 and 2 were launched, respectively in 1995 and 2007, with the objective of observing the Arctic sea ice, but through a special Canadian program, Hurricane Watch (Iris and Burger 2004), has provided excellent high resolution data of TC's, especially in the Caribbean and off the East coast of the USA (e.g. Vachon and Katsaros 1999). More of the background will be recounted in connection with each instrument being discussed.

In Sect. 2.2 we define the instruments that form the basis of our review of TC studies using microwave data. In Sect. 2.3 we give examples of observing TC's with some of the satellite sensors. In Sect. 2.4 emphasis is on practical applications of remote sensing in observing one particular typhoon, named MEGI, which crossed the South China Sea in October 2010. In Sect. 2.5 we summarize the impact of the many microwave instruments on TC research and forecasting and describe a few new instruments in advanced planning stages, a scanning altimeter, a TRMM follow on involving many satellites, and the need for coalitions of agencies to contribute to increased frequency of sampling the important variables. These advances are expected to enhance the value of microwave observations in the near future.

2.2 Instruments

We include here description of the instruments that we refer to in the examples of hurricane studies below to separate the technical details from the scientific use of the data they provide.

2.2.1 Passive Microwave Radiometry

Passive microwave sensing of the Earth started in the U.S.S.R. in the early 1960s. The first two satellites to carry four microwave radiometers into space were Kosmos-243 launched in 1968 and Kosmos-384 in 1970. The nadir-viewing trace radiometers operated at wavelengths $\lambda = 8.5$, 3.4, 1.35 and 0.8 cm. The Instantaneous Field Of View (IFOV) was 50×50 km (apogee) and 35×35 km (perigee) at $\lambda = 8.5$ and approximately 22×22 km (apogee) and 15×15 km (perigee) at other wavelengths. The possibilities of retrieving sea surface temperature, near-surface winds, total atmospheric water vapor content, total cloud liquid water content, rain intensity and sea ice parameters were documented by Basharinov et al. (1974). Many more passive microwave instruments have been flown, including early on the SEASAT and NIMBUS VII SMMR's mentioned in the introduction. They have become the work-horses of satellite remote sensing of tropical cyclones, being based on well-established technology.

Advantages of microwave remote sensing from space are due to combination of several factors. Sensors can penetrate through clouds and low-intensity precipitation and estimate the water surface temperature and near-surface wind speed practically over the whole ocean. Multi-frequency measurements allow retrieval both of integrated parameters such as total atmospheric water vapor content, total cloud liquid water content and rain rate (satellite multi-frequency radiometers—sensors) and the vertical profiles of air temperature and humidity (sounders). The molecular oxygen absorption band (50–70 GHz) and a line (118.75 GHz) are well suited for temperature sounding due to O_2 being uniformly distributed within the atmosphere. These advantages are realized due to the state of the art of the current radiometer construction providing highly stable calibration during long-term operation in space that is necessary, in particular, for climate change studies and for providing a dependable data source when tropical cyclones threaten.

Passive microwave radiometers have a strong heritage of instruments: the Kosmos-243 four channel nadir-looking nonscanning microwave radiometer, the Meteor-18 and Meteor-28 two-polarization nonscanning microwave radiometers, the Indian Bhaskara I and II, two-channel microwave radiometers, the DMSP Special Sensor Microwave Imager (SSM/I), the Tropical Rainfall Measuring Mission (TRMM) Microwave Imager (TMI), the Advanced Earth Observing Satellite-II (ADEOS-II) Advanced Microwave Scanning Radiometer (AMSR), the Advanced Microwave Scanning Radiometer–Earth Observing (AMSR-E), the NOAA Advanced Microwave Scanning Units (AMSU-A and AMSU-B), the Special Sensor Microwave Imager/Sounder (SSMI/S) and the multifrequency polarimetric microwave radiometer WindSat. These instruments measure horizontal (h) and/or vertical (v) polarized brightness temperatures $T_B^{V,H}$ over a range of frequencies, υ. WindSat also measures the cross-correlation of the vertical and horizontal polarizations. The cross-correlation terms represent the third and fourth parameters of the modified Stokes vector (e.g. Gaiser et al. 2004).

2.2.1.1 DMSP SSM/I

The SSM/I is a seven-channel, linearly polarized passive microwave radiometer aboard selected Defense Meteorological Satellite Program (DMSP) vehicles (Hollinger et al. 1990). The SSM/I is carried aboard the F8, F10, F11, F12, and F13 vehicles. Some of the channels on the F8 SSM/I are not operational. As an example orbital information for the F11 vehicle is as follows: maximum/minimum altitude: 878/841 km; inclination: 98.8°; period: 101.9 min and ascending equator crossing local time 18:11 (at launch). Other satellites except the F10 have similar orbits. The F10 did not achieve the desired orbit (more elliptical than desired), so orbital information given above is not applicable to the F10. Since 1987, the SSM/Is have proven to be stable, well-calibrated instruments whose data are useful for a variety of geophysical applications.

The SSM/I instrument consists of an offset parabolic reflector that is 61 × 66 cm fed by a seven-port horn antenna. The reflector and feed are mounted on a drum which contains the radiometers, digital data subsystem, mechanical scanning subsystem, and power subsystem. The drum assembly rotates about the axis of the drum. A small mirror and a hot reference absorber are mounted on the assembly. The instrument sweeps a cone around the satellite velocity vector so that the Earth incidence angle is always 53°. The SSM/I rotates continuously about an axis parallel to the local spacecraft vertical at 31.6 rpm. The SSM/I measures the upwelling brightness temperature over an angular section of 102.4° about the sub-satellite track. When looking in the forward direction of the spacecraft, the scan is directed from left to right with active scene measurements lying 51.2° about the forward direction. The resulting swath width is 1,400 km, which results in 24 h global coverage. The spacecraft sub-satellite point travels 12.5 km during the 1.9 s period. Data are recorded during the 102.4° of the cone when the antenna beam intercepts the Earth's surface (Hollinger et al. 1990).

Vertical and horizontal polarization measurements are taken for 19.4, 37.0 and 85.5 GHz channels and for the 22.2 GHz only at V-polarization. The channel IFOV varies with channel energy, position in the scan, along scan or along track direction and altitude of the satellite. The 85.5 GHz IFOV is the smallest with a 13 × 15 km, and the 19.4 GHz IFOV is the largest at 45 × 70 km. The 37.0 GHz IFOV is 30 × 38 km, and 22.2 GHz one is 40 × 60 km.

Because the 85 GHz IFOV is smaller, it is sampled twice as often, i.e., 128 times per scan. The sampling interval is 4.22 ms and equals the time for the beam to travel 12.5 km in the cross track direction. Radiometer data at the remaining frequencies are sampled every other scan with 64 uniformly spaced samples being taken and have an 8.44 ms interval. Scan A denotes scans in which all channels are sampled and Scan B denotes scans in which only 85.5 GHz data are taken. Thus, one data cycle consists of four 85.5 GHz scans and two scans of the 19.4, 22.2, and 37.0 GHz channels. The sensitivity of all channels is high and equal 0.4 K for 19.4 and 37.0 GHz and 0.8 K for the two other channels.

Remote Sensing System (RSS) has recently reprocessed the entire brightness temperature ($T_B^{V,H}$) data set using advanced calibration techniques, more rigorous quality control, and better geolocation. The new methods are applied to all the SSM/I channels, including 85 GHz. 21 years of SSM/I brightness temperatures are now available on hard disk arrays. The SSM/I time series consists of 6 satellites spanning the period from 1987 to the present, which in 2012 represents a total of 55 satellite years (http://www.ssmi.com/ssmi/ssmi_brightness_temperatures.html).

The data from the DMSP satellites are received and used at operational centers continuously for determination of position and intensity of TCs (http://www.nrlmry.navy.mil/TC.html), and for forecast of their evolution as well as for research (e.g. Ferraro 1997; Ferraro et al. 2005; Wentz and Spencer 1998; Mitnik et al. 2002; Chen et al. 2004; Liu et al. 2008a; Wimmers and Velden 2007).

2.2.1.2 SSMIS

The primary mission of the Special Sensor Microwave Imager/Sounder, SSMIS, instrument is to combine and extend the imaging and sounding capabilities of three previously separate DMSP microwave sensors: the SSM/T-1 temperature sounder, the SSM/T-2 moisture sounder, and the SSM/I. With improved temperature sounding capabilities, the SSMIS is capable of profiling the mesosphere, i.e. 10–0.03 mb. As such, the SSMIS is currently the only operational passive microwave sensor that can collect temperature measurements in the 40–80 km altitude range. In addition, SSMIS offers capabilities associated with radiometer channels having common fields of view, uniform polarizations, and fixed spatial resolutions across the active scene/scan sector, making it thus far the most complex and unique operational satellite passive microwave imager/sounder flown (Kunkee et al. 2008).

SSMIS is a 24-channel, passive microwave radiometer designed to obtain a variety of polarized atmospheric temperature, moisture, and land variables under most weather conditions. Channel frequencies range from 19 to 183 GHz and are obtained over a swath width of approximately 1,707 km. Detailed channel characteristics are provided in Table 2.1. As indicated in Table 2.1, the SSMIS channels correspond to four main categories of measurement parameters (Bell 2006): Lower Atmospheric Sounding (LAS), Upper Atmospheric Sounding (UAS), Environmental (ENV) and Imaging (IMA) data.

The SSMIS employs the same conical scan geometry and external calibration scheme as the heritage SSM/I instrument (Fig. 2.2). By employing the conical scan geometry of the SSM/I, the SSMIS maintains uniform spatial resolution, polarization purity, and common concentric Fields of View (FOV) for all channels across the swath (Kunkee et al. 2008).

Important information on SSMI and SSMIS can be find at web sites:

http://nsidc.org/data/docs/daac/ssmis_instrument/index.html

http://www.ncdc.noaa.gov/oa/rsad/ssmi/gridded/index.php ("Description of Global Gridded Products National Environmental Satellite, Data, and Information Service—NESDIS") and

http://www.ncdc.noaa.gov/oa/rsad/ssmi/swath/ ("SSMI and SSMIS Monitoring and Documentation", NESDIS).

Application of SSMIS data to tropical cyclone study is considered by Hawkins et al. (2008).

2.2.1.3 AMSU-A and AMSU-B

The Advanced Microwave Sounding Unit (AMSU) is derived from the Microwave Sounding Unit (MSU) which began service in 1978 on TIROS-N and continued on the NOAA 6 through 14 satellites. AMSU flies on the NOAA 15, launched 13 May 1998; NOAA 16, launched 21 September 2000; and NOAA 17, launched 24 June 2002.

Table 2.1 Channel characteristics of the SSMIS instrument

Channel	Center frequency (GHz)	3-db width (MHz)	Polarization	NEΔT (K)[a]	Sampling interval (km)[b]	Channel application
1	50.3	380	V	034	37.5	LAS
2	52.8	389	V	0.32	37.5	LAS
3	53.596	380	V	0.33	37.5	LAS
4	54.4	383	V	0.33	37.5	LAS
5	55.5	391	V	0.34	37.5	LAS
6	57.29	330	RCP[c]	0.41	37.5	LAS
7	59.4	239	RCP	0.40	37.5	LAS
8	150	1642(2)[d]	H	0.89	12.5	IMA
9	183.31 ± 6.6	1526(2)	H	0.97	12.5	IMA
10	183.31 ± 3	1019(2)	H	0.67	12.5	IMA
11	183.31 ± 1	513(2)	H	0.81	12.5	IMA
12	19.35	355	H	0.33	25	ENV
13	19.35	357	V	0.31	25	ENV
14	22.235	401	V	0.43	25	ENV
15	37	1,616	H	0.25	25	ENV
16	37	1,545	V	0.20	25	ENV
17	91.655	1418(2)	V	0.33	12.5	IMA
18	91.655	1411(2)	H	0.32	12.5	IMA
19	63.283248 ± 0.285271	1.35(2)	RCP	2.7	75	UAS
20	60.792668 ± 0.357892	1.35(2)	RCP	2.7	75	UAS
21	60.792668 ± 0.357892 ± 0.002	1.3(4)	RCP	1.9	75	UAS
22	60.792668 ± 0.357892 ± 0.0055	2.6(4)	RCP	1.3	75	UAS
23	60.792668 ± 0.357892 ± 0.016	7.35(4)	RCP	0.8	75	UAS
24	60.792668 ± 0.357892 ± 0.050	26.5(4)	RCP	0.9	37.5	LAS

[a] NEΔT for instrument temperature (0 °C) and calibration target (260 K) with integration times of 8.4 ms for channels 12–16; 12.6 ms for channels 1–7, 24; and 25.2 ms for channels 19–23 and 4.2 ms for channels 8–11, 17-18

[b] Along-scan direction sampling based on 833 km spacecraft altitude

[c] RCP denotes right-hand circular polarization

[d] Number of sub-bands is indicated by (n) next to individual 3-db width

AMSU has a finer vertical resolution with 15 channels for atmospheric temperature sounding. Using this new instrument, there are several techniques to estimate size, structure, intensity, wind speed in TCs, as well as to retrieve the global atmospheric temperature and humidity profiles from the Earth's surface to the upper stratosphere, at approximately 2-millibar pressure altitude. AMSUs are always situated on polar-orbiting satellites in sun-synchronous orbits and they cross the equator at the same local solar time at ascending and descending orbits. The AMSU has two sub-instruments, AMSU-A with 15 channels (23.8, 31.4, 89.0 GHz with V-polarization and 12 channels in oxygen absorption band between 50.3 and 57.6 GHz with H- or V-polarization) and AMSU-B with 5 channels (89.9, 150 GHz and three channels: 183.31 ± 1.00, 183.31 ± 3.00 and 183.31 ± 7.00 GHz within a vapor absorption line, all with V-polarization). Swath width is equal to 1,650 km and IFOV at nadir for a nominal spacecraft altitude of 833 km is 48 km for AMSU-A which is twice that of MSU and it is 16 km for AMSU-B. The AMSU-A antenna provides a cross-track scan, scanning ±48.3° from nadir with a total of 30 Earth

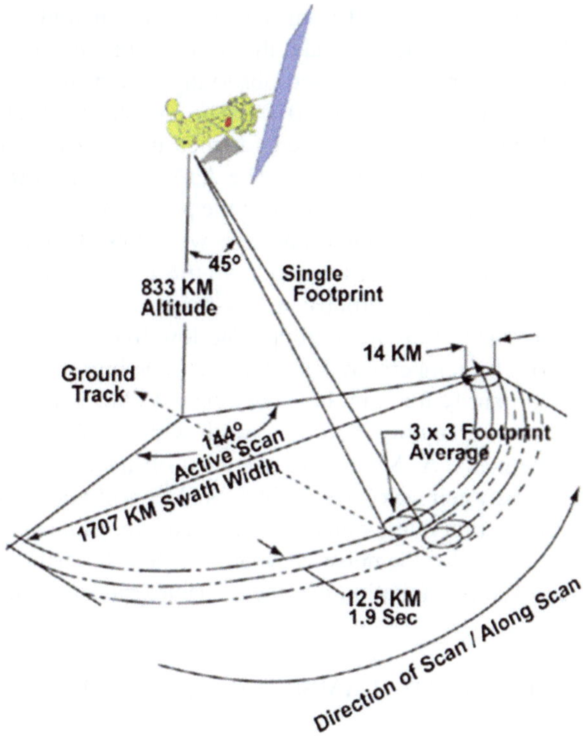

Fig. 2.2 *SSMIS Scan Geometry* (Poe et al. 2001). Schematic of the sampling by the scanning microwave/ imager sounder

fields-of view per scan line. The instrument completes one scan every 8 s. The AMSU-B antenna also provides a cross-track scan, scanning ±48.95° from nadir with a total of 90 Earth fields-of view per scan line. The instrument completes one scan every 2.66 s (Bennartz 2000; Kramer 2002).

AMSU applicability to retrieval of the oceanic and atmospheric parameters in the tropical zone and to TC analysis was shown by Knaff et al. (2000), Staelin and Chen (2000), Spencer and Braswell (2001), Demuth et al. (2004, 2006), and Bessho et al. (2006).

2.2.1.4 Suomi NPP, ATMS

The Suomi National Polar-orbiting Partnership or Suomi NPP, previously known as the National Polar-orbiting Operational Environmental Satellite System Preparatory Project (NPP) and NPP-Bridge is a NASA weather satellite which was launched on October 27, 2011 and was successfully inserted into its sun-synchronous low-Earth orbit. Three months after the satellite's launch it was named Suomi NPP after Verner E. Suomi http://en.wikipedia.org/wiki/Verner_E._Suomi, a meteorologist at the University of Wisconsin-Madison. The Suomi NPP is the first mission designed to collect critical data to improve weather forecasts in the short-term and increase our understanding of long-term climate change. In the next decade, the Suomi NPP will be the principal U.S. operational, polar-orbiting, space-based observing system for weather forecasting and climate monitoring (Lee et al. 2010). The Advanced Technology Microwave Sounder (ATMS) sensor is designed for measurement of water vapor content at various altitudes of the atmosphere. The ATMS is a cross-track scanner with 22 channels that combine all the channels of the preceding AMSU-A1, AMSU-A2, and AMSU-B sensors into a single package with considerable savings in mass, power and volume (http://npp.gsfc.nasa.gov/atms.html). The 22 channels of the ATMS are divided into two groups: a low-frequency (23–57 GHz) group, and a high-frequency (88–183 GHz) group. The low-frequency channels, 1 through 15, are primarily for temperature soundings and the high-frequency channels, 16 through 22, are primarily for humidity soundings. ATMS will provide vertical temperature and moisture soundings across a swath width of 2,300 km. In comparison to heritage instruments, ATMS has an extended off nadir view angle, which will result in no gaps between successive orbits. ATMS will also be capable of retrieving surface precipitation rates, water path estimates for rain and snow, and peak vertical wind. Recent studies demonstrate that ATMS will perform better than AMSU for these retrieved precipitation parameters (Surussavadee and Staelin 2008).

2.2.1.5 ADEOS-II AMSR and Aqua AMSR-E

The AMSR was launched onboard the Advanced Earth Observing Satellite-II (ADEOS-II, Midori) on December 14, 2002 and the AMSR-E was launched onboard the Aqua satellite on May 4, 2002. On Aqua it is abbreviated as AMSR-Earth Observation System or AMSR-E; it is similar to SSM/I. AMSR-E takes dual-polarization observations at the 6 frequencies: 6.925, 10.65, 18.7, 23.8, 36.5 and 89.0 GHz. AMSR has two additional channels with V-polarization at the frequencies of 50.3 and 52.8 GHz. From an altitude of 705 km radiometers measure the upwelling Earth brightness temperatures with a resulting swath width of 1,445 km (AMSR-E) and 1,600 km (AMSR). The nadir angle is fixed at 47.4° which results in an Earth incidence angle of 55.0° $\pm$ 0.3°. During a scan period the sub-satellite point travels 10 km. The IFOV for each channel is different (Table 2.2), however, Earth observations are recorded at equal sample intervals of 10 km (5 km for the 89 GHz channels) along the scan. The sensitivity and other characteristics of the channels are given in Table 2.2 (Kawanashi et al. 2003). AMSR adopts the external calibration scheme which is similar to that of SSM/I. The Hot Load is maintained at a physical temperature of about 300 K, while the Cold Sky Mirror introduces the cosmic background radiation (approximately 2.7 K). Unfortunately, the AMSR mission ended in October 2003 and AMSR-E ceased observations on October 4, 2011.

Near real-time browse images of precipitation, water vapor, and track of the latest tropical cyclone observed by PR/VIRS (TRMM) and AMSR-E (Aqua) from developing to decaying stage was available till October 4, 2011 via "JAXA/EORC

Table 2.2 Characteristics of the AMSR and AMSR-E channels

Center frequency (GHz)	6.925	10.65	18.7	23.8	36.5	89.0	50.3/52.8
Bandwidth (MHz)	350	100	200	400	1000	3000	200/400
Polarization	V/H	V/H	V/H	V/H	V/H	V/H	V
Sensitivity (K, target)	0.3	0.6	0.6	0.6	0.6	1.1	1.8/1.3
IFOV (km × km)	71 × 41	46 × 26	25 × 15	23 × 14	14 × 8	6 × 4	12 × 6
Sampling rate (km × km)	10 × 10	10 × 10	10 × 10	10 × 10	10 × 10	5 × 5	10 × 10
Integration time (ms)	2.6	2.6	2.6	2.6	2.6	1.3	2.6

50.3/52.8 channels were at ADEOS II AMSR only

Real-Time Monitoring for Tropical Cyclones" (http://sharaku.eorc.jaxa.jp/AMSR/relay/monitor.html); Japan Aerospace Exploration Agency (JAXA)/Earth Observation Research Center. Thus, Real-time monitoring for tropical cyclones is now available. The coverage of AMSR-E has been expanded to the global coverage (June 1, 2011). JAXA EORC provides access to online archive of images, movies and data of tropical cyclones observed by TRMM/Aqua/Midori-II satellites (http://sharaku.eorc.jaxa.jp/TYPHOON_RT/index.html).

Applications of AMSR/AMSR-E data to tropical cyclones were considered in many journal and conference papers (e.g. Wilheit et al. 2003; Mitnik and Mitnik 2003, 2005b, 2010, 2011; Mitnik et al. 2005a, 2009; Shi and Wang 2007; Yan and Weng 2008; Greenwald 2009; Shibata 2006, 2009).

2.2.1.6 Coriolis/WindSat

WindSat is a satellite-based multifrequency polarimetric microwave radiometer developed by the Naval Research Laboratory for the U.S. Navy and the National Polar-orbiting Operational Environmental Satellite System (NPOESS) Integrated Program Office (IPO). It is designed to demonstrate the capability of polarimetric microwave radiometry to measure the ocean surface wind vector from space. WindSat is the primary payload on the Air Force Coriolis satellite, which was launched on 6 January 2003. It is in an 840-km circular Sun-synchronous orbit. Using a conically-scanned 1.83 m offset parabolic reflector with multiple feeds, WindSat covers a 1,025 km active swath. The 10.7, 18.7, and 37.0 GHz channels are fully polarimetric; that is, they derive all four Stokes parameters by measuring the six principal polarizations. The 6.8 GHz channel is dual polarimetric (vertical and horizontal) and provides sea surface temperature as a secondary product. The 23.8 GHz channel is also dual polarimetric because its purpose is to correct for atmospheric water vapor which is unpolarized. Gaiser et al. (2004) describe this passive microwave wind sensing instrument (Table 2.3) which allows to separate effects of wind speed and direction and therefore provides information of the wind vector.

Obtaining wind speed and direction from a passive sensor was proposed in order to save the great expense of launching scatterometers. Several retrieval algorithms were developed (e.g., Bettenhausen et al. 2006; Smith et al. 2006). The Coriolis has been in operation for several years and the data are used operationally,

Table 2.3 Characteristics of the WindSat channels

Frequency (GHz)	Polarization	Bandwidth (MHz)	Incidence angle (deg)	IFOV(km × km)
6.8	V, H	125	53.5	40 × 60
10.7	V, H, ±45, L, R	300	49.9	25 × 38
18.7	V, H ± 45, L, R	750	55.3	16 × 27
23.8	V, H	500	53.0	12 × 20
37.0	V, H ± 45, L, R	2,000	53.0	8 × 13

in particular, to retrieve wind vectors in tropical cyclone areas including rain conditions (Adams et al. 2006; Meissner and Wentz 2006, 2008; Knaff et al. 2011; Mims et al. 2010; Yueh et al. 2006). The experience does not indicate that the quality of WindSat wind retrievals is equivalent to scatterometer wind vectors.

2.2.1.7 Global Change Observation Mission-W1 AMSR2

The Advanced Microwave Scanning Radiometer-2 (AMSR2) will be launched as a single mission instrument onboard on the first satellite of the Water Series of Global Change Observation Mission (GCOM-W1). "Shizuku" (meaning a "drop" or a "dew") was selected by the Japan Aerospace Exploration Agency (JAXA) as a GCOM-W1 nickname. GCOM is a series of JAXA Earth Observation Missions, successors to the ill-fated ADEOS-II satellite and to the Aqua Mission. GCOM is, together with the GPM Mission, Japan's contribution to the Global Earth Observation System of Systems (GEOSS). The mission of GCOM-W1 is to observe the water cycle.

AMSR2 is multi-frequency, total-power microwave radiometer system with V- and H-polarization channels for all frequencies. Frequency channel set is identical to that of AMSR-E except 7.3 GHz channel for Radio Frequency Interference (RFI) mitigation (Table 2.4). It has a digitization at 12 bits and a dynamic range from 2.7 to 340 K. AMSR2 has two-point external calibration with the improved HTS (hot-load) and performs deep space calibration maneuvers to check consistency between main reflector and CSM (Imaoka et al. 2010; Oki et al. 2010). Scan angle is nominally 55° with an offset 2 m parabolic antenna, the world's largest observation sensor aboard a satellite. The height of the rotating part is about 2.7 m and the weight is about 250 kg. The AMSR2 can keep rotating such a large and heavy antenna at a speed of one turn per 1.5 s for 24 h a day and more than 5 years without a minute of rest. It covers a swath width of 1,450 km. The AMSR2 can observe over 99 % of the Earth's area in just 2 days. The GCOM-W1 AMSR2 data will continue valuable time series of quantitative observations of tropical cyclones which are collected since 2002 by Aqua AMSR-E (http://sharaku.eorc.jaxa.jp/TYPHOON_RT/index.html).

GCOM-W1 was successfully launched on May 18, 2012 with the H-2A rocket. JAXA will take about 45 days to insert the satellite into an orbit ahead of Aqua;

Table 2.4 GCOM-W1AMSR2 characteristics

Center frequency (GHz)	6.925/7.3	10.65	18.7	23.8	36.5	89.0
Polarization	V/H	V/H	V/H	V/H	V/H	V/H
Band width (MHz)	350	100	200	400	1,000	3,000
IFOV (km × km)	35 × 62	24 × 42	14 × 22	15 × 26	7 × 12	3 × 5
Sampling rate (km × km)	10 × 10	10 × 10	10 × 10	10 × 10	10 × 10	5 × 5

that is it will be in the so-called A-train (L'Ecuyer and Jiang 2010; Lee et al. 2012) in order to participate in an Earth observation mission called the "A-Train" in cooperation with satellites of other countries while performing the initial functional confirmation for about 3 months. NASA and its international partners operate several Earth observing satellites that closely follow one after another along the same orbital "track". This coordinated group of satellites, constituting a significant subset of NASA's current operating major satellite missions, is called the Afternoon Constellation, or the A-Train, for short. The Mean Local Time of the Ascending Node (MLTAN) of the GCOM-W1 orbit will be between (OCO-2—259.5 s) and (Aqua—79.5 s). It has a sun synchronous orbit (A-Train orbit). Altitude 699.6 km (on Equator), inclination 98.2°, local sun time: 13:30 ± 15 min. Five satellites currently fly in the A-Train: GCOM-W1, Aqua, CloudSat, CALIPSO, and Aura. Orbiting Carbon Observatory 2 (OCO-2) will join the configuration in 2013.

2.2.2 Passive and Active Microwave Systems for Atmospheric Measurements

2.2.2.1 CloudSat

The CloudSat Mission was jointly developed by the NASA/JPL and the Canadian Space Agency. CloudSat's payload, the Cloud Profiling Radar (CPR), is the first spaceborne 94-GHz (W-band) radar and is contributing vertical cloud profiles (VCPs) over the globe (Stephens et al. 2002). CPR provides direct information on the vertical distribution of clouds and their optical and physical properties with unprecedented spatial and vertical resolution important for weather system analysis and forecasts (Stephens et al. 2002; Kim et al. 2011; Lee et al. 2012). Knowledge of the vertical cloud profiles, including parameters such as the cloud-top and cloud-bottom height and the cloud-layer thickness under various cloudiness conditions (e.g. semi-transparent/opaque clouds and multiple cloud layers), is necessary, in order to better understand complex cloud radiative forcing effects. The VCP affects atmospheric circulation by determining the vertical gradient of radiative heating/cooling and latent heating.

The Cloud Profiling Radar measures the returned backscattered energy by clouds as a function of height along the orbital track with a 240 m vertical range resolution between the surface and 30 km along its flight path with footprint size of 1.4 × 1.7 km (Stephens et al. 2002, Tanelli et al. 2008). Unfortunately, it is only capable of quantitatively profiling lightly precipitating cloud systems (Mitrescu et al. 2010). For higher precipitation rates, complications arising from increased extinction and multiple scattering make quantitative precipitation analysis almost impossible. At rain rate exceeding 30 mm/h that is typical for typhoon eyewall and rain bands radar signals backscattered by the sea surface disappear due to very high two-way attenuation. Despite these limitations CloudSat profiles can show precipitation features such as melting layers (as bright bands, Matrosov 2010), deep convective towers, and multiple cloud layers. The capability to distinguish between convective and stratiform precipitating systems is also important for TC study.

2.2.2.2 Tropical Rainfall Measuring Mission

Most records of precipitation over the sea are based on indirect estimates by remote sensing using the cloud brightness or cloud top temperature as indicators of deep and raining systems (e.g. Adler et al. 2003). Direct measurements of precipitation over oceans are almost non-existent, and even attempts to put rain gauges on ships and buoys have had little success due to flow distortion by the ship's structure. Buoys also have issues with contamination by sea spray, so today satellite estimates of precipitation are typically calibrated against coastal radar estimates of precipitation, which in turn are calibrated against ground based rain-gauge networks (Adler et al. 2003). In spite of these difficulties we do now have precipitation estimates based on inferences from the temperature and patterns of cold cloud tops and from microwave radiometer measurements of the cloud water particles and scattering by large ice particles. (These are airborne precipitation particles, which may or may not coincide with precipitation impacting on the sea surface). A method for estimating the ground precipitation combining various methods called CMORPH (which stands for CPC Morphing technique, where CPC is Climate Prediction Center), has been developed for routine use (Joyce et al. 2004). It is flexible, because it can combine various measurements and their algorithms by "morphing" them together.

A dramatic step forward in rain measurements over the oceans from space was provided by the Tropical Rainfall Measuring Mission, TRMM (Simpson et al. 1988; Kummerow et al. 1998; Kummerow et al. 2000) which carried a narrow-swath rain radar (215 km) and the TRMM Microwave Radiometer, TMI, with a 760 km swath width. It was launched in November 1997 and provided data for a decade. The satellite had a low-earth (402 km height), non-polar orbit that allowed it to sample low latitude regions at different times in the diurnal cycle during a month's time. It outlived its scheduled life-time and has continued operations at a lower altitude (350 km) with somewhat reduced functions. It has been used at operational centers when available, but was not intended to be an operational weather satellite, being that its revisit times for the same location takes a whole month. The goal was to obtain climatology of tropical and subtropical precipitation. One of its major contributions was its calibration of the microwave radiometer algorithms and the additional insight into storm structure that its vertical resolution provided. See further Sect. 2.3.5.

The TRMM Microwave imager: The TMI is a nine-channel passive microwave radiometer based upon the DMSP SSM/I. The key differences are the addition of 10.7-GHz channels with V- and H-polarizations and a frequency change of the water vapor channel from 22.235 to 21.3 GHz. This change off the center of the water vapor line was made in order to avoid saturation in the tropical orbit of TRMM. Additionally, this provides improving the accuracy of total water vapor retrieval (Mitnik and Mitnik 2006). Table 2.2 presents the performance characteristics of the nine TMI channels (Kummerow et al. 1998). The increased spatial resolution evident in Table 2.2 is due to the lower orbit of the TRMM satellite with respect to the DMSP rather than sensor differences (Table 2.5).

Combination of passive and active microwave sensing has proved its high efficiency for precipitation study. This approach will be advanced at the Global Precipitation Measurements (GPM) project. GPM is designed to make more accurate and frequent observation of tropical rainfall by expanding its observing areas to higher latitudes. GPM is composed of one core satellite and approximately eight constellation satellites. The core satellite carries a Dual-frequency Precipitation Radar (DPR) and a microwave radiometer will be used to measure precipitation structure and to provide a calibration standard for the constellation satellites with microwave radiometers. The core satellite is scheduled for launch in early 2014 and the low-inclination satellites for launch in November 2014 (http://science.nasa.gov/missions/gpm/).

2.2.3 *Variables Retrieved by Microwave Radiometers, Sounders, CloudSat and TRMM*

2.2.3.1 Ocean Surface

Several geophysical variables can be retrieved from the microwave sensors. Sea surface temperature was a very desirable quantity, because infrared signals emitted from the sea surface, already well established technology, were cut-off by clouds.

Table 2.5 TMI characteristics

Center frequencies (GHz)	10.65	19.35	21. 3	37.0	85.5
Bandwidth (MHz)	100	500	200	2000	3000
Polarization	V/H	V/H	V	V/H	V/H
Sensitivity (K)	0.63/0.54	0.50/0.47	0.71	0.36/0.31	0.52/0.93
IFOV (km × km)	63 × 37	30 × 18	23 × 18	16 × 9	7 × 5
Sampling interval (km × km)	13.9 × 9.1				13.9 × 4.6

The incidence angle of the antenna beam is 52.88 at the earth's surface and a swath width is 758.5 km. TMI swath covers precipitation radar (*PR*) swath that is equal to 215 km as shown in Fig. 3 (Kummerow et al. 1998)

This has resulted in biased observations because they could only be obtained outside storms. Microwaves at low frequencies (of the order of 5 and 6 GHz) penetrate with little absorption through clouds (less so through rain) and recent sensors allow to retrieve SST under cloud conditions and estimate ocean surface thermal variability caused by diurnal heating or wind action in the wake of hurricanes and typhoons (Wentz et al. 2000; Gentemann and Minnett 2008; Gentemann et al. 2008, 2010a, b; Reynolds et al. 2007, 2010; Shibata 2009).

Sea surface wind response by passive microwave radiometers was deemed to be caused by the increased emissivity of the sea surface when white caps form and there is foam on the sea surface. This indicates that the state of the sea, which is not always simply related to the mean wind speed, contributes some uncertainty in the inversion algorithms. However, on the open ocean much of the variability is averaged out by the relatively large footprints required for the lower frequencies used for wind speed (6.9–37 GHz) on above described radiometers SSM/I, TMI, SeaWinds, AMSU, AMSR-E (Sasaki et al. 1987; Krasnopolsky et al. 2000; Azis et al. 2005; Bessho et al. 2006; Shibata 2006; Boukabara and Weng 2008; Mitnik and Mitnik 2010, 2011). In fact, the physics of the measurement is not used in the calibration for wind speed, but conversion to geophysical information resorts to empirical fits between satellite microwave measurements and surface observations. The sophisticated use of the Stokes parameters in WindSat provides information of speed and direction from this instrument, but it has greater limitations compared to the scatterometers, due to atmospheric interference.

2.2.3.2 Atmospheric Integrated Water Vapor, Cloud Liquid Water Content and Precipitation

Satellite passive microwave data are unique in being able to provide large-scale maps of integrated water vapor, *V*, and integrated cloud liquid water, *Q*, independently of time of a day and under cloudy conditions. *V* and *Q* over the oceans have been successfully retrieved from satellite microwave measurements carried out by multichannel microwave radiometers on board "Kosmos-243", Nimbus 7, DMSP series satellites, TRMM, Aqua, ADEOS-II, WindSat, NOAA, TOPEX, Jason and several other satellites. Nevertheless despite comparatively long history of satellite passive microwave observations and their advantages, the interpretation of data from microwave sensors always had large errors under significantly changing environmental conditions (Wentz 1997) and has always been limited in ocean areas with heavy rain, which are typical for tropical cyclones and for the Intertropical Convergence Zone. To estimate water vapor and cloud liquid water in various marine weather systems and, especially, climatic trends from satellite passive microwave data, there is a strong need in having precise retrieval algorithms tuned. Many algorithms exist for retrieval of the integrated water vapor *V* and integrated cloud liquid water from satellite passive microwave measurements.

Some of the algorithms are based on purely statistical techniques, the starting point of which is a database containing both brightness temperatures $T_B^{V,H}(\nu_i)$ and the corresponding environmental conditions for global or for the specific regions such as the tropics. Such a method can easily be used for *V*, by collocating ship and small island radiosondes with satellite overpasses. The brightness temperatures can also be computed using a state of the art model of microwave radiative transfer in the atmosphere–ocean system. Due to the lack of any routine measurement of cloud liquid water, a reasonable approach is construction of cloud liquid water content profiles using the radiosonde profiles of air relative humidity and temperature, information on the amount and forms of clouds and generalized data on their liquid water content (Mitnik and Mitnik 2003). The inversion is then performed using a statistical method, with or without a priori choice of channels or functions of the channels. As the *V* retrieval is close to linear, a simple multi-linear regression on $T_B^{V,H}(\nu_i)$ has often been used.

The most well-known purely statistical algorithms for SSM/I are those of Alishouse et al. (1990a, b) for water vapor, which is a linear combination of T_B^V at 19, 22 and 37 GHz channels, with addition of a quadratic term in $T_B^V(22)$ to better account for the non-linearity, and for *Q*, as linear function of brightness temperatures $T_B^V(19)$, $T_B^H(19)$, $T_B^V(37)$ and $T_B^H(85)$.

Alishouse et al. (1990a) algorithm is used up to now to retrieve total water vapor content from SSM/I and AMSR-A brightness temperatures and get animations of global and regional *V*-fields (Wimmers and Velden 2007) showing, in particular, evolution of tropical cyclones and their interaction with other atmospheric pressure systems and the ocean (http://tropic.ssec.wisc.edu/real-time/mimic-tpw/global/main.html).

The other ones are physically-based, resting on the Radiative Transfer Equation (RTE) simulations (Mitnik 1969; Basharinov et al. 1974; Greenwald et al. 1993; Wentz 1992, 1997; Petty and Katsaros 1992; Liu and Curry 1993; Grody et al. 2001; Ferraro et al. 2005; Horváth and Davies 2007; Mitnik and Mitnik 2003, 2006, 2010; Meissner and Wentz 2006; O'Dell et al. 2008; Yan and Weng 2008; Weng et al. 1997, 2003, 2007). The established advantage of physically-based algorithms is their well developed theoretical basis and physical background of each step.

Neural Networks (NNs)-based algorithms can also be both statistically and physically-grounded. Their general point is that they don't require a *priori* knowledge of the transfer function, which can be of any non-linear character. This method has been used successfully in a large variety of remote sensing applications, including our specific case of *V*- and *Q*-parameter retrievals (Jung et al. 1998; Krasnopolsky et al. 2000; Mallet et al. 2002; Moreau et al. 2002; Bobylev et al. 2010).

Absorption spectrum by small cloud droplets is determined by the spectrum of the dielectric permittivity of fresh water $\varepsilon(\nu)$ that, in turn, depends on the temperature of cloud droplets t_{cl} (Mitnik 1978; Mitnik and Mitnik 2010; Mätzler et al. 2010). Uncertainties in ε values, especially for the temperatures below −20 °C increase the errors of both T_Bs simulation and the satellite-derived *Q* values. There is no information on vertical distribution of liquid water in thin clouds in the microwave observations (Ebell et al. 2010).

The uncertainty in the water vapor emission model is the dominant error source for the satellite-derived V values. An improved model of microwave absorption in the atmosphere will result in better retrievals of V, Q, W and sea surface emissivity, all of which are important applications of microwave remote sensing (Turner et al. 2009; Liu et al. 2011).

The emissivity of the calm sea surface is determined by the spectrum of the dielectric permittivity of saline water that, in turn, depends on its temperature and salinity (Liu et al. 2011). Experimental and theoretical research on the emissivity of the rough sea surface under variable environmental conditions continues. There remain discrepancies in published data and theories that emphasize the difficulties in performing accurate experiments and arrive at their theoretical description (Webster et al. 1976a; Sasaki et al. 1987; Aziz et al. 2005; Shibata 2004, 2006; Uhlhorn et al. 2007; Boukabara and Weng 2008; El-Nimri et al. 2010; Liu et al. 2011). Problems increase at high winds and intense wave breaking that is typical for the tropical cyclone areas (Shibata 2006; Uhlhorn et al. 2007; Yan and Weng 2008; Mims et al. 2010).

Typical values of total water vapor content in tropical cyclone areas change from 40 to 50 kg/m^2 at the periphery of a vortex and in a band of dry air frequently observed in TC circulation system to high values of 60–70 kg/m^2 in the central area and in spiral rain bands. The values of total cloud liquid water content can reach 2 kg/m^2 and more in the eye wall and in rain bands. Clouds with high Q-values are surrounded by clouds with the lower Q values that, in turn, are imbedded in the moist air mass. Typically the size of the area with the increased V-values is significantly larger than the area with $Q > 0.5$ kg/m^2. Retrieval of V and Q values at the areas of high atmospheric attenuation caused by water vapor, clouds and precipitation over the sea surface disturbed by storm winds is a complicated task. Nonetheless, great progress has been made as illustrated in Sect. 2.4.

2.2.4 Scatterometry

2.2.4.1 Satellite Instruments 1978–Present

Scatterometer instruments were developed to infer the surface wind vectors over the oceans from space. The principle of their operation was famously illustrated by the scatterometer on the SEASAT satellite in 1978 (e.g. Jones et al. 1982, and review by Katsaros and Brown 1991), but had earlier been proven by circle flights with a prototype radar. The relative position between radar look direction and the wind direction showed clear effects on the radar "backscatter", a diffuse radar return, different from reflection. The SEASAT satellite carried the well established visible and infrared sensors, as well as several experimental instruments including a Ku-band scatterometer, i.e. a radar system consisting of 3 "stick" antennae, whose returns depend on the roughness of the sea and from which wind vectors (wind speed and direction) can be inferred. (Full discussion of the physics and techniques as well as other issues in this section can be obtained from the excellent text by Robinson 2004). The SEASAT Ku-band scatterometer data allowed scientists to learn how to interpret radar returns from 3 stick antennae, as they consecutively viewed the same area on the sea surface under the advancing satellite. The algorithm should return wind speed and direction by inverting an empirical model relating measured scatterometer backscatter coefficients (in a matrix obtained by the 3 forward looking antenna and the 3 backward looking antenna observing the same pixel on the sea surface) and the comparison data on surface wind speed and direction. The geophysical model-function, GFM, is calibrated by use of in situ and/or output from atmospheric numerical models collocated in space and time with scatterometer measurements (e.g. Bentamy et al. 1999; Stoffelen 1999; Wentz et al. 2001). Since 1991 there have been several scatterometers in space, two from the European Space Agency, ESA, on the European Remote Sensing satellites 1 and 2, ERS 1/2, launched in 1991 and 1995, respectively. This ESA scatterometer, the Advanced Microwave Instrument (AMI) "sees" well through clouds since it is based on C-band radar technology, but such scatterometers are relatively insensitive to the small ocean waves at low wind speeds, while the Ku-band instruments are more adversely impacted by clouds and precipitation, but respond better to the small waves or roughness elements at low wind speeds. A Japanese–U.S. collaboration lead to the launch of Ku-band scatterometers, NSCAT (NASA scatterometer) on the Advanced Earth Observing Satellite (ADEOS) in 1996. Unfortunately, ADEOSI had an early demise, which lead to the fast launch in 1999 of QuikSCAT, a NASA satellite carrying the SeaWinds, a scanning scatterometer. This scatterometer was a copy of the instrument designed for ADEOS II, which was launched in 2002, Fig. 2.3. The SeaWinds on QuikSCAT was very successful (e.g. Ebuchi et al. 2002; Katsaros et al. 2001).

It collected data for a full 10 years allowing many new analysis techniques and algorithms to be tested. For example, Long (2004) achieved higher spatial resolution by sophisticated processing of the original data. When ADEOS II was flying its scatterometer, there was a period of a tandem sampling by two identical scatterometers, until ADEOS II also met an early demise (Liu et al. 2008b). The SeaWinds scatterometer has a continuous 1,800 km wide swath, including the nadir region, which has proven very important. One such instrument views the global ocean once every 12 h at 50° latitude, but less frequently in the tropics. The sampling with several scatterometers is highly desirable due to the rapidly changing wind field at any one location.

Since October 2007, the new Advanced Scatterometer, ASCAT, onboard MetOp-A, launched by EUMETSAT, is collecting data for an operational agency. EUMETSAT is the European organization for meteorological satellites for weather, climate and environmental applications. It provides valuable surface wind information with high spatial and temporal resolution over the global ocean using two C-band beams, each side of nadir. MetOp, Europe's newest weather satellite, was successfully launched on 17 September 2012. Metop-B will first follow and then replace its identical twin, Metop-A, in a polar, sun-synchronous orbit.

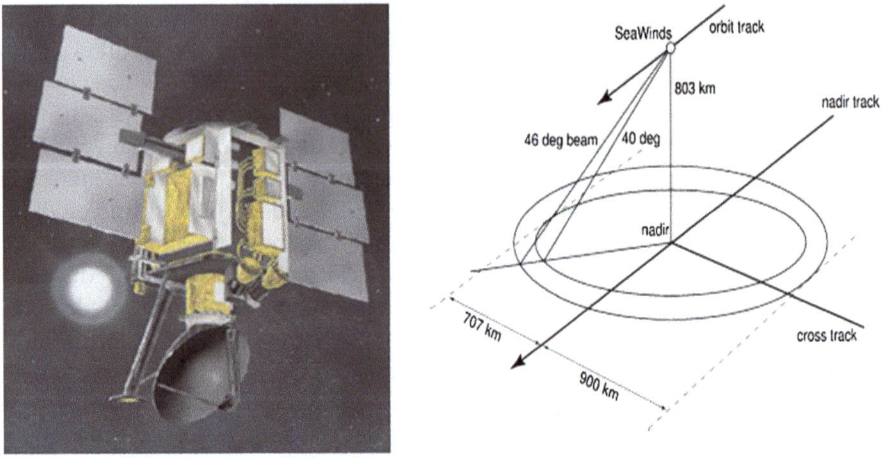

Fig. 2.3 SeaWinds on QuikSCAT satellite illustrating the sampling cone. *Left* The SeaWinds instrument on board QuikSCAT. The rotating disk antenna is 1 m in diameter (courtesy NASA/JPL). *Right* A schematic of the measurement geometry of the SeaWinds instrument on board QuikSCAT (from Freilich and Dunbar 2000). As indicated in the figure the far swath extends from 700 to 900 km on either side. The nadir swath is 499 km wide, centered on nadir

2.2.4.2 Wind Stress Versus Wind Speed

Interpretation and use of scatterometers data requires use of an atmospheric boundary layer model and understanding of the physics of the measurement. It was clear from the beginning that the backscatter signal being dependent on the sea surface roughness dominated by small waves, gravity and capillary waves, in a wavelength band that resonates with the radar's wavelength requires a direct relationship of these waves to the quantity of interest, which could be mean wind speed at a certain height or the wind stress at the sea surface. The short waves are generated by the local wind forcing of the ocean and are therefore related to the surface stress exerted by the atmosphere. A complexity arises due to some of the atmospheric stress being carried by the form drag due to larger waves, especially at high wind speeds. The conversion of the radar backscatter to a wind stress or wind speed is therefore rather complex and one "model function" may not be universally applicable. However, experience has shown that the current formulations for the different scatterometer systems perform quite well. They are strictly empirical fits and therefore depend on the quality of the "comparison data" from the sea surface.

Once we accept that the scatterometer measures wind stress, we then often need to convert it to a wind speed for many practical purposes, not least of which is for calibrating with the comparison data, which often is simply mean wind speed (and direction). That conversion involves use of a boundary layer model, since the wind speed at any one height also depends on the atmospheric stratification as defined in Monin–Obukhov theory (e.g. Businger et al. 1971). The equivalent Neutral wind speed is that speed in neutral atmospheric stratification that would give the same stress as the scatterometer measured. The convention is to translate the comparison data to the value of the "equivalent neutral wind" at 10 m height before the inversion of the backscatter signals take place (e.g. Liu and Tang 1996). Thus, it is this Equivalent Neutral Wind speed and direction that is the output of the model function. It often involves up to 3 "aliases", since the uncertainty in the signals, the comparison data and in the model function itself sometimes returns more than one plausible solution. Methods to eliminate the false aliases have been described (e.g. Woiceshyn et al. 1986). Currently the TOGA Coare3.0 boundary layer algorithm (Fairall et al. 2003; Bentamy et al. 2003) is used to convert the comparison data to 10 m neutral wind speed, when algorithms are tested and verified. However, sometimes the atmospheric stratification is not available, which makes the intercomparison inexact (W. T. Liu, personal communication, 2012).

The typical credentials for scatterometers have included wind speed accuracy of 2 m s^{-1} and wind direction within 20° (e.g. Jones et al. 1982; Ebuchi et al 2002). More complex concerns stem from the inherent limitation of the physics in that the wind speed measured is really the difference between the wind stress on the surface and the current vector in the sea (Liu et al. 2008a, b). The effects of the stratification in patterns of strong sea surface temperature gradients can also show up as variability in the surface wind as interpreted from the scatterometers (equivalent neutral 10 m wind, which can be directly converted to friction velocity). This variability can result from the atmospheric stress being transferred differently to the surface in stable and unstable sections of the field of view. For many application the 10 m neutral wind or the wind stress is the quantity of interest, but if the actual wind, say at 10 m height or any other height is needed, additional information on the atmospheric temperature gradient (and to a lesser extent the humidity gradient) must be furnished. The articles by Liu et al. (2008b) also discuss the role of inadequate sampling by just one scatterometer for many purposes and the opportunity to enhance the sampling by coordination between many satellite wind sensing systems.

2.2.5 Synthetic Aperture Radars, SARs

The active microwave sensors such as altimeters, scatterometers and Synthetic Aperture Radars (SARs) (Campbell 2002; CCRS 2009; Jackson and Apel 2004) are radars operating in the microwave region (1–30 GHz in frequency, 1–30 cm in wavelength) at different polarizations and incidence angles. Over the sea, the radar return depends, on the degree of development of the sea surface roughness (Valenzuela 1978). The roughness is composed of centimeter-decimeter water waves produced by the wind. Since the wind field has its own spatial pattern, which depends on its strength, on the thermodynamic characteristics at the air-sea interface and on the interaction between the wind flow and the orography near coastlines or islands, it generates certain spatial features. The radar backscatter reacts to the sea surface roughness. Therefore, the study of the characteristics of the radar backscatter provides information on the characteristics of the wind and of the Marine Atmospheric Boundary Layer (MABL, see e.g. Katsaros et al. 2000). In this layer, important exchanges of sensible and latent heat, CO_2 and momentum take place over a large spectrum of temporal frequencies and spatial scales. It is the wind stress that drives the surface waves and the ocean drift currents (e.g. Bentamy et al. 2003; Zecchetto 2010; Baklanov et al. 2011; Liu and Xie 2008).

At present, there are several satellites equipped with Synthetic Aperture Radar (SAR) instruments: the European ENVISAT launched in March 2002 (ESA 2002), the German TerraSAR-X launched by Deutsche Luft und Raumfahrt in June 2007 (Buckreuss et al. 2003), the Italian Constellation of Small Satellites for Mediterranean basin Observations (COSMOSkyMed or CSK), developed by the Agenzia Spaziale Italiana (ASI) (since June 2007) (Coletta et al. 2007), the Canadian commercial satellite RADARSAT-1 (launched in November 1995) and RADARSAT-2 (launched in December 2007) (Morena et al. 2004), the Indian RISAT launched in April 2012. Below only the Envisat Advanced SAR (http://envisat.esa.int) and RADARSAT-1 and-2 SAR instruments (Hurley 2010) operating at C-band will be considered, since they have wide swath mode with swath width of 400 and 500 km and SAR image can cover a significant part of a tropical cyclone. Spatial resolution is approximately 150 by 150 m for Envisat ASAR and 100 × 100 m for RADARSAT-1 and -2 SARs (Fig. 2.4). In the Wide Swath Mode, the ScanSAR technique is used, providing images of a wider strip (405 km)

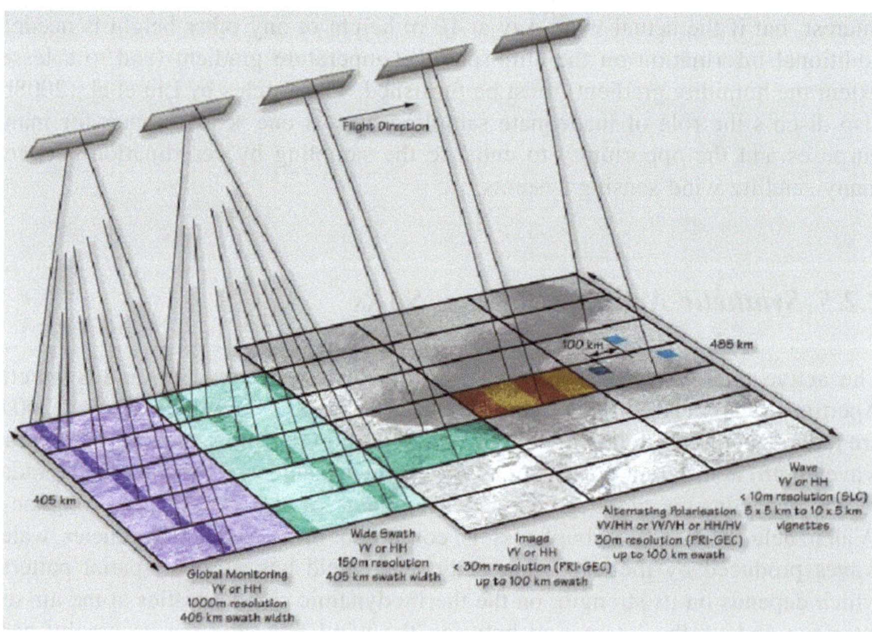

Fig. 2.4 Envisat ASAR operating modes

with medium-resolution (150 m) in HH or VV polarisation. The total swath consists of five subswaths and the ASAR transmits bursts of pulses to each of the subswaths in turn in such a way that a continuous along-track image is built up for each subswath.

The increased availability of satellite SAR images offers to scientists many opportunities to investigate the structure of the Marine Atmospheric Boundary layer, MABL, in tropical cyclones. Scientific literature about SAR images over the ocean has shown a variety of geophysical phenomena detectable by SAR (e.g. Jackson and Apel 2004), including the multi-scale structure in the atmospheric turbulence under high winds and the structure of the convective turbulence under low wind.

Satellite based Synthetic Aperture Radar (SAR) can operate day and night and in all weather condition. Radar signals penetrate clouds and precipitation that presents important advantage for tropical cyclone study compared to visible and infrared sensor such as AVHRR and MODIS. Furthermore, radars are truly sensing the surface and therefore observe the impact of the atmosphere on the ocean surface.

The high-resolution of SAR wind retrievals make these data suitable for understanding tropical cyclone wind and precipitation structure. Examples of high-resolution Envisat SAR images show the details which SARs can measure in examples from Northwest Pacific typhoons Aere and Songda in 2004 (Mitnik et al. 2005a), Talim and Matsa in 2006, Sinlaku in 2008 (Dagestad et al. 2010), Parma in 2009, Megi in 2010 (reported on in this chapter, Sect. 2.4) as well as Atlantic hurricanes such as Katrina in 2005 (Repucci et al. 2010).

2.2.6 Altimeters

Altimeters measure the height of the satellite above the sea surface, as the name implies. They obtain a narrow swath of 7 km width in the nadir track of the polar orbiting satellites that carry them. Satellite altimeters are designed to deliver relatively long revisit (10 days and more) and global views of sea surface height. The concept is well established since early missions GEOS-3 and SEASAT that date back to the 1970s. The launch of the TOPEX/Poseidon in 1992 provided the greatest impetus for satellite altimetry research in the 20th century. Its launch was followed by the Jason-1 (2001) and Jason-2 (2008). The European space Agency, ESA, satellites were launched in 1991, 1995, 2002, 2010 (ERS-1, ERS-2, ENVISAT, Cryosat-2) and the US Navy launched a series of Geosat satellites (1985 and 1998). Additional missions are planned in order to provide continuous coverage, e.g., AltiKa, Sentinel-3, HY-2, including new experimental concepts such as SWOT (Fu et al. 2009), which uses multiple frequencies and sampling by a scanning antenna. To date satellite altimetry has focused in the open ocean (Fu and Cazenave 2001), but interest in coastal observation is leading to new research (e.g. Vignudelli et al. 2011). An uninterrupted flow of altimeter data has been accumulating, contributing to the ability to address scientific and societal challenges related to the ocean. However, in some cases the various missions were designed without continuity in mind, with different observing strategies usually driven by their particular objectives of accuracy, spatial resolution and temporal revisit requirements. For example, the T/P and Jason series main objective was to generate the best estimates of sea level over time to serve climate monitoring, but these data can be also used to better understand the ocean circulation. As a result, data from the various satellite altimeters were processed independently. Nevertheless, the missions have common foundations concerning retrievals, orbits, geophysical corrections, data calibration and exploitation. Very useful and unique information on tropical cyclones derived by altimeters includes accurate wind speed and sea state data along their tracks the width of 5–10-km (Quilfen et al. 2010; 2011).

2.2.7 Airborne Radiometers for Surface Wind

Stepped Frequency Microwave Radiometer (SFMR) and Hurricane Imaging Radiometer (HIRAD), history and instrument descriptions. Estimating hurricane surface wind distributions and maxima is an operational requirement of the Tropical Prediction Center/National Hurricane Center (TPC/NHC), and emergency management decisions rely on coastal watches and warnings issued by NHC based partly on observed winds. Surface wind speed estimates by NHC are determined largely from extrapolated aircraft flight-level wind data. In 1997, Global Positioning System (GPS) dropwindsondes (Hock and Franklin 1999) first demonstrated the ability to provide in situ measurements of hurricane surface wind velocities, most importantly in the inner core, and recent work utilizing these measurements has improved the accuracy of extrapolations (Franklin et al. 2003).

Since 1984, the Stepped Frequency Microwave Radiometer (SFMR) has flown on at least one of two NOAA WP-3D research aircraft to estimate hurricane surface wind speeds (Uhlhorn and Black 2003). Beginning in 2004, a redesigned SFMR was flown on one WP-3D aircraft for operational surface wind speed measurements. After procedural testing during 2004, both WP-3Ds had operational SFMRs installed in 2005. The unusually active 2005 Atlantic hurricane season provided ample opportunity to evaluate the SFMR performance over the entire range of expected surface wind speeds (10–70 m/s). In particular, extreme wind speed (>60 m/s) measurements were obtained from flights into Saffir–Simpson (SS) category 5 Hurricanes Katrina and Rita.

Since hurricane reconnaissance began in 1947, numerous methods have been employed to estimate the distribution of surface winds in hurricanes. Sea state catalogs have provided a guide to determination of the wind speed (Black and Adams 1983). For many years surface winds have been estimated by flight-level measurements using various extrapolation algorithms (Miller 1958; Powell 1980; Powell et al. 1996). Maximum sustained winds have been estimated using pressure–wind relationships summarized by Harper (2002) and most recently described by Knaff and Zehr (2007). In studies prior to 1980, Basharinov et al. (1974), Nordberg et al. (1969), Ross and Cardone (1974) and Wilheit (1979) showed that passive microwave emissions from the sea surface are also strongly correlated with wind speed.

The concept for the first experimental SFMR was proposed by C. T. Swift at the University of Massachusetts Microwave Remote Sensing Laboratory (C. T. Swift 1976, personal communication) and built by the National Aeronautics and Space Administration's (NASA) Langley Research Center in 1978 (Harrington 1980). The SFMR design involved a single nadir-viewing antenna and receiver capable of making measurements of radio emission from the sea surface at four selectable frequencies between 4.5 and 7.2 GHz. The "stepping" procedure allowed for estimating the surface wind speed in hurricanes by correcting for rain-induced effects in the measurements, and therefore enabling recovery of the rain rate. The first measurements by the original SFMR were made from the National Oceanic and Atmospheric Administration (NOAA) WC-130 aircraft in Hurricane Allen in 1980 and reported by Jones et al. (1982), as well as Black and Swift (1984), Delnore et al. (1985), and Swift and Goodberlet 1992). By making assumptions about the vertical structure of the atmosphere together with SST measurements by a downward-looking airborne infrared radiometer, reasonable estimates of the ocean surface brightness temperature (TB) were made at 4.5, 5.0, 5.6, and 6.6 GHz. Wind speeds were then calculated assuming a linear increase in wind speed with TB, independent of frequency. Agreement between surface (20 m) winds extrapolated from the 1,500-m flight level and the SFMR estimates for independent flight legs were within 6–10 %. Despite the success in Allen, this instrument was never again flown into a hurricane.

The SFMR measures nadir brightness temperature (TB) at six C-band frequencies, and a retrieval algorithm uses a Geophysical Model Function (GMF) relating surface emissivity and wind speed to produce surface wind speed estimates along the flight track. Early emissivity/wind speed models used in microwave radiometry applications have been developed for winds <25 m/s (Goodberlet et al. 1989; Wentz et al. 1986; Wentz 1983; Webster et al. 1976a), or either relied upon or were validated against aircraft flight-level high wind data extrapolated to the sea surface (Uhlhorn and Black 2003; Tanner et al. 1987; Black and Swift 1984; Swift et al. 1984; Jones et al. 1981). These early methods resulted in considerable uncertainty about the retrieved hurricane surface wind speed, especially under extreme conditions.

A second SFMR was designed and built in 1982 under the supervision of Swift (Swift et al. 1986). The number of frequencies was expanded to six between 4.6 and 7.2 GHz, and the instrument integration time was reduced to less than 1 s, resulting in improved spatial resolution. A new retrieval algorithm was also implemented and described in Tanner et al. (1987). This instrument was flown on board the NOAA WP-3D in 1984 and during 12 flights during the 1985 hurricane season. The SFMR was further modified in 1986 and initially used for studies of sea ice structure (St. Germain et al. 1993). Using data obtained in Hurricanes Earl (1985), Gilbert (1988), and Hugo (1989), the empirical emissivity–wind speed relationships were refined to include winds over 60 m/s. With support from the Office of the Federal Coordinator for Meteorology (OFCM) the existing horn antenna was replaced with a dipole array antenna in 1993. The new antenna with a new set of six frequencies was flown in Hurricane Olivia (1994) and retrieved high quality wind estimates. Further funds were provided by OFCM for an upgrade of the SFMR's receiver, which allowed for increased calibration stability. The reconfigured SFMR (Goodberlet and Swift 1996) was first flown in Hurricane Jerry in 1995. Minor modifications were made to reduce background noise levels after the 1995 season.

The Stepped Frequency Microwave Radiometer represents a potentially significant advancement in the remote measurement of hurricane near-surface wind speeds, most notably at speeds >50 m/s. This improvement is due to both refined remote sensing technologies, and more accurate ground-truth data in the form of GPS dropwind-sonde wind speed measurements. The SFMR yields wind speed measurements that are overall within ±4 m/s rms of the dropwind-sonde-estimated surface wind. Figure 2.5 illustrates the cross-track scanning ability of the new Hurricane Imaging RADiometer (HIRAD) based on a WP-3D installation compared to the nadir-viewing SFMR. The figure also shows the sampling of the

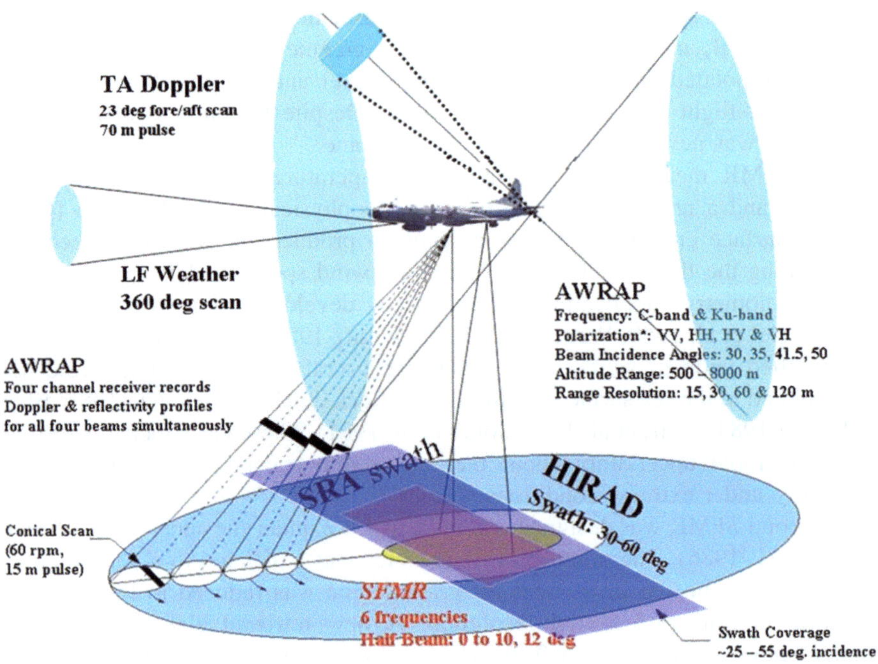

Fig. 2.5 Remote sensing instruments on the NOAA WP-3D aircraft. These include the 360° scanning lower fuselage (LF) for vertically integrated radar reflectivity mapping; the TAil (TA) Doppler radar for 3-D wind and reflectivity mapping (0.5–12 km altitude); stepped frequency microwave radiometer (SFMR—*yellow oval* swath) for surface wind and integrated rain rate; wideswath scanning radar altimeter (WSRA—*pink rectangular* swath) for directional surface wave spectra; hurricane imaging RADiometer (HIRAD), scanning SFMR system now installed on NASA global hawk; airborne wind and rain atmospheric profiling radar, now a dual-frequency, dual polarized system for wind and rain 3D profiles from flight level to surface. A scanning Doppler lidar system has recently been added for wind profiling in clear air

atmospheric wind field above the sea surface by the tail-mounted, X-band Doppler radar operation in a fore-aft scanning mode. Details of this important technology is not included in this review.

Surface winds in hurricanes have been estimated remotely using the Stepped-Frequency Microwave Radiometer (SFMR) from the NOAA WP-3D aircraft for the past 31 years. Since the use of the GPS dropwindsonde system in hurricanes was first initiated in 1997 as a 'gold standard' for verifying and calibrating SFMR winds, routine collocated SFMR and GPS surface wind estimates have been made. During the 1998, 1999, and 2001 hurricane seasons, a total of 249 paired samples were acquired and compared.

The SFMR equivalent 1-min mean, 10-m level neutral stability winds were found to be biased high by 2.3 m/s relative to the 10-m GPS winds computed from an estimate of the mean boundary layer wind. Across the range of wind speeds from 10 to 60 m/s, the RMSE was 3.3 m/s. The bias was found to be dependent on storm quadrant and independent of wind speed, a result that suggests a possible relationship between microwave brightness temperatures and surface wave properties. Tests of retrieved winds' sensitivities to sea surface temperature, salinity, atmospheric thermodynamic variability, and surface wind direction indicate wind speed errors of less than 1 m/s above 15 m/s.

For the first time, the NOAA/Aircraft Operations Center (AOC) flew Stepped Frequency Microwave Radiometers (SFMRs) on both WP-3D research aircraft for operational hurricane surface wind speed measurement in 2005. An unprecedented number of major hurricanes provided ample data to evaluate both instrument performance and surface wind speed retrieval quality up to 70 m/s (Saffir–Simpson category 5). To this end, a new microwave emissivity–wind speed model function based on estimates of near-surface winds in hurricanes by GPS dropwindsondes is proposed. For practical purposes, utilizing this function removes a previously documented high bias in moderate SFMR-measured windspeeds (10–50 m/s), and additionally corrects an extreme wind speed (>60 m/s) underestimate. The AOC operational SFMRs yield retrievals that are precise to within ±2 % at 30 m/s, which is a factor of 2 improvement over the NOAA Hurricane Research Division's SFMR, and comparable to the precision found here for GPS dropwindsonde near-surface wind speeds. A small (1.6 m/s), but statistically significant, overall high bias was found for independent SFMR measurements utilizing emissivity data not used for model function development. Across the range of measured wind speeds (10–70 m/s), SFMR 10-s averaged wind speeds are within 4 m/s (RMS) of the dropwindsonde near-surface estimate, or 5–25 % depending on speed. However, an analysis of eyewall peak wind speeds indicates an overall 2.6 m/s GPS low bias relative to the peak SFMR estimate on the same flight leg, suggesting a real increase in the maximum wind speed estimate due to SFMR's high-density sampling. Through a series of statistical tests, the SFMR is shown to reduce the overall bias in the peak surface wind speed estimate by order 50 % over the current flight-level wind reduction method and is comparable at extreme wind speeds.

The updated model function is demonstrated to behave differently below and above the hurricane wind speed threshold (~32 m/s), which may have implications for air–sea momentum and kinetic energy exchange. The change in behavior is at least qualitatively consistent with recent laboratory and field results concerning the drag coefficient in high wind speed conditions, which show a fairly clear "leveling off" of the drag coefficient with increased wind speed above ~30 m/s (Donelan et al. 2004; Powell et al. 2003). Finally, a composite analysis of historical data indicates that the earth-relative SFMR peak wind speed is typically located in the hurricane's right-front quadrant, which is consistent with previous observational and theoretical studies of surface wind structure.

Recent studies (Dunion et al. 2003; Franklin et al. 2003; Powell et al. 1999, 2003) indicate that boundary layer models used to extrapolate flight-level wind speeds (Powell 1980; Deardorff 1972; Cardone 1969; Blackadar 1965) show a tendency to underestimate surface wind speeds >50 m/s. Anecdotal evidence also suggested that the SFMR underestimated surface wind speeds in similar conditions. Utilizing the previous GMF, comparisons of SFMR-derived wind speeds with GPS surface-reduced, 0–500-m layer-averaged winds indicated no such tendency (Uhlhorn and Black 2003), but, in fact, very few in situ data were available at speeds above 50 m s^{-1}. However, when compared with surface-adjusted lowest 150-m layer-averaged winds obtained in 2004, a low bias in SFMR retrievals became apparent at extreme wind speeds. With the recent surge in particularly intense hurricanes, analyzing and correcting this anomalous behavior has become critical. During 2005, a large dataset of contemporaneous SFMR and GPS dropwindsonde observations was obtained, resulting in significant improvement to the empirically derived SFMR emissivity–wind speed GMF, most notably at extreme wind speeds. This new model function has been implemented operationally on NOAA WP-3D SFMRs beginning in 2006, and on SFMRs installed on Air Force Reserve Command (AFRC) WC-130J aircraft since 2007.

The observation of a linear increase in sea surface emissivity with wind speed at hurricane force is a shift in our understanding of sea surface radiometric properties. In practical terms, the previous quadratic emissivity–wind speed model consistently underestimated winds at speeds >50 m/s, and the new linear GMF corrects this anomaly. From a scientific standpoint, additional evidence is found for a distinct physical change in the sea surface when surface winds exceed ~30 m/s, and further, that momentum and kinetic energy exchanges are relatively reduced at high wind speeds (Donelan et al. 2004). This gain in understanding may, potentially, lead to improvements in air–sea exchange parameterizations, ultimately resulting in superior hurricane intensity forecasts.

2.3 Tropical Cyclones, a Few Examples

In this section we describe a couple of special uses of microwave remote sensing that have benefitted TC understanding. These examples are in addition to the sequence about typhoon MEGI from 2010, that we cover with microwave observations that have become routine for real-time ascertaining of a TC's size, track and intensity. We discuss some particular uses of passive microwave, TRMM PR profiling, the sequence of scatterometer winds in the early stages of a TC, altimeter data used to assess heat content of the upper ocean in advance of a storm, the amazing looks of hurricane eyes seen with RADARSAT 1 SAR and last the climatology of total rainfall from landfalling TCs based on TRMM's long term record.

2.3.1 Multi-sensor Studies of TCs

The most advanced approach for obtaining quantitative data on tropical cyclones (and other marine weather systems) is the use of observations from several satellites by different sensors. These data are used extensively for both operational applications and for research on TC physics. An excellent source of multisensor data is a website developed at the Naval Research Laboratory Marine Meteorology Division (MMD), Monterey, California, USA: http://www.nrlmry.navy.mil/TC.html (see also Hawkins et al. 2001). Another source of multisensor data on tropical cyclones is a web site developed by the Earth Observation Research Center of the Japan Aerospace Exploration Agency: http://sharaku.eorc.jaxa.jp/TYP_DB/index_e.shtml where tropical cyclone data observed by TRMM's Precipitation Radar, TMI and VIRS, Aqua/AMSR-E and Midori-II (ADEOS-II)/AMSR are available for the period 1997–2011. The multisensor approach with a focus on microwave techniques is in ever-growing use. It was described in many papers devoted to estimates of particular parameters (wind, temperature, precipitation, etc.) and to diagnose and study structure and evolution of individual tropical cyclones. In particular, a novel morphing algorithm called the Morphed Integrated Microwave Imagery at the Cooperative Institute for Meteorological satellite Studies (CIMES) (MIMIC) was introduced by Wimmers and Velden (2007) to improve visualization of tropical cyclones, which in turn enhanced the ability to analyze and forecast these events. The authors' approach was to blend sequential individual microwave images together through a "morphing" process, which is the digital manipulation of two or more images to make them appear to change into one another naturally. Kidder and Jones (2007) developed a process for blending the total water vapor content values retrieved from the AMSU instruments on three NOAA satellites, and the SSM/I instruments on three DMSP satellites by application of the Alishouse (1990a) algorithm. The data are then mapped to a projection useful to forecasters and composited for 12 h to make a global map. These maps are produced hourly using Data Processing and Error Analysis System (DPEAS) software and made available to forecasters online. (See also papers by Hawkins et al. 2001; Knaff et al. 2011; Prasad and Singh 2007). Below two examples of multi-sensor study of tropical cyclones.

2.3.1.1 Tropical Cyclone Sinlaku (2008)

Typhoon *Sinlaku* developed as a tropical depression in the western Pacific Ocean, east of the northern Philippine Islands on September 8, 2008 and strengthened to a tropical storm the same day. The storm's outer rain bands intensified seasonal monsoon rains, causing flooding on the Philippine island of Luzon. On passing the Philippines, Sinlaku strengthened to become a typhoon and attained Category 4 status on September 10, 2008. Central pressure dropped sharply to 935 mb at 12 UTC on September 10 and in a day began to increase slowly and reached 945 mb at 12 UTC on September 13. Maximum sustained wind was around 50–64 m/s.

Results of *Sinlaku* sensing taken by satellite active and passive microwave instruments on September 10 are shown in Fig. 2.6. An Envisat ASAR image was acquired at 01:30 UTC. The *Sinlaku* eye looks as a dark circle due to weak winds. Its diameter is approximately 30 km. The light tone area surrounding the eye and elongated to the north marks the storm wind zone. Spiral rain bands of various length and width as well as individual rain cells appear as negative and sometimes positive radar contrast against the background (Fig. 2.6a). Magnitude and sign of the contrast depend on both the sea surface roughness and characteristics of the rain. In turn, rain falling on the sea surface can increase or decrease its roughness and also cause a decrease of radar signals due to twofold attenuation in the atmosphere.

MetOp ASCAT scatterometer data reveal the large-scale features of the sea surface wind field: large zones of cyclonic circulation to the north and to the south of *Sinlaku*, the increase of wind towards the typhoon center, wind shadows to the east of Luzon and to the south of Taiwan (Fig. 2.6b, c).The typhoon area is characterized by high values of the total water vapor content (Fig. 2.6e). Total cloud liquid water content exceeds 1–1.5 kg/m^2 in a large circular area around the center as well as in rain bands (Fig. 2.6f) in which heavy rains are typical. Strong attenuation by rain and clouds hinders wind speed estimates with the use of passive microwave measurements. As a result, AMSR-E retrieved wind fields cover only the areas with lower atmospheric absorption (Figs. 2.6d, g). The large-scale features of these fields agree well with the scatterometers data (Fig. 2.6b, c).

ENVISAT ASAR image of *Sinlaku* is seen in Fig. 2.7, which illustrates the fine details provided by this high performance SAR.

2.3.2 Altimeters and Upper Ocean Heat Content

A valuable application of altimetry in regard to TC research and operations has been the incorporation of altimetric data to estimate the depth of the warm water available as a heat source for hurricane intensification in the Caribbean and Gulf of Mexico region (Mainelli et al. 2008; Goni et al. 2009) and in the western Pacific ocean (Lin et al. 2008). Assimilation of altimeter data into operational oceanographic models and the increased precision (of the order of 1 cm in sea surface height), makes this

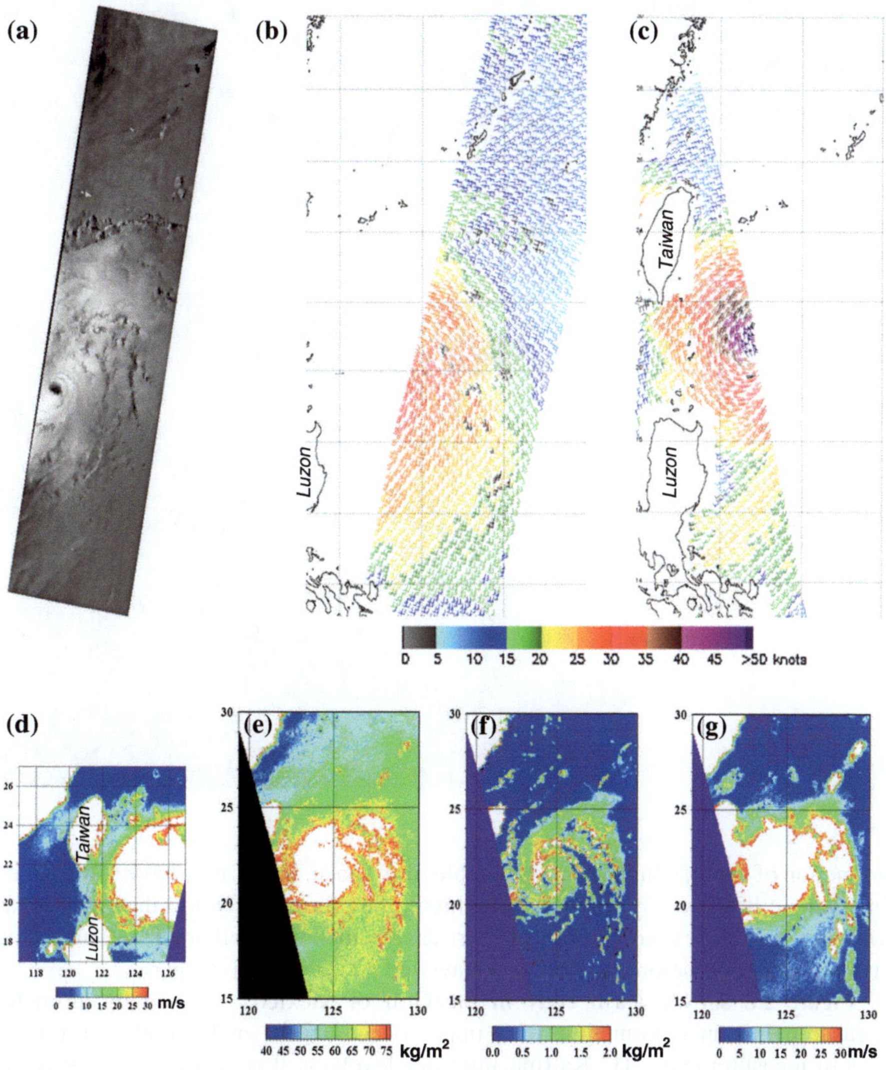

Fig. 2.6 Typhoon Sinlaku sensing by active and passive microwave sensors: Envisat ASAR image at 01:30 UTC (**a**), wind field derived from MetOp ASCAT at 01:37 UTC (**b**) and at 12:43 UTC (**c**) and from Aqua AMSR-E at 17:50 UTC (**d**) on 10 September. Fields of total atmospheric water vapor content (**e**), total cloud liquid water content (**f**) and wind speed (**g**) retrieved from Aqua AMSR-E measurements at 04:45 UTC on 11 September 2008

technology ready for operational use. EUMETSAT and NOAA have already established cooperation agreements with regard to altimetry. The method for using altimeter data when estimating the heat content in warm eddies ahead of the arrival of a TC, includes assimilation of the altimeter information into a numerical model of the upper ocean, which has been initialized with buoy data of the temperature structure as

Fig. 2.7 Spiral rain bands converging to Sinlaku eye on Envisat ASAR image acquired on 10 September 2008 at 01:30 UTC

a function of depth, which is not obtainable from space. Buoys may have to be placed in strategic locations by aircraft deployment, but this is well worth the expense, as was proven in the case of the Katrina hurricane, which was well predicted, including its intensification before landfall as it travelled across a warm eddy.

Figure 2.8 shows warm eddy in the Gulf of Mexico, revealed by altimeter measurements in conjunction with upper ocean model and initialization from in situ measurements. TC Katrina intensity levels as it passes over this eddy is illustrated with the size and color of the circles along the track.

2.3.3 SAR Observed Eyes of Hurricanes

The SAR of RADARSAT 1 and 2 is only 500 km wide in the so-called wide mode sampling. That means that it will not often observe a whole TC within its swath. Further, the Canadian Space Agency's agents, who operated the satellite sampling, needed requests for taking an image many days in advance of the satellite passing

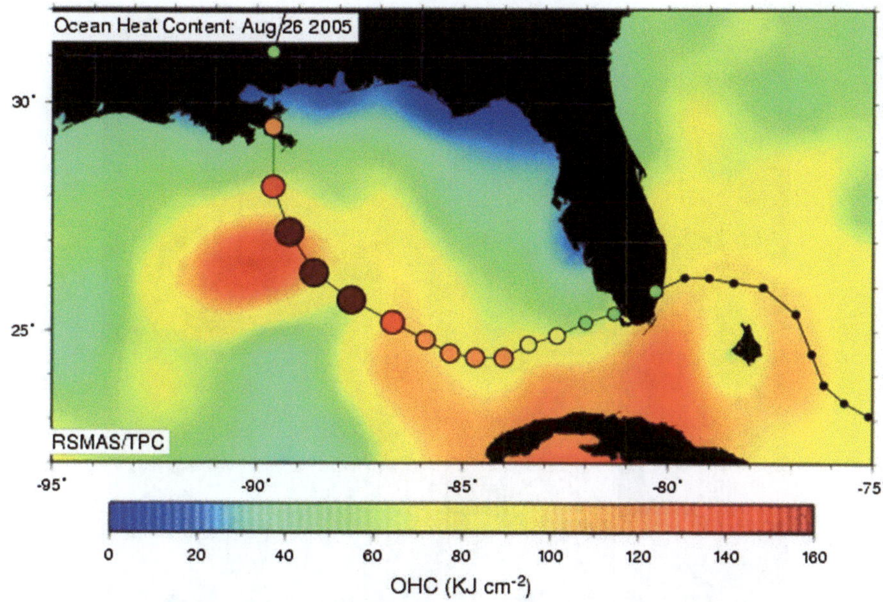

Fig. 2.8 Ocean heat content (OHC) for Hurricane Katrina used to derive OHC input to the statistical hurricane intensity prediction system (SHIPS) used operationally at the U. S. National Hurricane Center (NHC). Track of hurricane Katrina across the Gulf with *dots* a TS, *green* CAT1, *yellow circles* CAT2, *tan* CAT3, *red* CAT4 and *dark brown* CAT5. Rapid intensification began just after crossing the loop current boundary outlined in ocean heat content (OHC) from satellite altimeters as *yellow*, with deep *red* indicating a warm ring in the process of breaking off and moving westward in the Gulf. It is the movement away from this warm ring and onto the shallow, relatively cool shelf waters that accompanies Katrina' rapid weakening (Mainelli et al. 2008)

the desired location. Fortunately, smart operators pushed the button to take images of hurricane Mitch at category 5 in the Caribbean in the early days of RADARSAT 1 operations. A wonderful development followed in that the Canadian Space Agency developed a program called *Hurricane Watch*, in which the long lead times were waived and operators were able to catch the centers of many TC's (e.g. Iris and Burger 2004; Vachon et al. 2001). These images are haunting and interesting, but full interpretation of what exactly a SAR observes from space in high wind and heavy rain situations is not yet achieved. The images of TC eyes obtained during 3 hurricane season in Fig. 2.9 clearly show that the vortex may be distorted in ways that indicate mesoscale circulations or breakdown of the vortex.

Such features had been suspected in studies of damage on land, where the damage was higher in streak-like regions (Black, personal communication), which would be consistent with the higher wind speeds where a mesoscale circulation was added to the vortex wind speed. Breakdown of vortices have been found in laboratory and theoretical studies and similar features can sometimes be seen in visible images from geostationary satellites (Schubert et al. 1999; Kossin et al. 2002; Kossin and Schubert 2001, 2004; Montgomery et al. 2002).

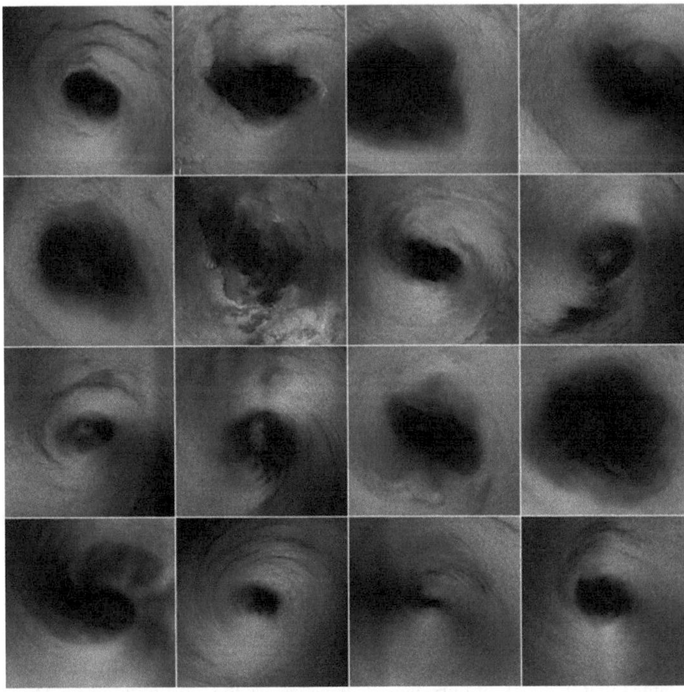

Danielle *31 Aug '98* **Dennis** *27 Aug '99* **Dennis** *29 Aug '99* **Dennis** *31 Aug '99*
Floyd *15 Sep '99* **Alberto** *17 Aug '00* **Florence** *13 Sep '00* **Dalila** *26 Jul '01*
Flossie *29 Aug '01* **Flossie** *1 Sep '01* **Erin** *11 Sep '01* **Erin** *13 Sep '01*
Felix *17 Sep '01* **Humberto** *27 Sep '01* **Juliet** *28 Nov '01* **Olga** *28 Nov '01*

Fig. 2.9 RADARSAT-1 SAR images of hurricane eyes (after Vachon et al. 2001) (© CSA, 1998, 1999, 2000, 2001)

2.3.4 Climatological Information on Precipitation From TRMM

A nice example of using TRMM's algorithms calibrated by its own rain radar and coastal radars is found in the climatological study of rainfall in tropical cyclones by Lonfat et al. (2004). The authors collected data from TRMM's TMI on precipitation associated with TC's for a 3 year period, January 1, 1998–December 31, 2001. They divided the 2121 instantaneous precipitation observations into 3 levels of storm intensity, tropical storm, Category 1–2 and 3–5 and separated the data for 6 ocean basins. They evaluated the precipitation in 10 km wide annuli around the storm center, out to a radius of 500 km, although in the outer regions their statistical sample was often too small due to the narrow swath of TMI. The study gave information about the asymmetry in the rainfall with respect to the storm track and as a function of storm intensity. They could calculate the total precipitation associated with each type of storm over its track. The data can be used to predict the rain falling on land after landfall and improve flood-risk forecasting. This subject has come a long way thanks to TRMM, which points the way to the next planned system described in Sect. 2.5. Stephens and Kummerow (2007) discuss the improvements in estimates that are possible when clouds and precipitation are analyzed in concert using CloudSat and TRMM information.

2.3.5 Airborne Passive Microwave Observations of TCs

As hurricane Katrina approached landfall in the state of Louisiana, USA, on August 28, 29 of 2005, the US instrumented aircraft, operated by the US Air Force for reconnaissance and for the Hurricane Research Division (HRD) of NOAA in Miami for both reconnaissance and research, were flying their characteristic patterns through the eye of this storm. It was a category 5 storm off shore with highest winds of 142 kt, but fortuitously the intensity declined to a category 3 just before landfall with surface winds estimated at 100 kt. Figure 2.10 is a display of the flight tracks, the estimated surface winds along the flight track by the SFMR on the HRD aircraft and by reduction of flight level winds at 700 mb to the surface from the W 3 130. Dropsonde wind speed estimates applying to 10 m height are also seen in these two figures. All 3 surface wind speed estimates show general agreement. Gridded wind fields based on these two data sets are also generated by the HWIND computer program operated by HRD and provided to the hurricane forecasters in near real time (Fig. 2.10e, f). The wind fields are similar, but there are also differences, even though the two wind fields apply at the same time. This indicates that the use of surface measurements with the SFMRs is desirable, but that inter comparisons and calibration efforts must continue.

2.4 Applications of Microwave Remote Sensing to Typhoon MEGI

Two of the authors of this chapter (LM and KK) experienced the outer regions of super typhoon Megi during the 2010 Pan Ocean Remote Sensing Conference (PORSEC), while in Keelung, Taiwan, during October 18–23, 2010. In particular, heavy rains in the northeast of Taiwan, on the Philippines and in China were connected with Megi. 45 persons died, 28 went missing. There were numerous landslides, roads and power lines were destroyed, the crop damaged, etc. General economic damage reached approximately 611 million US$. Several rainy days in

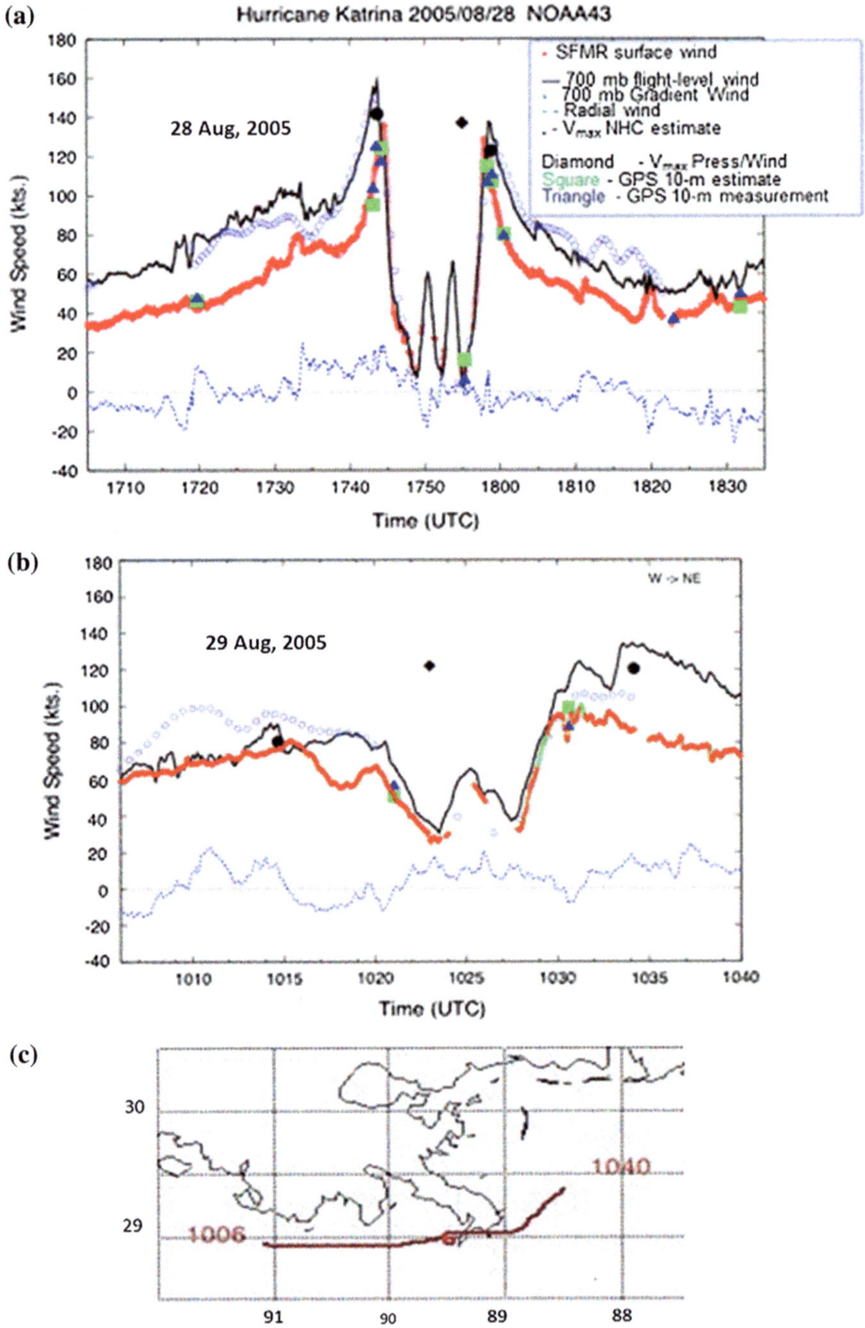
(a)
Hurricane Katrina 2005/08/28 NOAA43
28 Aug, 2005
SFMR surface wind
700 mb flight-level wind
700 mb Gradient Wind
Radial wind
V$_{max}$ NHC estimate
Diamond - V$_{max}$ Press/Wind
Square - GPS 10-m estimate
Triangle - GPS 10-m measurement
Wind Speed (kts.)
180
160
140
120
100
80
60
40
20
0
-20
-40
1710
1720
1730
1740
1750
1800
1810
1820
1830
Time (UTC)
(b)
29 Aug, 2005
W -> NE
Wind Speed (kts.)
180
160
140
120
100
80
60
40
20
0
-20
-40
1010
1015
1020
1025
1030
1035
1040
Time (UTC)
(c)
30
29
1006
1040
91
90
89
88

◀ **Fig. 2.10** WP-3D observations along west to east flight tracks in Hurricane Katrina on 28 and 29 August, 2005 (panels **a** and **b**), respectively, near peak intensity on the 28th and just prior to landfall on the 29th. Profiles of observed 700 mb wind and gradient wind speed computed from smoothed D-value (pressure minus radar altitude difference) profiles and SFMR surface wind speed. Also shown for comparison are dropsonde measured 10-m wind speeds, dropsonde estimated 10-m wind speeds (from mean boundary layer winds), NHC peak wind estimate (*black solid circles*) and NHC peak wind from pressure-wind relationship (*black diamond*). Panel **c** is the west to east flight track, just offshore, on the 29th. **d–g** H*WIND surface wind analysis centered on 0930 GMT 29 August. Panel **d** indicates analysis of surface wind estimated as 80 % of 700 mb flight-level measurement and panel **e** indicates surface wind analysis from SFMR surface wind measurements. Contours are in 5-kt intervals with *yellow–brown* transition indicating 50 kt and *red–magenta* transition indicating 90 kt. *White arrows* are estimated surface wind direction (flight level wind direction minus 25°). Panel **f** is the flight track and estimated surface wind barbs for the WC-130H aircraft flying at 700 mb and **g** is the WP-3D 650 mb flight track and surface wind barbs. Additional surface wind observations are indicated by the legend between panels

Keelung were caused by a spiral rain band of the typhoon. All participants of the conference watched *Megi*'s track every day. Moreover, the Internet allowed us to routinely obtain the data about Megi from several satellites: Aqua, Terra, TRMM, MetOp, Envisat and others which were processed with the use of both official and our original algorithms. These observations and ancillary information were included in the reports which were presented at a tropical storm session during PORSEC 2010. Research was continued in Russia by LM. This work, which has allowed a detailed quantitative description of *Megi* evolution is in part presented below. Fortuitously, a joint field program between the University of Taiwan and U.S. agencies (WHICH) was carried out in the South China Sea, using aircraft, ships and buoys during this same period. This program had as is goal to understand the structure and evolution of typhoons in this area and the influence of sea surface temperature patterns and upper ocean structures. One of us, PB, was in the field in charge of the aircraft program, and has been able to select representative example of microwave and other observations from the aircraft for the same period of time.

2.4.1 Megi Observed From Space

The Japan Meteorological Agency (JMA) marked a developing area of low pressure located southeast of Guam as a tropical depression on October 13. Early on that day, the Joint Typhoon Warning Center (JTWC) in Hawaii also classified the system as a tropical depression, giving it the identifier 15 W. Sea surface temperature exceeding 28 °C and high ocean heat content contributed greatly to strengthening of the depression. The low pressure system situated to the southwest of a subtropical ridge slowly tracked west-northwest towards the Philippines. Around 1200 UTC, the depression further intensified into a tropical storm, earning the name *Megi* from the JMA (Fig. 2.11).

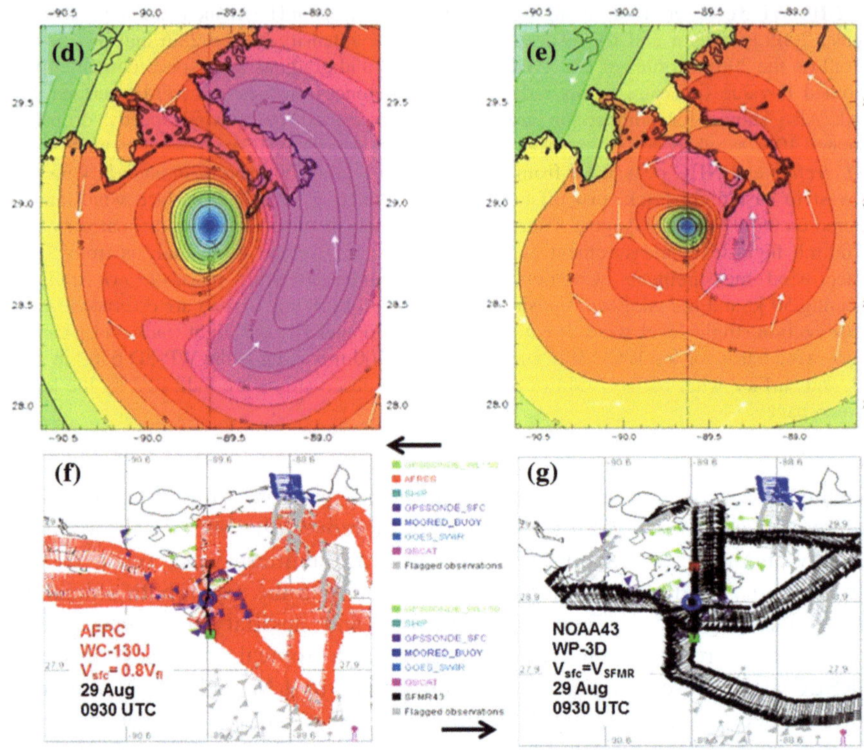

Fig. 2.10 continued

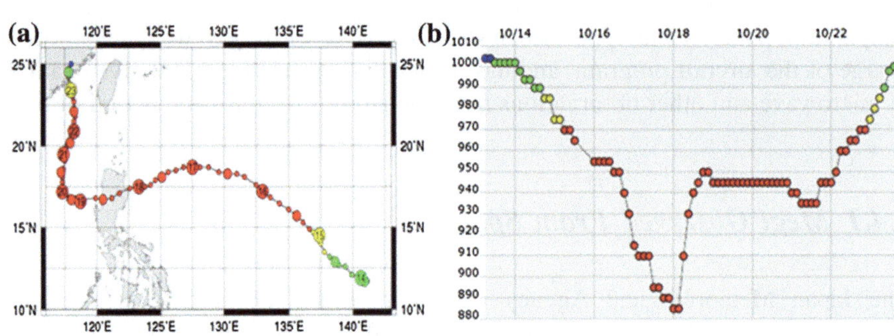

Fig. 2.11 Track (**a**) and central pressure (**b**) in super Typhoon Megi. The colors represent: *blue* tropical depression; *green* tropical storm; *yellow* severe tropical storm and *red* typhoon (http://agora.ex.nii.ac.jp/digital-typhoon/help/track.html.en)

Throughout the morning of October 14, satellite visible and infrared images showed that a dense overcast developed over the center of *Megi*. Later that day, an eye appeared on satellite imagery, resulting in the JTWC upgrading Megi to a minimal typhoon. *Megi* progressed northwestward around the periphery of the subtropical ridge, strengthening to a category 3 typhoon on October 16. Later in the day, the storm turned westward, and strengthened to a category 5 on the Saffir-Simpson Scale for Hurricane Classification (Table 2.6). Megi's cloud system consisting of well-developed spiral rain bands, a central circular dense cloud patch with a cloudless eye covered a large area to the east of Taiwan and Luzon (Fig. 2.12a). Measurements were obtained on 17 October at 04:55 UTC when central pressure dropped to 910 mb. The wind field derived from the Aqua AMSR-E microwave brightness temperatures with the use of an algorithm by Mitnik and Mitnik (2010, 2011) is shown in Fig. 2.12b. Rains and heavy clouds prevent wind speed retrieval in the areas marked by white color in Fig. 2.12b.

Fig. 2.12 Aqua MODIS visible image (**a**), Aqua AMSR-E-derived wind speed (**b**), total cloud liquid water content (**c**), and total water vapor content (**d**) obtained on 17 October at 04:55 UTC

The total cloud liquid water content Q and total water vapor content V in the typhoon area derived with the use of an algorithm by Mitnik and Mitnik (2003), Mitnik et al. (2009), and Bobylev et al. (2010) are shown in Fig. 2.12. A circular eye wall with a diameter of approximately 100 km and rain bands with $Q > 1.5$ kg/m^2 are clearly seen in Fig. 2.12c. The width of the eye wall is approximately 20 km. Within the eye, Q values decreased to 0.5 kg/m^2 and even lower. An area with high water vapor content (55–65 kg/m^2) surrounds the cloud field (Fig. 2.12d). V-values became still higher closer to heavy clouds and precipitation. V and Q retrieval errors are also higher here.

Satellite SAR images possess high spatial resolution. Typhoon Megi was within an Envisat ASAR swath on 17 October at 01:22 UTC (Fig. 2.13a). The brightness distribution of a SAR image of a tropical cyclone gives a detailed picture of the surface roughness which is determined mainly by the sea surface wind speed and direction and also rains since rain drops falling on the sea surface change its roughness characteristics.

Besides, heavy rains that are typical for eye wall and spiral bands of typhoons produce significant attenuation of SAR signals propagating from satellite to the sea surface and back from the sea surface to the satellite. An enlarged fragment of the image (Fig. 2.13b) shows the structure of the eye, eye wall and narrow bands spiraling to the eye. They appear with negative radar contrast against the background. The darker bands and patches with irregular edges (Fig. 2.13c, d) are indicators of the rain-induced two-way attenuation as well as sea surface roughness damping (Lines of squalls which frequently accompany heavy rains from convective clouds manifest themselves as narrow bright lines, Fig. 2.13e).

On 17 October Megi continued to strengthen. The central pressure decreased to 890 mb, maximum wind increased to 125 knots, the largest radius/diameter of storm wind was 220/440 km, and the largest radius/diameter of gale force wind reached 650/1,060 km. Current satellite sensors cannot estimate so strong winds due to physical and technical limitations. Maximum estimated winds derived from MetOp ASCAT data over the Megi central area did not exceed 25–27 m/s (Fig. 2.14).

However, currently airplane Stepped Frequency Microwave Radiometer (SFMR) has measured wind speeds up to 70 m/s in Atlantic tropical hurricanes (Uhlhorn et al. 2007). In Northwest Pacific typhoons SFMR-derived wind speed was 92 m/s—the highest SFMR wind ever measured. Figure 2.15 shows SFMR wind speed and rain rate retrievals as well as flight level wind versus *Megi* radius.

Table 2.6 Saffir–Simpson scale for hurricane classification

Strength	Winds (kts)	Pressure (mb)	Storm surge (m)	Damage
Cat. #1	65–82	>980	1.2–1.5	Minimal
Cat. #2	83–95	965–979	1.8–2.5	Moderate
Cat. #3	96–113	945–964	2.7–3.6	Extensive
Cat. #4	114–135	920–944	4.0–5.5	Extreme
Cat. #5	>135	<920	>5.5	Catastrophic

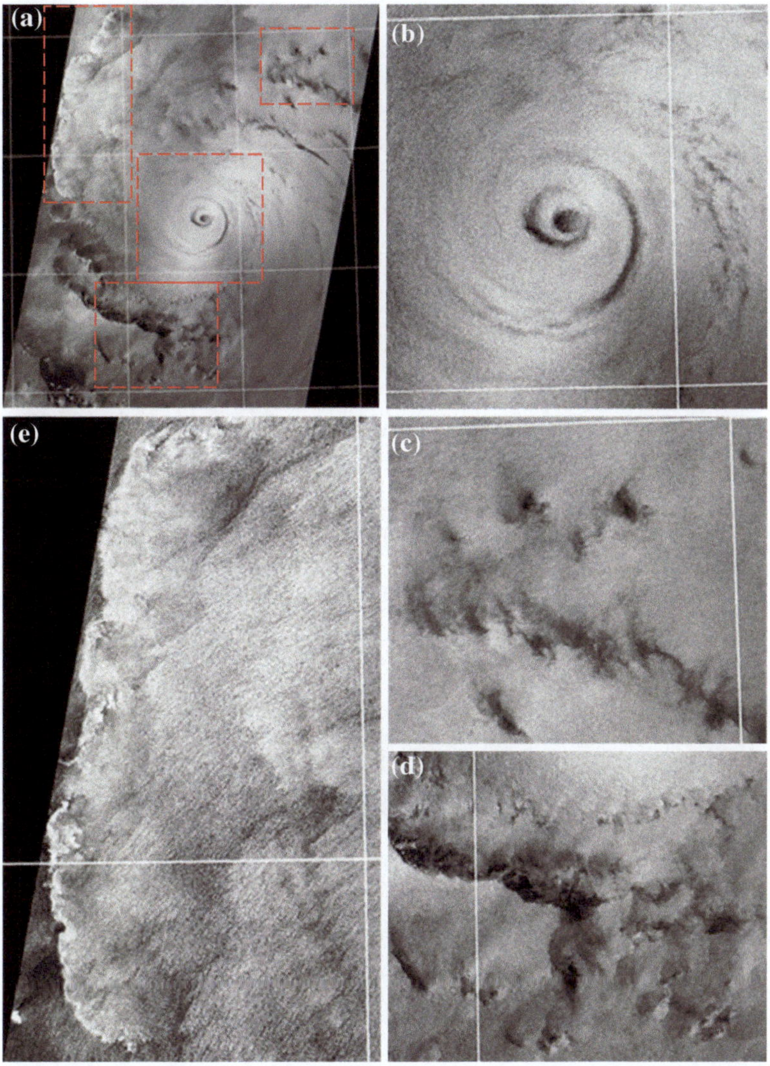

Fig. 2.13 Envisat ASAR image of typhoon Megi acquired on 17 October at 01:24 UTC (**a**) and enlarged fragments showing the signatures of eye and eye wall (**b**) rain cells and spiral rain bands near the wall (**c**) and (**d**) and squall lines (**e**). *Red dotted rectangles* in (**a**) mark the fragment boundaries (ESA©2010, ESA©2008)

Airborne SFMR observations were carried out at NE-SW pass across the *Megi* center on 17 October at 11:00–11:34 UTC. Wind speed reached 88 m/s and rain rate–40 mm/h.

On October 18, 2010, central pressure dropped to a minimum value of 885 mb. At approximately 07 UTC, the center of *Megi* crossed Luzon (Philippines) and continued westward movement. Central pressure began to rise and at 18 UTC

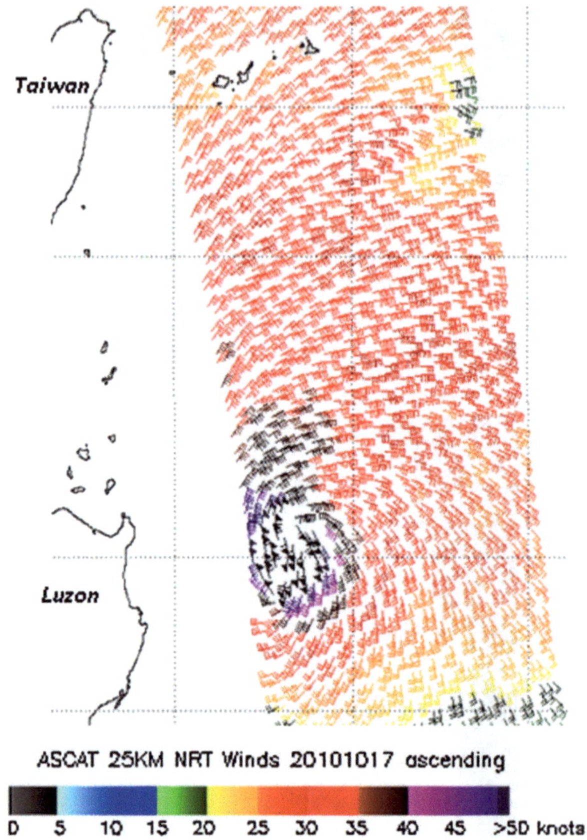

Fig. 2.14 MetOp ASCAT-derived wind field in typhoon Megi area on 17 October at 13:15 UTC when central pressure dropped to 895 mb. *Color scale* is wind speed in knots

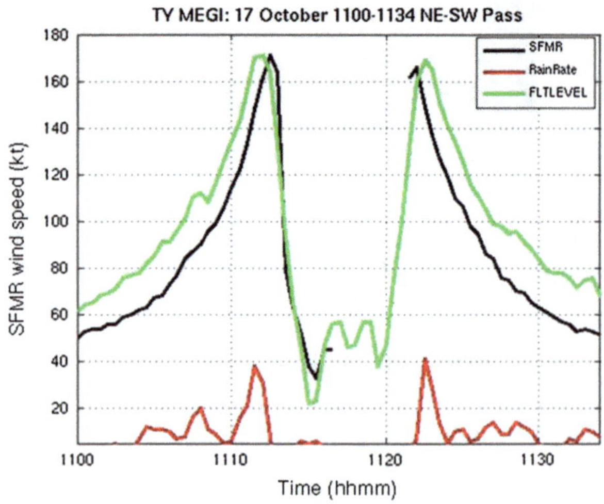

Fig. 2.15 Surface (*black*) and flight level (*green*) wind measured on October 17 during a traverse of super Typhoon Megi by the WC-130J aircraft using the stepped frequency microwave radiometer (SFMR) and flight level wind sensors. Vertically integrated rain rate (*red*) is also observed by the SFMR

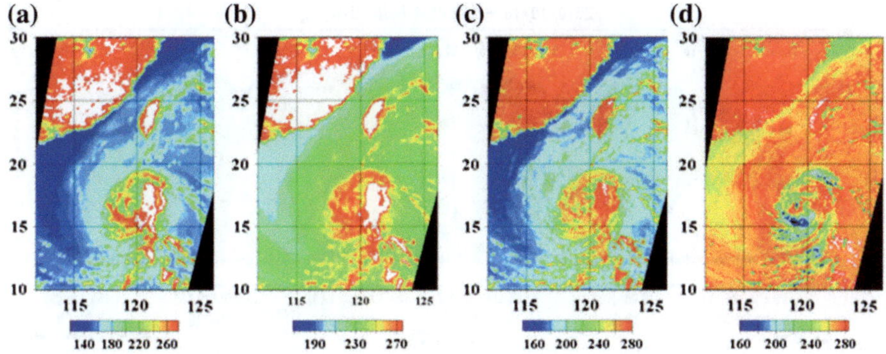

Fig. 2.16 Aqua AMSR-E brightness temperatures in Kelvin degrees of typhoon Megi at 18.7 GHz, H-pol (**a**), 23.8 GHz, H-pol (**b**), 36.5 GHz, H-pol (**c**) and 89.0 GHz, H-pol (**d**) measured on 18 October at 17:50 UTC

reached 950 mb. AMSR-E brightness temperatures at frequencies of 18.7, 23.8, 36.5 and 89.0 GHz with horizontal (H) polarization measured at 1750 UTC are shown in Fig. 2.16. A dark arrow marks the CloudSat track.

CloudSat's Cloud Precipitation Radar (CPR) operating at a frequency of 94 GHz and viewing in nadir crossed the eastern sector of the typhoon on a North–South track approximately when the *Megi* center was located over Luzon and central pressure increased to 945 hPa. CPR shows clouds at elevations higher than 16 km (Fig. 2.17). The blue areas along the top of the clouds indicate cloud ice. The wavy blue lines on the bottom center indicate intense rainfall. Where the solid line along the bottom disappears there is intense rainfall exceeding 30 mm/h. (http://esciencenews.com/topics/astronomy.space).

A moisture-saturated cloud band crossed Northern Taiwan (Fig. 2.18) and gave rise to torrential rains.

After traversing Luzon, the typhoon weakened to a category 2 but rapidly regained strength in the South China Sea where SST exceeded 30 °C, strengthening back to a category 4 on October 19. The eye diameter increased to approximately 120 km (Fig. 2.19). The circular eye wall had high contrast at all AMSR-E channels due to intense precipitation and high values of cloud liquid water content, Q. From analysis of brightness temperatures it follows that Q values exceeded 1.5–2 kg/m^2 in the eye wall and were less than 0.3–0.4 kg/m^2 inside the eye.

Megi slowed its forward speed due to the arrival of another trough over the South China Sea extending over the Taiwan Strait and breaking the ridge, turning the typhoon northwestward, then towards the north–northeast.

On 21 October *Megi* was observed by satellite SAR, Fig. 2.20. This time it was the German Aerospace radar satellite TerraSAR-X (http://www.dlr.de/en/desktopdefault.aspx/tabid-1/86_read-27445/). The swath width of TerraSAR-X is 100 km. The size of the dark central area (eye of *Megi*) is approximately 50 km, confirmed by Aqua AMSR-E and MODIS data, Fig. 2.20 and also by Fig. 2.21,

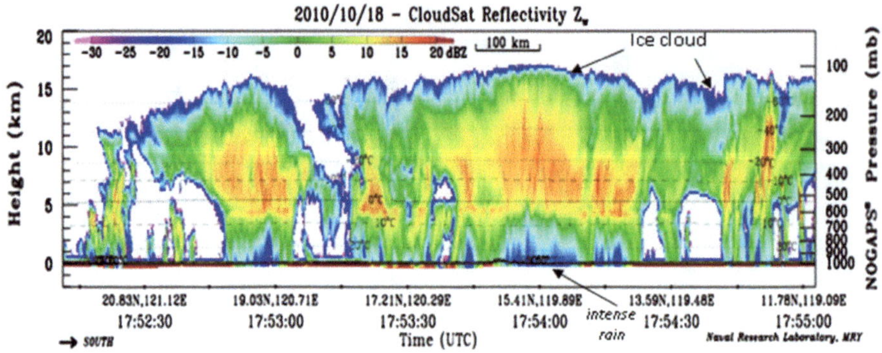

Fig. 2.17 CloudSat reflectivity of typhoon Megi cloudiness on 18 October at 17:54 UTC. The cloudsat precipitation radar (CPR) operating at a frequency of 94 GHz and viewing in nadir crossed the eastern sector of the typhoon on a North–South track at approximately when the Megi center was located over Luzon and central pressure increased to 950 mb. CPR shows clouds at elevations higher than 16 km. The *blue areas* along the *top of the clouds* indicate cloud ice. The *wavy blue lines on the bottom center* indicate intense rainfall. They cause strong attenuation of 94-GHz radar signals. When rainfall exceeds approximately 30 mm/h radar signals reflected by the sea surface disappear due to high two-way attenuation

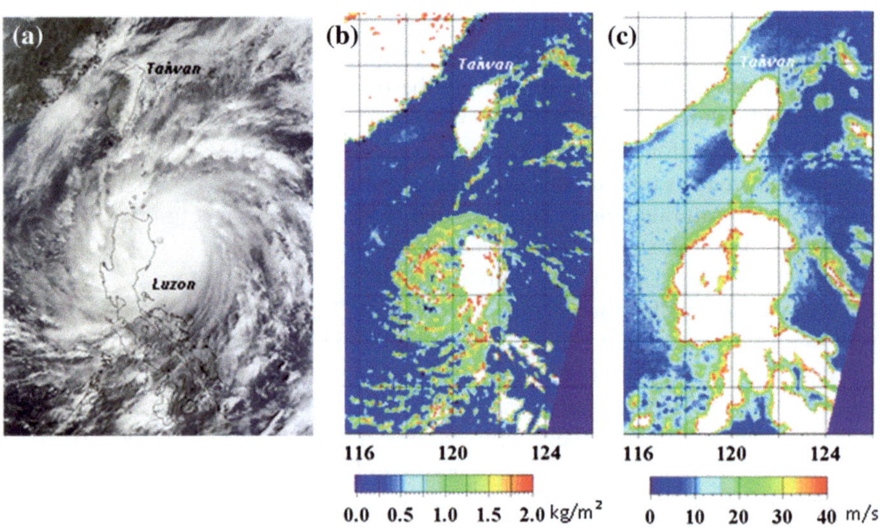

Fig. 2.18 Aqua MODIS visible image at 02:35 UTC (**a**), Aqua AMSR-E-derived total cloud liquid water content (**b**), and wind speed (**c**), at 17:38 UTC on 18 October, 2010. AMSR-E TBs allow retrieval of the cloud liquid water content and total water vapor content over the ocean and when rain rate is less than approximately 5 mm/h. That is why Taiwan, Luzon and China are shown by *white color*

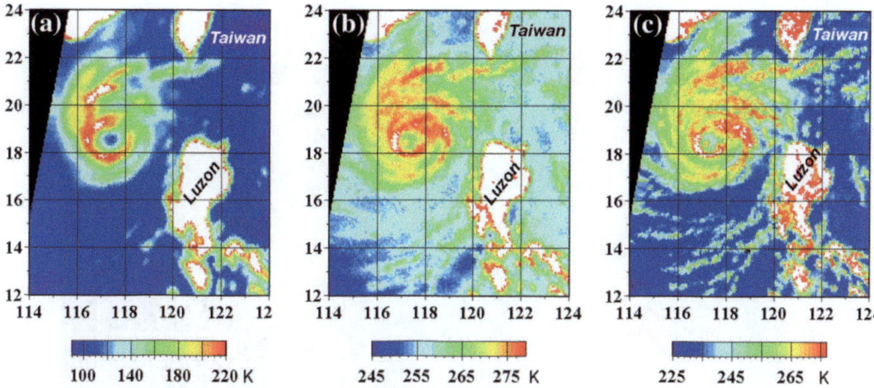

Fig. 2.19 Aqua AMSR-E brightness temperatures in Kelvin degrees taken on 20 October 2010 at 17:38 UTC and used in wind speed retrieval algorithm: (**a**) 10.7 GHz, H-pol; (**b**) 23.8 GHz, V-pol and (**c**) 36.5 GHz, V-pol. Central pressure $P_{min} = 948$ mb, $V_{max} = 100$ knots

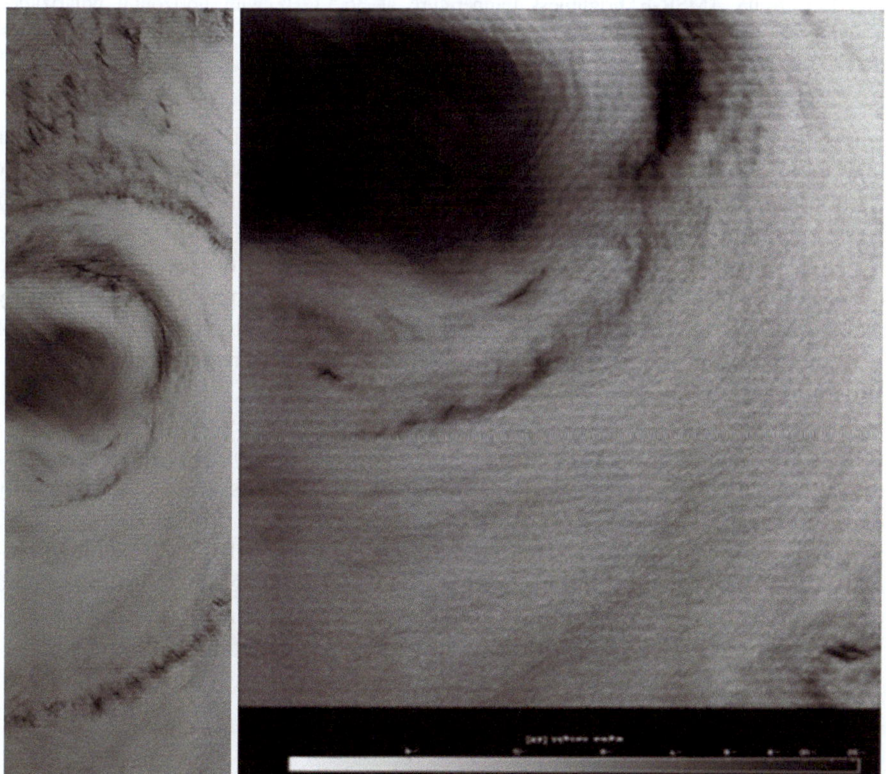

Fig. 2.20 TerraSar-X image of typhoon Megi at 22:06 UTC on 21 October 2010 over the Northern South China Sea. Pixel size of the ScanSAR mode is 8.25 m. Swath width is 100 km, and the scene covers a distance of 450 km (*left*) http://www.dlr.de/en/desktopdefault.aspx/tabid-6221/10233_read-27445/. The radiometric calibrated image (*right*) (from Lehner et al. (2011))

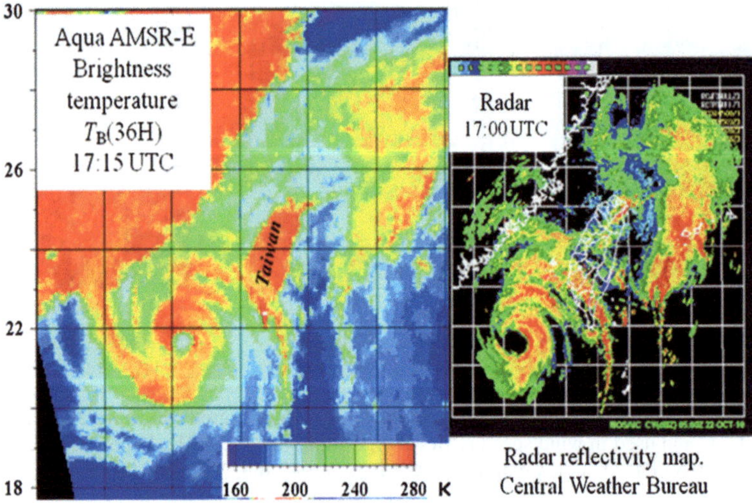

Fig. 2.21 Aqua AMSR-E brightness temperature at 36.5 GHz with horizontal polarization acquired at 17:15 UTC and radar reflectivity map produced by coastal radars of the Central Weather Bureau in Taiwan at 17:00 on 22 October 2010 (b)

whose times of capture enclose the TERRA-SAR image. It is significantly smaller than in an earlier image (Fig. 2.19), while the central pressure decreased only by 5 mb (See Fig. 2.11). Dark curvilinear bands and patches are due to scattering of 3-cm radiation by water droplets and hail.

On 22 October a Taiwan coastal radar traced *Megi*'s northward movement and changes in its spiral rain bands. The field of brightness temperature measured by Aqua AMSR-E and a radar reflectivity map acquired with a time difference of 15 min are in a good agreement, since both satellite and ground sensors are sensitive to water droplets in the atmosphere (Fig. 2.21).

Satellite measurements obtained by passive and active sensors prove their usefulness for quantitative research on the life cycle of tropical cyclones and for operational applications especially in data sparse areas. This was demonstrated by quasi-operational analysis of near real-time satellite data obtained over *Megi*, provided via the Internet and presented during PORSEC2010 in Keelung. Microwave sensors enhance overall knowledge and understanding of TC evolution, and can significantly improve forecasts of their tracks and intensity, as well as, dangerous terrain effects such as storm surges, heavy precipitation and flooding.

2.5 Summary and Perspectives

Over the past 4 decades remote sensing has developed to include passive and active microwave instruments, employing a wide range of frequencies and sampling techniques. Applied to TC research and operational diagnosis this work has had very significant impact. Together with frequent observation in the visible and infrared spectrum from geostationary and polar orbiting satellites, they have allowed new insights and development of theory and numerical models. Improvements in understanding of TC structures and their life cycle and in understanding the interactions of the microwave signals with the underlying sea surface and the atmospheric constituents have ensued—One might ask, where can we best go from here?—Much work is still needed for understanding the TC life cycle, especially rapid intensification. Economic stress has slowed down the replacement of many valuable sensors as they fail of old age. The enterprise of satellite remote sensing (and aircraft remote sensing as well) is an expensive undertaking, but not with the perspective of the disasters that occur, when coastal populations are attacked unawares by intense TC's. Such events were recorded especially before the satellite era (e.g. the New England storm of 1938, Galveson Texas, 1900 and 1915). Pielke and Landsea (1998) review the cost of the 14 most expensive hurricanes up to 1998 with normalization based on inflation, population density and level of wealth that has changed over time. The cost has kept on rising, but the loss of life has ben dramatically reduced when the populations are warned and evacuated. Hurricane Katrina's impact on Louisiana in 2005 was largely due to infrastructure failures and improper evacuations. The approaching hurricane and its intensity were known. In no region on the globe should surprises like the Galvston 1900 hurricane and the *Katrina* disaster happen today, but communicating the risks to the threatened populations and having transportation and shelters prepared can still be a problem.

For global watchfulness from space we need dependable, operational satellite systems with frequent sampling. This will require full international cooperation from inception of design and orbit planning, through development, operations and data distribution (Katsaros et al. 2011). Such data must be made available in near real time for optimal reward in improved warnings. In order to accomplish this in greater measure than we have today, scientists, politicians and institutional leaders must consider this holistic, international approach a priority. The uncertainty of research results, when observations remain punctual, and conclusions *per force* are anecdotal makes it difficult to gather folks into action and achieve adoption of good risk-aversion practices. It is, therefore, crucial to lay a solid foundation of high quality, routine and relevant data for the hoped-for international cooperation. Several international organizations are at work on these principles, notably the World Meteorological Organization (WMO) and various branches of the United Nations Educational, Scientific and Cultural Organization (UNESCO). The UNESCO efforts include education and training for capacity building in many countries where the need for better science training exists.

All along creative minds present us with new and innovative ideas for how to improve the measurements and the use of the information. Examples of sophisticated new systems include the scanning altimeter, Surface Water and Ocean Topography (SWOT, Fu et al. 2009; http://swot.jpl.nasa.gov/). Satellite altimetry is recognized as an essential component of the Global Earth Observation System of Systems. To date satellite altimetry has focused on the open ocean (Fu and

Cazenave 2001), but the coastal ocean has emerged as an important domain for these data. Only through a synergy from various altimeters can a thorough and complete characterization of mesoscale circulation be obtained. Coastal and mesoscale issues have gotten a new push with the proposed SWOT, which would provide better coverage by an altimeter system than the current nadir viewing ones. It would also have multiple frequencies providing better accuracy.

For the precipitation and water content, as vapor or cloud liquid water in storms of all kinds, a follow-on system to TRMM has been proposed, which would include many passive radiometers and a central rain radar to provide the inter-calibration; the Japanese Aerospace and Exploration Agency, JAXA, and NASA are playing important roles with other international partners included. A US National Research Council Report describes the GPM mission with focus on NOAA's role.

We could increase the revisit time of sampling the ocean surface wind readily, if all nations making scatterometer measurements coordinated their equator crossing times. Liu et al. (2008b) provide graphic illustration that 6 h revisits or better would be possible globally with 4 scatterometers, if orbits are ideally timed, but even two coordinated scatterometers already improve chances of catching circulation patterns during early developments of tropical depression within 12 h (even though many systems with early circulations do not develop to storms; see Liu et al. 2008b).

The judicious investment in new satellite systems from many of the technologies illustrated above, where the proof-of-concept is well established would seem the way forward. In view of possible increases in the frequency of severe tropical cyclones due to a warming climate (Webster et al. 2005; Emanuel 2005; Elsner et al. 2008; Knutson et al. 2010) and the large population centers in vulnerable coastal areas, it is urgent that the community of scientists drive such progress, while not stifling development of new and innovative sensors.

Acknowledgments The authors thank members of Satellite Oceanography Laboratory, POI FEB Russian Academy of Science for Aqua AMSR-E, Terra and Aqua MODIS and Envisat ASAR processing. We are grateful to Professor Hans Graber for carrying out SAR image collection under the auspices of the Canadian "Hurricane Watch" program and for permission by Dr. Paris Vashon for use of the RADARSAT images in Sect. 2.3.

This work is partially supported by a grant from the Office of Naval Research, N0001411WX21223, from NSF grant ATM0631685 to University of Miami, and from the Russian Fund of Basic Researchgrant 11-05–ophi-m-2011.

References

Adams, I.S., Hennon, C.C., Jones, L., Ahmad, K.A.: Evaluation of hurricane ocean vector winds from WindSat. IEEE Trans. Geosci. Remote Sens. **44**, 656–667 (2006)

Adler, R.F., Huffman, G.J., Chang, A., Ferraro, R., Xie, P., Janoviak, J., Rudolf, B., Schneider, U., Curtis, S., Bolvin, D., Gruber, A., Susskind, J., Arkin, P.: The version 2 global precipitation climatology project (GPCP) monthly precipitation analysis (1979–present). J Hydrometeor **4**, 1147–1167 (2003)

Alishouse, J.C., Snyder, S.A., Vongsathorn, J., Ferraro, R.R.: Determination of oceanic total column water vapour from the SSM/I. IEEE Trans. Geosci. Remote Sens. **28**, 811–816 (1990a)

Alishouse, J.C., Snider, J.B., Westwater, E.R., Swift, C.T., Ruf, C.S., Snyder, S.A., Vongsathorn, J., Ferraro, R.R.: Determination of cloud liquid water-content using the SSM/I. IEEE Trans. Geosci. Remote Sens. **28**, 817–822 (1990b)

Aziz, M.A., Raising, S.C., Asher, W.E., Rose, L.A., Gaiser, P.W., Horgan, K.A.: Effects of air–sea interaction parameters on ocean surface microwave emission at 10 and 37 GHz. IEEE Trans. Geosci. Remote Sens. **43**, 1763–1774 (2005)

Baklanov, A.A., Grisogono, B., Bornstein, R., Mahrt, L., Zilitinkevich, S.S., Taylor, P., Larsen, S.E., Rotach, M.W., Fernando, H.J.S.: The nature, theory, and modeling of atmospheric planetary boundary layers. Bul Amer Meteor Soc **92**, 123–128 (2011)

Basharinov, A.E., Gurvich, A.S., Egorov, S.T.: Radio Emission of the Earth as a Planet (Radioizluchenie Zemli kak Planeti). Nauka Publishing House, Moscow (1974). (188 pp., in Russian)

Bell, W.: A preprocessor for SSMIS radiances scientific description. NWPSAF-MO-UD-014, Version 1.0. Met Office, United Kingdom: EUMETSAT (2006)

Bennartz, R.: Optimal convolution of AMSU-B to AMSU-A. J Atm. Oceanic Technol. **17**, 1215–1225 (2000)

Bentamy, A., Queffeulou, P., Quilfen, Y., Katsaros, K.: Ocean surface wind fields estimated from satellite measurements. IEEE Trans. Geosci. Remote Sens. **37**, 2469–2486 (1999)

Bentamy, A., Katsaros, K.B., Mestas-Nuñez, A.M., Drennan, W.M., Forde, E.B., Roquet, H.: Satellite estimates of wind speed and latent heat flux over the global oceans. J. Climate **16**, 636–656 (2003)

Bessho, K., DeMaria, M., Knaff, J.A.: Tropical cyclone wind retrievals from the advanced microwave sounder unit (AMSU): application to surface wind analysis. J. Appl. Meteorol. **45**, 399–415 (2006)

Bettenhausen, M.H., Smith, C.K., Bevilacqua, R.M., Wang, N.-Y., Gaiser, P.W., Cox, S.: A nonlinear optimization algorithm for WindSat wind vector retrievals. IEEE Trans. Geosci. Remote Sens. **44**, 597–608 (2006)

Black, P.G., Adams, W.L.: Guidance for estimating surface winds based on sea state observations from aircraft and sea state catalog. NOAA Tech. Memo. FCM-G1-1983, 83 pp (1983)

Black, P.G., Swift, C.T.: Airborne stepped frequency microwave radiometer measurements of rainfall rate and surface wind speed in hurricanes. Preprints, Second Conf. on Radar Meteorology, Zurich, Switzerland, Amer Meteor Soc 433–438 (1984)

Blackadar, A.K.: A single-layer theory of the vertical distribution of wind in a baroclinic neutral atmospheric boundary layer, Final Rept. AFCRL-65-531, Department of Meteorology, Pennsylvania State University, p. 22 (1965)

Bobylev, L.P., Zabolotskikh, E.V., Mitnik, L.M., Mitnik, M.L.: Atmospheric water vapor and cloud liquid water retrieval over the Arctic Ocean using satellite passive microwave sensing. IEEE Trans. Geosci. Remote Sens. **49**, 283–294 (2010)

Boukabara, S.A., Weng, F.: Microwave emissivity over ocean in all-weather conditions: validation using WINDSAT and airborne GPS dropsondes. IEEE Trans. Geosci. Remote Sens. **46**, 376–384 (2008)

Buckreuss, S., Balzer, W., Muhlbauer, P., Werninghaus, R., Pitz, W.: The TerraSAR-X satellite project. Proc IGARSS **5**, 3096–3098 (2003)

Businger, J.A., Wyngaard, J.C., Izumi, Y., Bradley, E.F.: Flux-profile relationships in the atmospheric surface layer. J Atm. Sci. **28**, 181–189 (1971)

Campbell, A.B.: Radar Remote Sensing of Planetary Surfaces. Cambridge University Press, Cambridge (2002). (331 pp)

Cardone, V.J.: Specification of the wind distribution in the marine boundary layer for wave forecasting. Report TR-69-01, Geophysical Sciences Laboratory. New York University. Available from NTIS #AD 702-490, 137 pp (1969)

CCRS (2009) Tutorial: fundamentals of remote sensing, technical report, Canada Centre for Remote Sensing. http://www.ccrs.nrcan.gc.ca/resource/tutor/gsarcd

Chen, S.H., Vandenberghe, F., Petty, G.W., Bresch, J.F.: Application of SSM/I satellite data to a hurricane simulation. Q. J. Royal Meteorol. Soc. **130**, 801–825 (2004). doi:10.1256/qj.02.672004

Coletta, A., Angino, G., Battazza, F., Caltagirone, F., Impagnatiello, F., Valentini, G., Capuzi, A., Fagioli, S., Leonardi, R.: COSMO-SkyMed program: utilization and description of an advanced space EO dual-use asset. In: Proceedings of Envisat Symposium, 23–27 Apr 2007, ESA SP-636, Montreux, Switzerland (2007)

Dagestad, K.-F., Johannessen, J.A., Hansen, M.W., Mouche, A., Collard, F.: Wind retrieval over the ocean from combined use of synthetic aperture radar and weather forecast models. In: ESA Living Planet Symposium, Begen, Norway, 28 June–2 July. http://due.esrin.esa.int/stse/files/project/131-176-149-30_2010720123246.pdf (2010)

Deardorff, J.W.: Parameterization of the planetary boundary layer for use in general circulation models. Mon Wea Rev **2**, 93–106 (1972)

Demuth, J., DeMaria, M., Knaff, J.A., Vonder Haar, T.H.: Validation of an advanced microwave sounder unit (AMSU) tropical cyclone intensity and size estimation algorithm. J Appl Meteor **43**, 282–296 (2004)

Demuth, J., DeMaria, M., Knaff, J.A.: Improvement of advanced microwave sounding unit tropical cyclone intensity and size estimation algorithms. J. Appl. Climatol. **45**, 1573–1581 (2006)

Donelan, M.A., Haus, B.K., Reul, N., Plant, W.J., Stassnie, M., Graber, H.C., Brown, O.B., Satzman, E.S.: On the limiting aerodynamic roughness of the ocean in very strong winds. Geophys. Res. Lett. **31**, L18306 (2004)

Dunion, J.P., Landsea, C.W., Houston, S.H., Powell, M.D.: A reanalysis of the surface winds for Hurricane Donna of 1960. Mon. Weather Rev. **131**, 1992–2011 (2003)

Ebell, K., Löhnert, U., Crewell, S., Turner, D.D.: On characterizing the error in a remotely sensed liquid water content profile. Atmos. Res. **98**, 57–68 (2010)

Ebuchi, N., Graber, H.C., Caruso, M.J.: Evaluation of wind vectors observed by QuikSCAT/SeaWinds using ocean buoy data. J. Atmos. Oceanic Tech. **19**, 2049–2062 (2002)

Elsner, J., Kossin, P., Jagger, T.: The increasing intensity of the strongest tropical cyclones. Nature **455**, 92–95 (2008)

El-Nimri, S.F., Linwood Jones, W., Uhlhorn, E., Ruf, C., Johnson, J., Black, P.: An improved C-band ocean surface emissivity model at hurricane-force wind speeds over a wide range of Earth incidence angles. IEEE Geosci. Remote Sens. Lett. **7**, 641–645 (2010)

Emanuel, K.A.: Increasing destructiveness of tropical cyclones over the past 30 years. Nature **436**, 686–688 (2005)

ESA (2002) ASAR Product Handbook, Technical report, European Space Agency, Paris, France

Fairall, C.W., Bradley, E.F., Hare, J.E., Grachev, A.A., Edson, J.B.: Bulk parameterization of air-sea fluxes: updates and verification for the coare algorithm. J. Climate **19**, 571–591 (2003)

Ferraro, R.R.: SSM/I derived global rainfall estimates for climatological purposes. J. Geophys. Res. **102**, 16715–16735 (1997)

Ferraro, R.R., Weng, F., Grody, N.C., Zhao, L., Meng, H., Kongoli, C., Pellegrino, P., Qiu, S., Dean, C.: NOAA operational hydrological products derived from the advanced microwave sounding unit. IEEE Trans. Geosci. Remote Sens. **43**, 1036–1049 (2005)

Franklin, J.L., Black, M.L., Valde, K.: GPS dropwindsonde wind profiles in hurricanes and their operational implications. Wea. Forecast. **18**, 32–44 (2003)

Freilich, M.H., Dunbar, R.S.: Seawinds: Algorithm Theoretical Basis Document. Oregon State University Tech. Rep., Corvallis (2000). (Tech. Rep., 56 pp)

Fu, L.L., Cazenave, A.: Satellite Altimetry and Earth Sciences: A Handbook of Techniques and Applications. Academic Press, San Diego (2001). (463 pp)

Fu, L.L., Alsdorf, D., Rodriguez, E., Morrow, R., Mognard, N., Lambin, J., Vaze, P., Lafon, T.: The SWOT (Surface Water and Ocean Topography) Mission, OceansObs'09 white paper p. 17, available from http://www.oceanobs09.net/ (2009)

Gaiser, P.W., St Germain, K.M., Twarog, E.M., Poe, G.A., Purdy, W., Richardson, D., Grossman, W., Jones, W.L., Spencer, D., Golba, G., Cleveland, J., Choy, L., Bevilacqua, R.M., Chang, P.S.: The WindSat Spaceborne polarimetric microwave radiometer: sensor description and early orbit performance. IEEE Trans. Geosci. Remote Sens. **42**, 2347–2361 (2004)

Gentemann, C.L., Meissner, T., Wentz, F.J.: Accuracy of satellite sea surface temperatures at 7 and 11 GHz. IEEE Trans. Geosci. Remote Sens. **48**, 1009–1018 (2010a). doi:10.1109/TGRS.20092030322

Gentemann, C.L., Wentz, F.J., Brewer, M., Hilburn, K., Smith, D.: Passive microwave remote sensing of the ocean: an overview. In: Barale, V., Gower, J., Alberotanza, L. (eds.) Oceanography from Space, pp. 13–33. Springer, Heidelberg (2010b). (Revisited)

Gentemann, C.L., Minnett, P.J., LeBorgne, P., Merchant, C.J.: Multi-satellite measurements of large diurnal SST warming events. Geophys. Res. Lett. **356**, L22602 (2008). doi:10.1029/2008GL035730

Gentemann, C.L., Minnett, P.J.: Radiometric measurements of ocean surface thermal variability. J. Geophys. Res. **113** (2008). doi:10.1029/2007JC004540

Goni, G., DeMaria, M., Knaff, J., Sapson, C., Ginnis, I., Bringas, F., Mavume, A., Lauer, C., Lin, I.-I., Ali, M., Sandery, P., Ramos-Buargue, S., Kang, K., Mehra, A., Chassignet, E., Halliwell, G.: Applications of satellite-derived ocean measurements to tropical cyclone intensity forecasting. Oceanography **22**, 190–197 (2009). doi:10.5670/oceanog.2009.78

Goodberlet, M.A., Swift, C.T., Wilkerson, J.C.: Ocean surface wind speed measurements of the special sensor microwave/imager (SSM/I). J. Geophys. Res. **94**, 14547–14555 (1989)

Goodberlet, M., Swift, C.T.: Development of a second generation stepped frequency microwave radiometer for IWRS. Tech. Rep. Prosensing, Inc. (formerly Quadrant Engineering), Amherst, MA, 42 pp (1996)

Greenwald, T.J., Stephens, G.L., Vonder Haar, T.H., Jackson, D.L.: A physical retrieval of cloud liquid water over the global oceans using special sensor microwave/imager (SSM/I) observations. J. Geophys. Res. **98**, 18471–18488 (1993)

Greenwald, T.J.: A 2-year comparison of AMSR-E and MODIS cloud liquid water path observations. Geophys. Res. Lett. **36**, L20805 (2009). doi:10.1029/2009GL040394

Grody, N.C., Zhao, J., Ferraro, R., Weng, F., Boers, R.: Determination of precipitable water and cloud liquid water from the NOAA 15 AMSU. J. Geophys. Res. **106**, 2943–2953 (2001)

Harrington, R.F.: The development of a stepped frequency microwave radiometer and its application to remote sensing of the earth. NASA Technical Report, TM-81847, 169 pp (1980)

Harper, B.A.: Tropical cyclone parameter estimation in the Australian Region, Systems Engineering Australia Pty Ltd for Woodside Energy Ltd, Perth, May 2002, 83 pp (2002)

Hawkins, J.D., Lee, T.F., Turk, J., Sampson, C., Kent, J., Richardson, K.: Real-time Internet distribution of satellite products for tropical cyclone reconnaissance. Bull. Amer. Meteor. Soc. **82**, 567–578 (2001)

Hawkins, J.D., Turk, F.J., Lee, T.F., Richardson, K.: Observations of tropical cyclones with the SSMIS. IEEE Trans. Geosci. Remote Sens. **46**, 901–912 (2008)

Hock, T.F., Franklin, J.L.: The NCAR GPS dropwindsonde. Bull. Amer. Meteor. Soc. **80**, 407–420 (1999)

Hollinger, J.P., Pierce, J.L., Poe, G.A.: SSM/I instrument evaluation. IEEE Trans. Geosci. Remote Sens. **28**, 781–790 (1990)

Horstmann, J., Thompson, D.R., Monaldo, F., Graber, H.C., Iris, S.: Can synthetic aperture radars be used to estimate hurricane force winds? Geophys. Res. Lett. **32**, L22 801 (2005)

Horváth, Á., Davies, R.: Comparison of microwave and optical cloud water path estimates from TMI, MODIS, and MISR. J. Geophys. Res. **112**, D01202 (2007). doi:10.1029/2006JD007101

Hurley, J.: Operational review: RADARSAT-1 & -2. SEA SAR 2010. 25–29 Jan 2010, ESA ESRIN, Frascati (Rome) Italy. (http://earth.eo.esa.int/workshops/seasar2010/8_Hurley.pdf) (2010)

Imaoka, K., Kachi, M., Fujii, H., Murakami, H., Hori, M., Ono, A., Igarashi, T., Nakagawa, K., Oki, T., Honda, Y., Shimoda, H.: Global change observation mission (GCOM) for monitoring carbon, water cycles, and climate change. Proc. IEEE **98**, 717–734 (2010)

Iris, S., Burger, G.: RADARSAT-1: Canadian space agency hurricane watch program. Proc. IGARSS **4**, 2742–2745 (2004)

Jackson, C.R., Apel, J.R. (eds): Synthetic aperture radar (SAR) marine user's manual, NOAA NESDIS Office of Research and Applications, Washington DC. www.sarusersmanual.com (2004)

Jones, W.L., Schroeder, L.C., Boggs, D.H., Bracalente, E.M., Browb, R.A., Dome, G.J., Pierson, W.J., Wentz, F.J.: The geophysical evaluation of remotely sensed wind vectors over the ocean. J. Geophys. Res. **87**, 3297–3317 (1982)

Joyce, R.J., Janowiak, J.E., Arkin, P.A., Xie, P.: CMORPH: a method that produces global precipitation estimates from passive microwave and infrared data at high spatial and temporal resolution. J. Hydromet. **5**, 487–503 (2004)

Jung, T., Ruprecht, E., Wagner, F.: Determination of cloud liquid water path over the oceans from SSM/I data using neural networks. J. Appl. Meteor. **37**, 832–844 (1998)

Katsaros, K.B., Brown, R.A.: Legacy of the Seasat mission for studies of the atmosphere and air-sea-ice interactions. Bull. Amer. Met. Soc. **72**, 967–981 (1991)

Katsaros, K.B., Forde, E.B., Chang, P., Liu, W.T.: QuikSCAT facilitates early detection of tropical depressions in 1999 hurricane season. Geophys. Res. Lett. **28**, 1043–1046 (2001)

Katsaros, K.B., Bentamy, A., Bourassa, M., Ebuchi, N., Gower, J., Liu, W.T., Vignudelli, S.: Climate data Issues from and oceanographic remote sensing perspective. In: Tang, D. (ed.) Remote Sensing of the Changing Oceans. Springer, Berlin (2011). doi:10.1007/978-642-16541-2_1

Katsaros, K.B., Vachon, P., Black, P., Dodge, P., Uhlhorn, E.: Wind fields from SAR: could they improve our understanding of storm dynamics? Johns Hopkins APL Tech. Dig. **21**, 86–93 (2000)

Kawanishi, T., Sezai, T., Ito, Y., Imaoka, K., Tukeshima, T., Ishido, Y., Shibata, A., Miura, M., Inahata, H., Spencer, R.W.: The advanced microwave scanning radiometer for the earth observing system (AMSR-E), NASDA's contribution to the EOS for global energy and water cycle studies. IEEE Trans. Geosci. Remote Sens. **41**, 184–194 (2003)

Kidder, S.Q., Jones, A.S.: A blended satellite total precipitable water product for operational forecasting. J. Atmos. Oceanic Technol. **24**, 74–81 (2007)

Kim, S.-W., Chung, E.-S., Yoon, S.-C., Sohn, B.-J., Sugimoto, N.: Intercomparisons of cloud-top and cloud-base heights from ground-based Lidar, CloudSat and CALIPSO measurements. Int. J. Remote Sens. **32**, 1179–1197 (2011)

Knaff, J.A., Zehr, R.M., Goldberg, M.D., Kidder, S.Q.: An example of temperature structure differences in two cyclone systems derived from the advanced microwave sounding unit. Wea. Forecast. **15**, 476–483 (2000)

Knaff, J.A., Zehr, R.M.: Reexamination of tropical cyclone pressure wind relationships. Wea. Forecast. **22**, 71–88 (2007)

Knaff, J.A., DeMaria, M., Molenar, D.A., Sampson, C.R., Seybold, M.G.: An automated, objective, multiple-satellite-platform tropical cyclone surface wind analysis. J. Appl. Meteor. Climatol. **50**, 2149–2166 (2011)

Knutson, T., McBride, L., Chan, C., Emanuel, K., Holland, G., Landsea, C., Held, I., Kossi, J., Srivastava, A., Sugi, M.: Tropical cyclones and climate change. Nat. Geosci. **3**, 157–163 (2010)

Kossin, J.P., Schubert, W.H.: Mesovortices, polygonal flow patterns, and rapid pressure falls in hurricane-like vortices. J. Atmos. Sci. **58**, 2196–2209 (2001)

Kossin, J.P., McNoldy, B.D., Schubert, W.H.: Vortical swirls in hurricane eye clouds. Mon. Weather. Rev. **130**, 3144–3149 (2002)

Kossin, J.P., Schubert, W.H.: Mesovortices in Hurricane Isabel (2003). Bull. Amer. Meteor. Soc. **85**, 151–153 (2004)

Kramer, H.J.: Observation of the Earth and Its Environment. Survey of Missions and Sensors. 4th edn. Springer, Berlin (2002). ISBN 3-540-42388-5, XXIX + 1510 pp., 1 CD-ROM

Krasnopolsky, V.M., Gemmill, W.H., Breaker, L.C.: A neural network multiparameter algorithm for SSM/I ocean retrievals—comparisons and validations. Remote Sens. Environ. **73**, 133–142 (2000)

Kummerow, C., Barnes, W., Kozu, T., Shiue, J., Simpson, J.: The tropical rainfall measuring mission (TRMM) sensor package. J. Atmos. Oceanic Technol. **15**, 808–816 (1998)

Kummerow, C., et al.: The status of the tropical rainfall measuring mission (TRMM) after two years in orbit. J. Appl. Meteor. **39**, 1965–1982 (2000)

Kunkee, D.B., Poe, G.A., Boucher, D.J., Swadley, S.D., Hong, Y., Wessel, J.E., Uliana, E.A.: Design and evaluation of the first special sensor microwave imager/sounder. IEEE Trans. Geosci. Remote Sens. **46**, 863–883 (2008)

L'Ecuyer, T.S., Jiang, J.H.: Touring the atmosphere aboard the A-train. Phys. Today **7**, 36–41 (2010)

Lee, T.F., Bankert, R.L., Mitrescu, C.: Meteorological education and training using A-train profilers. Bull. Amer. Meteorol. Soc. **93**, 687–696 (2012)

Lee, T.F., Nelson, C.S., Dills, P., Riishojgaard, L.P., Jones, A., Li, L., Miller, S., Flynn, L.E., Jedlovec, G., McCarty, W., Hoffman, C., McWilliams, G.: NPOESS: next-generation operational global earth observations. Bull. Amer. Meteorol. Soc. **91**, 727–740 (2010)

Lehner, S., He, M. -X., Li, X. -M., Velotto D., Ren Y. -Z. Application of coastal monitoring using high resolution SAR. 2011 Dragon 2 Symposium, Praga Czesh Republic, 20–24 June (2011)

Lin, I.-I., Wu, C.C., Pun, I.-P.: Upper ocean thermal structure and the western North Pacific category 5 typhoons. Part I. Ocean features and category 5 typhoons' intensification. Mon. Weather Rev. **136**, 288–306 (2008)

Liu, G.-R., Chao, C.C., Ho, C.-Y.: Applying satellite-estimated storm rotation speed to improve typhoon rainfall potential technique. Wea. Forecast. **23**, 259–269 (2008a)

Liu, G., Curry, J.A.: Determination of characteristic feature of cloud liquid water from satellite microwave measurements. J. Geophys. Res. **98**, 5069–5092 (1993)

Liu, W.T., Tang, W.: Equivalent Neutral Wind. JPL Publication 96-17, JetPropulsion Laboratory, Pasadena, 16 pp. http://airsea.jpl.nasa.gov/publication/paper/Liu-Tang-1996-jpl.pdf (1996)

Liu, W.T., Xie, X.: Ocean-atmosphere momentum coupling in the Kuroshio extension observed from space. J. Oceanogr. **64**, 631–637 (2008)

Liu, W.T., Tang, W., Xie, X., Navalgund, R.R., Xu, K.: Power density of ocean surface wind from international scatterometers tandem missions. Int. J. Remote Sens. **29**, 6109–6116 (2008b)

Liu, Q., Weng, F., English, S.J.: An improved fast microwave water emissivity model. IEEE Trans. Geosci. Remote Sens. **49**, 1238–1250 (2011)

Lonfat, M., Marks, F.D., Chen, S.S.: Precipitation distribution in tropical cyclones using the tropical rainfall measuring mission (TRMM) microwave imager: a global perspective. Mon. Weather Rev. **132**, 1645–1660 (2004)

Long, D.G.: Reconstruction of high resolution ocean wind vectors from low resolution scatterometer measurements. Proc. SPIE vol. 5562 Image Reconstruction from Incomplete Data III, Bellingham, WA, SPIE, 196-207 (2004)

Mainelli, M., DeMaria, M., Shay, L.K., Goni, G.: Application of oceanic heat content estimation to operational forecasting of recent Atlantic category 5 Hurricanes. Wea. Forecast. **23**, 1–16 (2008)

Mallet, C., Moreau, E., Casagrande, L., Klapisz, C.: Determination of integrated cloud liquid water path and total precipitate water from SSM/I data using a neural network algorithm. Int. J. Remote Sens. **23**, 661–674 (2002)

Matrosov, S.Y.: CloudSat studies of stratiform precipitation systems observed in the vicinity of the Southern Great Plains atmospheric radiation measurement site. J. Appl. Meteor. Clim. **49**, 1756–1765 (2010)

Mätzler, C., Rosenkranz, P.W., Cermak, J.: Microwave absorption of supercooled clouds and implications for the dielectric properties of water. J. Geophys. Res. **115**, D23208 (2010). doi:10.1029/2010JD014283

Meissner, T., Wentz, F.J.: Ocean retrievals for WindSat: radiative transfer model, algorithm, validation. In: 9th Specialist Meeting Microwave Radiometry Remote Sensing Applications, Puerto Rico, USA, 2006. Available: http://www.remss.com/papers/meissner_and_wentz(2006).pdf (2006)

Meissner, T., Wentz, F.J.: Wind retrievals under rain for passive satellite microwave radiometers and its application to hurricane tracking. Proc. IGARSS (2008)

Miller, B.I.: The three-dimensional wind structure around a tropical cyclone. National Hurricane Research Project Rept. 15, U.S. Weather Bureau, 41 pp (1958)

Mims, A., Kroodsma, R., Ruf, C., McKague, D.: WindSat retrieval of ocean surface wind speed in tropical cyclones. In: Proceedings of IGARSS 2010. Hawaii, USA, pp. 1831–1834, 26–30 July 2010

Mitnik, L. M.: Algorithm for atmospheric total water vapor determination from satellite measurements of microwave radiometric emission Proc. USSR Hydrometeorological Cent. **50**, 94–102 (1969) (in Russian)

Mitnik, L. M.: Cloud investigation by microwave radiometric technique. RIHMI-WDC, Obninsk, 67 pp (1978) (in Russian)

Mitnik, L.M., Mitnik, M.L.: Retrieval of atmospheric and ocean surface parameters from ADEOS-II AMSR data: comparison of errors of global and regional algorithms. Radio Sci. **38**, 8065 (2003). doi:10.1029/2002RS002659

Mitnik, L.M., Mitnik, M.L.: Retrieval of total water vapor content and total cloud liquid water content over the ocean by microwave sensing from DMSP, TRMM, AQUA and ADEOS-II satellites. Inv. Earth Space **4**, 34–41 (2006). (in Russian)

Mitnik, L.M., Mitnik, M.L.: AMSR-E advanced wind speed retrieval algorithm and its application to marine weather systems. In: Proceedings of IGARSS 2010, Hawaii, pp. 3224–3227, 26–30 July 2010

Mitnik, M.L., Mitnik, L.M.: Retrieval algorithm of sea surface wind speed from AMSR-E microwave radiometer data and its application to weather system analysis in tropical zone. Curr. Probl. Earth Remote Sens. Space **8**, 297–303 (2011). (in Russian)

Mitnik, L., Chen, K.-S., Wang, J.-T., Mitnik, M., Hsu, M.-K.: Satellite microwave observations of typhoon Herb (1996) near Taiwan. Glob. Atmos. Ocean Syst. **8**, 19–39 (2002)

Mitnik, L.M., Mitnik, M.L., Gurvich, I.A.: Monitoring tropical cyclone evolution over the NW Pacific with Aqua AMSR-E and Envisat ASAR. In: Proceedings of 31st International Symposium Remote Sensing Environment (ISRSE). St. Petersburg, 20–24 June 2005. http://www.isprs.org/publications/related/ISRSE/html/papers/376.pdf (2005a)

Mitnik, M.L., Mitnik, L.M.: Oxygen channels of ADEOS-II AMSR: comparison of the measured and simulated brightness temperatures. In: Proceedings of 31st ISRSE, St. Petersburg. http://www.isprs.org/publications/related/ISRSE/html/papers/908.pdf (2005b)

Mitnik, L.M., Mitnik, M.L., Zabolotskikh, E.V.: Microwave sensing of the atmosphere-ocean system with ADEOS-II AMSR and Aqua AMSR-E. J. Remote Sens. Soc. Japan **29**, 156–165 (2009)

Mitrescu, C., L'Ecuyer, T., Haynes, J., Miller, S., Turk, J.: CloudSat precipitation profiling algorithm—model description. J. Appl. Meteor. Climatol. **49**, 991–1003 (2010)

Montgomery, M.T., Vladimirov, V.A., Denissenko, P.V.: An experimental study on hurricane mesovortices. J. Fluid Mech. **471**, 1–32 (2002)

Moreau, E., Mallet, C., Maxbboux, B., Badran, F., Klapisz, C.: Atmospheric liquid water retrieval using a gated experts neural network. J. Atmos. Ocean Technol. **19**, 457–467 (2002)

Morena, L.C., James, K., Beck, J.: An introduction to the RADARSAT-2 mission. Canadian J. Remote Sens. **30**, 221–234 (2004)

Njoku, E.G.: Passive microwave remote sensing of the earth from space—a review. Proc. IEEE **70**, 728–750 (1982). doi:10.1109/PROC.1982.12380

Nordberg, W., Conaway, J., Thaddeus, P.: Microwave observations of the sea state from aircraft. Q. J. Royal Meteorol. Soc. **95**, 408–413 (1969)

O'Dell, C.W., Wentz, F.J., Bennartz, R.: Cloud liquid water path from satellite-based passive microwave observations: a new climatology over the global oceans. J. Climate **21**, 1721–1739 (2008)

Oki, T., Imaoka, K., Kachi, M.: AMSR instrument on GCOM-W1/2: concepts and applications. In: Proceedings of IGARSS 2010 Honolulu, pp. 1363–1366, 25–30 July 2010

Petty, G.W., Katsaros, K.B.: The response of the SSM/I to the marine environment. Part I: An analytic model for the atmospheric component of observed brightness temperatures. J. Atmos. Oceanic Technol. **9**, 746–761 (1992)

Pielke, R.A., Landsea, C.W.: Normalized hurricane damages in the United States, 1923–1995. Wea. Forecast. **13**, 621–631 (1998)

Poe, G., St. Germain, K., Bobak, J., et al.: DMSP calibration/validation plan for the special sensor microwave imager sounder (SSMIS). Naval Research Laboratory, Washington, DC, USA (2001)

Powell, M.D.: Evaluations of diagnostic marine boundary-layer models applied to hurricanes. Mon. Weather Rev. **108**, 57–766 (1980)

Powell, M.D., Houston, S.H., Reinhold, T.A.: Hurricane Andrew's landfall in south Florida. Part I: Standardizing measurements for documentation of surface wind fields. Wea. Forecast. **11**, 304–328 (1996)

Powell, M.D., Black, P.G., Houston, S.H., Reinhold, T.A.: GPS sonde insights on boundary layer structure in hurricanes. In: Preprints, 23rd Conference on Hurricanes and Tropical Meteorology, Dallas, TX, American Meteorol. Soc. Boston, pp. 881–884, 10–15 Jan 1999

Powell, M.D., Vickery, P.J., Reinhold, T.A.: Reduced drag coefficient for high wind speeds in tropical cyclones. Nature **422**, 279–283 (2003)

Prasad, A.K., Singh, R.P.: Features of hurricane Katrina using multi sensor data. Intern. J. Remote Sens. **28**, 4709–4713 (2007). doi:10.1080/01431160500522668

Quilfen, Y., Chapron, B., Tournadre, J.: Satellite microwave surface observations in tropical cyclones. Mon. Weather Rev. **138**, 421–437 (2010). doi:http://dx.doi.org/10.1175/2009MWR3040.1

Quilfen, Y., Vandemark, D., Chapron, B., Feng, H., Sienkiewicz, J.: Estimating gale to hurricane force winds using the satellite altimeter. J. Atm. Oceanic Technol. **28**, 453–458 (2011)

Reynolds, R.W., Gentemann, C.L., Corlett, G.K.: Evaluation of AATSR and TMI satelite SST data. J. Climate **23**, 152–165 (2010). doi:10.1175/2009JCL132421

Reppucci, A., Lehner, S., Schulz-Stellenfleth, J., Brusch, S.: Tropical cyclone intensity estimated from wide-swath SAR images. IEEE Trans. Geosci. Remote Sens. **48**, 1639–1649 (2010)

Reynolds, R.W., Smith, T.M., Liu, C., Chelton, D.B., Casey, K.S., Schlax, M.G.: Daily high-resolution-blended analyses for sea surface temperature. J. Climate **20**, 5473–5496 (2007)

Ross, D.B., Cardone, V.: Observations of oceanic whitecaps and their relation to remote measurements of surface wind speed. J. Geophys. Res. **79**, 444–452 (1974)

Robinson, I.S.: Measuring the Oceans from Space: Theoretical Principles and Methods of Satellite Oceanography. Springer, Berlin (2004). (646 p)

Sasaki, Y., Asanuma, I., Muneyama, K., Naito, G., Suzuki, T.: The dependence of sea surface microwave emission on wind speed, frequency, incidence angle, and polarization over the frequency range from 1 to 40 GHz. IEEE Trans. Geosci. Remote Sens. **25**, 138–146 (1987)

Schubert, W.H., Montgomery, M.T., Taft, R.K., Guinn, T.A., Fulton, S.R., Kossin, J.P., Edwards, J.P.: Polygonal eyewalls, asymmetric eye contraction and potential vorticity mixing in hurricanes. J. Atmos. Sci. **56**, 1197–1223 (1999)

Shi, W., Wang, M.: Observations of a hurricane Katrina-induced phytoplankton bloom in the Gulf of Mexico. Geophys. Res. Lett. **34**, L11607 (2007). doi:10.1029/2007GL02972

Shibata, A.: AMSR/AMSR-E SST algorithm developments; removal of ocean wind effect. Italian J. Remote Sensing, **30/31**, 131–142 (2004)

Shibata, A.: A wind speed retrieval algorithm by combining 6 and 10 GHz data from advanced microwave scanning radiometer: wind speed inside hurricanes. J. Oceanogr. **62**, 351–359 (2006)

Shibata, A.: Algorithm development in retrieving sea surface temperature and sea surface wind speed. J. Remote Sens. Soc. Japan **29**, 167–173 (2009)

Simpson, J., Adler, R.F., North, G.R.: Proposed tropical rainfall measuring mission (TRMM) satellite. Bull. Amer. Meteor. Soc. **69**, 278–295 (1988)

Smith, C.K., Bettenhausen, M., Gaiser, P.W.: A statistical approach to WindSat ocean surface wind vector retrieval. IEEE Geosci. Remote Sens. Lett. **3**, 164–168 (2006)

Spencer, R.W., Braswell, W.D.: Atlantic tropical cyclone monitoring with AMSU-A: estimation of maximum sustained wind speeds. Mon. Weather Rev. **129**, 1518–1532 (2001). doi:10.1175/1520-0493(2001)129<1518:ATCMWA>2.0.CO;2

Staelin, D.H., Chen, F.W.: Precipitation observations near 54 and 183 GHz using the NOAA-15 satellite. IEEE Trans. Geosci. Remote Sens. **38**, 2322–2332 (2000)

Stephens, G.L., Vane, D.G., Boain, R.J., Mace, G.G., Sassen, K., Wang, Z., Illingworth, A.J., O'Connor, E.J., Rossow, W.B., Durden, S.L., Miller, S.D., Austin, R.T., Benedetti, A., Mitrescu, C., the CloudSat Science Team: The CloudSat mission and the A-train. Bull. Am. Meteorol. Soc. **83**, 1771–1790 (2002)

Stephens, G.L., Kummerow, C.D.: The remote sensing of clouds and precipitation from space: a review. J. Atmos. Sci. **64**, 3742–3765 (2007)

St. Germain, K.M., Swift, C.T., Grenfell, T.C.: Determination of dielectric constant of young sea ice using microwave spectral radiometry. J. Geophys. Res. **98**, 4675–4679 (1993)

Stoffelen, A.: A simple method for calibration of a scatterometer over the ocean. J. Atmosph. Oceanic Technol. **16**, 275–282 (1999)

Surussavadee, C., Staelin, D.H.: NPOESS precipitation retrievals using the ATMS passive microwave spectrometer. In: Proceedings of IGARSS 2008, Boston, MA, IEEE, vol. 5, pp. 570–573 (2008)

Swift, C.T., Goodberlet, M.A.: Passive microwave remote sensing of the ocean. Specialist Meeting on Microwave Radiometry and Remote Sensing Applications. Boulder, CO, U.S. DOC/NOAA/ERL/WPL, pp. 87–93 (1992)

Swift, C.T., Dehority, D.C., Tanner, A.B., McIntosh, R.E.: Passive microwave spectral emission from saline ice at C-band during the growth phase. IEEE Trans. Geosci. Remote Sens. **GE-24**, 840–848 (1986)

Tanelli, S., Durden, S.L., Pak, K.S., Reinke, D.G., Partain, P., Haynes, J.M., Marchand, R.T.: CloudSat's cloud profiling radar after two years in orbit: performance, calibration, and processing. IEEE Trans. Geosci. Remote Sens. **46**, 3560–3573 (2008)

Tanner, A.C., Swift, C.T., Black, P.G.: Operational airborne remote sensing of wind speeds in hurricanes. In: Preprints, 17th Conference on Hurricanes and Tropical Meteorology, Miami, FL, American Meteor Society. pp. 385–387 (1987)

Turner, D.D., Cadeddu, M.P., Löhnert, U., Crewell, S., Vogelmann, A.M.: Modifications to the water vapor continuum in the microwave suggested by ground-based 150-GHz observations. IEEE Trans. Geosci. Remote Sens. **47**, 3326–3337 (2009)

Uhlhorn, E.W., Black, P.G.: Verification of remotely sensed sea surface winds in hurricanes. J. Atmos. Oceanic Technol. **20**, 99–116 (2003)

Uhlhorn, E.W., Black, P.G., Franklin, J.L., Goodberlet, M., Carswell, J., Goldstain, A.S.: Hurricane surface wind measurements from an operational stepped frequency microwave radiometer. Mon. Weather Rev. **135**, 3070–3085 (2007)

Vachon, P.W., Katsaros, K.B.: Atmospheric cyclones from spaceborne SAR. Backscatter **10**, 14–19 (1999)

Vachon, P.W., Black, P.G., Dodge, P.P., Katsaros, K.B., Clemente-Colon, P., Pichel, W., McDonnell, K.: RADARSAT-1 Hurricane watch. In: Proceedings of IGARSS 2001, Sidney, Australia, 9–13 July 2001, on CD-ROM, 3 p

Valenzuela, G.: Theories for the interaction of electromagnetic and oceanic waves—a review. Bound Layer Meteorol **13**, 61–85 (1978)

Vignudelli, S., et al.: Satellite altimetry: sailing closer to the coast. In: Tang D. (ed.) Remote Sensing of the Chaning Oceans, Springer, Berlin (2011). doi:10.1007/978-3-16541-2_11

Webster, P.J., Holland, G.J., Curry, J.A., Chang, H.R.: Changes in tropical cyclone number, duration, and intensity in a warming environment. Science **309**, 1844–1846 (2005)

Webster, W.J., Wilheit, T.T., Ross, D.B., Gloersen, P.: Spectral characteristics of the microwave emission from a wind-driven foam covered sea. J. Geophys. Res. **81**, 3095–3099 (1976a)

Weng, F., Grody, N.C.: Retrieval of cloud liquid water using the special sensor microwave imager (SSM/I). J. Geophys. Res. **99**, 25535–25551 (1994)

Weng, F., Zhao, L., Poe, G., Ferraro, R., Li, X., Grody, N.: AMSU cloud and precipitation algorithms. Radio Sci. **38**, 8068–8079 (2003)

Weng, F., Grody, N.C., Ferraro, R.R., Basist, A., Forsyth, D.: Cloud liquid water climatology derived from the special sensor microwave imager. J. Climate **10**, 1086–1098 (1997)

Weng, F., Zhu, T., Yan, B.: Satellite data assimilation in numerical weather prediction models. Part II: Uses of rain-affected radiances from microwave observations for hurricane vortex analysis. J. Atm. Sci. **64**, 3910–3925 (2007)

Wentz, F.J.: A model function for ocean microwave brightness temperatures. J. Geophys. Res. **88**, 1892–1908 (1983)

Wentz, F.J., Mattox, L.A., Peteherych, S.: New algorithms for microwave measurements of ocean winds: applications to SEASAT and the special sensor microwave imager. J. Geophys. Res. **91**, 2289–2307 (1986)

Wentz, F.J., Spencer, R.W.: SSM/I rain retrievals within a unified all-weather ocean algorithm. J. Atm. Sci. **55**, 1613–1627 (1998)

Wentz, F.J.: A well-calibrated ocean algorithm for special sensor microwave/imager. J. Geophys. Res. **102**, 8703–8718 (1997)

Wentz, F.J.: Advanced algorithms for QuikScat and SeaWinds/AMSR. In: Proceedings of IGARSS 2001 Sydney, Australia (2001)

Wentz, F.J., Gentemann, C., Smith, D., Chelton, D.: Satellite measurements of sea surface temperature through clouds. Science **288**, 847–850 (2000)

Wentz, F. J., Smith, D. K., Mears, C. A., Gentemann, C. L.: Advanced algorithms for QuikScat and SeaWinds/AMSR. In: Proceedings of Geoscience and Remote Sensing Symposium, 2001. IGARSS '01. IEEE 2001 International, **3** (2001). doi:10.1109/IGARSS.2001.976752

Wilheit, T., Kummerow, C., Ferraro, R.: Rainfall algorithms for AMSR-E. IEEE Trans. Geosci. Remote Sens. **41**, 204–214 (2003)

Wilheit, T.T.: A model for the microwave emissivity of the ocean's surface. IEEE Trans. Geosci. Electron. **GE-17**, 244–249 (1979)

Wimmers, A.J., Velden, C.S.: MIMIC: a new approach to visualizing satellite microwave imagery of tropical cyclones. Bull. Amer. Meteor. Soc. **88**, 1187–1196 (2007)

Woiceshyn, P., Wurtele, M., Boggs, D., McGoldrick, L., Peteherych, S.: The necessity for a new parameterization of an empirical model for wind/ocean scatterometry. J. Geophys. Res. **91**, 2273–2288 (1986)

Yan, B., Weng, F.: Applications of AMSR-E measurements for tropical cyclone predictions, part I: retrieval of sea surface temperature and wind speed. Adv. Atm. Sci. **25**, 227–245 (2008)

Yueh, S.H., Wilson, W., Dinardo, S., Hsiao, S.V.: Polarimetric microwave wind radiometer model function and retrieval testing for WindSat. IEEE Trans. Geosci. Remote Sens. **44**, 584–596 (2006)

Zecchetto, S.: Ocean wind fields from satellite active microwave sensors. In: Imperatore, P., Riccio, D. (eds.) Geoscience and Remote Sensing, New Achievements. Intech, pp. 263–283 (2010)

Chapter 3
Typhoon Eye Observations Using SAR and MTSAT

A. K. Liu, Y.-H. Cheng, C.-R. Ho, S.-J. Huang and N.-J. Kuo

Abstract Typhoon eyes have been delineated from the smoother area in the Radarsat Synthetic Aperture Radar (SAR) images of ocean surface roughness and from the warmer area in the Multi-functional Transport Satellite (MTSAT) infrared images by using wavelet analysis. Case studies for different typhoons and environment have been investigated to demonstrate that SAR can be a powerful tool to help in typhoon tracking and prediction, especially at the ocean surface. It is found that the distance between the center locations of these typhoon's eyes, as determined by SAR and MTSAT, respectively, is quite significant (14–26 km) for all five cases. The result of large center distance between typhoon eyes at the cloud level from MTSAT data and on the ocean surface from SAR data implies that the eyewall shaft may be highly tilted and the vertical wind shear profile is more complex than generally expected. Some of the issues concerning the definition of typhoon eye and typhoon tracking/prediction have been identified and compared with other data sets. Also, the tilted structure and associated vertical wind shear, especially during typhoon turning and staggering, may be caused by the ocean feedback or island blocking effects.

Keywords Typhoon eye · SAR · MTSAT · Eyewall · Wavelet transform · Tracking

---

Y.-H. Cheng · C.-R. Ho · S.-J. Huang · N.-J. Kuo
Department of Environmental Informatics, National Taiwan Ocean University, Keelung, Taiwan

A. K. Liu (✉)
NASA Goddard Space Flight Center, Greenbelt, MD, USA
e-mail: lindatonyliu@yahoo.com

D. L. Tang and G. J. Sui (eds.), *Typhoon Impact and Crisis Management*, Advances in Natural and Technological Hazards Research 40,
DOI: 10.1007/978-3-642-40695-9_3,

3.1 Introduction

Satellite remote sensing with repeated coverage is the most efficient method to monitor and study ocean environment, marine productivity, and oil spills/pollution. The ability of meteorological satellite remote sensing for monitoring clouds, fronts, and typhoons has been amply demonstrated. One of the first and most important applications of satellite observations of the tropical cyclone has been the estimation of associated intensity of tropical cyclones with the temperature of the eye, cloud organization, and surrounding environment. For example, meteorological satellite images show changes in the cloud organization, central area of the cyclone, and rain bands. The Synthetic Aperture Radar (SAR) images of wind-related ocean features, such as surface waves, rain cells, tropical cyclones, internal lee waves, and katabatic wind have been studied recently (Hsu et al. 2010). The combined use of infrared and visible imagery, optical sensors, and SAR can provide frequent high-resolution coverage of the typhoon evolution.

The western North Pacific is the area with the highest seasonal frequency of tropical cyclones in the world, including storms of highest intensity (hurricanes and super-typhoons). These severe typhoons not only impact the coastal area through powerful winds and torrential rain, but also mix the ocean surface and cause upper ocean response along its passage. Cooling responses to a typhoon affect the upper ocean environment and provide feedback to the typhoon itself (Black and Holland 1995; Schade and Emanuel 1999). In addition to the cooling response induced by a typhoon, the passage of the typhoon also plays a key role in influencing the upper layer marine ecosystem (Lin et al. 2003; Babin et al. 2004; Walker et al. 2005). Thus, more understanding of the internal behavior of typhoons and their interactions with the surrounding atmosphere and underlying ocean proves to be the key for further improving the prediction of typhoon intensity change. Conventionally, the typhoon eye observation is from weather satellite using Infrared (IR) or Visible (VIS) wavelengths, which is the observation at cloud top with height ranging from 12 to 18 km. However, because of the vertical wind shear tilt, the location of typhoon eye on the ocean surface may be quite different from that inferred from cloud imagery. In general, the impact of typhoons on a coastal community is near the surface, whether over land, or the adjacent ocean. Therefore, typhoon information (e.g., location and intensity of strongest winds) near the ocean surface is critical for the typhoon tracking operation and forecast. Owing to its warm-core structure, the tropical cyclone's strongest winds are, in fact, located in the lowermost troposphere, typically near the top of a shallow boundary layer 500–1,000 m deep.

A typhoon monitoring program in Asia incorporating SAR along with conventional observations has been carried out at the National Taiwan Ocean University (NTOU) recently. The objectives are to study typhoon characteristics and to retrieve wind fields by a wavelet tracking technique using multiple sensors including SAR. The first project is focused on the tracking of typhoon's eye using satellite images of SAR from Radarsat-1 of Canadian Space Agency (CSA), and

MTSAT from Japan Meteorological Agency (JMA). A typhoon eye is a zone of relatively weak winds that exists in the center of the swirling vortex of a typhoon. Weather in the eye is relatively calm and free of clouds above a temperature inversion atop the boundary layer, with slowly swirling stratocumulus in the lower eye, the sluggish motion being attributable to mesovortices in the eye (Dunkerton 2010). The eye has also the warmest temperatures at the upper cloud levels. It is normally circular in shape, sometimes accompanied by a polygonal deformation of the adjacent eyewall, and may range in size from 8 to 200 km in diameter. Weatherford and Gray (1988a) used the aircraft dataset, collected by the United States Air Force WC-130 aircraft, to analyze the relationship between cyclone inner-core intensity and outer-core wind strength. Weatherford and Gray (1988b) found that outer-radius wind strength and inner-core intensity can vary greatly. But they didn't have enough information to measure the size of the typhoon eye. Brueske and Velden (2003) estimated the tropical cyclone's intensity and eye size from the NOAA-KLM series Advanced Microwave Sounding Unit (AMSU). The optimal horizontal resolution at nadir is 48 km for AMSU-A and 16.3 km for AMSU-B. The eyewall, a ring of cumulonimbus convection, surrounds the eye and contains the sharpest radial pressure gradient nearly coincident with strongest wind. Because the eyewall slopes outward with height, the eye is approximately an inverted, truncated cone. Conversely, the upper boundary of the sloping eyewall resembles the shape of a stadium or amphitheatre. The air aloft in the eye is clear, warm, and dry, separated by an inversion from more moist, usually cloudy air near the ocean surface. Willoughby (1998) reported that the typhoon eye contains two air masses separated by an inversion and the air below the inversion exchanges momentum and moisture with the ocean and mixes in complicated ways with air from the eyewall.

Microwave sensors have improved the detection of internal cyclone structure, such as the location of the eye, and morphology of precipitation in the adjacent eyewall, because those wavelengths are sensed through high clouds that sometimes obscure the eye in IR and VIS images. However, passive microwave sensors do not have enough horizontal resolution to estimate precisely the size and structure of typhoon eye area. For example, the Advanced Microwave Scanning Radiometer (AMSR) 89 GHz band has only a 6.25 km resolution. Therefore, tracking the calm areas on the ocean surface from high-resolution SAR imagery (with 25–50 m resolution) or the warmest areas on the cloud top from thermal IR imagery (with 1–4 km resolution) may provide a better detection and clearer description of a typhoon's eye. SAR, in particular, has the all-weather capability to "see" through the clouds for the signature and footprint of a typhoon on the ocean surface, in common with passive microwave sensors. The operating frequency of SAR sensors is typically lower than that of other microwave sensors designed to study clouds and precipitation; therefore SAR provides a much better depiction of ocean surface roughness. SAR images, however, are not completely immune from precipitation attenuation and surface splash effects, but since the typhoon eye is generally free of such effects, SAR imagery is particularly useful for studies of the eye (Dunkerton 2010).

In this note, first the SAR and MTSAT satellite data used are listed for reference, and cases selected for study are identified. Then, the typhoon eyes are delineated from the smoother area in SAR images and from the warmer area in MTSAT/IR images by using wavelet analysis. Four case studies for different typhoons and environment have been investigated to measure the center distance between the typhoon eye derived from SAR data and that from MTSAT data. As demonstrated by these case studies, SAR can be a powerful tool to help in typhoon tracking and prediction, especially at the ocean surface. Finally, some of the issues on the definition of typhoon's eye and typhoon tracking/prediction have been identified and are compared with other data sets. Also, the highly tilted eyewall structure and its associated vertical wind shear, especially during typhoon turning and staggering have been examined and are discussed here.

3.2 Data

The Multi-functional Transport Satellite (MTSAT) series fulfills a meteorological function for the Japan Meteorological Agency (JMA). MTSAT is a geostationary meteorological satellite providing imagery for the East Asia and the Western Pacific area. The orbit of MTSAT-1R is 35,800 km above the equator at 140° east longitude. MTSAT-1R is observing every hour in normal condition and half hour during severe weather. The MTSAT-1R meteorological sensor contains one visible and four infrared channels. The wavelength of Visible (VIS) channel is 0.55–0.90 μm and its resolution is 1 km at the sub-satellite point. The infrared channels include IR1 (10.3–11.3 μm), IR2 (11.5–12.5 μm), IR3 (6.5–7.0 μm), and IR4 (3.5–4.0 μm) and their resolution is 4 km.

The Center for Southeastern Tropical Advanced Remote Sensing (CSTARS) at the University of Miami has provided access for the scientific community to real-time, high-resolution satellite imagery since 2003. Currently, existing SAR satellites orbiting in space and accessible to CSTARS are the Canadian Space Agency's (CSA) Radarsat-1, the European Space Agency's ERS-2 and ENVISAT Advanced SAR. Other SAR satellites now in orbit include CSA's Radarsat-2 (with additional beam and polarization capabilities), PALSAR (Japanese Space Agency), TerraSAR-X (Germany/DLR), CRS (Chinese Remote Sensing Satellite) and Cosmo Sky-Med (Italian Space Agency). Radarsat-1 is a C-Band SAR that penetrates clouds and can image near the terminator in all weather conditions, and was launched in November 1995. The ScanSAR Wide product of Radarsat-1 used in this study is a geo-referenced ground coordinate multi-look image. The ScanSAR Wide product has a pixel size of 50 $\times$ 50 m with nominal image coverage of 500 km. Because SAR has the capability to penetrate clouds, it provides a new method to study typhoons with high-resolution observation near the ocean surface below the inversion layer.

The Moderate Resolution Imaging Spectrometer (MODIS) views the entire Earth's surface every 1–2 days, and its image collected within the typhoon area has been used for calibration and comparison with SAR and MTSAT data. Also, the Joint Typhoon Warning Center (JTWC) and JMA's archived best track data have been used to check the consistency of the typhoon's location. In this study, all typhoon images collected by Radarsat in conjunction with MTSAT with a relatively strong typhoon eye in the Western Pacific Ocean from 2003 to 2007 have been examined. The detailed image processing method is outlined in the next section for the case studies of typhoon eye.

3.3 Approach

Wavelet transforms are analogous to a Fourier transform but are localized in both frequency and time. A two-dimensional wavelet transform is a highly efficient band-pass data filter, which can be used to separate various scales of processes (Ruskai et al. 1992; Liu et al. 1997). For effective identification and tracking of common features such as the typhoon eye in a chosen image, a two-dimensional Mexican-hat wavelet transform is applied—to the sub-scene of image covering the typhoon eye—with a spatial scales corresponding to extracted features, while filtering out noise in the data (Wu and Liu 2003).

The Mexican hat wavelet is the second derivative of a Gaussian function, and the resulting transform is the Laplacian of a Gaussian smoothed function. Thus, zeroes correspond to the inflection points of the original function. The contours of zero crossing indicate the edges in the pattern of the input function. To perform the wavelet transform, first a suitable length scale is chosen, which corresponds to the horizontal scale of the Gaussian function. Then an edge detection procedure is carried out to determine the pixel locations of significant differentials so that the feature of interest can be delineated from the background (Wu and Liu 2003). A threshold can be chosen such that all amplitude changes above a minimum contrast level are detected as edges. However, in a noisy image, there is a tradeoff between (1) missing valid edges and (2) over-detection: that is, detecting false edges induced by noise. Several thresholds can be explored via trial and error and a final choice selected based on observation and experience (Liu et al. 2003).

A typhoon eye is a quasi-circular zone of relatively weak winds that exists in the center of the swirling vortex of a typhoon. For the SAR imaging mechanism, a weak wind area with less surface roughness has relatively low radar backscatter compared to the adjacent eyewall. Thus, the typhoon eye can be delineated in a SAR image as the darker (less rough) area with relatively low radar backscatter, and its location pinpointed by using wavelet analysis. A typhoon eye has also the warmest temperatures at the upper cloud levels. So, in a MTSAT/IR image, the warmer area also can be delineated by using wavelet analysis, and its position may be regarded as the typhoon's eye observed in the upper cloud field.

Once the typhoon's eye has been extracted from the image, its center location is calculated as the geometric center with respect to the delineated eye contours, one at lower levels (SAR) and another at upper levels (MTSAT-IR). Then the center displacement is estimated as the distance between the centers of typhoon eye as detected from SAR and MTSAT/IR data. In the estimation of center distance a spheroid coordinates has been used for better accuracy. Thus, the displacement of lower and upper centers represents the vertical tilt of the typhoon's eyewall shaft with increasing height.

3.4 Case Studies

A total of 5 cases of various typhoons in different years, locations, and conditions have been examined for the comparison of typhoon eyes between SAR and MTSAT/IR data collected almost at the same time. Also, the typhoon locations from the archived best tracks of JTWC and JMA have been checked for the calibration and validation along with other sensors, such as MTSAT/VIS, and MODIS. Some of the issues on the definition of typhoon's eye, the error analysis based on sensor resolution, and the resulting vertical wind shear are discussed in the Sect. 3.4.1.

3.4.1 Baseline Case of Typhoon Man-yi in July 2007

In this baseline case, Typhoon Man-yi was turning and approaching Japan on July 13, 2007. Figure 3.1 shows the images and zoom-in sub-scenes of Radarsat SAR and MTSAT/IR with the respective eye boundaries (red contours) delineated by wavelet analysis. In the MTSAT/IR image, the formation of a small eye (with warmer temperatures relative to the eyewall) and the expanded area of very cold cloud in bands around the eye indicate the increasing intensity. The typhoon regional track map, the contours and centers of eyes from the Radarsat SAR and MTSAT/IR images, respectively, are shown in Fig. 3.2. In the regional track map (Fig. 3.2a), the imaging location is indicated by a big arrow, and the typhoon location is tracked by days (big circle with numbers and different colors), with tick marks separated by a half day (median circle) and 3 h (small circles). Also, the typhoon best-tracks from the JTWC (green line) and JMA (black line) post-analysis are overlaid in Fig. 3.2b for reference. Then, the contours and centers of typhoon eyes derived from SAR (green dot), MTSAT/IR (blue dot) and MTSAT/VIS (red dot) images around 9:33 UTC are interpolated and plotted in Fig. 3.3 for comparison. At this stage the typhoon intensity from the ocean surface pressure field was beginning to decrease, and typhoon was also changing its direction of approach to Japan from the south. Notice that both centers of typhoon eye derived from the MTSAT/IR and MTSAT/VIS data are consistent and quite far from the eye center derived from SAR data, and the separation distance ia 19 km, approximately perpendicular to the track direction. Several cases have been checked between IR and VIS channels for MTSAT, and their respective eye

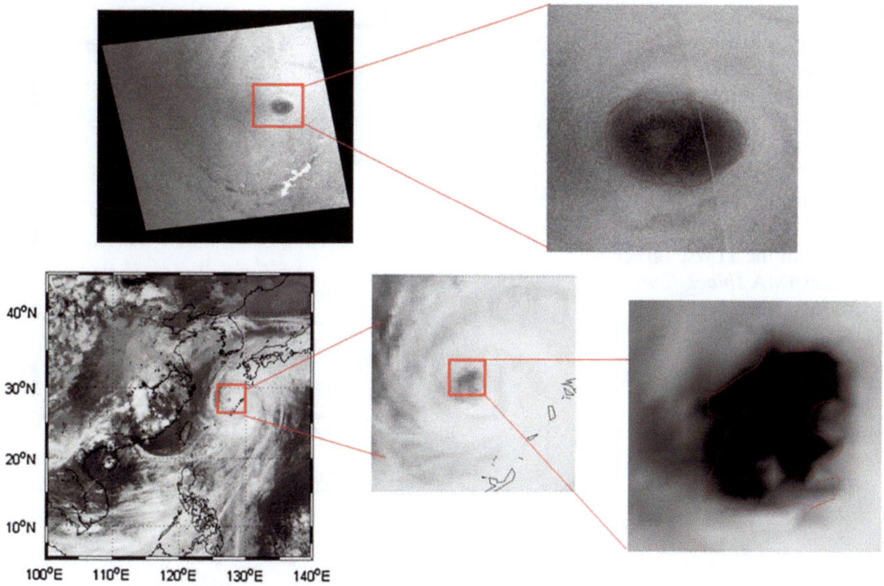

Fig. 3.1 Images and zoom-in sub-scenes of **a** Radarsat SAR (*upper panel*) and **b** MTSAT/IR (*lower panel*) collected on July 13, 2007 with the typhoon eye boundary (*red contours*) delineated by wavelet transform for Typhoon Man-yi

centers are always within 8 km (2 pixels of IR channel) probably due to the contaminations of small cloud patches in the visible channel with higher 1 km resolution as compared with IR channel with 4 km resolution.

In order to calibrate the accuracy of MTSAT/IR, MODIS images have also been used to check for consistency. First, the centers of typhoon eyes derived from MODIS/IR and MODIS/VIS data are found to be consistently within 4 km on July 12, and within 7 km on July 13. For comparison, the contours and centers of typhoon eyes derived from the MODIS/IR and MTSAT/IR images at 13:33 UTC are shown in Fig. 3.4 with a center separation distance of 12 km approximately. It is also found that the eye centers from SAR and MODIS data are closer to the typhoon tracks from JTWC and JMC than the MTSAT-derived eye center. Therefore, it is suspected that the MODIS data may have been used in the post-analysis of typhoon's best tracks by both JTWC and JMA in this case.

3.4.2 Typhoon Kajiki in October 2007

In this typical case, Typhoon Kajiki was over the open ocean and speeding up after beginning to re-curve on October 20, 2007. At this stage, the typhoon intensity was decreasing. Figure 3.5 shows the typhoon regional track map, the extracted

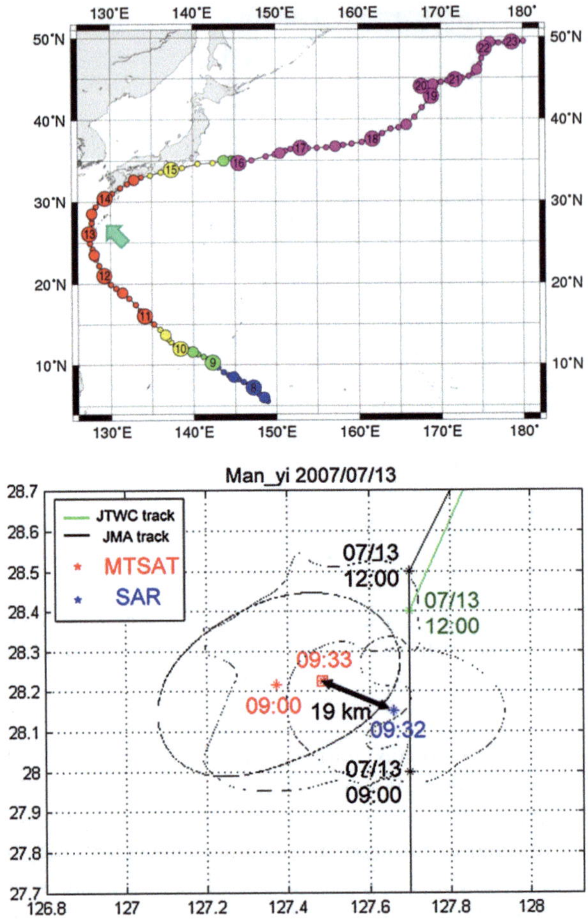

Fig. 3.2 **a** Regional typhoon track map, **b** extracted contours and centers of typhoon eyes from RADARSAT SAR, and MTSAT/IR images when Typhoon Man-yi turned and approached to Japan on July 13, 2007. The best typhoon tracks from the JTWC (*green line*) and JMA (*black line*) post-analysis of Typhoon Man-yi are overlaid

contours and centers of typhoon eyes derived from SAR and MTSAT/IR data for this case. The typhoon eye derived from SAR data was lagging and dragging behind that derived from MTSAT data with a distance of 14 km at 20:25 UTC while the typhoon was speeding up over the open ocean. However, the SAR-derived typhoon eye is right at the average center of both tracks from the JTWC (green line) and JMA (black line) post-analysis in this case. In general, the typhoon is speeding up due to the wind and pressure fields aloft. Therefore, the typhoon eye near the ocean surface below the inversion layer may experience less effect from the high level wind and so may be expected to drag behind the typhoon eye at the cloud level way above the inversion layer.

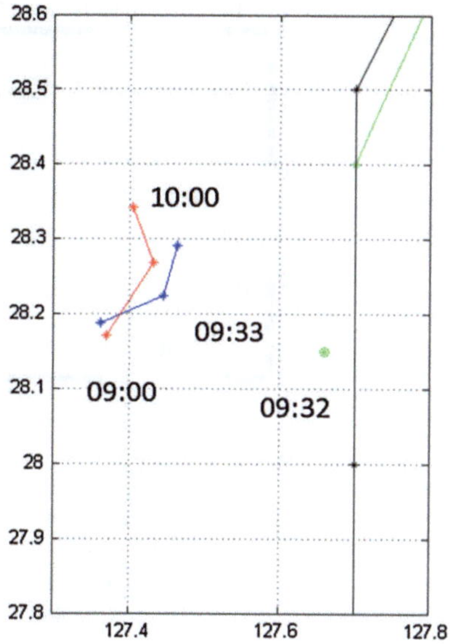

Fig. 3.3 Comparison of the centers of typhoon eyes derived from SAR (*green dot*), MTSAT/IR (*blue dot*) and MTSAT/VIS (*red dot*) images around 9:33 UTC on July 13, 2007. The best typhoon tracks from the JTWC (*green line*) and JMA (*black line*) post-analysis of Typhoon Man-yi are also overlaid for reference

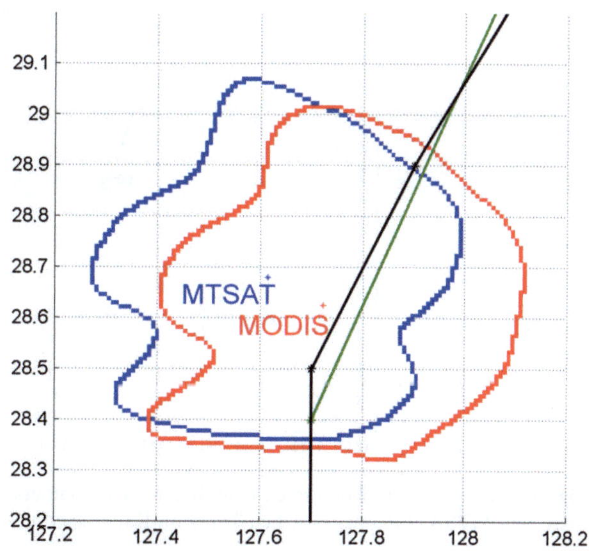

Fig. 3.4 Comparison of typhoon eye contours and centers derived from MODIS/IR and MTSAT/IR images at 13:33 UTC on July 13, 2007 with a center separation distance of 12 km approximately. The best typhoon tracks from the JTWC (*green line*) and JMA (*black line*) post-analysis of Typhoon Man-yi are also overlaid for reference

3.4.3 Typhoon Usagi in August 2007

In this case, Typhoon Usagi was headed towards Japan on August 1, 2007 as shown in the typhoon regional track map, the extracted contours and centers of typhoon eyes from SAR, and MTSAT/IR images in Fig. 3.6. The distance between typhoon eyes derived from SAR and MTSAT data is more than 15 km. At this stage, the typhoon intensity was quite steady based on the ocean surface pressure

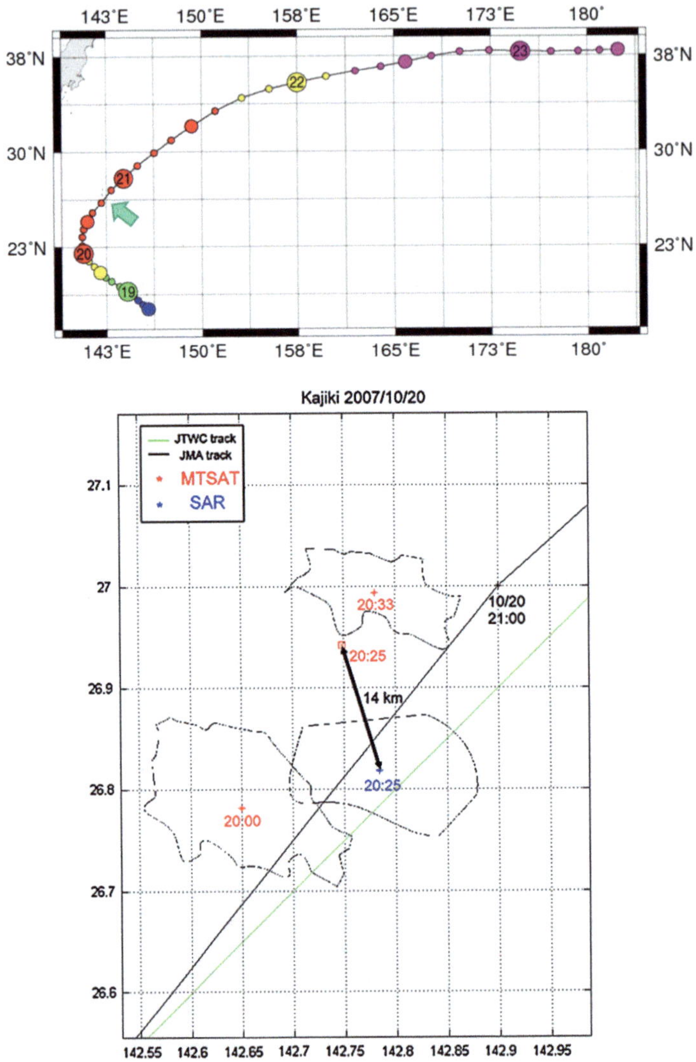

Fig. 3.5 **a** Regional typhoon track map, **b** extracted contours and centers of typhoon eyes from SAR, and MTSAT/IR images, when Typhoon Kajiki sped up on October 20, 2007

field. The typhoon eyes derived from SAR and MODIS data are close to the JTWC track, while the center of eye derived from MTSAT data is close to the JMA track. Notice that the typhoon eye at the cloud level derived by MTSAT data is lagging behind that derived by SAR data near the ocean surface. This is probably due to the blocking effect caused by island and mountains in the wind field at the cloud level, however, the local wind field was not much affected near the ocean surface below the inversion layer.

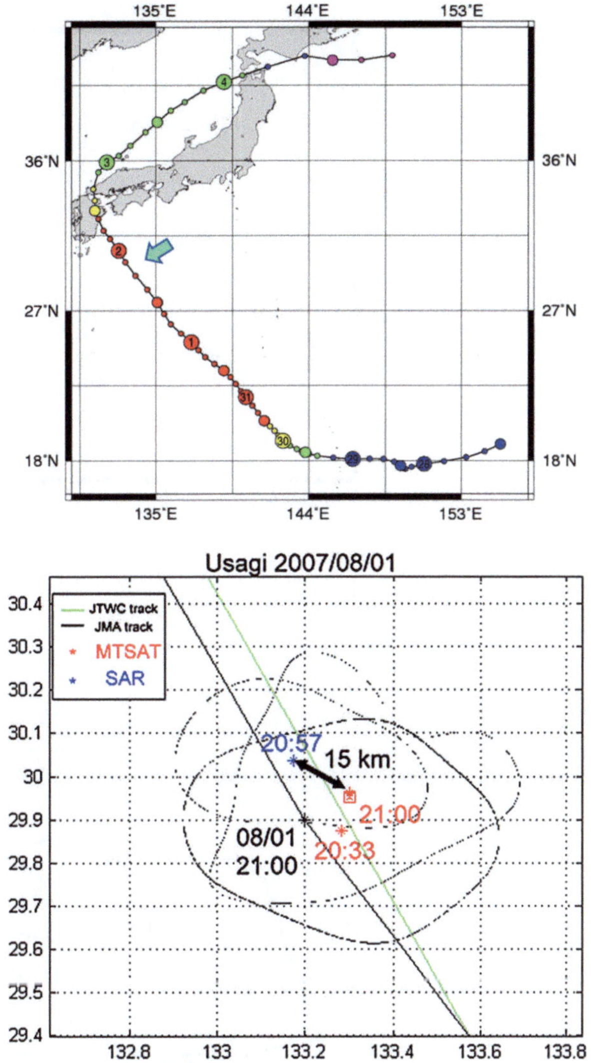

Fig. 3.6 **a** Regional typhoon track map, **b** extracted contours and centers of typhoon eyes from SAR, and MTSAT/IR images when Typhoon Usagi headed towards Japan on August 1, 2007

3.4.4 Typhoon Kirogi in October 2005

In this special interesting case, Typhoon Kirogi was turning and staggering on April 14 (Fig. 3.7) and then speeding up on April 15 (Fig. 3.8) as shown in the typhoon regional track map, the extracted contours and centers of typhoon eyes, from SAR and MTSAT/IR images, respectively. At this stage, the typhoon intensity was steady. Also, both typhoon eyes derived from SAR and MTSAT data follow the track from JMA post-analysis really well, but not very well the JTWC track on the 14th. When the typhoon was turning and staggering on the 14th, the

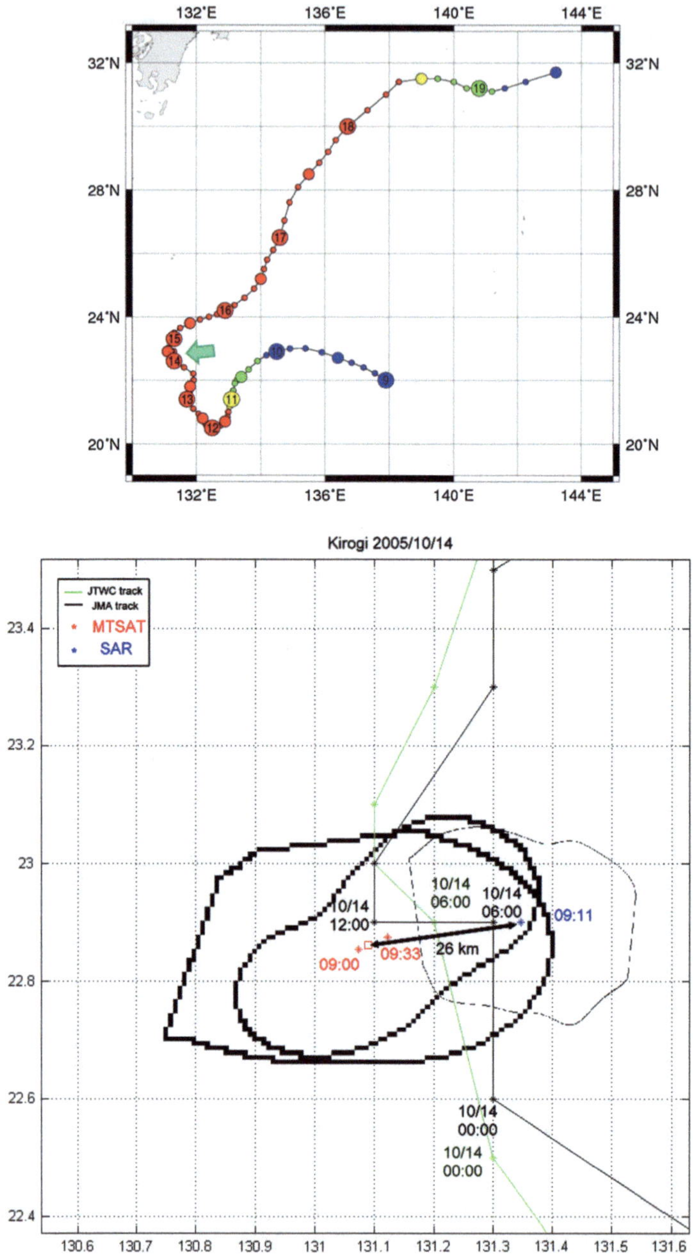

Fig. 3.7 **a** Regional typhoon track map, **b** extracted contours and centers of typhoon eyes from SAR, and MTSAT/IR images when Typhoon Kirogi turned and staggered on October 14, 2005

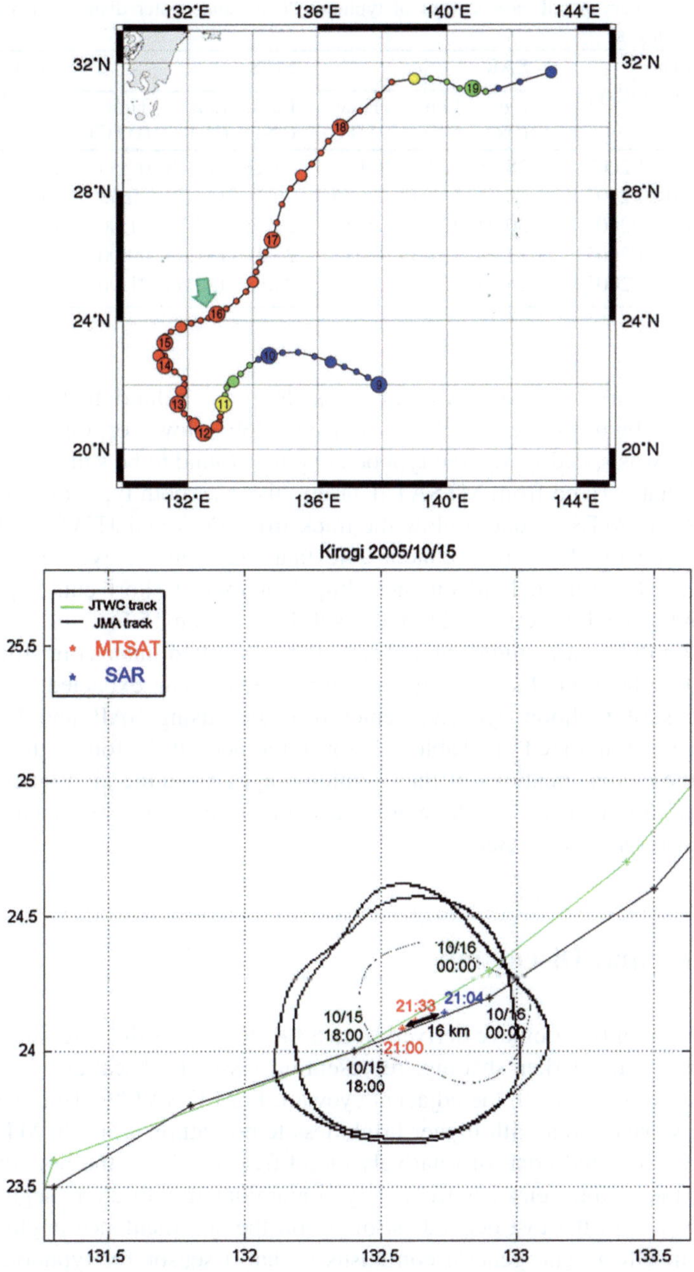

Fig. 3.8 **a** Regional typhoon track map, **b** extracted contours and centers of typhoon eyes from SAR, and **c** MTSAT/IR images when Typhoon Kirogi sped up on October 15, 2005

Table 3.1 Summary of all case studies of typhoon "eye" and center distance using SAR and MTSAT/IR images

Typhoon name	Date dd/mm/yyyy	SAR			MTSAT			Distance between eyes (km)
		Lat (°N)	Lon (°E)	Time (UTC)	Lat (°N)	Lon (°E)	Time (UTC)	
Man-yi	07/13/2007	28.15	127.66	09:32	28.23	127.48	09:33	19
Kajiki	10/20/2007	26.81	142.78	20:25	26.99	142.78	20:33	14
Usagi	08/01/2007	30.04	133.17	20:57	29.96	133.30	21:00	15
Kirogi	10/14/2005	22.90	131.34	09:11	22.87	131.09	09:00	26
Kirogi	10/15/2005	24.14	132.73	21:04	24.08	132.58	21:00	16

typhoon eye derived from SAR data was dragging behind that derived from MTSAT data by a distance of 26 km approximately. However, on the 15th when the typhoon was speeding up, the typhoon eye was found to be suddenly ahead, by 16 km, of that derived from MTSAT data. In this case, both typhoon eyes derived from SAR and MTSAT data follow the track from JMA and JTWC post-analysis really well on the 15th. It is conceivable that the typhoon eye near the ocean surface was dragging behind on the 14th, then took a short cut, bypassed the turning and moved directly to the new track location on 15th.

Based on these case studies and observation, the tilted inner-core structure and associated vertical wind shear may be more complex than expected. All of these case studies of typhoon eye and center distances using SAR and MTSAT/IR images are summarized in Table 3.1 for reference. It is found the distances between the center locations of these typhoon eyes are quite significant, ranging from 14 to 26 km. However, there are many important issues still needing further study as discussed in the Sect. 3.5.

3.5 Issues and Discussion

The definition of typhoon eye derived from SAR imagery is the smoother area near storm center with darker shading, representing lesser backscattering from radar relative to that underneath the adjacent eyewall. For MTSAT/IR data, the typhoon eye is the warmer area with higher brightness temperature, while in MTSAT/VIS data it is an inverted cone of relatively cloud-free air above the boundary layer. However, the visible sensor is frequently contaminated with cloud segments, and reflectivity within the eye depends strongly on the sun incidence angle owing to the stadium effect. The general consensus is that a sensor for typhoon intensity studies should be one which measures the wind field directly. Lacking direct measurements of surface wind, the next best alternative is the high-resolution SAR imagery based on the surface roughness inferred from radar backscattering of capillary waves on the ocean surface. Roughness increases with wind intensity, and the retrieval of wind speed requires a suitable geophysical model function. In a heavily precipitating area such as the hurricane eyewall, SAR images may contain artifacts owing to attenuation and surface splash; nevertheless, these images are well-suited to studies of the hurricane eye (Dunkerton 2010) which is relatively calm and free of such effects, as demonstrated here.

It is found that the typhoon eye derived from SAR always follow the typhoon tracks from the JTWC and JMA post-analysis really well. It is conceivable that these best tracks have already incorporated other satellite data (e.g. MODIS, AMSR, TRMM, …) and even ship/buoy data over the ocean for post-analysis. As demonstrated here, SAR can be a powerful tool for typhoon tracking and prediction, especially near the ocean surface below the inversion layer. As SAR data become increasingly abundant and computers faster, there is an urgent need for, and rising interest in, combining multiple sensors such as MTSAT and MODIS for typhoon tracking in near real-time applications. High-resolution SAR definitely can help in the operational use of MTSAT for improved location accuracy, with attendant improvements in short-term prediction and model assimilation of typhoon tracks over the open ocean, including storm surge forecasts relevant to coastal communities where the typhoon has its most serious impact.

The accuracy of our wavelet method for typhoon eye extraction is limited only by the persistence of the feature, the spatial resolution, and navigational accuracy of satellite data. Since the remapped SAR images have a pixel size of 50 m, they have no serious accuracy issue. However, the MTSAT/IR images have a pixel size of 4 km, so the derived eye may have geolocation error of O (4 km) due to sensor resolution. Parallax owing to the stadium effect may contribute a similar amount to location error in geostationary imagery (Dunkerton 2010). Since most of the eye contours are relatively axi-symmetric, the threshold used to delineate eye contour by wavelet analysis will not much affect our derived center location of typhoon eye. For an average eye size of 50 km in diameter, the maximum error is estimated to be O (8 km), that is, about 2 pixels. As the average size of the eye is usually much larger than the height of the typhoon, and the eyewall itself tilts radially outward with height, a slight tilt of the typhoon's eye/eyewall "inverted cone" would be hard to detect by human eyes. Therefore, only a combination of high-resolution instruments and objective analysis can offer a precise observation of the typhoon eye. However, the accuracy of eye location from MTSAT/IR data may still be a concern as compared with that derived from MODIS/IR data with 12 km eye center separation (Fig. 3.4). More detailed sensitivity study between MTSAT and MODIS images is warranted to determine the accuracy of MTSAT eye locations.

The large location offset in some of our case studies, ranging from 14 to 26 km, between typhoon eyes derived from MTSAT data at the cloud level and that from SAR data on the ocean surface implies that the eyewall shaft may be highly tilted and the associated vertical wind shear more complex than generally expected. This large offset is relatively small to compare with the average eye size of 50 km. However, it is comparable to the cloud level at an average height of 12 km, making the angle of tilt significant. Also, as mentioned above, the accuracy of eye location from MTSAT data may be a concern with an uncertainty of 12 km (3 pixels). For a small island such as Taiwan, 26 km—as found in the case 4—can be the distance between two cities! As shown in case 3, the typhoon eye derived from MTSAT data is lagging behind that derived from SAR data by 15 km probably due to island/mountain blocking effect. In this case, the typhoon could land on the coast almost half an hour earlier than the prediction based on the MTSAT data alone. Definitely, the high-resolution instantaneous observation by SAR (50 m) can be complementary to the more frequent observations (every 15 min) using MTSAT for real-time operation. Although lacking the temporal continuity of geostationary imagery, a solitary SAR image may afford some advance warning that the MTSAT eye location requires correction to account for position errors of O (10 km). Also, tilted and twisted vertical wind shear profile has to be considered in the meso-scale typhoon model for a better prediction and operation use on typhoon track and intensity. Based on these observations, the feedback from the ocean upwelling may be significant especially during typhoon turning and staggering as shown in the case 4 study.

Acknowledgments The authors would like to thank Norden Huang of the National Central University for his encouragement and technical advice. The Radarsat SAR data is provided from CSTARS station in the University of Miami under Typhoon Monitoring Program and copyrighted to CSA. The MTSAT-IR data is provided by the Central Weather Bureau of Taiwan. Technical and scientific edits of the submitted manuscript were provided by Tim Dunkerton of NorthWest Research Associates, Seattle WA. This work is performed during AKL's stay at NTOU as a visiting professor in 2010 supported by the National Research Council (NSC) of Taiwan.

References

Babin, S.M., Carton, J.A., Dickey, T.D., Wiggert, J.D.: Satellite evidence of hurricane-induced phytoplankton blooms in an oceanic desert. J. Geophys. Res. **109**, C03043 (2004). doi:10.1029/2003JC001938

Black, P.G., Holland, G.J.: The boundary layer of tropical Cyclone Kerry (1979). Mon. Weather Rev. **123**, 2007–2028 (1995)

Brueske, K.F., Velden, C.S.: Satellite-based tropical cyclone intensity estimation using the NOAA-KLM series Advanced Microwave Sounding Unit (AMSU). Mon. Wea. Rev. **131**, 687–697 (2003)

Dunkerton, T.J.: Microwave imagery and in situ validation of eye mesovortex structure in Hurricane Katrina (2005) at peak intensity. In: AMS 29th Conference on Hurricanes and Tropical Meteorology, Tucson AZ (2010)

Hsu, M.K., Wang, C.T., Liu, A.K., Chen, K.S.: Satellite Remote Sensing of Southeast Asia: SAR Atlas, p. 133. Tingmao Publish Co., Taipei, Taiwan (2010)

Lin, I., Liu, W.T., Wu, C.-C., Wong, G.T.F., Hu, C., Chen, Z., Liang, W.-D., Yang, Y., Liu, K.-K.: New evidence for enhanced ocean primary production triggered by tropical cyclone. Geophys. Res. Lett., **30**, 1718 (2003). doi:10.1029/2003GL017141

Liu, A.K., Peng, C.Y., Chang, S.Y.-S.: Wavelet analysis of satellite images for coastal watch. IEEE J. Ocean. Eng. **22**, 9–17 (1997)

Liu, A.K., Wu, S.Y., Zhao, Y.: Wavelet analysis of satellite images in ocean applications. In: Chen C.H. (ed.) Frontiers of Remote Sensing Information Processing, pp. 141–162, World scientific. (2003)

Ruskai, M. B., Beylkin, G., Coifman, R., Daubechies, I., Mallat, S., Meyer, Y., Raphael, L.: Wavelets and Their Applications. Jones and Bartlett Publishers, Boston, MA, USA (1992)

Schade, L.R., Emanuel, K.A.: The ocean's effect on the intensity of tropical cyclones: results from a simple coupled atmosphere ocean model. J. Atmos. Sci. **56**, 642–651 (1999)

Walker, N.D., Leben, R.R., Balasubramanian, S.: Hurricane-forced upwelling and chlorophyll a enhancement within cold-core cyclones in the Gulf of Mexico. Geophys. Res. Lett. **32**, L18610, (2005). doi:10.1029/2005GL023716

Weatherford, C.L., Gray, W.M.: Typhoon structure as revealed by aircraft reconnaissance. Part I: data analysis and climatology. Mon. Weather Rev. **116**, 1032–1043 (1988a)

Weatherford, C.L., Gray, W.M.: Typhoon structure as revealed by aircraft reconnaissance. Part II: structural variability. Mon. Weather Rev. **116**, 1044–1056 (1988b)

Willoughby, H.E.: Tropical cyclone eye thermodynamics. Mon. Weather Rev. **126**, 3053–3067 (1998)

Wu, S.Y., Liu, A.K.: Toward an automated ocean feature detection, extraction and classification algorithm for SAR imagery. Int. J. Remote Sens. **24**, 935–951 (2003)

Chapter 4
Satellite Passive and Active Microwave Methods for Arctic Cyclone Studies

E. V. Zabolotskikh, L. M. Mitnik, L. P. Bobylev and B. Chapron

Abstract Satellite passive microwave methods for the Arctic cyclone studies are considered. The data of Special Sensor Microwave Imager (SSM/I) onboard Defense Meteorological Satellite Program (DMSP) satellites and Advanced Microwave Scanning Radiometer—Earth Observing System (AMSR-E) onboard Aqua satellite and AMSR2 onboard GCOM-W satellite are shown to be quite informative for polar low studies. The methodology of polar low detection and tracking, based on analysis of the total atmospheric water vapor fields, retrieved from passive microwave satellite measurement data, is described. Several case studies of mesoscale low system are considered using passive microwave and other satellite data, including visible, infrared and synthetic aperture radar images and scatterometer wind fields.

Keywords Arctic mesoscale cyclones · Polar lows · Satellite passive and active microwave instruments · SSM/I · AMSR-E · AMSR2 · Envisat ASAR · MetOp scatterometer

---

E. V. Zabolotskikh (✉)
Satellite Oceanography Laboratory, Russian State Hydrometeorological University and Nansen Centre, Malookhtinsky Prospect 98, Saint Petersburg, Russia 195196
e-mail: liza@rshu.ru

L. M. Mitnik
V.I. Il'ichev Pacific Oceanological Institute (POI), Vladivostok, Russia

L. P. Bobylev
Nansen Centre, St. Petersburg, Russia

L. P. Bobylev
Nansen Centre, Bergen, Norway

B. Chapron
Department of Oceanography and Ecosystem Dynamics, IFREMER, Brest, France

D. L. Tang and G. J. Sui (eds.), *Typhoon Impact and Crisis Management*,
Advances in Natural and Technological Hazards Research 40,
DOI: 10.1007/978-3-642-40695-9_4,

4.1 Introduction

Arctic intensive cyclones, though not being as vigorous as tropical cyclones, nevertheless pose a serious threat to marine operations and to coastal communities when they make landfall. An especially distinguished class of Arctic cyclones is represented by the range of mesoscale lows with a horizontal length scale of less than 1,000 km. Such mesoscale vortices develop both in the Arctic and Antarctic poleward of the main polar front or other major frontal zones. In spite of a large number of research studies of these systems they are still a challenge to modelers and forecasters due to their very fast development and variety of underlying conditions. The term "mesocyclone" covers a very wide range of weather systems from insignificant, minor vortices with only a weak cloud signature and no surface circulation, to the very active maritime disturbances known as polar lows, which in extreme cases may have winds of hurricane force and bring heavy snowfall. Though such systems can be observed in both hemispheres, the Arctic polar lows are significantly more intense than Antarctic ones (Rasmussen and Turner 2003). And although it has been known for many years that strong small storms in high latitudes could appear with almost no precursor, it was only with the general availability of the imagery from polar orbiting weather satellites in the 1960s that it was realized that these phenomena were quite common.

This imagery indicated that the storms developed over the high latitude ocean areas mostly during winter and tended to decline rapidly over land or ice. The most intense polar lows develop over those ocean areas where the sea surface temperatures remain relatively high during outbreaks of cold air. Such conditions ensure strong, deep convection and the satellite imagery almost always shows cumulonimbus clouds associated with the most active polar low systems. Until recently mostly visible and infrared images, showing a wide range of cloud signatures associated with polar low development, have been used for their detection and study (Bader et al. 1995; Rasmussen and Turner 2003).

Small size and short lifetime of polar lows hinder their detection and tracking and complicate their modeling. Sparse synoptic observations in areas where polar lows frequently occur provide insufficient data for modeling and forecasting, though it should be noticed that the last decade is distinguished by the growing number of modeling studies which already use satellite products for initial and boundary conditions. The first modeling results were generally not successful because the atmospheric models did not have a resolution high enough to resolve the mesoscale systems, which often have a diameter of only a few hundred kilometers. There were also difficulties because of the poor representation of convection in many of the early models. But the latest generation of models with resolutions of 50 km and higher, and a good representation of physical processes, are having more success in simulating some polar mesoscale weather systems (Kusaka et al. 2005).

While the availability of visible and infrared imagery from polar orbiting satellites provided a major advance in the study of polar lows in the early years of meteorological satellites, these observations can capture only the structure of storm tops and some mesocyclones can be masked by higher clouds (Friedman et al. 1999). Starting with Seasat in 1978 and with Kosmos-1500 in 1983, Spaceborne Synthetic Aperture Radar (SAR) and Real Aperture Radar (RAR) systems have provided the scientific community with high and medium resolution images of the ocean surface. The high spatial resolution of radar systems, combined with their ability to "see" through clouds independently of the time of day and their ability to retrieve wind fields allow them to register the high wind speeds accompanying polar lows (Sikora et al. 2000). These instruments enable deep insight into the internal properties of polar lows, mark the accurate location of the atmospheric fronts and polar low centers near the sea surface; and indicate the presence of small-scale organized variations of surface wind at various scales (Mitnik et al. 1996). The most serious limitation of radar imagery is its scarcity which hampers using this instrument in polar low studies especially considering how rapidly these storms develop.

4.2 Capabilities of Passive Microwave Instruments Related to Polar Low Studies

Among instrumental data satellite passive microwave data remain an invaluable source of regularly available remotely sensed data, as opposed to occasional SAR images. They are independent of clouds, as opposed to infrared and visible imagery. The value of the satellite passive microwave sensors for the Arctic cyclone studies can hardly be overestimated since they are among those instruments which can provide the quantitative estimation of a whole set of geophysical parameters. Moreover, regularity and high temporal resolution in the polar regions make them the most appropriate for study of the most intense polar lows developing. Satellite passive microwave data make possible retrievals of such important geophysical parameters of the atmosphere–ocean system as sea surface wind speed, atmospheric water vapor content, cloud liquid water content and precipitation rate, atmospheric temperature and humidity profiles (Mitnik 2009).

Among satellite passive microwave instruments two groups of sensors can be distinguished. The first group includes passive microwave sounders, providing three dimensional maps of the atmospheric temperature and humidity. These instruments allow detecting the warm core of a polar low by mean of analyzing cyclone air temperature in the lower troposphere. The Microwave Sounding Units (MSU) operating on NOAA polar-orbiting platforms were the principal sources of satellite temperature profiles from late 1978 to the early 2000s. Beginning in 1998, a series of follow-on to the MSUs instruments, the Advanced Microwave Sounding Units (AMSUs), began operation. The AMSU instruments are similar to the MSUs, but they make measurements using a larger number of channels, thus sampling the atmosphere in a larger number of layers, including a number of layers higher in the stratosphere. The AMSU instrument proved to be a promising source of remotely sensed data for study of polar lows since it was a considerable advancement over its predecessor, possessing much improved resolution in both the horizontal and vertical directions. Thus, the study, described by (Moore and Vonder Haar 2003) illustrates the utility of AMSU-A data in polar-low research and forecasting by means of examination of a polar low that occurred in the Labrador Sea on 17–18 March 2000. The warm core of the polar low was identified by investigation of the brightness temperature T_B fields at the frequency of 53.6 GHz having 48 km horizontal resolution at nadir as compared with 110 km for MSU.

The presence of a warm core or lack of that, has dynamical implications, because the formation of a warm core in a polar low is generally associated with an intensification of surface winds (Rasmussen et al. 1993). Therefore, this approach allows for a straightforward and time-efficient examination of the microwave data that, nonetheless, provides insight into the polar-low structure and intensity. The ability of AMSU to identify reliably the warm core of a polar low illustrates its utility not only in the tracking of polar lows but also in the forecasting of some parameters such as surface wind speed. The informational content of AMSU is not limited to that of the atmospheric temperature field. Data at 50.3 and 89.0 GHz provide information about the total water vapor content Q and total cloud liquid water content V variations as well as about the sea ice, and the brightness temperatures at frequencies in the area of a strong water vapor resonance line at 183.3 GHz serve for retrieval of vertical profiles of atmospheric moisture. The great potential of satellite passive microwave cross-track sounders is associated with their ability to reconstruct the vertical structure of the cyclones. And here the availability of reliable retrieval techniques allowing correctly estimate atmospheric air temperature and moisture profiles gets the first priority in usage of these data. A comparatively new instrument—Special Sensor Microwave Imager/ Sounder (SSMI/S) onboard Defense Meteorological Satellite Program (DMSP) F16, F17 and F19 satellites the first of which was launched on 18 October 2003 combines in itself the capabilities of both imager and sounder allowing the geophysical parameter field and—simultaneously—vertical cyclone structure reconstruction. SSMI/S is a 24 channel 21-frequency, linearly polarized passive microwave radiometer system. The appearance of more sophisticated microwave instruments such as the Advanced Technology Microwave Sounder (ATMS) sensor with 22 microwave channels in the 23–183 GHz range onboard the Suomi National Polar-orbiting Partnership or Suomi NPP satellite launched on October 27, 2011 provides the opportunity to apply these methods to data with ever-increasing resolution across a swath width of 2,300 km (Wagner et al. 2011). Horizontal resolution of ATMS sensor is approximately 75 km for 23.8 and 31.4 GHz channels, 32 km for channels used for air temperature profile retrieval (52.8–57.6 GHz) and 16 km for atmospheric moisture retrieval channels (166–190 GHz).

Well calibrated satellite passive microwave imagers, starting with Special Sensor Microwave/Imager (SSM/I) onboard DMSP series satellites, and continued by Advanced Microwave Scanning Radiometer—Earth Observing System (AMSR-E) onboard the Aqua satellite (till October 2011) and AMSR2 instrument onboard Japan's GCOM-W1 satellite (launched in May 2012) (Mitnik et al. 2008), have been providing a set of quantitative parameters inside polar lows. The demonstration of newly found capabilities for Arctic cyclone studies, based on atmospheric water vapor field analysis, based on AMSR-E data, is given by (Bobylev et al. 2011). In this study a new methodology for detection and tracking of the Arctic lows was developed, based on satellite passive microwave data. Using this methodology implies the completion of two consecutive steps: atmospheric total water vapor content field reconstruction from passive microwave measurements of SSM/I (SSMI/S) and/or AMSR-E/AMSR2 instruments, and identification of the lows by means of vortex structure detection in these fields. The application of such an approach is most attractive in the polar regions characterized by low water vapor and cloud liquid water values, allowing accurate parameter estimation as opposed to tropical areas (Bobylev et al. 2010). Polar regions also have frequent revisits by SSM/I and AMSR-E/AMSR2. Over these areas of most frequent polar low occurrence 3–4 images per day from each instrument can be used for the studies.

4.3 Multi-sensor Approach

The most comprehensive approach for satellite-based studies of polar lows is one that incorporates data from multiple satellite platforms. Satellite data can provide a great deal of data on the sea surface characteristics and on the form and composition of the clouds associated with mesoscale cyclones/polar lows from which information can be inferred about the air masses, sea surface wind and physical processes involved in the formation of the systems. Visible and infrared imagery shows the cloud structure and locations of the lows and provides data on the cloud-top temperatures. If a nearby radiosonde profile is available or an estimate of the atmospheric profile is obtained from a numerical model, then the actual height of the cloud tops can be estimated. The usage of satellite passive microwave data and identification of polar lows from water vapor content fields, described before, proved to be very effective especially with simultaneously retrieved sea surface wind field. Total atmospheric water vapor content and total cloud liquid water content, retrieved from passive microwave data, are particularly valuable since they provide information on the whole atmospheric column and not just for the top of the cloud. Scatterometer data provide additional regular estimation of the sea surface winds improving temporal resolution of the wind data. SAR images, when available, can add invaluable information about the high-resolution near-surface wind field, precise locations and structure of the atmospheric fronts and polar low centers as well as about the sea ice presence of and structure of the marginal ice zone. All these types of data allow an estimation of the spatial distribution of ocean surface and atmospheric parameters and size of the lows, although these cannot easily be related to the surface pressure pattern without other data being available. Most comprehensive study of the polar lows—from their appearance to destruction requires a multi-sensor approach with combined use of data from various sensors taking advantage of the unique features provided by each of the different sensors. Though all aforementioned remote sensing sensors are capable of detecting a polar low, each of them suffers from various deficiencies. Taken together, they provide most complete knowledge about a mesoscale low: the highest temporal resolution and all available parameter estimation.

4.4 Case Studies

Below we present several examples demonstrating the usage of several types of satellite data, for polar low studies.

Figures 4.1 and 4.2 show two polar low developments in the Barents Sea on 7–8 January 2009 as imaged by SSM/I and AMSR-E retrieved atmospheric water vapor fields and confirmed by Envisat ASAR images. It can be seen that AMSR-E data provided a much more detailed structure of the polar lows than the SSM/I data did due to its higher spatial resolution. To improve SSM/I parameter field resolution advanced techniques can be developed using measurements at a higher frequency (85.5 GHz, both polarization). The atmospheric fronts are seen in Figs. 4.1 and 4.2 as dark narrow bands and patches in the AMSR-E retrieved water vapor fields, having higher resolution than SSM/I retrieved water vapor fields displayed in the same figures. These features coincide with the bands and patches

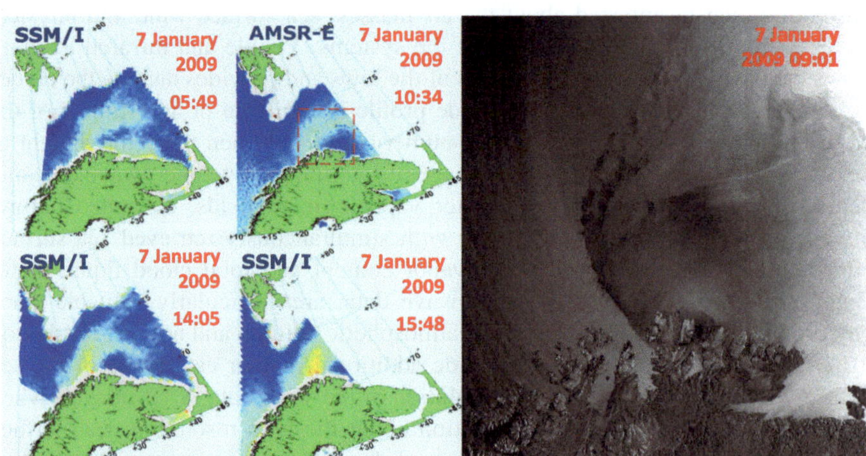

Fig. 4.1 The process of polar low evolution during 10 h on 7 January 2009 observed by SSM/I and AMSR-E-retrieved atmospheric water vapor content fields and Envisat ASAR. *Red dashed rectangle* on AMSR-E field for 10:34 UTC marks the boundary of ASAR image

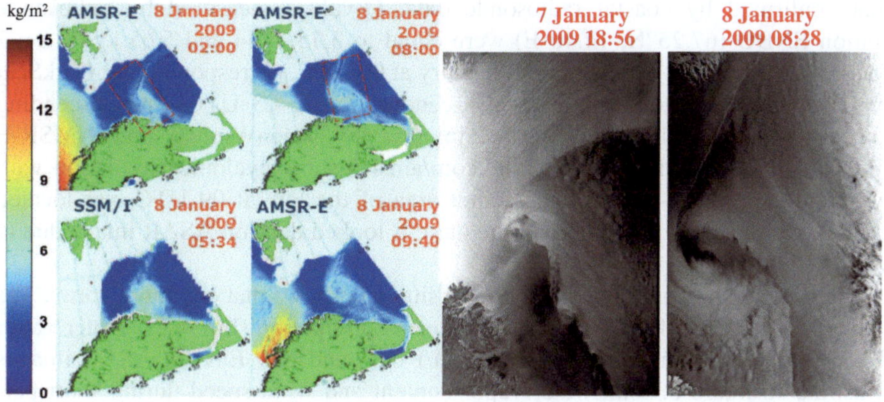

Fig. 4.2 The process of polar low evolution during 13 h on 7–8 January 2009 observed by SSM/I and AMSR-E-retrieved atmospheric water vapor content fields and Envisat ASAR. *Red dashed rectangle* on AMSR-E field for 8 January at 02:00 UTC marks the boundary of ASAR image on 7 January 2009 at 18:56 UTC. *Red dashed rectangle* on AMSR-E field for 8 January at 08:00 UTC marks the boundary of ASAR image on 8 January 2009 at 08:28 UTC

of the total cloud liquid water content located in the atmospheric front areas. The location of the fronts at the sea surface are clearly detected on Envisat ASAR images acquired on 7 January at 09:01 and at 18:54 UTC and on 8 January at 08:28 UTC (Figs. 4.1 and 4.2). In particular, a cold air outbreak playing a key role in a polar low development manifests itself as bands of alternating brightness to the west and northwest from the center of the low on both images acquired on 7 January. These bands are due to the organized variations of the sea surface wind and caused by roll convection in the marine boundary layer of the atmosphere. Their orientation corresponds to the northern surface winds. Cloud streets are an indicator of roll convection on the NOAA AVHRR, Terra and Aqua MODIS thermal infrared images (not shown). Very likely that this process is also accompanied by Q-variations, however, spatial smoothing hinders their indication in water vapor field. To the east of the front (a dark line) winds blow from the east-northeast that follows from orientation of cyclonically-inflected low-contrast bands on ASAR image (Fig. 4.1). Wind speed here is lower than in the area of cold air outbreak. Just the presence of clouds with $V \geq 0.1–0.2\ kg/m^2$ in the atmospheric front zones in combination with the sharp changes in wind speed and direction are responsible for the appearance of the dark patches and bands in Figs. 4.1 and 4.2. However, the flow of dry air directed to the center of a polar low also can contribute to the decreased Q-values retrieved from satellite microwave data. Such flow was found by direct airborne measurements during polar low study (Wagner et al. 2011). Satellite data near ice edge in the area of the cold air outbreak (Fig. 4.1) and the radiosonde-derived Q-values are in a good agreement. At Bear Island station (a red dot in Figs. 4.1 and 4.2) Q varied from 1.1 to 2.04 kg/m^2 on 7 January. Satellite-derived total water vapor content increases downflow that confirmed by coastal radiosonde data. Q-values measured by radiosonde station ENBO (67.25°N, 14.40°E) were equal to 3.75, 3.44 and 5.65 kg/m^2 on 7 January at 00 and 12 UTC and on 8 January at 00 UTC, correspondingly. QuikScat wind fields derived on 7 January at ascending ($\approx$02–03 UTC) and descending ($\approx$18–19 UTC) orbits (ftp://podaac.jpl.nasa.gov/OceanWinds/) and AMSR-E retrieved wind speed fields (ftp.ssmi.com/amsre) show cyclonic circulation with maximum speed 25–27 m/s. The center area of the low at 09 UTC was located partly over the sea and partly over the land. It looked dark on ASAR image due to weak wind.

Analysis of the various satellite and in situ data shows that one more vortex was formed to the north from the first one after its dissipation (Fig. 4.2). Water vapor fields retrieved from AMSR-E and SSM/I data and two Envisat ASAR images illustrate structure of total water vapor content and wind speed during short period—from 18:54 02 UTC on 7 January till 09:40 UTC on 8 January.

The synergistic use of the data from various sensors for this polar low study is presented in Fig. 4.3. Collocated in time AMSR-E retrieved atmospheric water vapor vortex structure shown in Fig. 4.3a, MODIS infrared (channel 31—10.78–11.28 μm) imaged cloud vortex structure Fig. 4.3b and ASAR-retrieved wind field. Figure 4.3c clearly indicates the existence and characteristics of a polar low (wind speeds are greater than 20 m/s). Sharp changes in the MODIS infrared image brightness reflect the differences between cloud top temperature and the sea surface temperature.

The second polar low, developed in the Barents Sea west of Novaya Zemlya near the ice edge, can serve as one more example of a so called spiraliform mesoscale cyclone. The spiraliform system seen in Fig. 4.4 on two MODIS images initially formed north of the ice edge near Novaya Zemlya over the open water in the evening of 4 March 2010 as a result of the cold air outbreak from the northern

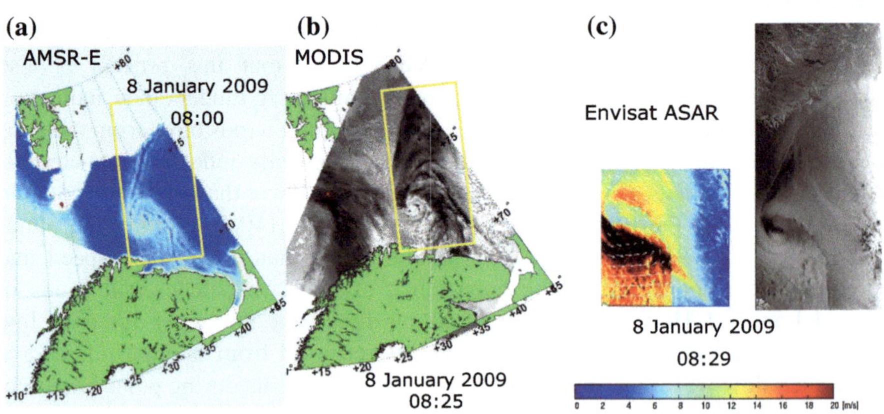

Fig. 4.3 **a** Collocated in time AMSR-E retrieved atmospheric water vapor. **b** MODIS infrared (channel 31—10.78–11.28 μm) imaged cloud structure and **c** ASAR image and corresponding retrieved wind field associated with the polar low event. *Yellow rectangle* on AMSR-E field and MODIS image marks the boundary of ASAR image on 8 January 2009 at 08:28 UTC

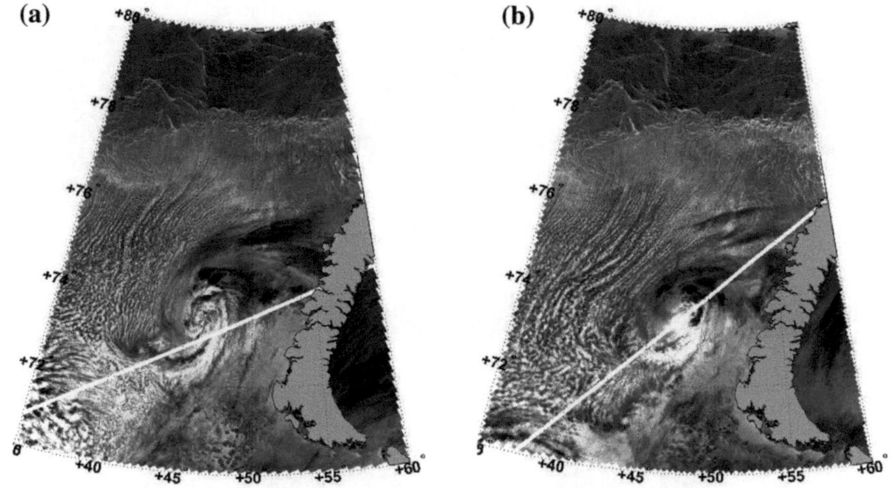

Fig. 4.4 MODIS infrared (channel 31—10.78–11.28 μm) images of a polar low west of Novaya Zemlya, taken on 5 March 2010 at: **a** 08:40 UTC, **b** 16:25 UTC

ice and from there moved southward, where it dissipated over the ice edge in the morning of 6 March 2010. No indication of the system exists afterwards.

Spiraliform systems are characterized by one or more spiral bands of convective clouds around the circulation centre. Spiraliform mesoscale cyclones occasionally have an eye-like cloud-free or nearly cloud-free area at the centre of the low (Rasmussen and Turner 2003). The form of the low shown in Fig. 4.4 is typical of many spiral structured polar lows at high latitudes. Since the air in polar regions where the spiral polar lows generally develop is conditionally unstable, the clouds along the frontal zones are predominantly of a convective nature.

Exploring Aqua AMSR-E data total atmospheric water vapor content, total cloud liquid water content and sea surface wind fields have been assembled. The polar low development as imaged by total atmospheric water vapor content fields is presented in Fig. 4.5.

The water vapor field associated with the polar low clearly shows the atmospheric front entering the central part of the low from the north-east. Characteristics of the fronts, which are formed within the boundary layer of the atmosphere, play an important role in the development of many mesoscale cyclones and polar lows in the Arctic region (Drüe and Heinemann 2001).

Such intensive mesocyclones form not only over the Arctic but, for instance, over the Japan Sea. The Japan Sea is not a northern sea geographically; however, it is a region of formation of intense mesoscale cyclones which are very fast-developing and fast-moving. The cold air outbreaks from the Asian continent and the thermal regime of the sea favors intense convective activity that manifests itself in the formation of mesoscale cloud rolls, cells and intense mesoscale cyclones (Mitnik et al. 2011). A typical intense mesocyclone over the Japan Sea on 9 February 2008 is shown in Fig. 4.6. Visible Aqua MODIS image shows the

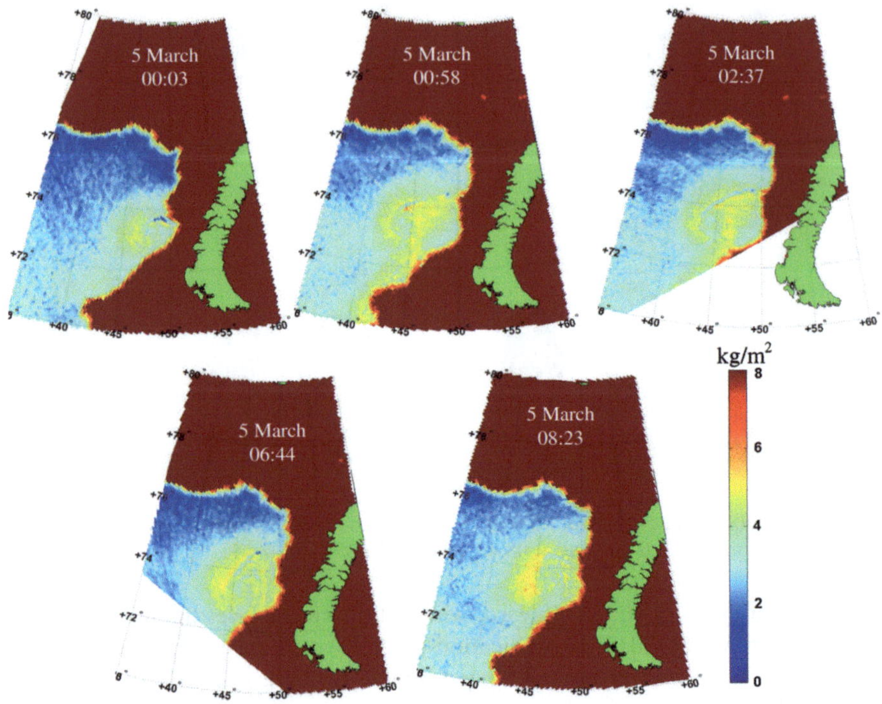

Fig. 4.5 Polar low that developed on 4–5 March in the Barents Sea, manifesting itself in total atmospheric water vapor content vortex structure, derived with neural network algorithm from Aqua AMSR-E measurement data

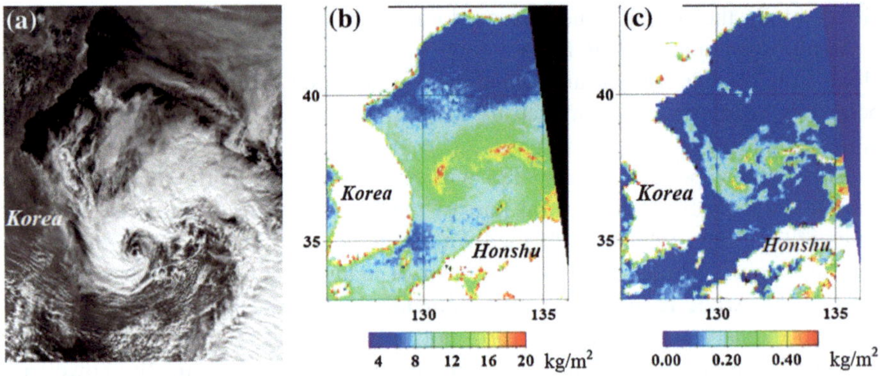

Fig. 4.6 MODIS visible image (**a**), total water vapor content (**b**) and total cloud liquid water content (**c**), retrieved from Aqua AMSR-E at 04:40 UTC on 9 February 2008

mesocyclone at the stage of its maximum development (Fig. 4.6a). Spiral cloudiness and a cloudless eye in the center with the diameter of approximately 15 km surrounded by a cloud wall are clearly identified.

4.5 Conclusion

A great diversity of initial conditions and forcing mechanisms, interacting with each other, can be responsible for the Arctic polar low rise and development. These mechanisms and conditions are not well defined due to the lack of conventional observations in the region (Yanase and Niino 2007). Taking into account the fact that polar lows are frequently missed in surface analysis maps and that most of the current models have insufficient temporal and spatial resolution for polar low detection, observational studies, including usage of various remote sensing data, remain the only reliable source of regular and more complete information about these dangerous phenomena.

Low spatial resolution is today the only factor limiting usage of satellite passive microwave radiometers. When a mesocyclone has a relatively small diameter ($\sim$100 km and less) it can hardly be caught on a passive microwave image. Water vapor fields retrieved from AMSR-E measurement data has 10 km resolution based on the resolution of re-sampled pixels for those channels which are used in the retrievals. This is a great advance compared to the 25 km resolution of water vapor fields retrieved from SSM/I data. Otherwise satellite passive microwave imager data, such as SSM/I, SSMI/S, AMSR-E and AMSR2, present an excellent tool for detection, tracking and study of these weather phenomena by means of retrieval of atmospheric columnar water vapor fields followed by their analysis. So far vortex structures in water vapor fields are detected by visual analysis but further advances are associated with the development of the automated procedure for such detection.

Radar data (scatterometers and SARs) should also be used when possible since they provide wind field information so valuable for polar low analysis. The high spatial resolution of SAR systems enable the study of small—scale organized variations of surface winds associated with polar low development, whereas scatterometer provide low resolution but regular wind field data allowing the confirmation of high wind events.

Microwave radiometers of the last generation—past AMSR-E and new AMSR2—have low frequency channels, allowing wind speed retrievals under precipitation conditions. Such a possibility is the most important since the surface wind speed is one of the main polar low characteristics and the greatest danger to ships, which often are small fishing vessels. The greater progress in using this possibility is anticipated in connection with higher spatial resolution of future radiometers.

Acknowledgments The study has been performed under the cooperation in frame of EC EuRuCAS Project #295068. Funding for this research was also provided by the Mega-grant of the Russian Federation Government to support scientific research under the supervision of leading scientist at RSHU, No. 11.G34.31.0078. This work is partly supported by RFBR Project 11-05-ophi-m-2011. Aqua AMSR-E data were provided by the Japan Aerospace Exploration Agency (JAXA) within the cooperation between the JAXA and the POI FEB RAS (project GCOM-1RA F10).

References

Bader, M.J., Forbes, G.S., Grant, J.R., Lilley, R.B.E., Waters, A.J.: Images in Weather Forecasting. A Practical Guide for Interpreting Satellite and Radar Imagery. Cambridge University Press, Cambridge (1995)

Bobylev, L.P., Zabolotskikh, E.V., Mitnik, L.M., Mitnik, M.L.: Atmospheric water vapor and cloud liquid water retrieval over the Arctic Ocean using satellite passive microwave sensing. IEEE Trans. Geosci. Remote Sens. **48**, 283–294 (2010)

Bobylev, L.P., Zabolotskikh, E.V., Mitnik, L.M., Mitnik, M.L.: Arctic polar low detection and monitoring using atmospheric water vapor retrievals from satellite passive microwave Data. IEEE Trans. Geosci. Remote Sens. **49**, 3302–3310 (2011)

Drüe, C., Heinemann, G.: Airborne investigation of arctic boundary-layer fronts over the marginal ice zone of the Davis Strait. Bound. Layer Meteorol. **101**, 261–292 (2001)

Friedman, K.S., Pichel, W.G., Clemente-Colon, P., Li, X.: A study of polar lows with synthetic aperture radar and GOES imagery. In: Proceedings of the IEEE International Geoscience and Remote Sensing Symposium, IGARSS 2009, pp. 1978–1980 (1999)

Kusaka, H., Kataniwa, S., Tanaka, H.L., Kimura, F., Hara, M.: Numerical simulation of polar low development over the Japan Sea using the WRF model. In: Proceedings of the Joint WRF/MM5 User's Workshop, pp. 19–22 (2005)

Mitnik, L.M.: Mesoscale atmospheric vortices in the Okhotsk and Bering Seas: results of satellite multisensor study. Influence of Climate Change on the Changing Arctic and Sub-Arctic Conditions, NATO Science for Peace and Security Series C: Environmental Security, pp. 37–56 (2009)

Mitnik, L.M., Gurvich, I.A., Pichugin, M.K.: Satellite sensing of intense winter mesocyclones over the Japan Sea. In: 2011 IEEE International of Geoscience and Remote Sensing Symposium (IGARSS), pp. 2345–2348 (2011)

Mitnik, L.M., Hsu, M.K., Mitnik, M.L.: Sharp gradients and mesoscale organized structures in sea surface wind field in the regions of polar vortex formation. Global Atmos. Ocean Syst **4**, 335–361 (1996)

Mitnik, L.M., Mitnik, M.L., Gurvich, I.A.: Severe weather study in middle and high oceanic latitudes using Aqua AMSR-E. In: IEEE International of Geoscience and Remote Sensing Symposium, IGARSS, pp. V–330 (2008)

Moore, R.W., Vonder Haar, T.H.: Diagnosis of a polar low warm core utilizing the advanced microwave sounding unit. Weather Forecast. **18**, 700–711 (2003)

Rasmussen, E., Turner, J., Twitchell, P.F.: Report of a workshop on applications of new forms of satellite data in polar low research. Bull. Am. Meteorol. Soc. **74**, 1057–1073 (1993)

Rasmussen, E.A., Turner, J.: Polar Lows: Mesoscale Weather Systems in the Polar Regions. Cambridge University Press, Cambridge (2003)

Sikora, T.D., Friedman, K.S., Pichel, W.G., Clemente-Colón, P.: Synthetic aperture radar as a tool for investigating polar mesoscale cyclones. Weather Forecast. **15**, 745–758 (2000)

Wagner, J.S., Gohm, A., Dörnbrack, A., Schäfler, A.: The mesoscale structure of a mature polar low: Simulations and airborne measurements. Q. J. Roy. Meteorol. Soc. **137**, 1516–1531 (2011) doi:10.1002/qj.857

Yanase, W., Niino, H.: Dependence of polar low development on baroclinicity and physical processes: an idealized high-resolution numerical experiment. J. Atmos. Sci. **64**, 3044–3067 (2007)

Chapter 5
Numerical Simulation and Forecasting Techniques for Tropical Cyclones in the South China Sea

Qilin Wan, Fangni Lin, Jinnan Yuan, Weiyu Ding, Gad Levy and DanLing Tang

Abstract This chapter introduces a numerical simulation and forecasting technique for tropical cyclones in the South China Sea, and describes a technique to initialize model simulation of typhoons that improves typhoon formation and forecasting in the simulation. It includes three-dimensional variational data assimilation and the curvature-modification linear balance equation. In order to determine the quality of the initial field in numerical models, a relocation method for the initial vortex is adopted as to decrease the error that is caused by bias in a tropical cyclone initial location in tropical cyclone models, and a bogus data assimilation (BDA) method is designed to assimilate typhoon bogus tangent wind into the initial wind field, and to update the typhoon's vortex structure. Doppler radar data, satellite infrared brightness temperature data, tropical rainfall measuring mission (TRMM) retrieved rainfall rate data, and other satellite observations are integrated into the background field by means of the GRAPES-3D-VAR (Global/regional assimilation and prediction system three-dimensional variational assimilation) system, so that the initial structure of the typhoon is closer to reality. It is shown through statistical analysis that the Tropical and Marine Meteorology (Guangzhou) operational forecasting system possesses effective forecasting skill for tropical cyclones.

Q. Wan (✉) · J. Yuan · W. Ding
Institute of Tropical and Marine Meteorology, China Meteorological Administration, Guangzhou, China
e-mail: qlwan@grmc.gov.cn

F. Lin
Guangzhou Central Meteorological Observatory, Guangdong Meteorological Administration, Guangzhou, China

G. Levy
North West Research Associates, Seattle, WA, USA

D. L. Tang
LTO, South China Sea Institute of Oceanology, Chinese Academy of Sciences, Guangzhou, China

D. L. Tang and G. J. Sui (eds.), *Typhoon Impact and Crisis Management*, Advances in Natural and Technological Hazards Research 40,
DOI: 10.1007/978-3-642-40695-9_5,

Keywords TC numerical prediction · Variational data assimilation · Typhoon initialization techniques · Radar data assimilation · Satellite observational data · Operational performance

5.1 Introduction

Great attention has been paid to tropical cyclone (TC) forecasting techniques by meteorologists worldwide because the damage caused by TCs is often catastrophic. China is one of the countries most affected by TCs, and improvements in TC forecasting techniques play an important role in disaster prevention as well as in economic and social developments.

Numerical models are currently and for the foreseeable future the primary method of TC forecasting. Numerical predictions represent a typical initial value problem in mathematical physics. Model integration requires an initial value, which is determined by a specific data assimilation scheme applied to meteorological observations. As a result, data assimilation has a special importance in numerical prediction.

One of the important factors in TC forecasting by numerical models is a high quality initial field. When there are insufficient observational data to forecast TCs, a so called "bogus" typhoon (Iwasaki et al. 1987; Mathur 1991; Kurihar et al. 1993; Leslie and Holland 1995; Wang et al. 1996a, b; Wang 1998; Zhou 1998; Meng et al. 2001) is introduced to the initial field, which, to a certain extent, improves the initial structure of TCs. In effect, artificial (bogus) data are introduced to the initial field to better represent the typhoon. However, applying bogus data independently ignores the original TC structure information in the background field. The BDA (Bogus data assimilation) method generates the initial structure of a typhoon using bogus data, as well as observational data. This method has the advantages of both bogus data and variational assimilation methods. However, the initial vortex location of a tropical cyclone in the background field regularly deviates from the observed position (Yan et al. 2004), even after variational assimilation is applied, which may lead to large errors in the predicted TC track. Therefore, it is necessary to relocate the initial vortex in order to reduce errors caused by inaccurate location of the TC. Application of satellite and radar data plays a positive role in improving the initial field. For instance, TRMM data can reflect precipitation well, and radial wind and echo intensity from radar show the progress of short-wave weather systems accurately and continuously. Assimilation of such remote sensing data assists in obtaining a reasonable initial field and is useful for prediction as well.

This chapter introduces a numerical assimilation and forecasting technique for TCs in the South China Sea. Section 5.2 describes the development of the typhoon initialization technique, including the framework for the three-dimensional variational assimilation system (GRAPES 3D-Var), curvature-modification linear balance equation, typhoon relocation technique, Bogus Data, and BDA scheme. Section 5.3 describes the application of Radar data. Section 5.4 describes the application of satellite data. Lastly, Sect. 5.5 introduces a numerical simulation of a typhoon in an operational run.

5.2 Typhoon Initialization Technique

5.2.1 Three-Dimensional Variational Assimilation

The major development objective of the GRAPES data assimilation is to establish effective satellite and remote sensing data assimilation in order to improve the quality of numerical predictions. The GRAPES also employs a variational assimilation scheme.

5.2.1.1 Basic Formulation

In the numerical prediction model **x** is a forecast variable, which represents the state of the atmosphere, called the "system state variable". The superscript **b** denotes the background field (the assimilation scheme prediction values from a previous time step), while the superscript **a** denotes the analysis field (the result of the assimilation of observations). Atmospheric state **x**, that are changing with time can be written as:

$$\mathbf{x}(t) = M(\mathbf{x}(t_0)) + \boldsymbol{\eta}(t) \tag{5.2.1}$$

where M is the forecasting model, and $\boldsymbol{\eta}(t)$ is the model error. t and t_0 refer to the prediction time and the initial time of the model.

Below, $\mathbf{y}^o$ is an observed value. The relation between the observed value and state value is:

$$\mathbf{y}^o(t) = H(\mathbf{x}(t)) + \varepsilon \tag{5.2.2}$$

Here, H is called the "observation operator", and represents the physical relation between observed values and the atmospheric state. There are different expressions for data from different observational equipment, which should be known. ε is the observation error.

Suppose that observations are carried out within $[t_0, t_1]$. The goal of assimilation is to determine the model state $\mathbf{x}(t_0)$ at moment t_0 such that the deviation between the values obtained from the model within $[t_0, t_1]$ and actual observations is minimized. $\mathbf{x}^a(t_0)$ must be found such that the following objective function J is minimized (Parrish and Derber 1992):

$$J(\mathbf{x}^{\mathrm{a}}(t_0)) = \frac{1}{2}\left[\mathbf{x}^{\mathrm{a}}(t_0) - \mathbf{x}^{\mathrm{b}}(t_0)\right]^{\mathrm{T}}\mathbf{B}^{-1}\left[\mathbf{x}^{\mathrm{a}}(t_0) - \mathbf{x}^{\mathrm{b}}(t_0)\right] + \frac{1}{2}\sum_i [\mathbf{y}^{\mathrm{o}}(t) - H(\mathbf{x}^{\mathrm{a}}(t))]^{\mathrm{T}}\mathbf{R}^{-1}[\mathbf{y}^{\mathrm{o}}(t) - H(\mathbf{x}^{\mathrm{a}}(t))] \tag{5.2.3}$$

Here, **B** and **R** are the background error and observation error covariance matrices. Superscript **T** denotes the matrix transpose and $\mathbf{x}^{\mathrm{a}}(t), t_1 \geq t \geq t_0$ is decided by the prediction model:

$$\mathbf{x}^{\mathrm{a}}(t) = M(\mathbf{x}^{\mathrm{a}}(t_0)) \tag{5.2.4}$$

This section only considers observations at a single moment, so the second item on the right of (5.2.3) involving a sum with respect to time is ignored. Thus only three dimensional variational assimilation is considered, and the problem is simplified:

$$J(\mathbf{x}^{\mathrm{a}}) = \frac{1}{2}\left[\mathbf{x}^{\mathrm{a}} - \mathbf{x}^{\mathrm{b}}\right]^{\mathrm{T}}\mathbf{B}^{-1}\left[\mathbf{x}^{\mathrm{a}} - \mathbf{x}^{\mathrm{b}}\right] + \frac{1}{2}[\mathbf{y}^{\mathrm{o}} - H(\mathbf{x}^{\mathrm{a}})]^{\mathrm{T}}\mathbf{R}^{-1}[\mathbf{y}^{\mathrm{o}} - H(\mathbf{x}^{\mathrm{a}})] \tag{5.2.5}$$

where **B** is a huge matrix and it is difficult to compute. It is also difficult to solve (5.2.5) directly, and a series of variables transforms are needed to find the minimum of J.

The use of the variational method for assimilating observations has a wide range of applications within the field of atmospheric science. This method can integrate an initial field and observations according to certain objective criteria, in order to get information that cannot be obtained from observations directly as it allows for dynamical consistency between mass and motion and the advection of information into data-sparse regions.

5.2.2 The Curvature-Modification Linear Balance Equation

It is necessary to consider the equilibrium constraint relationship between the horizontal wind field and the geopotential height field in a variational assimilation system. Taking geostrophic equilibrium, or linear balance equations, as wind and pressure constraints, with the adjoint operator is appropriate for high latitudes (Coriolis parameter, f, is large) and for large-scale systems. The ideal wind and pressure constrains for a strong vortex system in low latitudes are the nonlinear balance equations, however. Since each observation operator should be first transformed into the tangent linear operator and then a linear adjoint is sought, if the nonlinear balance equation was applied to small-scale systems, the rationale for its linearization would be complex. Therefore, a convenient, effective wind and pressure constrains for strong vortex systems in a variational assimilation system is needed.

5.2.2.1 The Curvature-Modification Linear Balance Equation

The transformation of a wind vector between natural coordinates and Cartesian coordinates (as shown in Fig. 5.1) is as follows:

$$\begin{aligned} u &= V\cos(a+90) = -V\sin a \\ v &= V\sin(a+90) = V\cos a \end{aligned} \tag{5.2.6}$$

In (5.2.6) u and v are the two components of the wind vector in Cartesian coordinates, V is the wind speed, a is the angle between the normal vector to the flow $\vec{n}$ and the x-axis of the Cartesian plane.

In natural coordinates:

$$V = R_S \frac{\partial a}{\partial t} \tag{5.2.7}$$

Above, R_S is the radius of curvature (not considering the difference in curvature between the streamline and trace). For a cyclonic vortex $R_S > 0$, and for an anticyclonic vortex $R_S < 0$.

Taking the partial derivative with respect to time of (5.2.6) and (5.2.7) yields:

$$\begin{aligned} \frac{\partial u}{\partial t} &= -\frac{V}{Rs}v - \frac{\partial V}{\partial t}\sin a \\ \frac{\partial v}{\partial t} &= \frac{V}{Rs}v + \frac{\partial V}{\partial t}\cos a \end{aligned} \tag{5.2.8}$$

The wind and pressure equilibrium constraints in Cartesian coordinates are written:

Fig. 5.1 Conversion of wind vector in natural coordinate system. The *curve* is the flow line *s* is the tangential vector, *n* is the normal vector, *a* is the angle between the vector *n* and the x-axis, and *Rs* is the radius of curvature

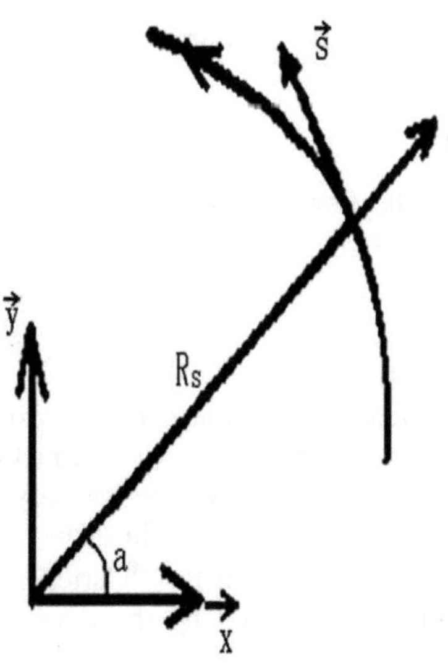

$$\begin{aligned}\frac{\partial u}{\partial t} - fv &= -\frac{\partial \phi}{\partial x}\\ \frac{\partial v}{\partial t} + fu &= -\frac{\partial \phi}{\partial y}\end{aligned} \tag{5.2.9}$$

For a strong vortex system (in this case for a strong vortex system with large-curvature vorticity), the last term on the right of (5.2.8) is omitted, and the other term on the right is substituted into (5.2.9) resulting in:

$$\begin{aligned}-\left(f + \frac{V}{Rs}\right)v &= -\frac{\partial \phi}{\partial x}\\ \left(f + \frac{V}{Rs}\right)u &= -\frac{\partial \phi}{\partial y}\end{aligned} \tag{5.2.10}$$

Defining the curvature modification of the Coriolis parameter as

$$\tilde{f} = f + \frac{V}{Rs} \tag{5.2.11}$$

From (5.2.10):

$$\nabla^2 \phi = \tilde{f}\varsigma + \vec{k} \cdot (\nabla \tilde{f} \times \vec{V}) \tag{5.2.12}$$

Equation (5.2.12) is defined as the "Curvature modified linear balance equation", and is consistent with the linear balance equation in this form. It will degenerate into this form when the flow line curvature radius is large.

Equation (5.2.12) can be rewritten as:

$$\nabla^2 \phi = \tilde{f}\nabla^2 \psi + \nabla \tilde{f} \cdot \nabla \psi \tag{5.2.13}$$

Here, ψ is the stream function.

Vorticity in natural coordinates can be written as:

$$\varsigma = \frac{V}{Rs} + \frac{\partial V}{\partial n} \tag{5.2.14}$$

The first term on the right of (5.2.14) acts as the curvature vorticity. From (5.2.14):

$$\frac{V}{Rs} = \varsigma - \frac{\partial V}{\partial n} = \nabla^2 \psi - \frac{\partial^2 \psi}{\partial^2 n} \tag{5.2.15}$$

Here, $\vec{n}$ is the normal vector to the streamline, and is the gradient vector of the stream function ψ as well. Equation (5.2.15) can be defined in Cartesian coordinates, and the curvature vorticity and radius of curvature can be calculated from the stream function. This can be used to determine the curvature-modification Coriolis term $\tilde{f}$ in the "Cartesian" coordinate system. The streamline curvature radius determined by the curvature-modification linear balance function is the

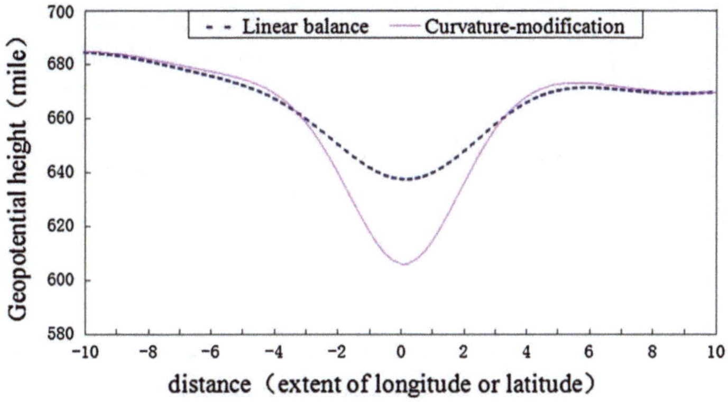

Fig. 5.2 Distribution of height field with different wind-pressure balance constraint relationships. *Dashed line* linear balance; *Solid line* curvature-modification linear balance

same as that determined by the geostrophic balance in a circular vortex system: Equal to the distance from a designated point to the center of the circle. This facilitates the variational assimilation process.

Figure 5.2 illustrates the meridional distribution (through the vortex center) of the GRAPES-3D-VAR analyzed height field, applying both the linear balance function and the curvature-modification linear balance equation as wind-pressure balance constraint relations. This figure shows that the low-pressure center analyzed with the curvature-modification linear balance equation was significantly deeper than that of the linear balance function. The geopotential height field in the vortex center decreased by 40 % resulting in lower central pressure and stronger vortex. This shows that the curvature-modification linear balance equation introduced here, and its application in variational assimilation had a great impact on the cyclone vortex analysis results.

Figure 5.3 describes the balance between the Coriolis force, the centrifugal force, and the pressure gradient force in the GRAPES-3D-VAR system when applying the linear balance function and curvature-modification linear balance equation as wind-pressure balance constraint relations. In low pressure vortex systems, the gradient force approximately equaled the Coriolis plus centrifugal force (gradient wind balance). As seen in the figure, curvature-modification linear wind-pressure constraints in GRAPES-3D-VAR were better than the linear balance constraints in gradient wind balance scenarios. Because there was a non-equilibrium component in the analyzed field, the curvature-modification linear balance pressure gradient force was not exactly equal to the sum of the Coriolis force and the centrifugal force. However, it is clearly shown that employing the linear balance equation as a balance constraint relation does meet the gradient wind balance conditions for strong cyclonic vortex systems, while applying the curvature-modification linear balance equation as the wind-pressure balance constraint relation has better results. This shows that using the

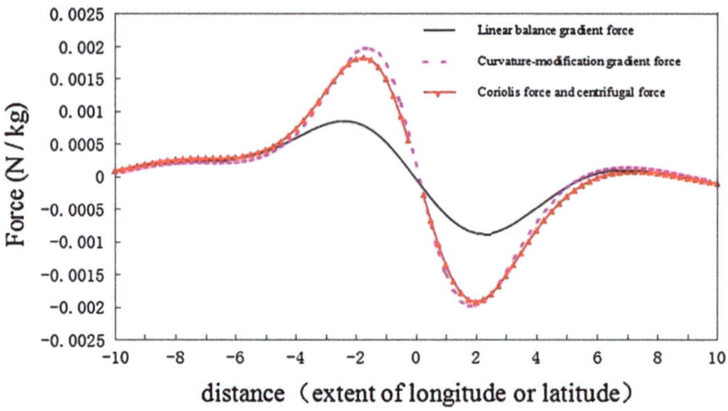

Fig. 5.3 Meridional distributions of forces upon different wind-pressure field balance constraints. *Solid triangles* the Coriolis force with centrifugal force; *solid line* linear balance pressure gradient force; *Dotted line* curvature-balance pressure gradient force

curvature-modification linear balance equation as a wind-pressure balance constraint relation in the variational assimilation scheme could effectively improve the relation between wind and pressure in the analysis field.

In conclusion, using the curvature-modification linear balance equation as the wind-pressure balance constraint relation in the three-dimensional variational assimilation system resulted in better performance and features, and improved the quality of strong wind vortex system analysis.

5.2.3 Typhoon Relocation Technique

A small deviation in the initial vortex position used for tropical cyclone numerical prediction may lead to relatively large error. Some tropical cyclone model initialization schemes tend to remove the initial vortex and insert bogus data (Leslie and Holland 1995; Wang et al. 1996b; Wang 1998; Huang et al. 2006). In this case, the initial vortex in the background field retains some valuable information however, the bogus data create a completely artificial vortex. Complete removal of the initial vortex may bring about inconsistency between bogus data and the environment, which will impact prediction outcomes. Therefore, it is necessary to relocate the initial vortex to reduce errors due to inaccurate position of the vortex.

To relocate the initial vortex in the background (or analysis) field, the following scheme was designed:

First, the initial position, based on the 850 hPa height field and the radius of the initial vortex, was set to be the radial distance r_0, where the average angular wind reached a maximum and monotonically decreased past a threshold (in this section, 3 m/s) for the first time.

Second, the large-scale environmental field *hE* and the vortex circulation *hV* were isolated from the background field *h* (e.g. the surface pressure, temperature, humidity and wind field), such that:

$$h = h_E + h_V \tag{5.2.16}$$

The separation method for vortex circulation was based on the twice-smoothing filtering technique by Kurihara et al. (1990, 1993), which involves separating the disturbance field via meridional and zonal filtering and then separating the vortex circulation from the disturbance field via cylindrical filtering. Meridional and zonal filtering were based upon a three point smoothing function, with smoothing coefficient:

$$a = \frac{1}{2}\left(1 - \cos\frac{2\pi}{m}\right)^{-1} \tag{5.2.17}$$

m was set to: $m = 2, 3, 4, 2, 5, 6, 7, 2, 8, 9, 2$ to carry out smoothing first along the zonal, and then, the meridional direction, to get the filtered disturbance field *hD*.

Vortex circulation was separated by cylindrical filtering for the disturbance field:

$$h_V(r,\theta) = [1 - E(r)][h_D(r,\theta) - \bar{h}_D(r_0)] \tag{5.2.18}$$

where $\bar{h}_D(r_0)$ is the angular average of the disturbance field *hD* in r_0, and the cylindrical filter function is:

$$E(r) = \frac{\exp\left[-(r_0 - r)^2/l^2\right] - \exp\left[-(r_0)^2/l^2\right]}{1 - \exp\left[-(r_0)^2/l^2\right]} \tag{5.2.19}$$

where *l* is the control parameter of *E* (*r*). In this section $l = r_0/5$.

The large-scale environmental field was determined according to (5.2.16) with the vortex circulation:

$$h_E = h - h_V \tag{5.2.20}$$

Finally, the relocated background was fixed after the vortex circulation determined with the filter was inserted into the large-scale environmental field over the tropical cyclone's observed position.

Comparison of the background before relocation at 1,000 hPa height and the wind field of tropical cyclone "Chanchu" on May 14, 2006 00 UTC shows the circumfluence after relocation, indicating that the circulation remained unchanged while the center of the tropical cyclone was corrected to its observed position (Fig. 5.4).

It can also be seen that differences between the relocated-background and the original background only exist near the center of the tropical vortex, meaning that changes only occurred within the tropical cyclone circulation, while elements of the surrounding environment remained the same (Fig. 5.5).

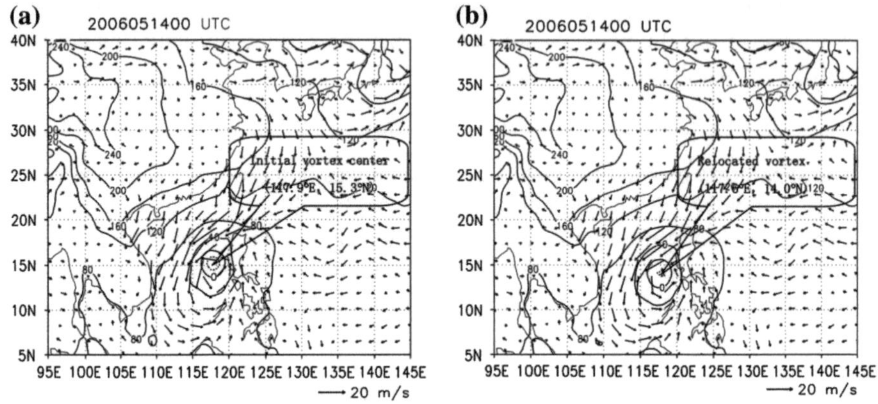

Fig. 5.4 Tropical cyclone "Chanchu" 1,000 hPa height fields (units: gpm) and wind field (units: m/s) on May 14, 2006 at 00 UTC. **a** Initial background field. **b** Relocated background field

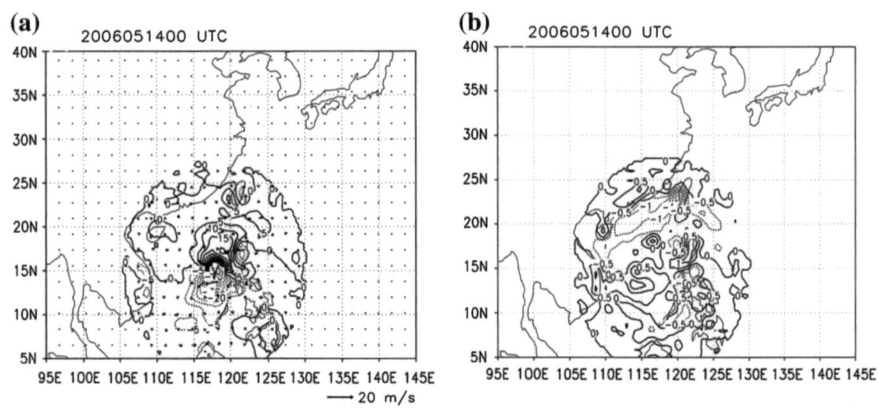

Fig. 5.5 May 14, 2006, 00 UTC 1,000 hPa the difference in the background field before and after relocation for tropical cyclone "Chanchu". **a** High field (units: gpm) and wind field (units: m/s). **b** Temperature (units: K)

5.2.4 Bogus Typhoon

5.2.4.1 Bogussing Scheme

The bogus typhoon scheme (hereafter "bogus typhoon") introduced here is an operational bogus typhoon from the Guangzhou Institute of Tropical and Marine Meteorology (Wang et al. 1996a). This scheme was adapted to the structure and features of South China Sea and western Pacific typhoons. It uses the observed location of the typhoon center, typhoon strength, typhoon fresh gale radius, and the initial move to construct an asymmetric bogus typhoon satisfying static equilibrium, gradient wind equilibrium, and quasi-thermodynamic equilibrium, which is consistent with the surrounding environment and convective parameterization of the model. The construction process included the following steps:

1. Calculation of the environmental parameters

Sea-level pressure, temperature, and moisture were calculated at each level in the environmental atmosphere around the area of the typhoon center. As the South China Sea typhoon itself was weak, with sparse ocean data, lower levels of the environmental atmosphere were stably stratified, making it difficult to produce convection. When this occurred, these levels were adjusted to neutral stratification.

2. Construction of a sea level pressure and height field in the typhoon area

The radial distribution of axisymmetric bogus typhoon sea level pressure $p(r)$ was obtained from the empirical formula:

$$p(r) = p_E - \Delta p\left[1 + \left(\frac{r}{R_0}\right)^2\right]^{-0.5} \tag{5.2.21}$$

where p_E is sea level pressure of the ambient atmosphere, Dp is the difference between environmental and typhoon central pressure, and R_0 is a scale parameter, computed using the fresh gale radius, gradient winds, and static relationship:

$$\frac{v^2}{r} + fv - \frac{\partial \phi}{\partial r} = 0 \tag{5.2.22}$$

$$\phi(r) = \phi_E + \Delta\phi_0\left[1 + \left(\frac{r}{R_0}\right)^2\right]^{-0.5} \tag{5.2.23}$$

$$\frac{\Delta\phi_0}{\Delta p} = -\frac{RT_E}{p_E} \tag{5.2.24}$$

In some cases, atmospheric pressure dropped so steeply that the solution for R_0 was too small, and R_0 and the typhoon central pressure had to be adjusted using controlled conditions:

$$\Delta p \times R_0 = \Delta p \times R_0$$

Here, "$'$" refers to the adjusted value. R_0' was adjusted to 50 km.

By Eq. (5.2.23), a potential sea level height difference was constructed as well:

$$D(r, p_s) = \phi(r) - \phi_E \tag{5.2.25}$$

The Sea surface temperature (T_E) and pressure (p_E) of the environment were increased by moist adiabatic lifting to the condensation level, and then by dry adiabatic lifting above, to find the cloud temperature, cloud top, and cloud base. An anticyclone was constructed at the cloud top:

$$D(r, p_t) = -B\Delta\phi_0\left[1 + \left(\frac{r}{R_0}\right)^2\right]^{-0.5} \tag{5.2.26}$$

Here, p_t is the cloud top pressure and B is the high-level anticyclone intensity factor, given by the typhoon strength. It was assumed that the anticyclone disappeared in the stratosphere (taking $p_{mid} = 70$ hPa).

$$D(r, p_{mid}) = 0 \tag{5.2.27}$$

Height deviation $D(r, p)$ away from the typhoon center was determined by:

$$D(r, p) = \alpha(r)D(0, p) + \beta(r) \tag{5.2.28}$$

where $D(0, p)$ is the height deviation of the typhoon center, computed by the hydrostatic equation using the typhoon center temperature (see below). Substitution of (5.2.25)–(5.2.27) into (5.2.28) results in expressions for $\alpha(r)$ and $\beta(r)$.

3. Construction of the warm-core structure and the tangential wind speed

The typhoon center vertical temperature profile was determined by:

$$T(0, p) = C[T_c(p) - T_E(p)] + T_E(p) \tag{5.2.29}$$

where $T_c(p)$ is the cloud temperature. Parameter C from cloud base to cloud top and from cloud top to the stratosphere can be obtained by (5.2.25)–(5.2.27), respectively through the hydrostatic equation (dp = − ρgdz).

From the gradient wind Eq. (5.2.22), the tangential wind can be obtained:

$$v_\theta = \left(\frac{-f + \sqrt{f^2 + \frac{4}{r}\frac{\partial\phi}{\partial r}}}{2}\right) \cdot r \tag{5.2.30}$$

4. Construction of relative humidity

Given a saturated bogus humidity field with a certain weighting increasing right under the environmental field, the weight coefficient was determined according to location relative to the typhoon center:

$$\begin{cases} w = 1 & r < R_8 \\ w = \cos\left(\frac{\pi}{2} \times \frac{r - R_8}{R_8}\right) & R_8 < r < 2R_8 \\ w = 0 & r > 2R_8 \end{cases} \tag{5.2.31}$$

where R_8 is fresh gale radius and relative humidity is:

$$RH = w \times 100\,\% \tag{5.2.32}$$

5. Secondary circulation and the initial move

Neglecting vertical and radial advection terms, the momentum equation, where steady state friction is assumed, expressed in cylindrical coordinates, is:

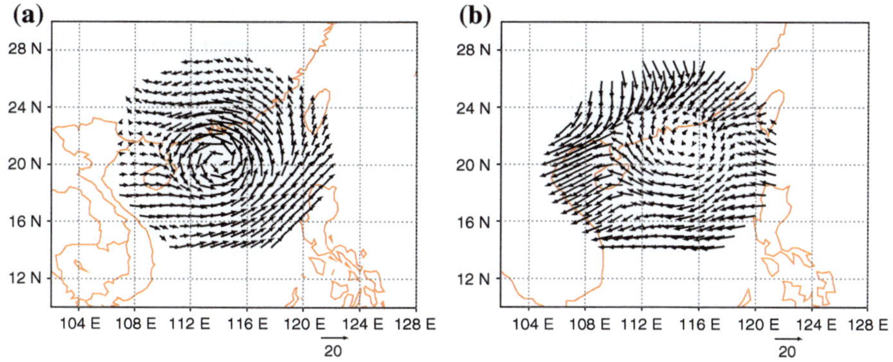

Fig. 5.6 Bogus typhoon. **a** 850 hPa. **b** 200 hPa

$$-\frac{v_\theta^2}{r} - fv_\theta + \frac{\partial \phi}{\partial r} + C_d \left|\vec{V}\right| v_r = 0 \tag{5.2.33}$$

$$\frac{v_r v_\theta}{r} + fv_r + C_d \left|\vec{V}\right| v_\theta = 0 \tag{5.2.34}$$

where v_r and v_θ represent the radial and tangential wind components, respectively, and C_d is the drag coefficient.

The formation of an asymmetric structure was based on the initial velocity calculated at the latitude and longitude of the typhoon center in two continuous moments and added to the wind field of the bogus typhoon. Figure 5.6 illustrates a bogus typhoon at 850 and 200 hPa using this scheme.

5.2.4.2 Structure of a Revised Bogus Typhoon

The bogus typhoon introduced before successfully described the structures of most tropical cyclones and has achieved a positive impact on TC operational numerical prediction in recent years (Wang et al. 1996a). This scheme can keep the majority of tropical cyclone track forecasts reliable. However, some other tropical cyclone track forecasts remain poor. So a question arises: "is the typhoon bogussing scheme consistent with the truth in poor forecasting cases or does it actually cause large forecast errors in such cases?" the answer was clear, as discussed below.

The track forecast error of TC 0012 on August 28, 2000 at 08 LBT was large for most forecasting methods. Figure 5.7a, b describe the typhoon bogussing scheme's tropical cyclone height field and temperature field structure at that time, which was close to the general conceptual bogus structure of tropical cyclones, but different from the information received from TOVS satellite observations of this tropical cyclone (Fig. 5.8a, b). The biggest difference in the height field was that the upper-level height field for the typhoon bogussing scheme had a strong positive anomaly, while the positive anomaly of the TOVS data was relatively weak. Additionally, the

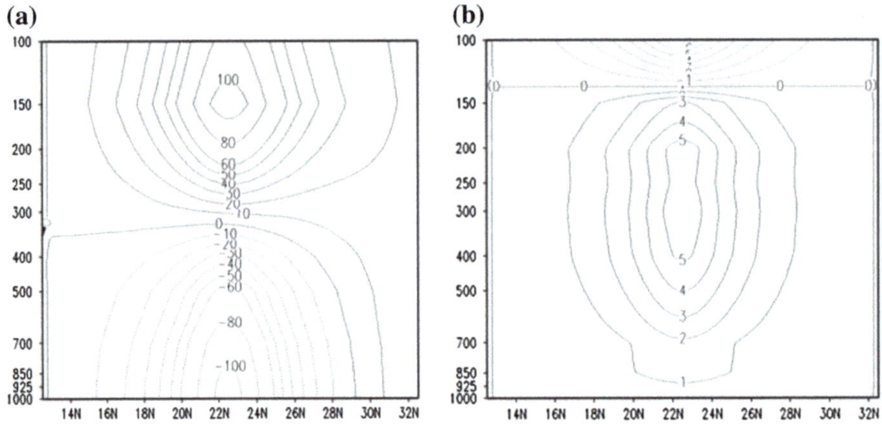

Fig. 5.7 The profile of 0012 TC BOGUS at 2000.08.28.08(BJT). **a** Height. **b** Temperature

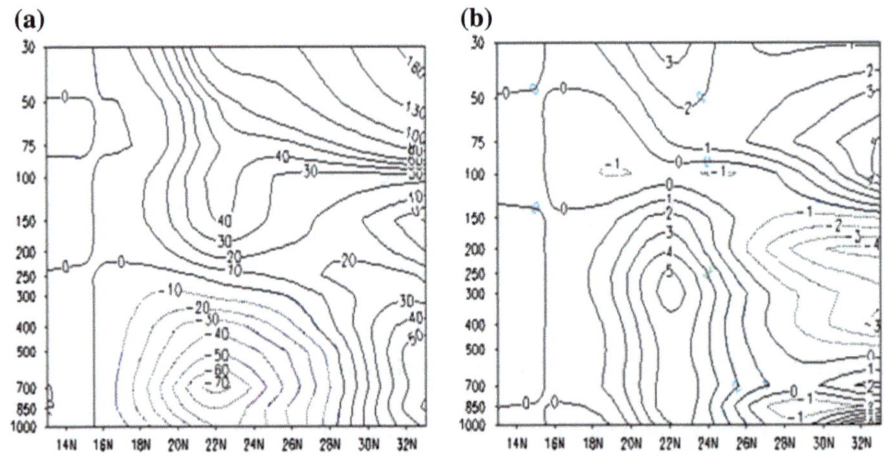

Fig. 5.8 The profile of 0012 TC at 2000.08.28.15(BJT). **a** Height. **b** Temperature

horizontal asymmetry of the TOVS data in the height field was more significant than that of bogus data. The biggest difference in temperature occurred in the high-level temperature field negative anomaly. That of the typhoon bogussing scheme was strong, while the TOVS data temperature negative anomaly in the high-level field was not obvious. The typhoon bogus data were constructed using a general conceptual model, and the actual structure of tropical cyclones was obviously different. This requires further study to improve tropical cyclone forecasting accuracy.

Figure 5.9a, b show the revised structure of the bogus typhoon after consulting the TOVS observed tropical cyclone structure (Fig. 5.8a, b). Comparing Fig. 5.8a, b and 5.7a, b shows that the revised structure of the typhoon bogus data had conspicuous changes in characteristics, approaching the actual structure of the tropical cyclone

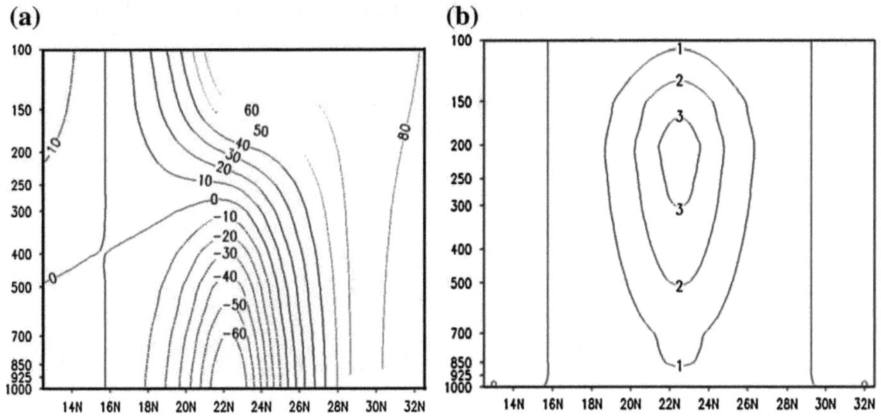

Fig. 5.9 The profile of 0012 TC modified bogus data at 2000.08.28.08(BJT). **a** Height. **b** Temperature

(TOVS described structures). Reduction in the high-level height field positive anomaly center and the horizontal asymmetry were more obvious, while high level temperature negative anomaly disappeared. Further analysis showed that the revised bogus typhoon was in good agreement with the environment.

5.2.4.3 Improvement of the Bogus Radial Circulation

TBB (Black-Body Temperature) data, used to define the convective heat of the TC, was based on the Davison (1992) scheme. The convection strength in the TC circulation can be expressed as:

$$Qcon = f(TBB)\,h(p) \tag{5.2.35}$$

$$f(TBB) = Qmax(TBB - Tb)/(Tt - Tb) \tag{5.2.36}$$

where $f\,(TBB)$ expresses the TBB intensity control factor, the largest convection heating rate in the vertical layer; $Qmax = 25$ K/day, $Tb = 273$, $Tt = 193$, and $h\,(p)$ is the vertical distribution of convective heating, determined by the difference between cloud temperature and ambient temperature.

Despite a complex asymmetrical TC structure, separating TC circulation into a non-axially symmetric component and an axially symmetric component is widely used in constructing bogus data in current operational models. In accordance with previous work, the non-axially symmetric component of the horizontal wind speed was uniform, and had no effect on the radial circulation of the TC. The axially symmetric component however, was the most important component in determining the TC circulation features. Here we focus on the axially symmetric component, to introduce radial circulation matched by convective clouds (satellite TBB).

The TC control equations of the axially symmetric component can be expressed as follows:

$$\frac{\partial V}{\partial t} + U\frac{\partial V}{\partial r} + W\frac{\partial V}{\partial z} + \frac{UV}{r} + fU = \boldsymbol{F}_V \tag{5.2.37}$$

$$\frac{\boldsymbol{V}^2}{r} + fV = \tilde{\theta}\frac{\partial \Lambda}{\partial r} \tag{5.2.38}$$

$$\frac{\partial \theta}{\partial t} + U\frac{\partial \theta}{\partial r} + W\frac{\partial \theta}{\partial z} = \frac{Q}{\Lambda} + \boldsymbol{F}_\theta \tag{5.2.39}$$

$$\frac{\partial q}{\partial t} + U\frac{\partial q}{\partial r} + W\frac{\partial q}{\partial z} = \boldsymbol{L}_q + \boldsymbol{F}_q \tag{5.2.40}$$

$$\frac{\partial \Lambda}{\partial z} = -\frac{g}{\theta} \tag{5.2.41}$$

$$\frac{\partial r\tilde{\rho}U}{r\partial t} + \frac{\partial \tilde{\rho}W}{\partial z} \approx 0 \tag{5.2.42}$$

$$\Lambda = \boldsymbol{C}_p\left(\frac{\boldsymbol{p}}{\boldsymbol{P}_0}\right)^{R/C_p} \tag{5.2.43}$$

In the above equations, U is the radial velocity, W is the vertical velocity, V is the tangential wind speed of the TC, and other symbols are as commonly defined in meteorology. The " ~ " indicates the background state of the weather elements (horizontal uniform non-disturbance state). We define:

$$U = -\frac{1}{\tilde{\rho}}\frac{\partial \tilde{\rho}\psi}{\partial z}\ ;\ W = -\frac{1}{r}\frac{\partial r\psi}{\partial r} \tag{5.2.44}$$

where ψ is the stream function of the meridional profile, the diagnostic equation of which can be derived by Eqs. (5.2.35)–(5.2.43):

$$A\frac{\partial^2 \psi}{\partial Z^2} - (B + B')\frac{\partial^2 \psi}{\partial zr} + C\frac{\partial^2 \psi}{\partial \boldsymbol{r}^2} + E\frac{\partial \psi}{\partial z} + F\frac{\partial \psi}{\partial r} + G\psi = D \tag{5.2.45}$$

In formula (5.2.45):

$$A = \left(f + \frac{2V}{r}\right)\left(f + \frac{V}{r} + \frac{\partial V}{\partial r}\right);\ B = \left(f + \frac{2V}{r}\right)\frac{\partial V}{\partial z};\ B' = \frac{g}{\tilde{\theta}}\frac{\partial \theta}{\partial r};\ C = \frac{g}{\tilde{\theta}}\frac{\partial \theta}{\partial z};$$

$$E = \frac{2B}{r} + AN^2 + \frac{\tilde{C}}{g}\left[\left(f + \frac{2V}{r}\right)\left(f + \frac{V}{r}\right) + \frac{\boldsymbol{V}^2}{\boldsymbol{r}^2}\right];\ F = \frac{1}{g}B'\tilde{C} + \frac{C}{r} - BN^2;$$

$$G = \frac{1}{gr}B'\tilde{C} - \frac{C}{\boldsymbol{r}^2} + \boldsymbol{N}^2\frac{3B}{r} + \boldsymbol{N}^2\frac{\tilde{C}}{g}\left[\left(f + \frac{2v}{r}\right)\left(f + \frac{v}{r}\right) + \frac{\boldsymbol{V}^2}{\boldsymbol{r}^2}\right];\quad \tilde{C} = \frac{g}{\tilde{\theta}}\frac{\partial \tilde{\theta}}{\partial z};$$

$$N^2 = \frac{1}{\tilde{\rho}}\frac{\partial \tilde{\rho}}{\partial z};\ D = \frac{g}{\tilde{\theta}}\frac{\partial}{\partial r}\left[\frac{Q}{\Lambda} + \boldsymbol{F}_\theta\right] - \frac{\partial}{\partial z}\left[\left(f + \frac{2V}{r}\right)\boldsymbol{F}_V\right] + \frac{\tilde{C}}{g}\left(f + \frac{2V}{r}\right)\boldsymbol{F}_V$$

From (5.2.45), we can see that as long as the tangential wind and potential temperature distribution are determined, the coefficients (A, B, C, etc.) on the left hand side of the equation are known, and the forces on the radial circulation are heat and dissipation. In fact, the vast majority of existing schemes for the construction of bogus typhoons only construct a tangential wind and potential temperature (or potential height, as geopotential height and potential temperature satisfy a hydrostatic balance) disturbance vortex. This means that, based on the existing bogus typhoon, and if the heat and dissipation field can be acquired, (5.2.45) can be used to diagnose the corresponding radial circulation. Thus changing a vortex with only tangential wind, into a three-dimensional wind field vortex with radial wind, vertical wind speed, and tangential wind fields satisfied. Considering that diabatic heating in the TC is mainly convective heating, according to Eqs. (5.2.35) and (5.2.36), the symmetry of the radial circulation of the bogus typhoon can be determined using the convection heating derived from TBB data.

Making use of (5.2.35)–(5.2.45) in the bogus typhoon formula, (5.2.21)–(5.2.24) and (5.2.29), with the addition of radial and vertical wind speed, transformed the bogus typhoon scheme, which had only one-dimensional tangential wind field, into one with a three-dimensional wind field with tangential, radial, and vertical wind speeds (in the radial coordinate system).

5.2.5 Bogus Data Assimilation

The cost function was defined as:

$$J(X) = J_b + J_{wind} + J_H + J_{RH} \tag{5.2.46}$$

Specifically, $X = (u, v, H, T, RH)^T$ is the model analyzed field at initialization, J_b, J_{wind}, J_H, J_{RH} are the cost function about background state, wind observation, geopotential height observation and water vapor observation respectively. They are expressed as follows:

$$J_b = \frac{1}{2}(x - x_b)^T B^{-1}(x - x_b) \tag{5.2.47}$$

$$J_{wind} = \frac{1}{2}\left(wind - wind_{bogus}\right)^T O_{wind}^{-1}\left(wind - wind_{bogus}\right) \tag{5.2.48}$$

$$J_H = \frac{1}{2}\left(H - H_{bogus}\right)^T O_H^{-1}\left(H - H_{bogus}\right) \tag{5.2.49}$$

$$J_{RH} = \frac{1}{2}\left(RH - RH_{bogus}\right)^T O_{RH}^{-1}\left(RH - RH_{bogus}\right) \tag{5.2.50}$$

Here, x_b is the model background state. B, O_{wind}, O_H, O_{RH} are the covariance matrix of the background errors, wind observation errors, geopotential height observation errors and relative humidity observation errors. "Wind" contains u and v components. H and RH are height and relative humidity, and the subscript "bogus" refers to the items constructed with the bogussing scheme.

The observed variables and the model background were calculated via the observation operator, as in formula (5.2.51):

$$y_o = H(x_b) \tag{5.2.51}$$

Specifically, y_o are observation variables which are expressed with the subscript "bogus" in Eqs. (5.2.46)–(5.2.48). If y_o and x_b were the same physical quantity, the observation operator H was just interpolation. In previous studies, the wind speed constructed by a bogussing scheme was usually assimilated as observational data. The bogus tangential wind and radial wind had no corresponding construction method, however. The construction of the radial wind did not take the wind field of the large-scale circulation into account. Instead, the background field included most of this information. If the wind field constructed via bogus data were directly assimilated as observational data, errors in the radial wind would be introduced into the system as well. If only the tangential wind was assimilated, while the radial wind was removed, the radial wind would be zero, which is not true. The BDA scheme in this article was based on the wind characteristics of the bogus data, and the tangential wind was assimilated with no constraints on radial wind. The formula used for tangential wind was:

$$Wind_{\tan g} = aU + bV \tag{5.2.52}$$

where a and b are the projection coefficients of the wind speed (U and V) in the tangential direction. The relation of the background wind field (U and V) and the bogus tangential wind, was established using formula (5.2.52) as an observation operator, which replaced $wind_{bogus}$ in formula (5.2.48). As a result, the bogus wind field distribution would alter the background field distribution of the tangential wind by adjusting the background wind field U and V components, and the radial component was not affected directly. The height field and the water vapor distribution constructed by the bogussing scheme were assimilated as observations by Eqs. (5.2.49) and (5.2.50).

For TC "Imbudo" (0307), the landfall path prediction was significantly improved by the BDA scheme, and was more consistent with observations (Fig. 5.10) than other methods.

The precipitation forecast at landfall, when using the BDA scheme, was close to reality. Figure 5.11a illustrates the predicted 3 h cumulative rainfall distribution of typhoon Imbudo from 00 to 03 a.m. from the 24 h BDA scheme. Figure 5.11b is the TRMM satellite retrieval precipitation distribution at that time. These results show that employing the BDA scheme for predicting the structure of the TC spiral rain-band resulted in an outstanding forecast.

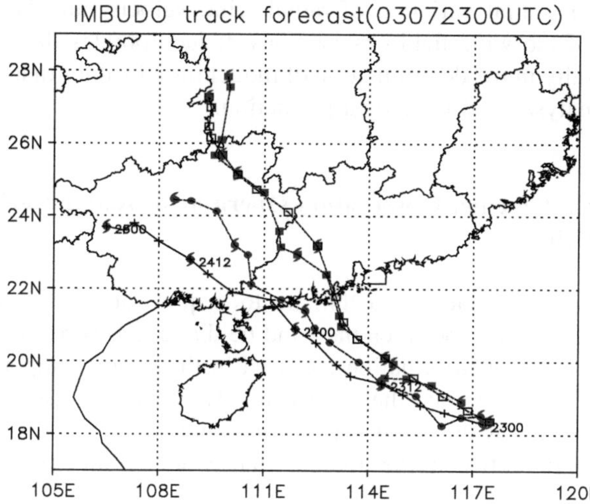

Fig. 5.10 Typhoon track marked with *asterisks*. *Solid dots* for BDA, *hollow boxes* for the bogus test, and *solid squares* for the control test

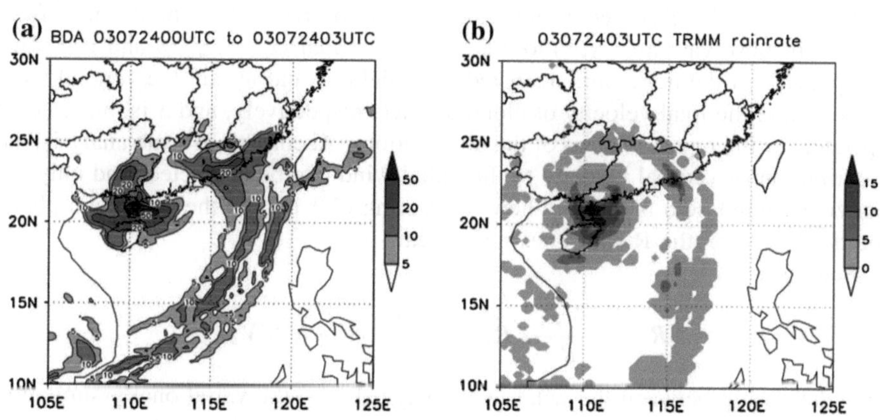

Fig. 5.11 Distribution of rainfall during landing. **a** BDA scheme, unit: mm. **b** TRMM satellite retrieval precipitation rate distribution, unit: mm h^{-1}

5.3 The Application of Radar Observational Data

5.3.1 The Variational Assimilation Technique for the Retrieval of Wind from Doppler Radar Data

The prediction skill of a mesoscale NWP system depends on the efficient utilization of Doppler information. Here, a technical scheme is put forward that acquires the atmospheric wind field by incorporating into the assimilation system the temporal and spatial variation of radial wind velocity and Doppler echoes, which are part of the wind-field observations. It incorporates the radar-retrieved wind field integrally in the determination of model initial values that can be used in both synoptic analyses and numerical prediction.

5.3.1.1 Technical Scheme: Observation Operator of Radar-Derived Wind Field

When using the 3-D variational technique to acquire an analyzed atmospheric wind field, the relationship between the wind field and radar observations needs to be known. Two types of information are used in this work, temporal and spatial variation of radial velocity and the radar echo. They will be discussed separately here. Below, u, v, and w are wind vector components.

The relationship between the radial wind velocity and a wind vector is expressed by:

$$S(V_R) = \frac{x - x_c}{R} u + \frac{y - y_c}{R} v + \frac{z - z_c}{R} (w - w_d) \tag{5.3.1}$$

Here, V_R is the observed radial wind velocity by radar, R is the distance from the site of observation (x, y, z) to the site of the radar (x_c, y_c, z_c), u and v are two components of the horizontal analyzed wind field, w and w_d are the vertical wind velocity and the final velocity of cloud droplets respectively, and S is the selected operator for the representation of the observations. According to scale analysis, the contribution of vertical velocity to the radial wind can be neglected, and only the horizontal wind field is retained here. Projecting (5.3.1) onto the horizontal plane, the third term on the R.H.S. does not appear.

Setting:

$$\boldsymbol{R}_x = \frac{x - \boldsymbol{x}_c}{R}, \ \boldsymbol{R}_y = \frac{y - \boldsymbol{y}_c}{R}, \ \boldsymbol{V}_{sR} = S(\boldsymbol{V}_R)$$

the relationship between the wind field analyzed with the variational assimilation technique and the radar observations as expressed in (5.3.1) is transformed to:

$$\boldsymbol{V}_{sR} = \boldsymbol{R}_x u + \boldsymbol{R}_y v \tag{5.3.2}$$

Continuous changes in the intensity of radar echoes can reveal the movement of cloud clusters with wind. For relatively short spans of time, the echo intensity of clouds can be assumed to be conserved (Qiu and Xu 1992; Laroche and Zawadzki 1994; Xu et al. 1994a, b; Shapiro et al. 1995; Xu and Qiu 1995; Qiu and Xu 1996; Qiu et al. 2000). For mesoscale assimilation, when local changes in echo intensity are attributed to cloud advection, the following simplified conservation equation for echo intensity is used to locate and study the relation between the wind field and radar observations. The equation can be written as:

$$\frac{\partial S(I)}{\partial t} = -\frac{\partial S(I)}{\partial x}u - \frac{\partial S(I)}{\partial y}v \tag{5.3.3}$$

Here, I is the intensity of the radar echo, u and v are the two components of the horizontal wind field, S is the selected operator for the representation of observations, and other symbols follow conventional meteorological meanings.

Equation (5.3.3) indicates the dimension of the wind speed. For this purpose,

$$\boldsymbol{I}_{sD} = \sqrt{\left(\frac{\partial S(I)}{\partial x}\right)^2 + \left(\frac{\partial S(I)}{\partial y}\right)^2}$$

$$\boldsymbol{I}_{sx} = -\frac{\partial S(I)}{\partial x}\Big/\boldsymbol{I}_{sD} \quad \boldsymbol{I}_{sy} = -\frac{\partial S(I)}{\partial y}\Big/\boldsymbol{I}_{sD} \quad \boldsymbol{V}_{st} = \frac{\partial S(I)}{\partial t}\Big/\boldsymbol{I}_{sD}$$

are introduced. Then:

$$\boldsymbol{V}_{st} = \boldsymbol{I}_{sx}u + \boldsymbol{I}_{sy}v \tag{5.3.4}$$

Here, $\boldsymbol{V}_{st}$ is the temporal and spatial variation of radar echo, which equals the component of the wind field in the direction of the gradient vector, indicating the velocity of echo movement with wind. If the gradient at a given point is orthogonal to the wind vector, no movement of the echo will be evident there, and $\boldsymbol{V}_{st}$ equals zero. If the gradient is parallel to the wind vector, the echo can be seen moving at wind velocity, and $\boldsymbol{V}_{st}$ equals the wind speed. Here, $\boldsymbol{V}_{st}$, the new "measured quantity of radar", is called the "seeable wind speed" of radar. It is evident that the seeable wind velocity has definite physical meaning, which helps determine observational error and control the quality of data.

Compared to the direct use of (5.3.3), advantages of employing the relation between the analyzed wind field and radar observations as expressed in (5.3.4) are listed below. For areas where there is significant spatial change in echo intensity (where the given threshold value for the horizontal gradient is satisfied), the local change in echo intensity within a short duration is affected by advection more than cloud clusters; the "seeable wind speed" exists and accurately depicts the wind field, which is then incorporated into the assimilation system. For areas where there is insignificant spatial change in echo intensity, the local change in echo intensity is caused primarily by cloud clusters, and does not reflect information about the wind field. No "seeable wind speed" can be calculated using (5.3.4), and this case is treated as a missing observation. This automatically prevents data incapable of describing the wind field from being included in the assimilation system. In contrast, the application of (5.3.3) does not rid the system of the effects of cloud clusters, as it does not filter out inadequate data.

In our work, the operator S follows the Barnes interpolation-filter scheme (Barnes 1964) for radar measurements at different times and locations to obtain grid-point values for a given time and spatial resolution. It is also necessary to run selective filtering for representative scales. The weighting function for the interpolation-filter scheme is given by:

$$W(\Delta x, \Delta y, \Delta z, \Delta t, \alpha, \beta, \gamma) = e^{-4\left(\frac{(\Delta x^2+\Delta y^2)}{\alpha^2}+\frac{\Delta z^2}{\beta^2}+\frac{\Delta t^2}{\gamma^2}\right)} \tag{5.3.5}$$

Here, $\Delta x, \Delta y, \Delta z, \Delta t$ are the spatial distance and temporal deviation from designated grid-points, and α, β, γ are selection parameters of horizontal, vertical, and temporal scales, respectively.

The variation and assimilation scheme with the radar wind-field data put forward in this work takes $\boldsymbol{V}_{sR}$, the radial wind velocity, and $\boldsymbol{V}_{st}$, the "seeable wind velocity" as observed quantities to determine the atmospheric wind field. From (5.3.2) to (5.3.4), the vectors of transformation operators for the two observed quantities are respectively:

$$\boldsymbol{R} = \begin{bmatrix} \boldsymbol{R}_x \\ \boldsymbol{R}_y \end{bmatrix} \tag{5.3.6}$$

$$\boldsymbol{I}_s = \begin{bmatrix} \boldsymbol{I}_{sx} \\ \boldsymbol{I}_{sy} \end{bmatrix} \tag{5.3.7}$$

R is the radial unit vector of the radar and I_s is the unit vector of the echo intensity (which has been selected for representation) in the horizontal gradient.

From (5.3.2) to (5.3.4), it is seen that the selected V_{sR} and $\boldsymbol{V}_{st}$ show a linear relationship with the assimilated wind field. Thus, the corresponding adjoint operator is the transposition matrix for (5.3.6) and (5.3.7).

The proposed scheme can be run operationally and can be combined naturally and directly with other types of observations. This makes it useful for the generation of initial values for NWP. The scheme is also applicable to existing data from the new-generation Doppler weather radars in China.

5.3.2 An Improved Method for Radar Seeable Velocity in Three-Dimensional Variational Assimilation

Direct assimilation of radar radial wind and "seeable velocity" (SV) had a significant effect, but the process of the vertical wind component was extremely simple and it only preserving the horizontal components made it difficult to receive accurate information on the wind field. Especially in regions with heavy rainfall, some discrepancies arise, as discussed below.

1. Calculating the terminal velocity of raindrops

The relationship between reflectance, I (units: dBz), and rainwater mixing ratio, qr, was calculated using:

$$I = 43.0963 + 17.5 \log_{10}(\rho q_r), \ \rho q_r > b \tag{5.3.8}$$

From the above:

$$\rho q_r = be^{cI},\ I > Ic \tag{5.3.9}$$

where, ρ is air density, qr is the rainwater mixing ratio, $b = 0.0034463089$, $c = \rho$ 0.13157629, and $Ic = 5$ dBz.

With (5.3.9) and the Marshall-Palmer raindrop spectral distribution, the relationship between the terminal velocity of raindrops and the radar echo can be derived:

$$W_d = 5.4a\, b^{0.125} e^{0.125cI},\ I > I_c \tag{5.3.10}$$

where, $a = \left(p_0/_{\bar{p}}\right)^{0.4}$, and $e = 2.718$

W_d can reach 5 m/s or so at lower levels in regions of heavy rainfall. This is 10 times the characteristic value of the vertical velocity in mesoscale systems and cannot be ignored.

2. Radar wind analysis technique

Assimilation of the radar radial wind and SV can be expressed as follows:

$$J(x) = J_b + J_{Vr} + J_{Vs} \tag{5.3.11}$$

where, $X = (u, v)^T$ is the analysis field at initialization. The three terms on the right hand side of (5.3.11) are the objective functions of the background field, the radial wind, and the SV. The specific formulas are:

$$J_b = \frac{1}{2}(x - x_b)^T B^{-1}(x - x_b) \tag{5.3.12}$$

$$J_{Vr} = \frac{1}{2}(V_r - V_{ro})^T O_{Vr}^{-1}(V_r - V_{ro}) \tag{5.3.13}$$

$$J_{Vs} = \frac{1}{2}(V_S - V_{So})^T O_{Vs}^{-1}(V_S - V_{So}) \tag{5.3.14}$$

where V_r and V_s represent the radar radial wind and SV, respectively. The observation operator *H*, establishes the relation with the analysis variables. The subscript *o* indicates radar data.

$$V_r = H_{V_r}(x_b)$$

$$V_s = H_{V_s}(x_b)$$

We will next discuss the radial wind and SV observation operators. The Doppler radar radial wind and wind vector relationship is:

$$V_r = \frac{x - x_c}{R}u + \frac{y - y_c}{R}v + \frac{z - z_c}{R}(w - w_d) \tag{5.3.15}$$

where, Vr is the observed radar radial wind, R is the distance between the point of observation (x, y, z) and the radar (x_c, y_c, z_c), u and v are the horizontal analysis wind field, and w and W_d are the atmospheric vertical velocity and terminal velocity of raindrops.

Wan et al. (2005) preserve only the horizontal wind component. However, the terminal velocity of raindrops in mesoscale weather systems is of such great magnitude that it should be considered even while ignoring the vertical velocity component when deriving the radial wind observation operator:

$$V_r = V_r + R_z W_d = R_x u + R_y v \tag{5.3.16}$$

here, $R_x = \frac{x - x_c}{R}$, $R_y = \frac{y - y_c}{R}$, $R_z = \frac{z - z_c}{R}$.

Generally, the vertical velocity was less than the terminal velocity of raindrops. It can be inferred that the value that the vertical displacement (distance) multiplied the vertical velocity was smaller, so it could be neglected. According to estimates based on scale analysis, the eigenvalues of echo intensity in vertical advection were less than those of horizontal advection by at least one order of magnitude. To observe the role of the terminal velocity of raindrops in the original formula for SV, the vertical velocity term is omitted, and the simplified echo balance equation can be expressed as follows:

$$\frac{\partial I}{\partial t} = -\frac{\partial I}{\partial x}u - \frac{\partial I}{\partial y}v + 1.125\frac{\partial I}{\partial z}W_d \tag{5.3.17}$$

Eliminating the dimension of echo intensity in (5.3.17) so that only the dimension of wind speed remains, leads to the modified formula for SV:

$$V_s = \frac{1}{I_s}\frac{\partial I}{\partial t} + 1.125 I_z W_d = I_x u + I_y v \tag{5.3.18}$$

Here, $I_x = -\frac{1}{I_s}\frac{\partial I}{\partial x}, I_y = -\frac{1}{I_s}\frac{\partial I}{\partial y}, I_z = -\frac{1}{I_s}\frac{\partial I}{\partial z}$, and $I_s = \sqrt{\left(\frac{\partial I}{\partial x}\right)^2 + \left(\frac{\partial I}{\partial y}\right)^2 + \left(\frac{\partial I}{\partial z}\right)^2}$, are the magnitude of the dimensionless echo gradient components (I_x, I_y, I_z), and the echo intensity gradient (I_s). The revised value of SV from (5.3.18) with only the echo advection term on the right is $1.125 I_z W_d$. This means that local clouds moving with the background wind field are expressed by SV, when considering the terminal velocity of raindrops. It has the dimension of wind speed, and expresses the advection of local variation in radar echo which responds to the horizontal atmospheric wind field. Furthermore, SV was quite easy to calculate using the radar echo intensity at two consecutive times.

From (5.3.16) to (5.3.18) the radial wind observation and SV operators are:

$$H_{V_r} = \begin{vmatrix} R_x \\ R_y \end{vmatrix} \tag{5.3.19}$$

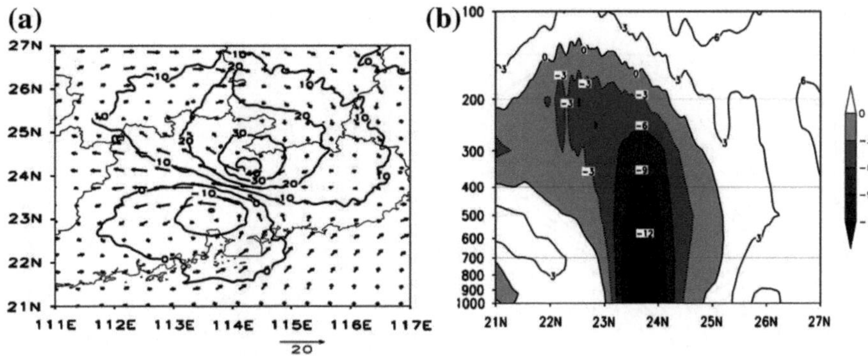

Fig. 5.12 Difference between simulated observation and free integration. **a** Horizontal distribution of the 700 hPa geopotential height (gpm) and wind (m/s). **b** The vertical-zonal section of average meridional wind (m/s) at 113° E–115° E

$$H_{V_s} = \begin{vmatrix} I_x \\ I_y \end{vmatrix} \tag{5.3.20}$$

The analysis above modifies the radial wind and SV observation operators. Comparison of the two observation operators shows that they are of the same form, the only difference being the u and v wind factor. Therefore, radial velocity and SV can share a data assimilation module in the assimilation process.

Using NCEP/NCAR 1 × 1° grid analysis data from June 25, 2008 at 00 p.m. (UTC) as the background field, we employed the GRAPES-3D-VAR assimilation system to assimilate radial wind and SV data using both the improved and original schemes. Figure 5.12a illustrates the difference (by subtraction) in horizontal distribution between the simulation and free integration on the 700 hPa geopotential height and wind fields. The average meridional wind difference field in the 113° E–115° E vertical—zonal section is shown in Fig. 5.12b below (values less than zero are shaded).

Figure 5.13a shows the horizontal distribution of the potential height increment field. There is only one center over the entire region of radar data. The wind field and the potential height field correspond to anti-cyclonic circulation. The horizontal wind increases up to 6 m/s. Figure 5.13b shows that the eastward wind increment appearing between the two centers of the typhoons in the vertical wind field increment of the original field, sloped southward with height as well. However, the center was higher in the lower (500–700 hPa) level, and failed to reflect the fact that intensity decreased with height. There is almost no radar data above 300 hPa because, due to the automatic station control, increments approached 0 m/s. Improved scheme results are shown in Fig. 5.14, demonstrating that the vertical distribution of the westerly wind increment in the 23° N–24° N section slopes southward with respect to altitude and weakens gradually. The center intensity is stronger than in the original method as well, approaching 5 m/s. The distribution of

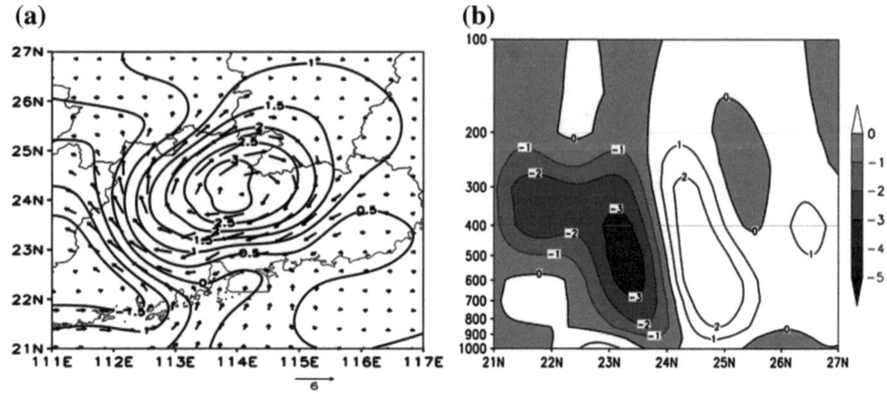

Fig. 5.13 Incremental approach of old method. **a** horizontal distribution of 700 hPa geopotential height (contour, gpm) and wind (vector, m/s). **b** The vertical-latitude section of average meridional wind (m/s) at 113° E–115° E

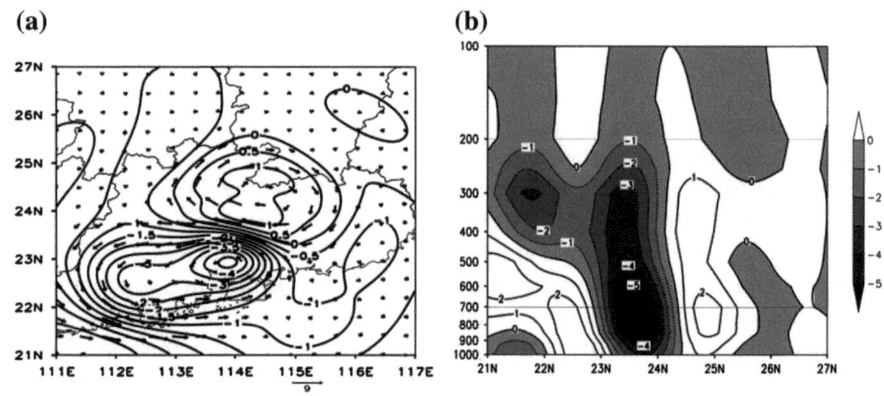

Fig. 5.14 Incremental approach of upgrade method. **a** Horizontal distribution of 700 hPa geopotential height (contour, gpm) and wind (vector, m/s). **b** The vertical-latitude section of average meridional wind (m/s) at 113° E–115° E

the potential height increment has a dipolar distribution, and is positive or negative to the north and south of the center, with wind strength increment as high as 9 m/s.

Assimilation results of both methods reflected the structure and distribution of the difference field (figure omitted) to some extent. This shows that they have the ability to describe the atmospheric wind field. Comparison of Figs. 5.12, 5.13 and 5.14 demostrates that when considering the effect of the terminal velocity of raindrops, the distribution of horizontal wind increment appeared in the structure of vortices, and was consistent with the "real atmosphere". The error in the position of the center was small, and the vertical changes in meridional wind were also close to those of the "real atmosphere."

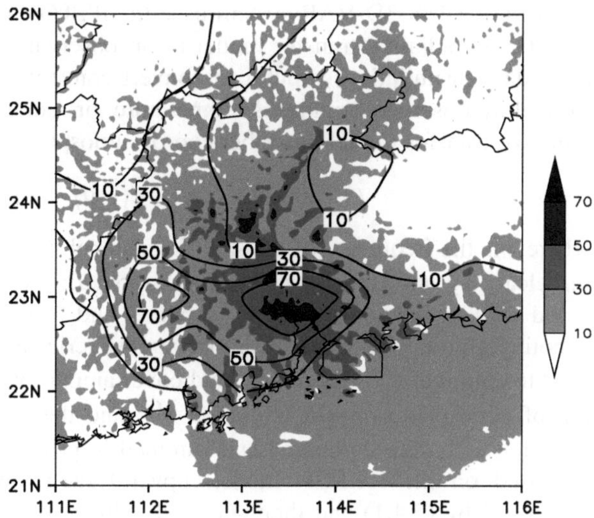

Fig. 5.15 Six hour precipitation (mm) of observed by ground station and simulation

Six hours of observed cumulative precipitation (contours), from June 25 2008 (06–12UTC) is shown in Fig. 5.15. Assimilation of radar information better simulated the distribution of precipitation (shaded area), and had a significant effect on the precipitation (compared with no assimilation of radar information).

5.4 The Application of Satellite Data

5.4.1 The Numerical Simulation of Satellite Infrared Channel Brightness Temperature

The numerical simulation of satellite brightness temperature was used as a way of testing simulation results, and was also an important part of the direct assimilation of satellite radiation data. The direct assimilation of radiation and satellite brightness temperature data commonly involves adopting a radiative transfer model as the observation operator, to simulate different channels of satellite brightness temperature (Wang 1995). As a result, radiative transfer model simulation capabilities will directly affect the assimilation results. Due to lack of three-dimensional cloud information, most of the direct assimilation of satellite data was limited to clear sky conditions, yet infrared data was significantly influenced by clouds. When cloud cover in tropical areas was high due to typhoons, rainstorms, and other extreme weather, such processes were inseparable. Considering only the radiation data under clear sky transmission, a large number of satellite observations in the tropics, in particular infrared channel data, were not suitable for assimilation.

The GRAPES-3D-VAR system used the RTTOV (fast radiative transfer model for TOVS) radiative transfer model as an observation operator for satellite radiation data assimilation, which can convert conventional meteorological elements to satellite observed brightness temperature, or radiation values. The atmospheric radiative transfer model describes top of atmosphere radiation with:

$$L(v,\theta) = (1-N)L^{\mathrm{Clr}}(v,\theta) + NL^{\mathrm{Cld}}(v,\theta) \tag{5.4.1}$$

where v is the frequency of the satellite channel, θ is the observation angle, L^{Clr} is the clear sky atmospheric upward radiation, L^{Cld} is the upwards radiation from cloud tops to the top of the atmosphere, and N is a number between 0 and 1 denoting cloud cover. The clear sky atmospheric radiative transfer equation was used to compute L^{Clr} when the cloud cover amount was 0. Otherwise, the cloud to top of atmosphere upward radiation L^{Clr} and the clear sky atmospheric upward radiation L^{Clr} were computed and summed in proportion.

Based on the RTTOV model, upward radiation in each cloud level was calculated by (5.4.1) and then summed, which converted the calculation process for translucent cloud radiation into a linear summation process for radiation and black body cloud radiation in clear conditions. The input parameters for the RTTOV model included: the atmospheric temperature profile, the absorbing gas profile, clouds, and surface features. The absorbing gas profile included water vapor (H_2O), ozone (O_3), and carbon dioxide (CO_2). The vertical cloud profile included cloud water content, cloud ice water content, cloud amount, and total cloud amount. Cloud top pressure was provided as an input parameter as well, when the simulated upward radiation at the top of the atmosphere was less than that during cloudy conditions. In recent years, with the development of meteorological microphysical process modeling technology, the regional mesoscale meteorological model has gained the capability of forecasting three-dimensional cloud structures. Making use of simulated three-dimensional cloud structure and other physical information, the RTTOV radiative transfer model was used to simulate radiation brightness temperature under cloudy conditions.

5.4.2 Three-Dimensional Variational Data Assimilation of TRMM Rain Rate Data

Because tropical regions are dominated by convective precipitation, an extension of the KUO-cumulus parameter scheme (Kuo 1965) was used as the observation operator to assimilate TRMM rain rate in the GRAPES-3D-VAR system.

An extension of the KUO-cumulus parameterization scheme, the FSU (Florida State University) convective parameter scheme, was used as an observation operator, to associate satellite retrieval rain rate and model variables. The precipitation formula used was:

$$R = I_1(1+\eta)(1-b) \tag{5.4.2}$$

In which $I_L = -\frac{1}{g}\int_{P_t}^{P_b} w\frac{\partial q}{\partial p}dp$ is the water vapor supply. P_t and P_b are cloud top and cloud bottom pressure respectively, $I_l\eta$ represents the mesoscale moisture supply, and η and b are the average vertical velocity and the 700 hPa relative vorticity in the air column. The input parameters for the convective parameter scheme included the atmospheric pressure (p), temperature (T), vertical velocity (ω), specific humidity (q) and the 700 hPa vorticity of each vertical layer. The precipitation observation operator was defined with the vertical velocity (ω) and the specific humidity (q) as the adjusted variables in the assimilation system:

$$y = H(\omega, q) \tag{5.4.3}$$

Here, the operator H is the convective parameter scheme defined by formula (5.4.2). The vertical velocity term in the observation operator can be calculated with the horizontal wind field (u, v) and the surface pressure. The 700 hPa vorticity can also be calculated from the horizontal wind field. The formula (5.4.3) can be rewritten as:

$$y = H(u, v, q) \tag{5.4.4}$$

Solving the partial derivative with respect to wind and specific humidity of the observation operator H gives the corresponding tangential linear operator H' and its adjoint operator H'^T. The code transpose method, commonly used in variational assimilation, was used for the construction of the adjoint operator. The observation operator H, was computed directly using the program code, while the conjugate transpose method was used to compile the program code to compute the adjoint operator. The wind field (u, v), water vapor field (q), and temperature distribution (T) of the initial field were then computed using a minimization algorithm. The physical balance relationship in the GRAPES-3D-VAR system was determined when the objective function J reached minimum. After application of the above method, the vertical structure of the moisture convergence in the model background field was adjusted by the observation operator and the structure of the model initial field was closer to that of the observed region of precipitation.

Figure 5.16a shows the 24 h cumulative precipitation for the control test (without TRMM data), and shows only a small area with more than 50 mm of precipitation near the typhoon center. After assimilation of TRMM data, the precipitation in the typhoon center increased significantly (Fig. 5.16b). The size of the region with 50 mm or more precipitation increased, and was closer to the 24-hour cumulative precipitation detected by TRMM (Fig. 5.16c). Therefore, the assimilation of TRMM data improved the 24 h cumulative rainfall forecast near the typhoon center, and was closer to satellite observations of the precipitation structure.

A control test and assimilation test for the 72 h forecast are shown in Fig. 5.17. The control test displays several distinct southward deflections in the typhoon path, which did not actually occur. Assimilation tests did not show this phenomenon.

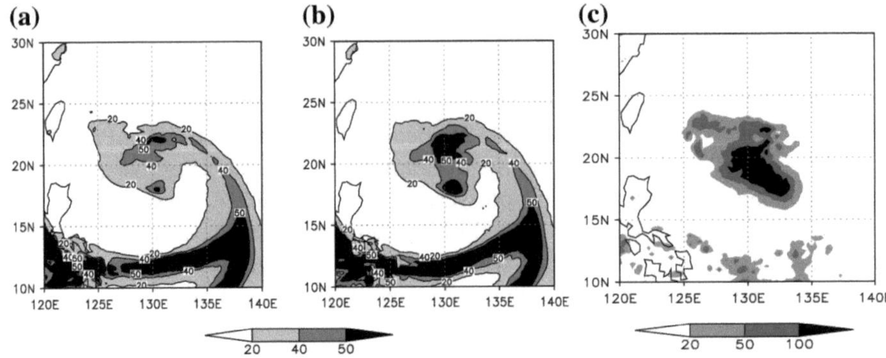

Fig. 5.16 24 h precipitation forecast. **a** Control test. **b** Assimilation test. **c** TRMM observation. Units: mm

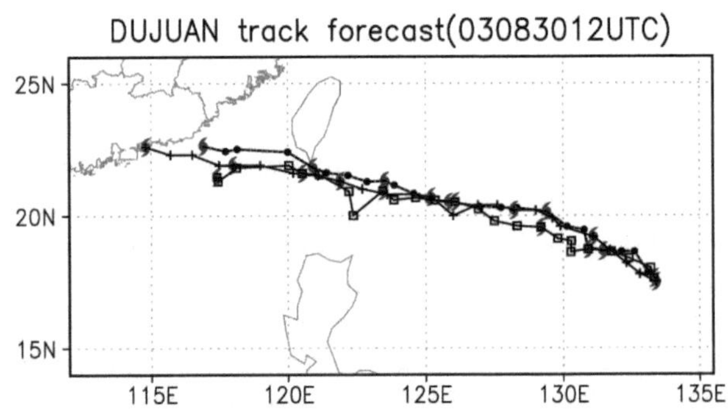

Fig. 5.17 Typhoon DUJUAN's track forecast. *Stars* observation, *open squares* control test, *solid dots* assimilation test

5.5 Typhoon Numerical Simulation and Operational Performance

5.5.1 Precise Typhoon Prediction

Based on vortex relocation technology, and using a high-resolution mesoscale model (with inner horizontal grid resolution of 2 km), the May 17, 2006 0000 UTC 24 h track forecast for tropical cyclone "Chanchu (0601)" was simulated. The simulated track (Fig. 5.18) and maximum wind speed (Fig. 5.19) were close to observations. Additionally, the precise structure distribution of surface wind and precipitation were simulated. There was conspicuous wind shear associated with heavy rain (Figs. 5.20, 5.21).

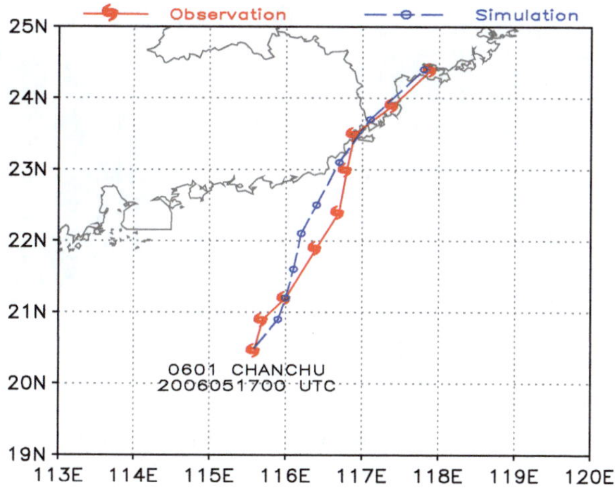

Fig. 5.18 Simulated track and observed track for "Chanchu" (0601)

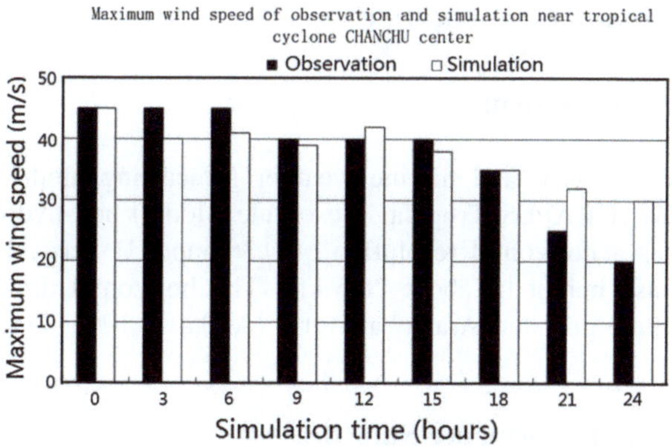

Fig. 5.19 24 h simulated and observed maximum wind speed for "Chanchu"

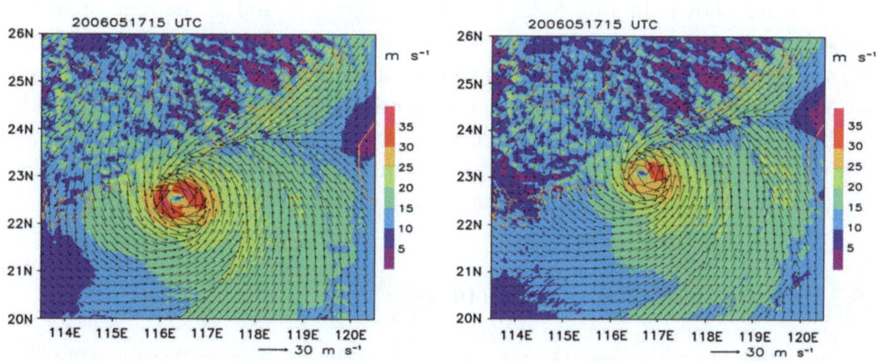

Fig. 5.20 Simulated surface wind at 17,1500 UTC and 17,1800 UTC

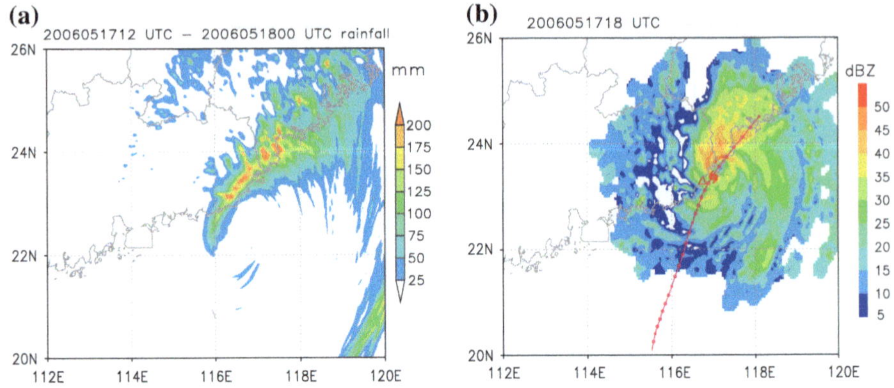

Fig. 5.21 Simulated 17,1200 UTC–18,0000 UTC cumulative precipitation (**a**) and observed radar reflectivity (**b**). *Red curve* the TC track, *TC symbol* TC center location, *hollow triangle* the radar location

5.5.2 Operational Performance

5.5.2.1 Operational System

The Guangzhou tropical and marine weather forecasting model, based on the GRAPES_TMM (GRAPES Tropical Mesoscale Model) non-hydrostatic equilibrium model, has a horizontal resolution of 0.36° and 31 vertical layers, with a model tropopause height of about 28.5 km. The horizontal dimension uses an equidistant latitude grid and Arakawa-C grid (Arakawa 1966) (Table 5.1).

5.5.2.2 Operation Performance Statistics

Assessments of typhoon track forecasts in 2009 (provided by the Shanghai Typhoon Institute) are shown in Fig. 5.22, in which objective and numerical prediction methods are compared to the Japanese global model (sample number above column). The assessments show that the Guangzhou numerical model relative error of prediction for 24 h was less than that of other methods, except the Jiangsu probability and the Shanghai ensemble models, while the relative errors of prediction for 48 and 72 h were only larger than the Shanghai ensemble. This means that the Guangzhou numerical model has excellent performance in typhoon path prediction.

5.5.2.3 Operational Case: Fanapi (1011)

In 2010, tropical cyclone no. 11 "Fanapi," formed in the western Pacific at 20:00 on September 15, and then strengthened into a super typhoon at 16:00 on September 17.

Table 5.1 Guangzhou tropical and marine weather forecasting model operational system. China

Analysis and assimilation	3D-VAR isobaric analysis (ground, ship, radiosonde, and ATOVS data) cloud motion wind information, etc. using a simple background error covariance
Typhoon initialization	The relocation technique and bogus typhoon technology are introduced into the numerical model through the three-dimensional variational method and value initialization
Dynamic frame	Completed pressure and non-static equilibrium, semi-implicit- and semi-Lagrangian differencing scheme, with isothermal atmosphere deducting and improved water vapor positive definite advection scheme, a height terrain following coordinates, with a Charney-Phillips jump layer in the vertical
Physical processes	Micro-physics (WSM6), long-wave radiation (RRTM), short-wave radiation (ECMWF), the surface layer (M–O similarity theory), the boundary layer (MRF), land surface process (SLAB), cumulus convection (KFETA)
Boundary conditions	Data from GFS, T639 and T213
The model forecast region	81.6° E–160.8° E; 0.8° N–50.5° N
Forecast times and range	Twice daily (00,12 UTC), and 120 h, respectively
The tropical storm report conditions	When numbered tropical cyclones move into the domain (90.0° E –155.0° E, 5.0° N–50.0° N), the model will automatically add a Bogus typhoon, and issue typhoon forecast products

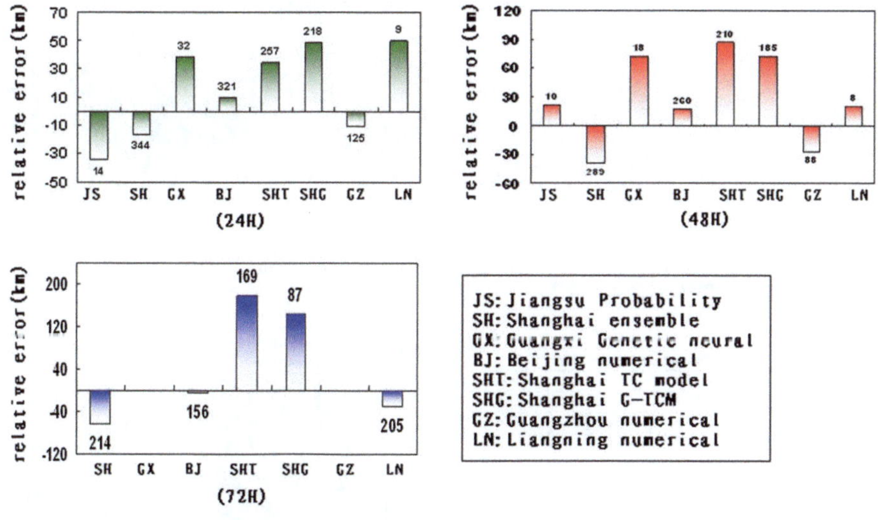

Fig. 5.22 Assessment of typhoon track forecasts for 24, 48, and 72 h

It was the most powerful typhoon to affect China. The operational performance system predicted both the track (Fig. 5.23) and the precipitation (Fig. 5.24) of Fanapi perfectly. The average 72 h path error is shown in Fig. 5.25. The average 24 h error was less than 100 km. The 48 and 72 h average errors were both less than 120 km reflecting a successful simulation.

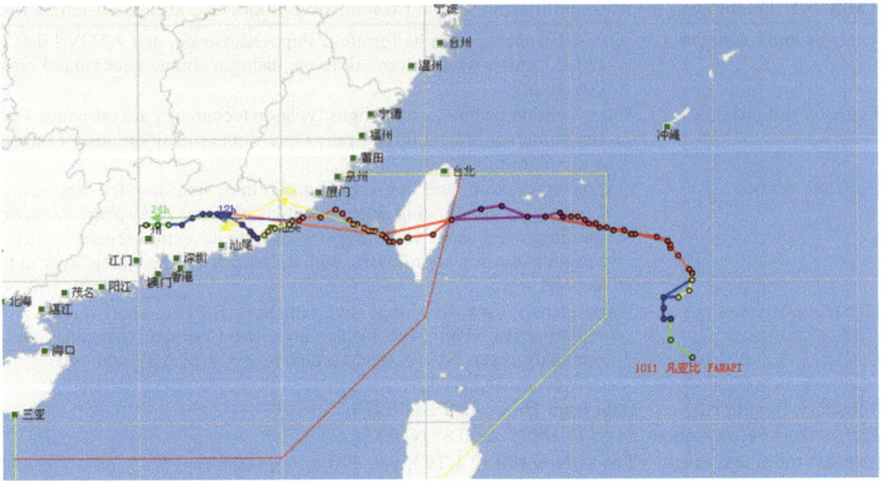

Fig. 5.23 Track forecast and observation of Fanapi (1011)

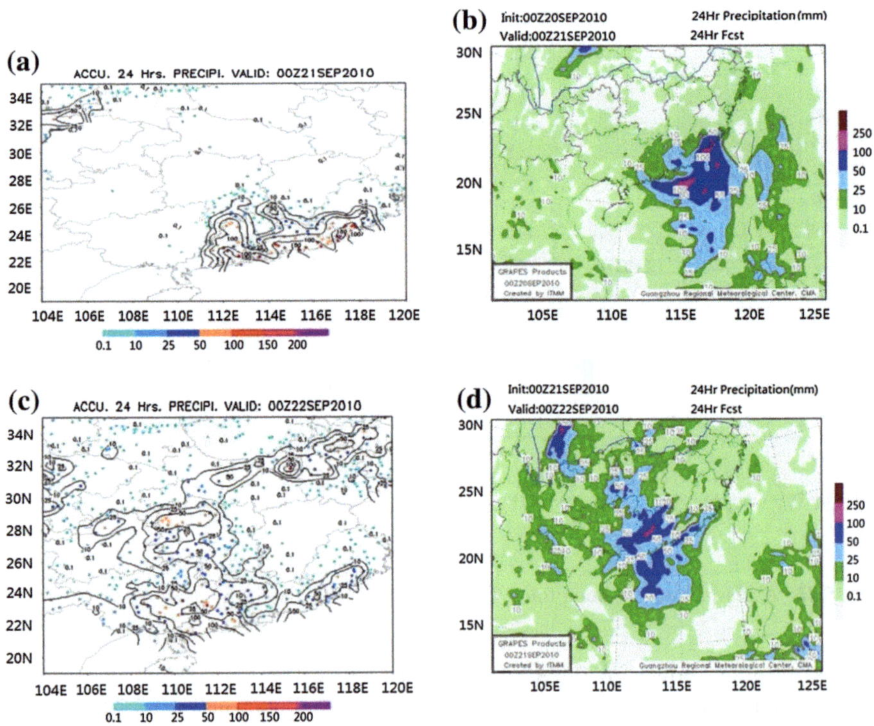

Fig. 5.24 Operational forecast (*right*) and observation (*left*), 24 h cumulative rainfall of Fanapi (1011)

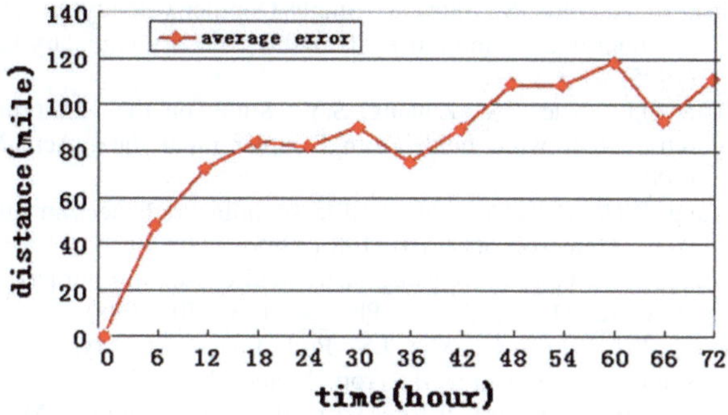

Fig. 5.25 72 Average errors of Typhoon Fanapi (1011) track forecast

References

Arakawa, A.: Computational design for long-term numerical integration of the equations of atmospheric motion. J. Comput. Phys. **1**, 119–143 (1966)

Barnes, S.L.: A technique for maximizing details in numerical weather map analysis. J. Appl. Meteor. **3**, 396–409 (1964)

Davidson, N.E., Puri, K.: Tropical prediction using dynamical nudging, satellite-defined convective heat sources, and cyclone bogus. Mon. Wea. Rev. **120**(11), 2501–2522 (1992)

Huang, X.G., Fei, J.F., Lu, H.C.: A contrast test of the methods to remove the analyzed typhoon. J. Appl. Meteorol. Sci. **17**(1), 81–86 (2006)

Iwasaki, T., Nakano, H., Sugi, M.: The performance of a typhoon track prediction model. J. Meteor. Soc. Japan. **85**, 555–570 (1987)

Kuo, H.L.: On formation and intensification of tropical cyclones through latent heat release by cumulus convection. J. Atmos. Sci. **22**, 40–63 (1965)

Kurihara, Y., Bender, M.A., Tuleya, R.E., Ross, R.J.: Prediction experiments of hurricane Gloria (1985) using a multiply nested movable mesh model. Mon. Wea. Rev. **118**(10), 2185–2198 (1990)

Kurihara, Y.M., Bender, M.A., Ross, R.J.: An initialization scheme of hurricane models by vortex specification. Mon. Wea. Rev. **121**, 2030–2045 (1993)

Laroche, S., Zawadzki, I.: A variational analysis method for retrieval of three-dimensional wind field from single- Doppler radar data. J. Atmos. Sci. **51**, 2664–2684 (1994)

Leslie, L.M., Holland, G.J.: On the bogussing of tropical cyclones in numerical model: a comparison of vortex profiles. Metror. Atmos. Phys. **56**, 101–110 (1995)

Mathur, M.B.: The national meteorological center's quasi-Lagrangian model for hurricane prediction. Mon. Wea. Rev. **119**, 1419–1447 (1991)

Meng, Z.Y., Xu, X.D., Chen, L.S.: A method to study the tangential wind profile for strong tropical cyclone and its contribution to abnormal tropical cyclone track simulation. Chin. J. Atmos. Sci. **25**(2), 193–199 (2001)

Parrish, D.F., Derber, J.C.: The National Meteorological Center's spectral statistical-interpolation analysis system. Mon. Wea. Rev. **120**, 1433–1466 (1992)

Qiu, C.J., Xu, Q.: A simple adjoint method of wind analysis for single-Doppler data. J. Atmos. Oceanic. Technol. **9**, 588–598 (1992)

Qiu, C.J., Xu, Q.: An improvement on the simple conjugate method for retrieving the wind fields from single-Doppler data. Q. J. Appl. Meteorol. **7**, 421–430 (1996)

Qiu, C.J., Yu, J.X., Xu, Q.: Use of Doppler-radar data in improving short-term prediction of mesoscale weather. Acta Meteorol. Sinica. **58**(2), 244–249 (2000)

Shapiro, A., Ellis, S., Shaw, J.: Single-Doppler velocity retrievals with Phoenix II data: clear air and microburst wind retrievals in the planetary boundary layer. J. Atmos. Sci. **52**, 1265–1287 (1995)

Wan, Q.L., Xue, J.S., Zhuang, S.Y.: Study on the variational assimilation technique for the retrieval of wind fields from Doppler radar data. Acta Meteorol. Sinica. **63**(2), 129–145 (2005)

Wang, Z.H.: Direct use of satellite sounding radiances in numerical weather prediction. Q. J. Appl. Meteorol. **6**(1), 101–108 (1995)

Wang, Y.Q.: On the bogusing of tropical cyclone in numerical model: the influence of vertical structure. Meteor. Atmos. Phy. **85**, 153–170 (1998)

Wang, K.L., He, A.G., Xue, J.S.: Preliminary test of typhoon trace numerical prediction for the south china sea area. J. Trop. Meteorol. 44–51 (1996a) (Research on typhoon, rainstorm operational numerical forecast method and technique. Meteorological Press)

Wang, G.M., Wang, S.W., Li, J.J.: A bogus typhoon scheme and its application to a movable nested mesh model. J. Trop. Meteorol. **12**(1), 9–17 (1996b)

Xu, Q., Qiu, C.J.: Adjoint-method retrievals of low-altitude wind fields from single-Doppler reflectivity and radial-wind data. J. Atmos. Oceanic Technol. **12**, 1111–1119 (1995)

Xu, Q., Qiu, C.J., Yu, J.X.: Adjoint-method retrievals of low-altitude wind fields from single-Doppler reflectivity measured during Phoenix II. J. Atmos. Oceanic Technol. **11**, 275–288 (1994a)

Xu, Q., Qiu, C.J., Yu, J.X.: Adjoint-method retrievals of low-altitude wind fields from single-Doppler wind data. J. Atmos. Oceanic Technol. **11**, 579–585 (1994b)

Yan, J.H., Ding, W.Y., Xu, J.P.: "Non-bogussing" initialization for tropical cyclone numerical prediction and its features. Q. J. Appl. Meteorol. **15**(5), 513–522 (2004)

Zhou, J.L.: U.S. tropical cyclone research and forecasting- a research priority of U.S. Weather Research Program, research and analysis on the trends of the United States development of meteorological science and technology, pp. 78–91. Meteorological Press, Beijing, (1998)

Part II
Typhoon Impacts on Environments

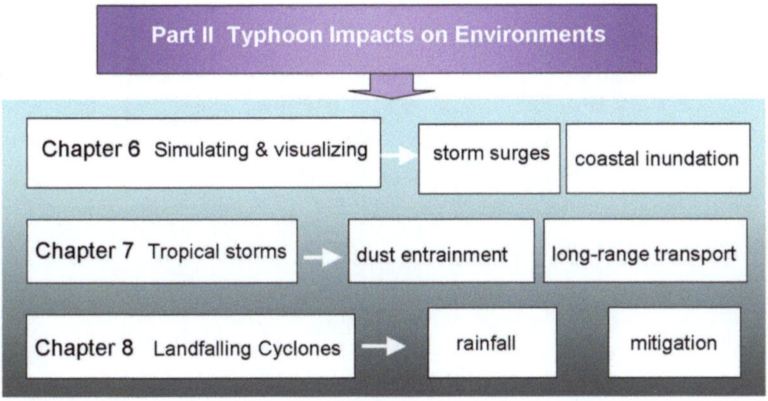

Chapter 6
Predicting and Visualizing Storm Surges and Coastal Inundation: A Case Study from Maryland, USA

Ming Li, Xiaohong Wang and Peng Jia

Abstract Many low-lying coastal regions are vulnerable to both chronic hazards associated with inundation by sea-level rise, and episodic storm surges generated by hurricanes and typhoons. Using Maryland's coast as an example, we provide an overview of a recent effort in the development of a state-of-the-art coastal inundation prediction system. We use a suite of atmospheric and hydrodynamic models to obtain an ensemble forecast of storm surge and overland inundation. Advanced graphic software such as ArcGIS and Google Earth is used to generate high-resolution images and animations of inundation in flood-prone areas. Such an end-to-end inundation prediction system can be applied to any coastal region. Given the accelerating sea-level rise and projected increases in the frequency and intensity of extreme weather events in a warming climate, we discuss how sea-level rise, changing tidal ranges and storm surges combine together to generate dangerously high surges in coastal regions.

Keywords Storm surge · Inundation · Hurricane · Typhoon · Tropical cyclone · Geographic information systems · Google earth · Graphic visualization

Submitted to Book on Remote Sensing of Typhoon Impacts and Crisis Management.

M. Li (✉) · P. Jia
Horn Point Laboratory, University of Maryland Center for Environmental Science,
2020 Horn Point Road, Cambridge, MD 21613, USA
e-mail: mingli@umces.edu
URL: http://www.hpl.umces.edu/faculty/li

X. Wang
Department of Mathematics and Computer Science, Salisbury University,
1101 Camden Avenue, Salisbury, MD 21801, USA

D. L. Tang and G. J. Sui (eds.), *Typhoon Impact and Crisis Management*,
Advances in Natural and Technological Hazards Research 40,
DOI: 10.1007/978-3-642-40695-9_6,

6.1 Introduction

In the United States, approximately 50 % of the population lives within 50 miles of the coast. The human presence in the coastal zone drives an economic engine that produces more than one-third of the U.S. gross national product. The physical infrastructure in the U.S. Gulf of Mexico and Atlantic coastal regions alone is worth about \$3 trillion (Marra et al. 2007). Globally, three-quarters of the world population now lives within 50 km of the sea. The vulnerability of coastal communities to the impacts of natural hazards is well known. Storm surge induced by tropical cyclones is one of the major threats to the life and property of coastal regions. On average, roughly 5 tropical cyclones every 3 years would strike the U.S. coastline, causing 50–100 casualties and billions dollars of property damage (http://www.nhc.noaa.gov/HAW2/english/basics.shtml).

Coastal regions are not only vulnerable to flooding due to episodic storm events but also to chronic hazards associated with climate change and sea level rise. Global sea level rose at a rate of 1.6 ± 0.2 mm year^{-1} from 1961 to 2003 and 3.4 ± 0.1 mm year^{-1} from 1993 to 2007 (Domingues et al. 2008). Climate models predict that the rate of sea-level rise will further accelerate in the 21st century (Meehl et al. 2007), with global-mean increases by 2,100 of 0.5–1.4 m, depending on greenhouse gas-emission scenarios and climate-model sensitivity (Rahmstorf 2007). Currently, sea level is rising at a rate near the upper end of the projections (Church et al. 2011).

In this paper we provide a review of recent progress in predicting and visualizing coastal inundations due to storm surges and sea-level rise. We use Maryland, USA. as an example of low-lying coastal regions prone to flooding and draw heavily from our own work in modeling Chesapeake Bay. Storm surges and coastal inundations are active research topics that have attracted wide attention in recent years. We apologize in advance for omitting many interesting studies in this short review.

6.2 Maryland's Low-Lying Coastal Areas: A Case Study

Maryland's coast is vulnerable to both chronic hazards associated with inundation by sea-level rise, and episodic storm surges generated by hurricanes/tropical storms and nor'easters. Due to its geography and geology, the Chesapeake Bay region is considered the third most vulnerable to storm surge and coastal inundation in the U.S., behind Louisiana and southern Florida.

As demonstrated by Hurricane Isabel (2003), Maryland's coast is extremely vulnerable to storm surges. With a semi-enclosed geometry, Chesapeake Bay can trap and amplify storm surges under certain conditions (Boicourt 2005; Li et al. 2006, 2007; Shen et al. 2006a, b, c). Bays' enclosed reaches offer protection from tropical storms that pass on the ocean side. Northeast-to-northerly winds in the western semicircle of a cyclone drive water away from the bay's head, even though they may cause dangerous sea-level setup and storm surges on the open coast. In contrast, when a storm passes on the land side, the confined nature of the bay becomes a liability and storm surges exceed those on the open coast. This shift from protection to vulnerability arises because southeasterly winds in the right-front quadrant of the storm blow water into the bay and pile water up against its head. This interesting scenario happened in September 2003 when Hurricane Isabel made landfall at the Outer Banks of North Carolina and moved north on the west side of Chesapeake Bay, creating widespread flooding in several populated areas including Washington, D.C., Baltimore, Annapolis and Maryland's Eastern Shore (Fig. 6.1). A similar scenario happened in August 1933 when Hurricane Chesapeake-Potomac caused extensive flooding over the Eastern Shore of Maryland and Virginia, and damaged agricultural fields by salt contamination. Over the past couple of decades, tropical storms have shown elevated activity on the U.S. East Coast: 16 storms have pounded Chesapeake Bay since 1996, including Hurricane Fran (1996), Floyd (1999), Isabel (2003), Ernesto (2006), Irene and Lee (2011). This pattern is expected to continue and may increase due to global warming (Goldenberg et al. 2001; Emanuel 2005; Webster et al. 2005). Nor'easters can also cause extensive flooding, as demonstrated during Nov. 2009 when the surge heights approached those of Isabel in southern Bay.

With the projected increases in the frequency and severity of storms and accelerating sea level rise, Marylanders and their properties are facing ever-increasing

Fig. 6.1 **a** Satellite Isabel (2003) at landfall. Photos of flooded streets (**b**, **c**) in the Greater Washington D.C. Area

risks. Especially vulnerable are those on the Maryland's Eastern Shore: Talbot, Dorchester, Wicomico, Worcester and Somerset counties (Fig. 6.2). Slope is a primary variable controlling the magnitude and range of sea-level rise impact over time. In Maryland's Eastern Shore where elevation change may only be as much as one foot per mile, gradual submergence of a large geographic area is quite likely over time. Land inundation due to sea-level rise is already occurring along low-lying coastal areas in Dorchester and Somerset Counties (IAN 2008). Two to three feet of additional sea-level rise will result in a dramatic intensification of coastal flood events and submerge thousands of acres of tidal wetlands and low-lying lands. Sea-level rise and storm surges also pose a significant threat to resources and infrastructure in Maryland's coastal zone. As growth and development continues in coastal areas, these impacts are likely to escalate. Figure 6.2 provides a graphical illustration of low-lying land areas in Maryland that are likely be subject to coastal inundation and flooding over the next 100 years.

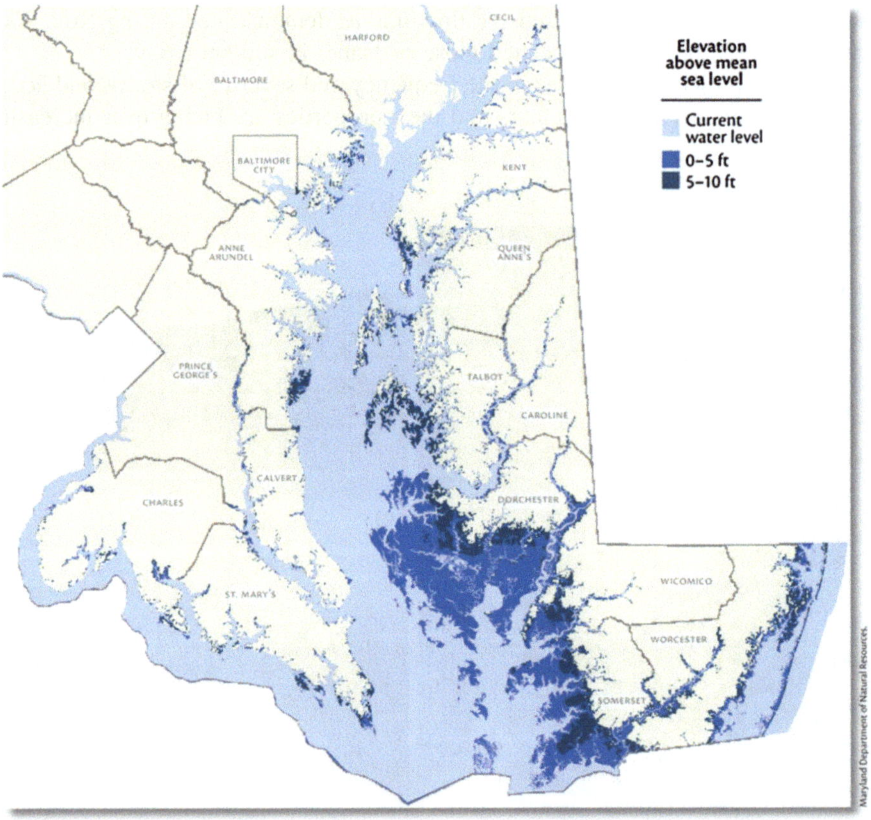

Fig. 6.2 Sea-level rise vulnerability in coastal areas of Maryland, calculated using LIDAR elevation data (from IAN 2008). Chesapeake Bay is shown as the *light blue area*

To address the coastal inundation issue confronting Maryland, we recently participated in the development of Chesapeake Inundation Prediction System (CIPS). CIPS is a collaborative effort involving meteorologists, oceanographers, hydrologists, economists and information technologists. We use a suite of atmospheric and hydrodynamic models to develop an ensemble forecast of storm surge and inundation with graphic visualization of model results including forecast uncertainty (Stamey et al. 2007). Advanced graphic software such as ArcGIS is employed to create high-resolution animations of inundation in flood-prone areas, which are used to quantify the economic impacts. CIPS provides an end-to-end, state-of-the-art, quantifiable storm surge and inundation forecast tool, and can be applied to other coastal regions in the U.S. and around the world.

6.3 Coupled Atmospheric-Hydrodynamic Models

Storm-surge models include atmospheric and oceanic submodels. The wind and pressure fields obtained from the atmospheric model are used to drive the hydrodynamic model for making storm-surge predictions. The atmospheric model ranges from simple parametric vortex models to mesoscale atmospheric forecasting models. Parametric surface winds are estimated either by assuming an idealized stationary, symmetric tropical cyclone with the observed path, surface pressure drop, and radius of maximum wind (Holland 1980; Peng et al. 2004), or by the planetary boundary layer model (Scheffner and Fitzpatrick 1997). Its operational utility has been limited by its sensitivity to errors in input parameters, such as the storm track, intensity and size (Rappaport et al. 2009). NOAA/Hurricane Research Division (HRD) has developed more accurate hurricane winds in real time (Powell et al. 1998; Houston et al. 1999); these winds are based on all available surface wind observations from buoys, coastal-marine automated observation platforms, ships, and other surface facilities. However, the HRD surface winds are only available prior to the hurricane's landfall.

Several high-resolution mesoscale atmospheric models have been used to make forecasts for tropical storms over coastal regions. The MM5 model is an earlier version of the mesoscale atmospheric model. A nested-grid (36/12/4 km) version of the MM5 has been used in real-time daily forecast mode at the University of Maryland (UMD) to provide hurricane prediction. Figure 6.3a–c show the MM5 forecast for the wind field during Hurricane Isabel. After making landfall at the Outer Banks in North Carolina, Isabel moved along a track to the west of Chesapeake Bay. The southeasterly winds in the right-front quadrant of the storm were the dominant winds over the Bay. The MM5 model predicted reasonably well the trajectory and intensity of Hurricane Isabel as well as the other meteorological fields (Li et al. 2006). The surface winds at a mid-Bay station appear to be slightly overpredicted, but the model accurately captured the shifts in the wind direction from the southwesterly to southeasterly during the passage of Isabel.

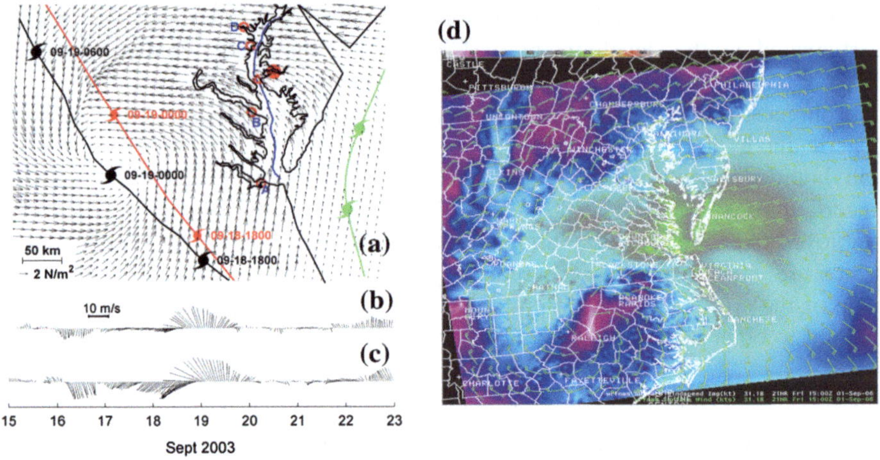

Fig. 6.3 **a** MM5-predicted surface wind stress over Chesapeake Bay during Hurricane Isabel (September 2003), superposed by observed (*red*) and predicted (*black*) Isabel's track and a hypothetical track for a storm moving on the ocean side of the Bay (*green*). *Red* open *circles* are used for tidal gauge stations (A-Hampton Road, VA; B-Lewisetta, MD; C-Annapolis, MD; D-Baltimore, MD), a *red* triangle for CBOS mid-Bay buoy and a red solid circle for a mid-Bay weather station. Comparison of observed (**b**, **c**) predicted horizontal wind vectors at the weather station. **d** WRF forecast of surface winds during Tropical Storm Ernesto

The Weather Research and Forecasting (WRF) model is a next-generation mesoscale numerical weather prediction system. It features multiple dynamical cores, a 3-dimensional variational (3DVAR) data assimilation system, and a software architecture allowing for computational parallelism and system extensibility. National Weather Service at Wakefield, Virginia runs a high-resolution (4 km) WRF model tailored for the Chesapeake Bay region (Fig. 6.3d). The model is able to resolve convection and produce detailed banding structures in tropical systems. The Regional Atmospheric Modeling System (RAMS) is another state-of-art mesoscale atmospheric forecasting model. Weatherflow Inc. employs a fine-resolution (2 km) RAMS to make operational forecasts for the Chesapeake Bay region. Global weather forecasting models such as Global Forecast System (GFS) and North American Mesoscale model (NAM) may be used to set up the initial and boundary conditions for the regional models.

Various 2D/3D hydrodynamic models have been developed and used to simulate storm surges, including the SLOSH (Sea, Lake, and Overland Surges from Hurricanes, Jelesnianski et al. 1992), ADCIRC (Luettich et al. 1992; Westerlink et al. 1992), ELCIRC (Stamey et al. 2007; Wang et al. 2008), SELFE (Shen and Gong 2009), CH3D (Sheng et al. 2010), Princeton Ocean Model (POM, Peng et al. 2004) and Finite Volume Coastal Ocean Model (FVCOM, Weisberg and Zheng 2008). These models use orthogonal curvilinear grids (SLOSH and POM), non-orthogonal curvilinear grids (CH3D), or unstructured triangular grids (ADCIRC and FVCOM). While a storm surge model usually simulates the wind-driven and pressure-induced surge and tide, wave-induced surge can be simulated by including the effects of waves on storm surge via one-way coupling or two-way coupling (e.g. Sheng et al. 2010) between a storm-surge model and a wave model such as SWAN. A few commonly used storm-surge models (ADCIRC, SELFE, SLOSH, FVCOM) are currently being compared in the SURA super-regional modeling testbed (http://testbed.sura.org).

Although unstructured models have the superior ability to fit complex coastlines easily, Li et al. (2006, 2007) and Zhong et al. (2010) used the structured-grid ROMS model to simulate storm surges in Chesapeake Bay. Regional Ocean Modeling System (ROMS) is a state-of-the-art regional ocean model (e.g. Shchepetkin and McWilliams 2005). It incorporates advanced modeling features such as high-order advection schemes, accurate pressure gradient algorithms, several subgrid-scale parameterizations, atmospheric, oceanic, and benthic boundary layers, biological modules and data assimilation (http://marine.rutgers.edu/po/). ROMS has found wide-ranging applications, including basin-scale ocean circulation (e.g. Haidvogel et al. 2000), shelf circulation (Marchesiello et al. 2003) and estuarine circulation (e.g. Warner et al. 2005; Li et al. 2005; Li and Zhong 2009; Zhong and Li 2006). Gridding techniques such as GRIDEN (Driscoll and Vavasis 1998; Zhong et al. 2010) and composite-domain method (Warner et al. 2010) provide reasonable approaches to refine grid resolutions in local regions for inundation predictions. The ROMS model was coupled to the MM5 model to predict the storm surge from Hurricane Isabel and the associated wind-driven currents. Figure 6.4a provides a 3D view of the sea-level distribution over the Bay at 0400 LST 19 September. The alignment of southeasterly winds with the long fetch of the lower Potomac River created the largest surge in Washington, DC, which reached 2.7 m above normal high tide. Sea levels in the northern Bay were also rising rapidly at this time. The observed temporal evolution of sea levels at 4 selected tidal stations was well captured by the model (Fig. 6.4b). The Root-Mean-Square (RMS) error averaged over 8 stations in the Bay is 0.13 m. The model's predictive skill as defined in Warner et al. (2005) has a high score of 0.96. It is worth noting that the storm surges reached 2.2 m at Baltimore and 2.0 m at Annapolis (Fig. 6.4b), causing flooding there.

To simulate overland inundation caused by storm surges, we have incorporated a simple wetting-and-drying scheme provided by ROMS. The formulation is based upon the concept of a 'critical depth' (D_{crit}) criterion (cf. Zhang et al. 2004; Oey 2005). As the model progresses, the total depth ($h + \eta$) is compared to D_{crit}. If $(h + \eta) < D_{crit}$, a 'flux blocking' algorithm is imposed to prevent transport out of that cell. Water can flow into any cell at any time, but the water cannot flow out if the total depth is less than D_{crit}. Cells become rewet if water flows back from adjacent cells. In our application, we choose $D_{crit} = 0.2$ m (Zhong et al. 2010). Recent model simulations suggest that storm-surge predictions are sensitive to bottom stress parameterization (Weisberg and Zheng 2008). Bottom stress is usually prescribed using a quadratic drag law. In 2D model, the drag coefficient C_D is prescribed as a constant or a depth-dependent function (cf. Daily and Harleman 1966; Luettich et al. 1992). The roughness height is related to sea bed types and

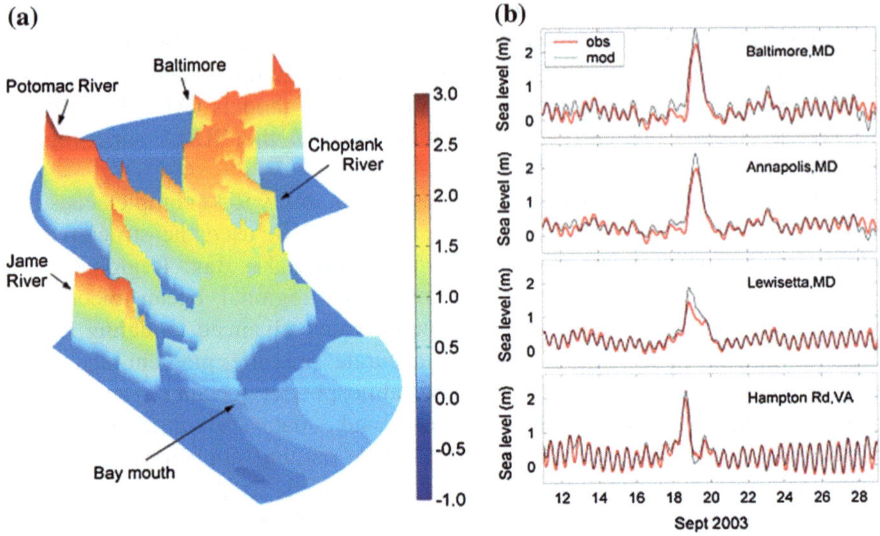

Fig. 6.4 ROMS validation for Hurricane Isabel: **a** 3D peak water levels at high storm surge; **b** observed (*red*) versus predicted (*black*) water levels at four tidal gauge stations

bed forms (Soulsby 1997). Different values may need to be used for intertidal zones and inundated shallow regions featuring salt marshes or other vegetative surfaces (Nicolle and Karptchev 2007).

6.4 Ensemble Forecasts

The magnitudes of storm surges are determined primarily by meteorological forcing, such as storm intensity, path, spatial and temporal scales, and topographic parameters such as the width and slope of continental shelf, geometry and character of local coastal and shelf features. For a semi-enclosed Bay such as Chesapeake Bay, Zhong et al. (2010) conducted a series of numerical sensitivity experiments for the storm surge generated by Hurricane Isabel. They found that small errors in the predicted hurricane parameters may lead to large errors in the storm surge prediction. Errors in the hurricane track and intensity mainly affect the model prediction on the surge height whereas errors in the translation speed change the prediction of not only the surge height, but also its arriving time and duration of high water. The surge height is more sensitive to the wind forcing in the upper Bay than in the lower Bay due to different response mechanisms in the two regions.

Ensemble forecasting has been shown to improve the weather forecasts and provide a means of conveying uncertainty. We have extended the ensemble forecasting technique to storm-surge predictions during the CIPS project. We use regional atmospheric models as well as global GFS and NAM models to generate an ensemble of atmospheric forcing fields: GFS, NAM, WRF-GFS, WRF-NAM, RAMS and MM5. The ROMS hydrodynamic model is forced by this ensemble of wind and pressure fields to produce an ensemble of storm surge forecasts.

We have tested the ensemble forecasting capability by comparing the model predictions of two historical storms (Hurricane Isabel in 2003 and Tropical Storm Ernesto in 2006) against observed sea levels. Figure 6.5 shows a comparison of sea level time series at several tidal stations between the observations and three model simulations: RAM, WRF-NAM (regional WRF forced with NAM boundary and initial conditions), WRF-GFS (regional WRF forced with GFS boundary and initial conditions). Because each atmospheric model produces slightly different forecasts for the wind and atmospheric pressure fields, the hydrodynamic model predicts a range of storm surge heights: some higher than the observed values while the others are lower. The forecasted surge heights fall into a narrow range at some stations (such as Cambridge, MD and Annapolis, MD) but spread over a wider range at other stations (such as CBBT, VA and Hampton, VA). We have also investigated how different models produce different predictions for overland inundations in a local region.

Figure 6.6 shows a comparison of the inundated area over the Eastern Shore of Maryland among six different models: MM5, RAM, WRF-NAM, WRF-GFS, GFS and NAM. We have included two coarse-resolution global weather forecasting models GFS and NAM for comparison. The inundated area is similar among

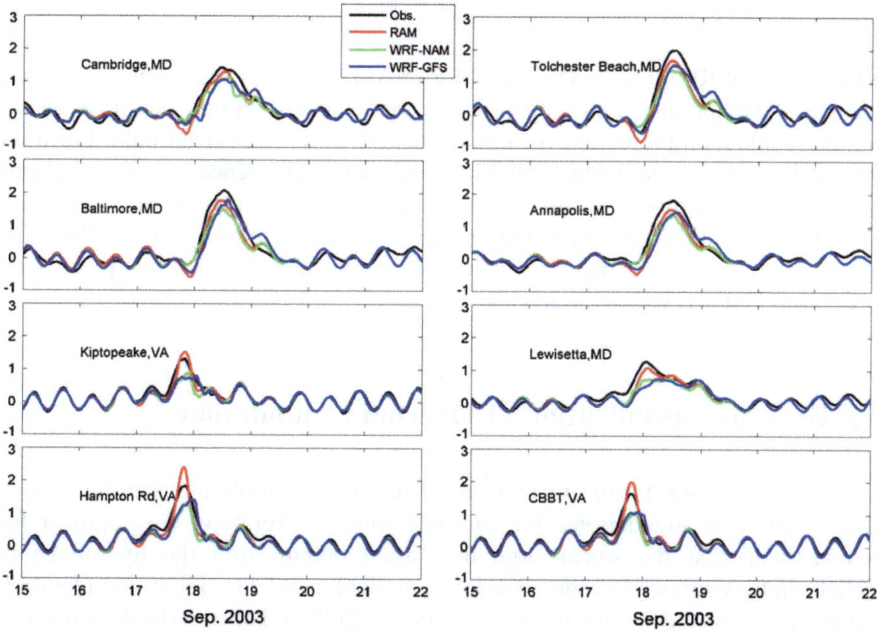

Fig. 6.5 Ensemble predictions of storm surges at tidal gauge stations during Hurricane Isabel

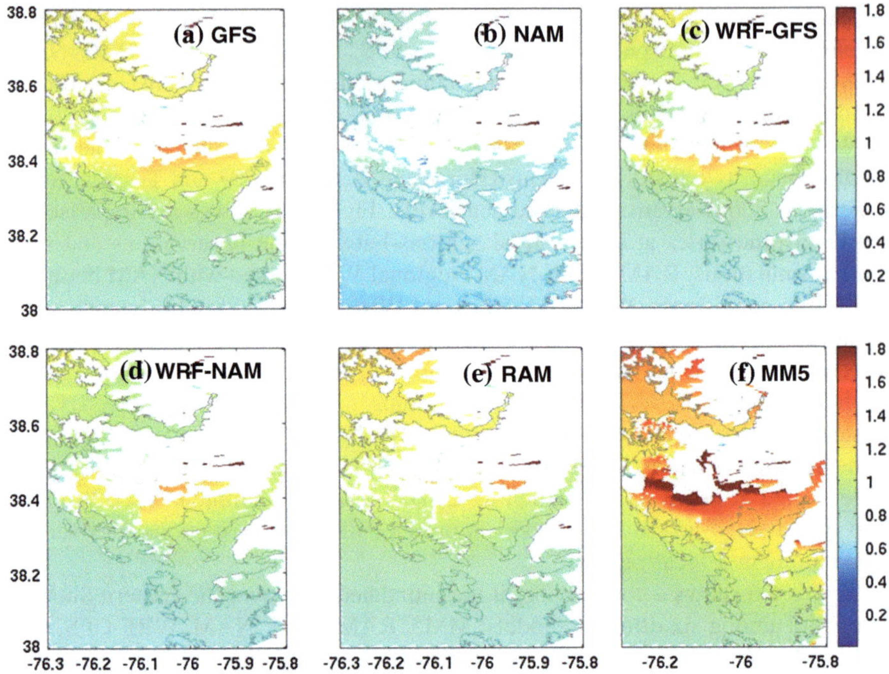

Fig. 6.6 Ensemble predictions of overland inundations over the Eastern Shore of Maryland during Isabel: **a** GFS; **b** NAM; **c** WRF-GFS; **d** WRF-NAM; **e** RAM; **f** MM5, all in unit of meters

WRF-NAM, WRF-GFS, RAM and GFS. NAM predicted least inundation area whereas MMS predicted the most. These model inter-comparisons have shown that the inundation prediction is highly sensitive to detailed wind field. For a flat area such as the Eastern Shore of Maryland, small differences in sea level predictions lead to substantial differences in the overland inundation. By conducting ensemble simulations, we are able to produce a range of predictions for the storm surges and overland inundations. Such information can help emergency managers to develop better response strategies.

6.5 Graphic Visualization of Overland Inundation

ArcGIS and Google Earth are two powerful software tools that can be used to display overland inundations due to storm surges. Prototype development has demonstrated that the storm-surge inundation output from the hydrodynamic model can be visualized to show the location, depth, and duration of inundation in static and animated products (Stamey et al. 2007). The overland inundation visualization was done at a spatial resolution of less than a city block ($\sim$50 m), vertical resolution of less than one foot ($\sim$30 cm or less), with a sequential time step of 1 h or less. Figure 6.7a shows the predicted inundation of Old Town Alexandria, Virginia during Hurricane Isabel.

GIS provides a convenient environment for viewing, querying and analyzing data, and facilitates coastal planning and emergency management. The water-level data from the hydrodynamic model are imported and overlaid on GIS raster and vector data layers. Specially, raw high-resolution LIDAR digital elevation data are processed onto a regular grid which is used as the basis for visualization. The water levels from the hydrodynamic model cells are then interpolated onto this grid. Inundation depths are obtained by subtracting the land elevation from the water level in each grid cell. The result is a wet/dry profile of inundation. To improve the accuracy of the profile, a layer representing buildings is used as a mask if one is available for the local jurisdiction. The resulting grid is then converted into vector polygons and stored in the geodatabase. The polygons in the geodatabase can be visualized by ArcGIS or be converted into Google Earth KML files.

Inundation polygon layers and desktop GIS software such as ArcGIS are suitable tools for GIS professionals or trained emergency managers. However, there is also a need for a simpler delivery vehicle for dissemination to a wider audience such as the general public. Google Earth provides such as a tool and is free of charge. User-generated layers can be displayed onto the Google Earth client by importing KML files. One can publish the PostGIS polygon objects as KML and convert timestamp fields in the PostGIS records into the corresponding Time Stamp elements in Google Earth. Hence one can show animated inundation in specific local regions using Google Earth. Google Earth also allows users to add and view 3D buildings, thus enabling 3D views of inundations over buildings and structures. We have used ArcGIS and Google Earth to visualize inundations over Cambridge, Dorchester County during Hurricane Isabel (Fig. 6.7b and c). Both still images and animations were generated to display the area extent, duration and water levels in the city.

6.6 Effects of Sea-Level Rise on Coastal Inundation

Coastal regions around Chesapeake Bay are not only affected by episodic storm surges but also affected by sea-level rise. Global sea level rose a rate of 1.6 ± 0.2 mm year^{-1} from 1961 to 2003 and 3.4 ± 0.1 mm year^{-1} from 1993 to 2007 (Domingues et al. 2008). Due to the greater rate of absolute sea level rise in the middle latitudes of the Northwest Atlantic Ocean (Church et al. 2008) and regional land subsidence associated with the post-glacial rebound (Nerem et al. 1998), the sea level rise over Chesapeake Bay is particularly large. Tide-gauge records in Chesapeake Bay reveal that sea levels increased by 3–4 mm per year

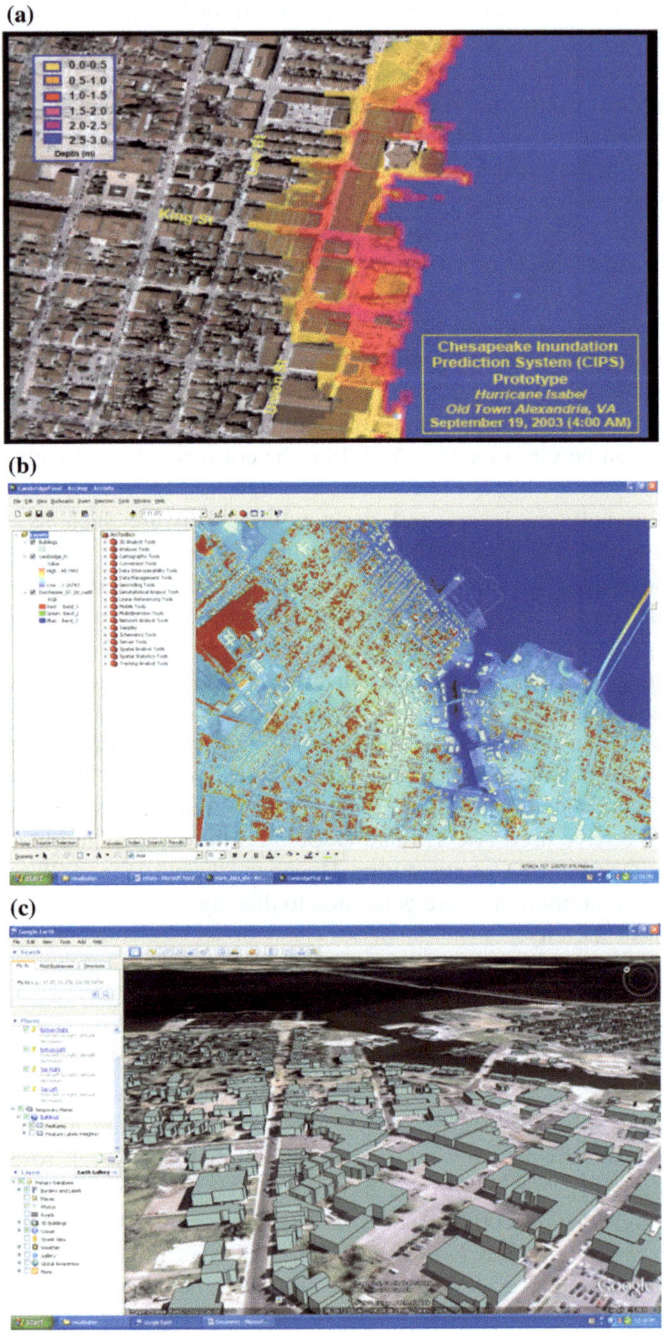

Fig. 6.7 Static GIS visualization of peak flooding in **a** Old Town Alexandria, Virginia (from Stamey et al. 2007) and **b** Cambridge, Maryland during Hurricane Isabel. **c** 3D visualization using Google Earth

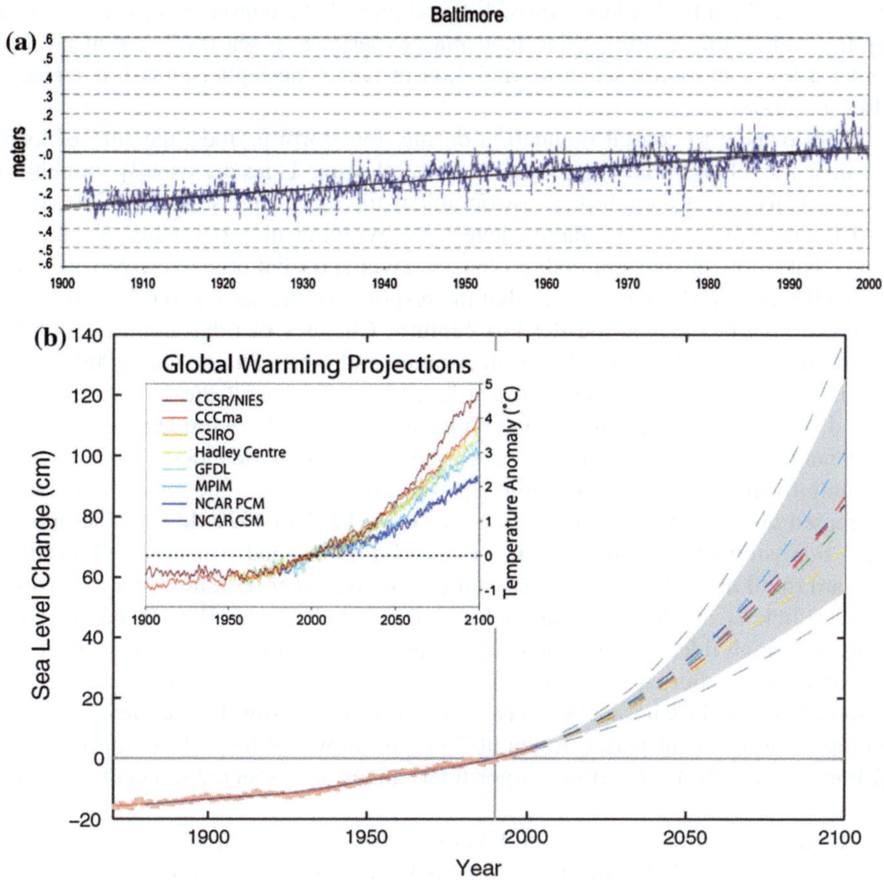

Fig. 6.8 **a** Time series of sea level at the tidal gauge station of Baltimore in the 20th century. **b** Global climate model predictions for the global-mean sea-level rise in the 21st century. The insert shows the model predictions for the global-mean temperature

over the 20th century (Zervas 2001), which was nearly twice that of the global average over the same time period (Fig. 6.8a). For example, the sea level at the tidal-gauge station in Baltimore, Maryland rose by over 30 cm in the 20th century.

Climate models predict that the rate of sea-level rise will accelerate in the 21st century (IPCC 2007; Meehl et al. 2007), with global-mean increases by 2,100 of 0.5–1.4 m, depending on greenhouse gas-emission scenarios and climate-model sensitivity (Fig. 6.8b, Rahmstorf 2007). Currently, sea level is rising at a rate near the upper end of the projections (Church et al. 2011). Moreover, it is recognized that a catastrophic melt of ice sheets could occur and lead to 6 m rise in global sea levels (Bindschadler 2008), but robust methods for quantifying it are not yet available. The Scientific and Technical Working Group of Maryland's Commission on Climate Change assessed the 2007 IPCC global sea-rise projections, along with regional land subsidence variables, and provided a conservative estimate that by the end of this century, Maryland may experience a sea-level rise of 0.83 m under a lower-emission scenario, and 1.04 m under the higher-emission scenario (Boesch 2008).

With respect to coastal inundation during non-storm periods, the sum of local mean sea level and tidal amplitude, is more relevant. Tides are usually thought of as stationary, even though their astronomical forcing changes slowly (Cartwright and Eden 1973), and tidal datum levels are evolving in geographically variable ways that imply changes in both global sea level and tidal processes. A new study by Müller et al. (2011) has shown that the response of the oceans to tidal forces has changed significantly during the last century. Changes of tidal amplitude and/or phase have taken place in both North Atlantic and North Pacific (Jay 2009). Flick et al. (2003) examined long-term sea-level records around the U.S. coasts and found significant changes in the tide range, either in the diurnal tide range or mean tide range. For example, at San Francisco, the diurnal tide range increased by 64 mm from 1900 to 1998, while at Wilmington, N.C., the mean tide range increased at a rate of 542 mm per century from 1935 to 1999. Inside Chesapeake Bay, tidal stations in its upper part, notably Baltimore and Annapolis, showed an upward trend in tidal range whereas stations on the lower bay showed a downward trend (Flick et al. 2003). Semi-enclosed Bays such as the Bay of Fundy have extremely high tides because their natural resonance periods are close to those for semidiurnal or diurnal tides (e.g. Garrett 1972). At the current sea level, the resonant period in Chesapeake Bay is about 48 h. Raising the sea level by 1 m shortens the resonant period to about 36 h and moves it toward the diurnal tides (Zhong et al. 2008). Therefore, larger tidal ranges are expected to occur at higher mean sea levels.

Recent studies suggest that sea-level rise may greatly amplify storm surges in shallow coastal regions. Smith et al. (2010) examined the potential impact of sea level rise on coastal surges in southeast Louisiana. In shallow wetland or wetland-fronted areas of moderate peak surges (2–3 m), however, the surge levels increase by as much as 1–3 m beyond the sea level rise. Their study highlights the fact that surge generation and propagation over shallow areas are nonlinear processes and sensitive to changes in the water depth. Other studies have shown strong nonlinear interactions between storm surges and tides. For example, Horsburgh and Wilson (2007) showed that larger surges are usually encountered around 3–5 h before the high tidal water in the North Sea. Therefore, the cumulative effects of sea level rise, tides and storm surges on coastal inundations are not simple linear additions. Strong nonlinear interactions between them can produce unexpectedly high water levels in shallow estuaries such as Chesapeake Bay. Future research is warranted to understand the nonlinear interactions between sea-level rise, tides and storm surges.

6.7 Conclusions and Discussions

Using the low-lying areas in Maryland, U.S.A. as an example of coastal regions prone to storm surges and coastal inundations, we have discussed how sea-level rise, changing tidal ranges and storm surges may lead to extremely high sea levels in a warming climate. We have also described the recent development of an end-to-end coastal inundation forecasting system that uses a suite of advanced regional atmospheric and hydrodynamic models to generate an ensemble forecast and employs the state-of-art software ArcGIS and Google Earth to produce high-resolution images and animations of overland inundations.

Many coastal regions in China are facing the same inundation risks due to sea level rise and storm surges. As a result of rapid urbanization, the Chinese coastal zone has become a region with highly developed economy and dense population. This coastal zone is inhabited by 42 % of the Chinese population and produces 51 % of the gross domestic product in China. For example, Shanghai, the largest city of China, lies in the Changjiang River Delta. During the period between 1978 and 2007, the mean sea level in Shanghai has risen 115 mm, as a result of global sea-level rise and local land subsidence (Xue et al. 2005). This large relative sea level rise, together with recent increases in the intensity and frequency of storms, has placed Shanghai at high risks of flooding and inundations. Computer models have been developed to evaluate these risks (Hu et al. 2007; Guo et al. 2009; Yin et al. 2011). Similarly, the southern Guangdong Province faces the high risks of typhoon-induced storm surges since much of the population inhabits low lying coastal regions. Each year over 3–4 named typhoons and tropical storms hit the coastal areas of Guangdong Province. With the continued urbanization, storm surges are a matter of great concern, particularly for the Pearl River Delta, where the physical geography (the southern part of the delta lies between −0.3 and 0.4 m relative to the sea level), and the urban development of the region render it extremely vulnerable (Zhang 2009; Zhang et al. 2011).

Given the same inundation issues facing the U.S. and Chinese coastal coasts, there is a great merit to conduct parallel investigations and develop collaborative projects between the American and Chinese scientists. Many numerical models of the atmosphere and ocean are open-source community models and can be applied to any coastal region around the world. By sharing the resources and experiences, we will all be better positioned to mitigate the impacts of storm surges and inundations in coastal regions.

Acknowledgments We thank Dr. Danling Tang for inviting us to attend the cross-strait workshop on typhoon impacts and crisis managements. We also thank Harry Wang, John Billet, Bill Boicourt, Kevin Seller, Liz Smith and others for their collaborations in developing the CIPS inundation prediction system. Michael Scott and Mick Tyndall's help with ArcGIS and Google Earth is greatly appreciated. This work was supported by grants from NOAA (NA07NOS4730214) and NSF (OCE-082543). This is UMCES contribution number 4731.

References

Bindschadler, R.: Why predicting West Antarctic ice sheet behavior is so hard: what we know, what we don't know and how we will find out. In: McCracken, M., Moore, F., Topping Jr, J.C. (eds.) The Likelihood and Character of Large and Disruptive Climate Change, pp. 75–80. EarthScan, London (2008)

Boesch, D.F. (ed.): Global Warming and the Free State: Comprehensive Assessment of Climate Change Impacts in Maryland. Report of the Scientific and Technical Working Group of the Maryland Commission on Climate Change. University of Maryland Center for Environmental Science, Cambridge, Maryland (2008)

Boicourt, W.C.: Physical response of Chesapeake Bay to hurricanes moving to the wrong side: refining the forecasts. In: Sellner, K.G. (ed.) Hurricane Isabel in Perspective, pp. 39–48. Chesapeake Research Consortium, CRC Publication 05-160, Edgewater, MD (2005)

Cartwright, D., Eden, A.C.: Corrected table of tidal harmonics. Geophys. J. Roy. Astron. Soc. **33**, 253–264 (1973)

Church, J.A., White, N.J., Aarup, T., Wilson, W.S., Woodworth, P.L., Domingues, C.M., Hunter, J.R., Lambeck, K.: Understanding global sea levels: past, present and future. Sustain. Sci. **3**, 9–22 (2008)

Church, J.A., Gregory, J.M., White, N.J., Platten, S.M., Mitrovica, J.X.: Understanding and projecting sea level rise. Oceanography **24**(2), 130–143 (2011)

Daily, J.W., Harleman, D.R.F.: Fluid Dynamics, pp. 297–298. Addison-Wesley Publishing Company Inc., Reading (1966)

Domingues, C.M., Church, J.A., White, N.J., Gleckler, P.J., Wijffels, S.E., Barker, P.M., Dunn, J.R.: Improved estimates of upper-ocean warming and multi-decadal sea-level rise. Nature **453**, 1090–1093 (2008)

Driscoll, T.A., Vavasis, S.A.: Numerical conformal mapping using cross-ratios and Delaunay triangulation. SIAM J. Sci. Comput. **19**(6), 1783–1803 (1998)

Emanuel, K.A.: Increasing destructiveness of tropical cyclones over the past 30 years. Nature **436**, 686–688 (2005)

Flick, R.E., Murray, J.F., Asce, L.: Trends in United States tidal datum statistics and tide range. J. Waterw. Port Coast. Ocean Eng. **129**, 155–164 (2003)

Garrett, C.: Tidal resonance in the Bay of Fundy and Gulf of Maine. Nature **238**, 441–443 (1972)

Goldenberg, S.B., Landsea, C.W., Mestas-Nunez, A.M., Gray, W.M.: The recent increase in Atlantic hurricane activity: causes and implications. Science **293**, 474–478 (2001)

Guo, Y., Zhang, J., Zhang, L., Shen, Y.: Computational investigation of typhoon-induced storm surge in Hangzhou Bay, China. Estuar. Coast. Shelf Sci. **85**, 530–536 (2009)

Haidvogel, D.B., Arango, H.G., Hedstrom, K., Beckmann, A., Malanotte-Rizzoli, P., Shchepetkin, A.F.: Model evaluation experiments in the North Atlantic Basin: simulations in nonlinear terrain-following coordinates. Dyn. Atmos. Oceans **32**, 239–281 (2000)

Holland, G.J.: An analytical model of the wind and pressure profiles in hurricanes. Mon. Weather Rev. **108**, 1212–1218 (1980)

Horsburgh, K.J., Wilson, C.: Tide-surge interaction and its role in the distribution of surge residuals in the North Sea. J. Geophys. Res. **112**, C08003 (2007). doi:10.1029/2006JC004033

Houston, S.H., Shaffer, W.A., Powelland, M.D., Chen, J.: Comparisons of HRD and SLOSH surface wind fields in hurricanes: implications for storm surge modeling. Weather Forecast. **14**, 671–686 (1999)

Hu, K., Ding, P., Ge, J.: Modeling of storm surge in the coastal waters of Yangtze Estuary and Hangzhou Bay. China J. Coast. Res. **1**(50), 527–533 (2007)

IAN.: Comprehensive Strategy for Reducing Maryland's Vulnerability to Climate Change Phase I: Sea-Level Rise and Coastal Storms. Report of the Maryland Commission on Climate Change Adaptation and Response Working Group, p. 38 (2008)

IPCC: Summary for policymakers. In: Solomon, S., Qin, D., Manning, M., Chen, Z., Marquis, M., Averyt, K.B., Tignor, M., Miller, H.L. (eds.) Climate Change 2007: The Physical Science Basis. Contribution of Working Group I to the Fourth Assessment Report of the Intergovernmental Panel on Climate Change. Cambridge University Press, Cambridge (2007)

Jay, D.A.: Evolution of tidal amplitudes in the eastern Pacific Ocean. Geophys. Res. Lett. **36**, L04603 (2009). doi:10.1029/2008GL036185

Jelesnianski, C.P., Chen, J., Shaffer, W.A.: SLOSH: Sea, Lake, and Overland Surges from Hurricanes. NOAA Technical Report NWS 48, p. 71 (1992)

Li, M., Zhong, L.: Flood-ebb and spring-neap variations of stratification, mixing and circulation in Chesapeake Bay. Cont. Shelf Res. **29**, 4–14 (2009)

Li, M., Zhong, L., Boicourt, W.C.: Simulations of Chesapeake Bay estuary: sensitivity to turbulence mixing parameterizations and comparison with observations. J. Geophys. Res. **110**, C12004 (2005). doi:10.1029/2004JC002585

Li, M., Zhong, L., Boicourt, W.C., Zhang, S., Zhang, D.-L.: Hurricane-induced storm surges, currents and destratification in a semi-enclosed bay. Geophys. Res. Lett. **33**, L02604 (2006). doi:10.1029/2005GL024992

Li, M., Zhong, L., Boicourt, W.C., Zhang, S., Zhang, D.-L.: Hurricane-induced destratification and restratification in a partially-mixed estuary. J. Mar. Res. **65**, 169–192 (2007)

Luettich, R.A., Westerlink, J.J., Scheffner, N.W.: ADCIRC: An Advanced Three dimensional Circulation Model for Shelves, Coasts and Estuaries. Report 1: Theory and Methodology of ADCIRC-2DDI and ADCIRC-3DL. Dredging Research Program Technical Report DRP-92-6, US Army Engineers Waterways Experiment Station, Vicksburg, Mississippi (1992)

Marchesiello, P., McWilliams, J.C., Shchepetkin, A.: Equilibrium structure and dynamics of the California current system. J. Phys. Oceanogr. **33**, 753–783 (2003)

Marra, J., Allen, T., Easterling, D., Fauver, S., Karl, T., Levinson, D., Marcy, D., Payne, J., Pietrafesa, L., Shea, E., Vaughan, L.: An integrating architecture for coastal inundation and erosion program planning and product development. Mar. Technol. Soc. J. **41**(1), 24–37 (2007)

Meehl, G.A., Stocker, T.F., Collins, W.D., Friedlingstein, P., Gaye, A.T., Gregory, J.M., Kitoh, A., Knutti, R., Murphy, J.M., Noda, A., Raper, S.C.B., Watterson, I.G., Weaver, A.J., Zhao, Z.-C.: Global climate projections. In: Solomon, S., Qin, D., Manning, M., Chen, Z., Marquis, M., Averyt, K.B., Tignor, M., Miller, H.L. (eds.) Climate Change 2007: The Physical Science Basis. Contribution of Working Group I to the Fourth Assessment Report of the Intergovernmental Panel on Climate Change. Cambridge University Press, Cambridge (2007)

Müller, M., Arbic, B.K., Mitrovica, J.X.: Secular trends in ocean tides: observations and model results. J. Geophys. Res. **116**, C05013 (2011). doi:10.1029/2010JC006387

Nerem, R., van Dam, T., Schenewerk, M.: Chesapeake Bay subsidence monitored as wetlands loss continues. Eos Trans. Am. Geophys. Union **79**, 149 (1998)

Nicolle, A., Karpytchev, M.: Evidence for spatially variable friction from tidal amplification and asymmetry in the Pertuis Breton (France). Cont. Shelf Res. **27**, 2346–2356 (2007)

Oey, L.-Y.: A wetting and drying scheme for POM. Ocean Model. **9**, 133–150 (2005)

Peng, M., Xie, L.L., Pietrafesa, L.J.: A numerical study of storm surge and inundation in the Croatan–Albemarle–Pamlico Estuary system. Estuar. Coast. Shelf Sci. **59**(1), 121–137 (2004)

Powell, M.D., Houston, S.H., Amat, L.R., Morisseau-Leroy, N.: The HRD real-time hurricane wind analysis system. J. Wind Eng. Ind. Aerodyn. **77**, 53–64 (1998)

Rahmstorf, S.: A semi-empirical approach to projecting future sea-level rise. Science **315**, 368–370 (2007)

Rappaport, E.N., Franklin, J.L., Avila, L.A.L.A.: Advances and challenges at the National Hurricane Center. Weather Forecast. **24**, 395–419 (2009)

Scheffner, N.W., Fitzpatrick, P.J.: Real-time predictions of surge propagation. In: Spaulding, M.L., Blumberg, A.F. (eds.) Estuarine and Coastal Modeling 1997, pp. 374–388. ASCE, 1998, Reston, VA (1997)

Shchepetkin, A.F., McWilliams, J.C.: The regional oceanic modeling system: a split-explicit, free-surface, topography-following-coordinate ocean model. Ocean Model. **9**, 347–404 (2005)

Shen, J., Gong, W.: Influence of model domain size, wind directions and Ekman transport on storm surge development inside the Chesapeake Bay: a case study of extratropical cyclone Ernesto. J. Mar. Syst. **75**, 198–215 (2009)
Shen, J., Gong, W., Wang, H.V.: Water level response to 1999 hurricane Floyd in the Chesapeake Bay. Cont. Shelf Res. **26**, 2484–2502 (2006a). doi:10.1016/j.csr.2006.07.02
Shen, J., Zhang, K., Xiao, C.C., Gong, W.: Improved prediction of storm surge inundation using a high-resolution unstructured grid model. J. Coast. Res. **22**(6), 1309–1319 (2006b). doi:10.2112/04-0288.1
Shen, J., Wang, H.V., Sisson, M., Gong, W.: Storm tide simulation in the Chesapeake Bay using an unstructured grid model. Estuar. Coast. Shelf Sci. **68**(1–2), 1–16 (2006c)
Sheng, Y.P., Zhang, Y., Paramygin, V.A.: Simulation of storm surge, wave, and coastal inundation in the Northeastern Gulf of Mexico region during Hurricane Ivan in 2004. Ocean Model. **35**, 314–331 (2010)
Smith, J.M., Cialone, M.A., Wamsley, T.V., McAlpin, T.O.: Potential impact of sea level rise on coastal surges in southeast Louisiana. Ocean Eng. **37**, 37–47 (2010)
Soulsby, R.L.: Dynamics of Marine Sands, p. 272. Thomas Telford Publishers, London (1997)
Stamey, B.H., Wang, V., Koterba, M.: Predicting the next storm surge flood. Sea Technol. **8**, 10–15 (2007)
Wang, C.-F., Wang, H.V., Kuo, A.Y.: Mass conservative transport scheme for the application of the ELCIRC model to water quality computation. J. Hydraul. Eng. **134**(8), 1166–1171 (2008)
Warner, J.C., Geyer, W., Lerczak, J.: Numerical modeling of an estuary: a comprehensive skill assessment. J. Geophys. Res. **110**(C5), C05001 (2005)
Warner, J.C., Geyer, W.R., Arango, H.G.: Using a composite grid approach in a complex coastal domain to estimate estuarine residence time. Comput. Geosci. **36**, 921–935 (2010)
Webster, P.J., Holland, H.J., Curry, J.A., Chang, H.R.-R.: Changes in tropical cyclone number, duration, and intensity in a warming environment. Science **309**, 1844–1846 (2005)
Weisberg, R.H., Zheng, L.: Hurricane storm surge simulations comparing three-dimensional with two-dimensional formulations based on an Ivan-like storm over the Tampa Bay, Florida region. J. Geophys. Res. **113**, C12001 (2008). doi:10.1029/2008JC005115
Westerlink, J.J., Luettich, R.A., Baptista, A.M., Scheffner, N.W., Farrar, P.P.: Tide and storm surge predictions using a finite element model. J. Hydraul. Eng. **118**, 1373–1390 (1992)
Xue, Y., Zhang, Y., Ye, S.J., Wu, J.C., Li, Q.F.: Land subsidence in China. Environ. Geol. **48**, 713–720 (2005)
Yin, J., Yin, Z., Hu, X., Xu, S., Wang, J., Li, Z., Zhong, H., Gan, F.: Multiple scenario analyses forecasting the confounding impacts of sea level rise and tides from storm induced coastal flooding in the city of Shanghai, China. Environ. Earth Sci. **63**, 407–414 (2011)
Zervas, C.: Sea Level Variations of the United States, 1854-1999, NOAA Technical Report NOS CO-OPS 36 (2001)
Zhang, J.: A vulnerability assessment of storm surge in Guangdong Province, China. Hum. Ecol. Risk Assess. **15**, 671–688 (2009)
Zhang, Y.-L., Baptista, A.M., Myers, E.P.: A cross-scale model for 3D baroclinic circulation in estuary-plume-shelf systems: I. Formulation and skill assessment. Cont. Shelf Res. **24**, 2187–2214 (2004)
Zhang, Q., Zhang, W., Chen, Y.D., Jiang, T.: Flood, draught and typhoon disasters during the last half-century in the Guangdong province, China. Nat. Hazards **57**, 267–278 (2011)
Zhong, L., Li, M.: Tidal energy fluxes and dissipation in the Chesapeake Bay. Cont. Shelf Res. **26**, 752–770 (2006)
Zhong, L., Li, M., Foreman, M.G.G.: Resonance and sea level variability in Chesapeake Bay. Cont. Shelf Res. **28**, 2565–2573 (2008)
Zhong, L., Li, M., Zhang, D.-L.: How do uncertainties in hurricane model forecasts affect storm surge predictions in a semi-enclosed bay? Estuar. Coast. Shelf Sci. **90**, 61–72 (2010)

Chapter 7
Influence of Tropical Storms in the Northern Indian Ocean on Dust Entrainment and Long-Range Transport

V. Ramaswamy

Abstract Ninety five tropical cyclonic events (tropical storms, depressions and cyclones) between 2001 and 2010 were studied to determine their impact on dust outbreaks and long-range transport over the northern Indian Ocean and south Asia. In addition to the winter and summer Shamal Winds, tropical cyclones are an important mechanism of dust entrainment and transport of dust in this region. Elevated dust levels were observed in the northern Arabian Sea during most tropical cyclone events. During the study period, fifteen tropical cyclones migrated close to the dust source areas leading to major dust storms. Anti-clockwise winds associated with these storms were observed to entrain dust and transport it mostly towards the west or south-westerly direction. Tropical cyclones and storms, located further away from dust source areas, significantly alter the dispersal pathways of dust plumes raised by other mechanisms. The Northern Bay of Bengal cyclone events are shown to aid advection of dust plumes from southwest Asia and Thar Desert over highly populated regions of the Indian Subcontinent. Tropical cyclones also play an important role in dispersal of fine-mode aerosols over South Asia and formation of complex aerosol-dust mixtures.

Keywords Tropical cyclones · Mineral dust · Northern Indian Ocean · Aerosols · Dust storms

7.1 Introduction

Tropical cyclones and accompanying strong winds and heavy rainfall cause extensive damage along coastal regions. They normally originate close to the equator where sea surface temperatures are above 26.5 °C and move westwards and eventually towards higher latitudes. As the cyclones migrate pole-wards

---

V. Ramaswamy (✉)
National Institute of Oceanography, Dona Paula, Goa 403004, India
e-mail: rams@nio.org; rams.goa@gmail.com

D. L. Tang and G. J. Sui (eds.), *Typhoon Impact and Crisis Management*,
Advances in Natural and Technological Hazards Research 40,
DOI: 10.1007/978-3-642-40695-9_7,

towards the semi-arid regions and deserts located near the Horse-Latitudes, the cyclonic winds can potentially cause major dust outbreaks. However, there are not many studies directly linking tropical cyclones to dust storms. Study of this linkage is important as mineral dust together with anthropogenic aerosols strongly perturb the radiation energy balance of the earth-atmosphere system, and have significant impacts on various terrestrial and marine ecosystems (Arimoto 2001; McTanish and Strong 2007; Ramanathan et al. 2001a; Rastogi and Sarin 2005; Tegen et al. 1996). The nutrients and toxins associated with dust can have a significant impact on ocean chemistry and primary productivity and its capacity to absorb atmospheric carbon dioxide (Duce et al. 1991; Jickells et al. 2005). Furthermore, the intensity and frequency of tropical cyclones are expected to rise in the northern Indian Ocean (Muni Krishna 2009) and for the first time in the Arabian Sea a super-cyclone, Gonu, was recorded in 2007.

Because of the northerly migration of the ITCZ during summer, much of India and Bay of Bengal receives copious rain during the SW monsoon period between June and September. The winter NE monsoon (December–February) is characterized by dry light winds. The Thar Desert in NW India and region west of it, including SW Asia, the Middle-East and Saharan Africa, receives scanty rain and constitutes the larges contiguous source of dust in the world (Goudie and Middleton 2006). Dust from these regions is transported over Middle-East and the northern Indian Ocean mainly by the so called "forty day" summer Shamal Winds and the numerous 3–5 day winter Shamal Winds (Ackerman and Cox 1988, 1989). In addition, convective events and local winds like Aandhi, Loo, Khamsin, etc. cause dust storms during spring-summer (Middleton 1986). The northern Indian Ocean and surrounding landmass is affected by several tropical cyclones which occur mainly before or after the summer monsoon. Though it is known that cyclone winds can cause dust storms, there are not many specific studies raise showing effect of tropical cyclones on dust entrainment and transport. One such study is the effect of the severe tropical cyclone Sidr in 2008 on dust storms in the Thar region (Badarinath et al. 2009). In the present study, the importance of episodic events like tropical cyclones and storms on uplift and dispersal of aerosols in the northern Indian Ocean and surrounding countries between 2001 and 2010 is discussed.

7.2 Data and Methods

7.2.1 Study Area

The entire northern Indian Ocean comes under the influence of the seasonally reversing monsoons. Heating of the Indian subcontinent and the Tibetan Plateau during spring-summer initiates a monsoonal circulation (Das 1968). The Inter-tropical convergence zone (ITCZ) which lies around 10° south latitude during January, migrates northward and reaches a maximum northerly extent in July covering almost the entire Arabian Sea and the Indian subcontinent (Fig. 7.1).

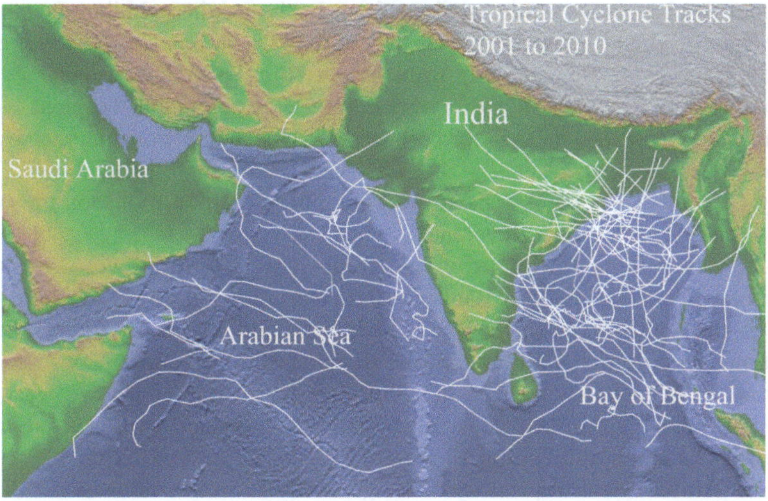

Fig. 7.1 Tracks of tropical cyclones and storms over the northern Indian Ocean between 2001 and 2010

Depressions and tropical cyclones often form in the vicinity of the monsoon trough. The development of tropical cyclones depends on SST's above 26.5 °C, and low wind shear in the upper troposphere. Since a minimum of Coriolis force is required, tropical cyclones do not form over the equator. They also require pre-existing disturbances in atmospheric circulation for initiation. During summer, the sub-tropical jet stream over north India disintegrates and reforms north of the Himalayas over central Asia and around this time a tropical easterly jet stream (TEJ) builds up over peninsular India extending up to Africa. The area south of the ITCZ has winds blowing predominantly from the ocean with consequent high precipitation. During spring-summer a quasi-permanent high is set up at the head of Saudi Arabia and eastern Mediterranean Sea. The strong pressure gradient between the quasi-permanent Arabian High and Asian Low sets up persistent northerlies in the Persian Gulf Area with strong dust laden winds, called Shamal, between May and end of July (Goudie and Middleton 2006). The intensity of the Shamal wind is enhanced by the presence of the Zagros Mountains.

The 3–5 days winter-Shamal Winds are episodic events due to passage of cold fronts between November and March. During this period, the polar jet migrates to lower latitudes and its proximity to the sub-tropical jet enhances disturbances in the Persian Gulf region. Dust storms can occur in the front and rear of the cold fronts. Dust events are seen mainly over Iraq, Iran and Afghanistan region with not much activity being recorded over the Arabian Peninsula. The SW Asian region including Baluchistan, Afghanistan, Iran and Iraq, are characterized by mountain chains and endorheic lakes which derive their water from snowmelt and occasional rain. Winds, intensified by mountain gaps, blowing over dry lake beds and deserts make this region very prone to dust storms. The dust is normally dispersed by the prevailing weak NE monsoon winds.

About 4–12 storms and depressions form in the northern Indian Ocean every year. They originate mostly between 5°N and 15°N latitude in the south eastern part of the Bay of Bengal and Arabian Sea (Fig. 7.1). Few tropical cyclones originating in the west Pacific cross over into the Bay of Bengal. Tropical cyclones of hurricane intensity in the northern Indian Ocean are conspicuous by their absence over areas overlaid by the TEJ because of high vertical wind shear. Though a number of depressions are formed during the SW monsoon they do not develop into severe tropical cyclones.

In the Bay of Bengal, winter cyclonic events move west-northwestwards and hit the Indian Peninsula and Sri Lanka coasts. During April–May, cyclonic events form in the south and central Bay which move initially west, then north and sometimes recurve to the NE and hit the Bangladesh or Myanmar Coast. After the TEJ has set, depressions form in the central or North Bay and move west-northwestwards and hit the east coast of India. Post-monsoon cyclones are mainly formed in the south and central Bay and hit the east coast of India.

In the Arabian Sea, tropical cyclones generally form in southeast Arabian Sea and adjoining central Arabian Sea in the months of May and between October and December. Some of the cyclones that originate in the Bay of Bengal travel across the Indian Peninsula and emerge in the Arabian Sea as low pressure areas. They frequently intensify into cyclonic storms. Most of the storms in Arabian Sea move in west-northwesterly direction towards Arabian Coast in the month of May and in a northerly direction towards India–Pakistan Coast in the month of June. In other months, they generally move initially northwest then north and finally recurve towards northeast affecting the north-eastern part of the Arabian Sea. A few tropical cyclones also move west north westwards towards the Arabian coast.

7.2.2 Classification of Tropical Cyclones

Tropical cyclonic events in the northern Indian Ocean are classified as depressions, storms or cyclones depending on the pressure deficiency and associated wind speeds. A seven tier classification adopted by the Indian Meteorological Department (IMD) is given in Table 7.1 (Anon 2012). The size of a tropical cyclone over northern Indian Ocean varies from 50 to 100 km radius to nearly 2,000 km with an

Table 7.1 Classification of tropical cyclones in the northern Indian Ocean (Anon 2012) adopted by the Indian Meteorological Organisation (IMD) and used in this study

System	Pressure deficiency hPa	Associated wind speed Kmph
Low pressure area	1.0	<32 (8.8)
Depression (D)	1.0–3.0	32–50 (8.8–13.9)
Deep depression (DD)	3.0–4.5	51–59 (13.9–16.4)
Cyclonic storm (CS)	4.5–8.5	60–90 (16.4–25)
Severe cyclonic storm (SCS)	8.5–15.5	90–119 (25.0–33.0)
Very severe cyclonic storm (VSCS)	15.5–65.6	119–220 (33.0–61.1)
Super cyclonic storm	>65.6	>220 (61.1)

average of 300–600 km. In the northern Indian Ocean tropical cyclones progress along the sea surface at a rate of 300–500 km a day. The lifetime of a tropical cyclone in the northern Indian Ocean is normally between 3 and 5 days though a few can last up to 10 days (Table 7.2).

7.2.3 Dust Source Areas

The largest contiguous dust source region in the world is the region that extends from NE Africa through the Sahara, Arabian and SW Asia up to the Thar Desert in India (Prospero et al. 2002). The dust from the Sahara, west of the Bodele depression in Chad, is transported predominantly towards the west across the Atlantic towards the Caribbean Islands or northwards across the Mediterranean towards Europe (Prospero 1999; Pye 1987). Amongst African sources only dust from the Nubian Desert and Horn of Africa is transported eastward. The Middle-East and SW Asian dust is transported mainly towards the east (Badarinath et al. 2010). It is seen that the disturbed soils of Iraq, SW Asia and south Asia are major contributors of dust in recent years (Tegen and Fung 1995; Givehchi et al. 2013). The principal dust sources contributing to sediments in the Arabian Sea and region eastward can be classified into six regions.

1. Syrian Desert (Al Jazirah) including Mesopotamia and An-Nafud province which includes the flood plains of the Tigris-Euphrates river Shat-al-Arab and Kuzestan plains of Iran. The desert pavements of the Ad Dahana in Saudi Arabia.
2. The Rub Al Khali, Oman depression and the coastal plains of Saudi Arabian Peninsula.
3. SW Asia which includes the endorheic basins of, Sistan e Baluchistan (Helmand Basin), Mashkel Basin, Jaz Muraian Basin, Dast e Kavir, Lut Basin, Ras Koh, Kerman, and coastal Makaran range. Smaller contributions are from the coastal plains bordering the Persian Gulf, with minor contributions from north Afghanistan and semi-arid regions of Central Asia.
4. Thar Desert including the Thal and Cholistan regions, the Indus and Punjab plains.
5. Nubian and East Libyan deserts, the coastal plains adjoining the Red Sea including the Tokar delta.
6. Horn of Africa including Somalia.

7.2.4 Tropical Storms and Cyclone Data

Data on number of cyclones and storms and related parameters such as maximum wind speed, minimum pressure, tropical cyclone tracks etc. were collected from a number of sources including the Indian Meteorology Department (IMD) reports (http://www.imd.gov.in/section/nhac/dynamic/cyclone.htm), and archives of cyclone and storms (http://weather.unisys.com/hurricane) and are summarized in

Table 7.2 Details of ninety five cyclone events (tropical cyclones, storms and depressions) in the northern Indian Ocean between January 2001 and December 2010 and their effect on entrainement and dispersal of mineral dust and aerosols

S. No	Start of event	End of event	Duration of event	Number and name of event	Intensity of event	Remarks
			Days	Popular name in brackets		
1	7-Dec-2010	8-Dec-2010	1	8B	DD	Dust Plume from Makaran to Oman driven by counter-clockwise winds Fine-mode aerosols from India-Pakistan pushed towards Western Arabian Sea
2	2-Nov-2010	12-Nov-2010	10	7B (Jal)	SCS	Fine-mode aerosols from Indus plains pushed towards Western Arabian Sea
3	19-Oct-2010	23-Oct-2010	4	6B (Gir)	VSCS	Fine-mode aerosols from India-Pakistan pushed towards Western Arabian Sea. Dense Haze over Gangetic plains being dispersed
4	13-Oct-2010	16-Oct-2010	3	5B	DD	Dust plumes observed along Makaran Coast. Haze driven from Indian subcontinent towards western Arabian Sea
5	7-Oct-2010	9-Oct-2010	2	4B	D	Dust plumes in Persian gulf, Gulf of Oman and northern Arabian Sea
6	30-May-2010	7-Jun-2010	8	3A (Phet)	VSCS	Cyclone Phet formed in the central Arabian Sea, moved towards Oman and caused a dust event in the Persian Gulf region, then veered towards NE Arabian Sea causing a major dust event in the Thar region
7	19-May-2010	23-May-2010	4	2A (Bandu)	CS	Two cyclones Bandu and Laila formed in the western Arabian Sea and Western Bay of Bengal simultaneously. Major dust events in the deserts adjacent to Red Sea and Persian Gulf. Dust over northern Arabian Sea
8	17-May-2010	21-May-2010	4	1B (Laila)	SCS	Same as above
9	10-Dec-2009	15-Dec-2009	5	8B (Ward)	CS	Counter-clockwise winds of cyclone Ward off Sri Lanka pushes fine-mode particles from Indo-Gangetic plains towards the western Arabian Sea
10	9-Nov-2009	11-Nov-2009	2	7A (Phyan)	CS	Cyclone Phyan in the eastern Arabian Sea pushes fine-mode particles from the Indus Plains and dust from Makaran pushed towards Oman

(continued)

Table 7.2 (continued)

S. No	Start of event	End of event	Duration of event	Number and name of event	Intensity of event	Remarks
			Days	Popular name in brackets		
11	5-Sep-2009	7-Sep-2009	2	6B	DD	Cyclone 6B strikes Bangladesh towards end of SW monsoon. Dust from Afghanistan Margo Desert pushed towards northern Arabian Sea
12	20-Jul-2009	21-Jul-2009	1	5B	DD	Dust from Gulf of Aden, Persian Gulf and Red Sea countries especially Tokar delta traceable to eastern Arabian Sea. Effect of Shamal winds enhanced due to cyclone 5B in the Bay of Bengal
13	25-Jun-2009	25-Jun-2009	1	4A	D	Counter clockwise winds of cyclone 4A in NW Arabian Sea pushing dust from Makaran towards Gulf of Aden
14	23-Jun-2009	24-Jun-2009	1	3A	D	Effects of this cyclone in Bay of Bengal cannot be distinguished from those of 4A in Arabian Sea
15	23-May-2009	26-May-2009	3	2B (Aila)	SCS	Cyclone in northern BOB causing a major dust event in Thar region with a dust plume trailing eastward down the Gangetic plains. Smaller dust events along the Makaran coast contributing to dust over the northern Arabian Sea
16	14-Apr-2009	17-Apr-2009	3	1B (Bijli)	CS	Cyclone in northern BOB pulling dust plumes from the Makaran coast towards the eastern Arabian Sea. Dust storms initiated due to frontal events in the Mesopotamian and Makaran region. Dust storms also seen in the Thar region
17	4-Dec-2008	7-Dec-2008	3	10B	DD	Counter-clockwise winds of Cyclone 10B off Sri Lanka pushes fine-mode particles from Indo-Gangetic plains towards western Arabian Sea
18	25-Nov–2008	27-Nov–2008	2	9B (Nisha)	CS	Cyclone in the SW Bay of Bengal pushes dust from Indian Subcontinent towards central Arabian Sea. Fine-Aerosols mix with dust from Makran Coast
19	13-Nov-2008	16-Nov-2008	3	8B (Khaimuk)	CS	Numerous dust plumes from Makaran coast being pulled southward towards central Arabian Sea and mixing with fine-mode aerosols from Indian subcontinent

(continued)

Table 7.2 (continued)

S. No	Start of event	End of event	Duration of event	Number and name of event	Intensity of event	Remarks
			Days	Popular name in brackets		
20	25-Oct-2008	27-Oct-2008	2	7B (Rashmi)	CS	Dust plumes from Makaran coast being pulled towards northern Arabian Sea and mixing with fine-mode aerosols from Indian subcontinent
21	19-Oct-2008	22-Oct-2008	3	6A	DD	Cyclone in Gulf of Aden creating dust storms in the Persian Gulf and Red Sea region
22	15-Sep-2008	19-Sep-2008	4	5B	DD	SW monsoon still active so major dust events seen in Mesopotamian and SW Asian region. Cyclone 5B in Bay of Bengal together with a disturbance in NW Arabian Sea leading to huge dust plume in the Persian Gulf being pulled into the Gulf of Oman and northern Arabian Sea
23	9-Aug-2008	9-Aug-2008	1	4B	D	Cyclone in the Bay of Bengal causing dust to be pulled right across the Arabian Sea towards Peninsular India
24	16-Jun-2008	18-Jun-2008	2	3B	D	Cyclone in the Bay of Bengal causing dust to be pulled right across the Arabian Sea towards Peninsular India and up to the western Bay of Bengal
25	5-Jun-2008	6-Jun-2008	1	2A	D	Cyclone 2A off Oman deflected huge dust plumes from Mesopotamian and Arabia peninsula towards Gulf of Aden and western Arabian Sea. Dust plumes also seen in Red Sea
26	27-Apr-2008	3-May-2008	6	1B (Nargis)	VSCS	Large dust event is Persian Gulf region due to Shamal event. The dust plumes pulled by cyclone Nargis across the northern Arabian Sea and over India. Dust events also seen over Thar region
27	11-Nov-2007	16-Nov-2007	5	12B (Sidr)	VSCS	Dust from Central and SW Asia Thar dust and fine-mode aerosols pulled down the Gangetic plains
28	27-Oct-2007	2-Nov-2007	6	11A	DD	Cyclone 11A along the Oman coast raises dust along the Red Sea coast and pulls Makaran Dust and Indus Haze towards northern Arabian Sea
29	27-Sep-2007	29-Sep-2007	2	10B	D	Not much effect because of stronger depression in Arabian Sea

(continued)

Table 7.2 (continued)

S. No	Start of event	End of event	Duration of event	Number and name of event	Intensity of event	Remarks
			Days	Popular name in brackets		
30	21-Sep-2007	24-Sep-2007	3	9B	D	Event 9B in the Bay of Bengal and a smaller depression in the north-eastern Arabian Sea occurring simultaneously. Persian Gulf and SW Asian dust pulled towards northern Arabian Sea
31	5-Aug-2007	7-Aug-2007	2	8B	DD	Dust entrained by Shamal winds over Persian Gulf and SW Asia and Red Sea pulled across the Arabian Sea
32	4-Jul-2007	9-Jul-2007	5	7B	DD	Cyclone in Bay of Bengal pulls dust across the Arabian Sea to Peninsular Indian and eastern Bay of Bengal
33	28-Jun-2007	30-Jun-2007	2	6B	DD	Cyclonic storm over the central Bay of Bengal Red pulls dust from Middle-East and SW Asia across the Arabian Sea
34	25-Jun-2007	26-Jun-2007	1	5A (Yemyin)	CS	Cyclone over Bay of Bengal crosses India and moves over to the northern Arabian Sea causing major dust storms over Persian Gulf
35	21-Jun-2007	23-Jun-2007	2	4B	DD	Dust in Red Sea, Gulf of Aden and Persian Gulf traceable across the central Arabian Sea
36	1-Jun-2007	7-Jun-2007	6	3A (Gonu)	VSCS	Super-Cyclone Gonu created major dust events in Middle-East and SW Asia
37	13-May-2007	14-May-2007	1	2B (Akash)	CS	Dust over northern Arabian Sea. Dust over Thar Desert pulled down Gangetic plains
38	3-May-2007	4-May-2007	1	1B	D	Minor dust events in northern Arabian Sea and Indo-Gangetic plains
39	29-Oct-2006	30-Oct-2006	1	12B (Ogni)	CS	Minor increase in dust over northern Arabian Sea from Makaran Coast
40	28-Sep-2006	30-Sep-2006	2	11B	D	Not much effect seen due to this cyclone as it occurred towards end of the SW monsoon
41	21-Sep-2006	24-Sep-2006	3	9B	D	Effect of this cyclone could not be ascertained due to stronger cyclone in Arabian Sea
42	21-Sep-2006	24-Sep-2006	3	10A (Mukda)	SCS	Major dust storms due to this cyclone in SW Asia with dust blown towards SW direction due to counter-clockwise winds

(continued)

Table 7.2 (continued)

S. No	Start of event	End of event	Duration of event	Number and name of event	Intensity of event	Remarks
			Days	Popular name in brackets		
43	3-Sep-2006	4-Sep-2006	1	8B	D	Effect of this cyclone could not be ascertained due to stronger cyclone in Arabian Sea
44	29-Aug-2006	1-Sep-2006	3	7B	D	A frontal system passes over SW Asia creating dust events. Dust plumes pulled eastwards by cyclone towards northern Arabian Sea and India
45	16-Aug-2006	18-Aug-2006	2	6B	D	High AOD over western Arabian Sea and Gulf of Oman. Extensive cloud cover western and central Arabian Sea prevents dust from being transported further eastward
46	12-Aug-2006	13-Aug-2006	1	5B	D	Two low pressure centers in Bay of Bengal and Arabian Sea. Dust from Middle-East and SW Asia pulled towards eastern Arabian Sea
47	2-Aug-2006	5-Aug-2006	3	4B	DD	Large dust plumes seen over northern, and central Arabian Sea, Red Sea and Persian Gulf
48	2-Jul-2006	5-Jul-2006	3	3B	DD	Large dust events seen in Red Sea and Persian Gulf due to Shamal winds. The deep depression in the Bay helps in transport of dust from Gulf of Aden to central Arabian Sea
49	25-Apr-2006	29-Apr-2006	4	2B (Mala)	VSCS	The severe tropical cyclone pulls down dust from the Thar region down the Gangetic plains and central India
50	13-Jan-2006	14-Jan-2006	1	1A	DD	Deep depression in the SE Arabian Sea blows fine-mode aerosols towards central Arabian Sea
51	15-Dec-2005	22-Dec-2005	7	12B	DD	Winter cyclone near Sri Lanka pushes aerosols from Indio-Gangetic plain towards northern Arabian Sea
52	6-Dec-2005	10-Dec-2005	4	11B (Fanoos)	CS	Winter cyclone near Sri Lanka pushes aerosols from Indus plains towards northern Arabian Sea

(continued)

Table 7.2 (continued)

S. No	Start of event	End of event	Duration of event	Number and name of event	Intensity of event	Remarks
			Days	Popular name in brackets		
53	28-Nov-2005	2-Dec-2005	4	10B (Baaz)	CS	Winter tropical storm in southern Bay of Bengal pushes aerosols from Indo-Gangetic plain towards northern Arabian Sea
54	20-Nov-2005	22-Nov-2005	2	9B	D	Winter depression in southern Bay of Bengal pushes aerosols from Indo-Gangetic plain towards northern Arabian Sea
55	26-Oct-2005	29-Oct-2005	3	8B	DD	Winter depression in southern Bay of Bengal pushes aerosols from Indo-Gangetic plain and minor dust from Makaran coast towards northern Arabian Sea
56	2-Oct-2005	3-Oct-2005	1	Name 3B	D	Depression in northern Bay disperses fine-mode aerosols over north India
57	17-Sep-2005	21-Sep-2005	4	7B (Pyarr)	CS	Cyclonic storm in Bay of Bengal pulls dust from Makaran coast to northern Arabian Sea and from Red sea and PG towards western Arabian Sea
58	14-Sep-2005	16-Sep-2005	2	6A	D	Two cyclones over Arabian Sea and Bay of Bengal affects dust plumes over Persian Gulf and Makaran coast
59	12-Sep-2005	16-Sep-2005	4	5B	D	Same as above
60	29-Jul-2005	31-Jul-2005	2	4B	DD	Deep depression over central Bay pulls dust from Saudi Arabian Peninsula across Arabian Sea into peninsular India
61	27-Jun-2005	5-Jul-2005	8	3B	D	Depression over central Bay pulls dust raised by Shamal winds over Saudi Arabia, Red Sea Area and SW Asia across Arabian Sea into peninsular India
62	21-Jun-2005	22-Jun-2005	1	2A	D	Depression over northeastern Arabian Sea causes a dust events over Makaran coast and also pulls dust from Persian Gulf into the northern Arabian Sea
63	13-Jan-2005	17-Jan-2005	4	1B (Hibaru)	CS	Winter storm SE of Sri Lanka pushes fine-mode aerosols from India to eastern Arabian Sea

(continued)

Table 7.2 (continued)

S. No	Start of event	End of event	Duration of event	Number and name of event	Intensity of event	Remarks
			Days	Popular name in brackets		
64	8-Jan-2005	10-Jan-2005	2	Name 01B	D	Western disturbance creates minor dust event in Makaran coast. Winter depression SE of Sri Lanka pulls the dust towards central Arabian Sea
65	29-Nov-2004	2-Dec-2004	3	10A (Agni)	SCS	A winter frontal system creates a major dust event over SW Asia and Thar region. The counter clockwise winds associated with cyclone Agni pushes the dust towards northern Arabian Sea and Oman
66	2-Nov-2004	7-Nov-2004	5	9B	DD	Deep depression over Bay along with small depression in western Arabian Sea pushes fine-mode aerosols from Indo-Gangetic plains towards northern Arabian Sea
67	4-Nov-2004	7-Nov-2004	3	Name 04A	D	Same as above
68	10-Oct-2004	10-Oct-2004	1	No name	D	Two cyclones one in the Bay and other in the NE Arabian Sea. The Arabian Sea depression is closer to dust sources and deflects the dust from SW Asia and Makaran coast towards Oman. Fine-mode aerosols from Indus plains also deflected towards SW
69	7-Oct-2004	8-Oct-2004	1	8B	D	Same as above
70	2-Oct-2004	4-Oct-2004	2	7B	D	Two simultaneous cyclones one in the Bay and other in the Arabian Sea. The Arabian Sea cyclone Onil caused a minor dust event along Makaran Coast
71	30-Sep-2004	4-Oct-2004	4	6A (Onil)	SCS	Same as above
72	12-Sep-2004	15-Sep-2004	3	5B	D	Dust over Afghanistan Helmand Basin deflected towards northeastern Arabian Sea by depression in Bay
73	11-Jun-2004	14-Jun-2004	3	4B	DD	Deep depression in northern Bay pulls Oman and Afghan dust eastward towards India

(continued)

Table 7.2 (continued)

S. No	Start of event	End of event	Duration of event	Number and name of event	Intensity of event	Remarks
			Days	Popular name in brackets		
74	10-Jun-2004	13-Jun-2004	3	3A	DD	Counter clockwise winds of a deep depression in Eastern Arabian Sea pushes mineral dust from SW Asia and Thar towards western Arabian Sea and Oman
75	16-May-2004	19-May-2004	3	2B	VSCS	A pre-SW monsoon very severe cyclone in Bay of Bengal influences the Shamal dust event in Saudi Arabian and pulls the dust plume towards the eastern Arabian Sea
76	5-May-2004	10-May-2004	5	1A	SCS	A large pre-monsoon cyclone in eastern Arabian Sea pushes dust from Thar Desert towards Arabian Sea
77	11-Dec-2003	16-Dec-2003	5	7B	SCS	A combination of frontal system in SW Asia and cyclone in southwestern Bay of Bengal pulls huge amounts of dust towards SE Arabian Sea
78	12-Nov-2003	15-Nov-2003	3	6A	SCS	A frontal system in SW Asia creates a dust plume. The cyclone in southern Arabian Sea has weak influence on this plume. Dust plume pulled towards northern Arabian Sea
79	26-Oct-2003	28-Oct-2003	2	5B	DD	A post monsoon cyclonic event crosses the Bay of Bengal. Fine aerosols from India pushed towards Arabian Sea and Bay of Bengal
80	6-Oct-2003	9-Oct-2003	3	4B	D	Depression in the northwestern Bay disperses fine aerosols from Indian sub-continent over Arabian Sea. Miner dust plumes along Makaran coast dispersed towards northern Arabian Sea
81	27-Aug-2003	28-Aug-2003	1	3B	D	SW monsoon depression in central Bay pulls dust from Middle-East towards central Arabian Sea
82	25-Jul-2003	28-Jul-2003	3	2B	DD	SW monsoon depression in central Bay pulls dust from Middle-East towards eastern central Arabian Sea
83	10-May-2003	19-May-2003	9	1B	VSCS	Convective winds create major dust storms over SW Asia and Thar Desert which is transported towards the south or easterly direction. Counter clockwise winds due to a severe cyclonic event in the central Bay disperses this dust over the Arabian Sea

(continued)

Table 7.2 (continued)

S. No	Start of event	End of event	Duration of event	Number and name of event	Intensity of event	Remarks
			Days	Popular name in brackets		
84	21-Dec-2002	25-Dec-2002	4	6B	CS	Cyclonic storm east of Sri Lanka disperses fine aerosols over India towards Arabian Sea
85	23-Nov-2002	28-Nov-2002	5	5B	CS	Cyclonic storm in central Bay disperses fine aerosols over India towards Arabian Sea
86	10-Nov-2002	12-Nov-2002	2	4B	SCS	Cyclonic storm in northwestern Bay disperses fine aerosols over India towards Arabian Sea
87	22-Oct-2002	23-Oct-2002	1	3B	D	A local weather disturbance creates a dust storm over western Indo-Gangetic plains and Thar region. Counter-clockwise winds of depression disperses the plume over northeastern Arabian Sea
88	11-May-2002	12-May-2002	1	2B	DD	Two simultaneous cyclones in the western Arabian Sea off Oman and the eastern Bay of Bengal. Since early May is a high dust season due to local convective winds, dust is raised at many locations, but mainly in western Gangetic plains Makaran coast and W Asia. The two cyclones disperse the dust plume over the northern Arabian Sea and Bay of Bengal
89	6-May-2002	10-May-2002	4	1A	CS	Same as above. Even though the cyclone is close to the Oman margin no major dust storm is seen along Oman coast, probably because of coarse sandy material and crust formation along the coast
90	11-Nov-2001	12-Nov-2001	1	6B	D	Counter clockwise winds of cyclone in Arabian Sea on 8th October creates a major dust plume over Makaran and SW Asian region. The dust is dispersed towards the SW direction over western Arabian Sea. After this cyclone dies down, two cyclones in quick succession form over the Bay of Bengal and disperse the remaining dust over Indo-Gangetic plains and NE Arabian Sea. Depression in western Bay disperses fine mode aerosols from India towards eastern Arabian Sea

(continued)

Table 7.2 (continued)

S. No	Start of event	End of event	Duration of event	Number and name of event	Intensity of event	Remarks
			Days	Popular name in brackets		
91	14-Oct-2001	16-Oct-2001	2	5B	CS	Same as above
92	8-Oct-2001	10-Oct-2001	2	4A	CS	Same as above
93	24-Sep-2001	27-Sep-2001	3	3A	CS	No major effect of this cyclone seen dust or aerosol transport
94	12-Jun-2001	13-Jun-2001	1	2B	D	Dust in SW Asia due to local convective winds dispersed down Indo-Gangetic plains by depression in the northern Bay of Bengal
95	21-May-2001	29-May-2001	8	1A	VSCS	Very severe cyclone in eastern Arabian Sea disperses dust from SW Asia and Thar region towards northern and western Arabian Sea

Table 7.2. The names and event numbers are according to a classification adopted by the Indian Meteorological Department.

7.2.5 Satellite Data and Images

Satellite images from a number of polar orbiting satellites like Oceansat II, SeaWifs, MODIS (Terra and Aqua) as well as geostationary satellites like Eumetsat and Kaplana were used for this study. The MODIS (Terra and Aqua) aerosol optical depth (AOD) at 550 nm and Angstorm exponent were obtained from the Giovanni archives. AOD was determined for the northern Arabian Sea for an area located between Lat 18°N–25°N and Longitude 55°E–72.5°E (Fig. 7.2) for

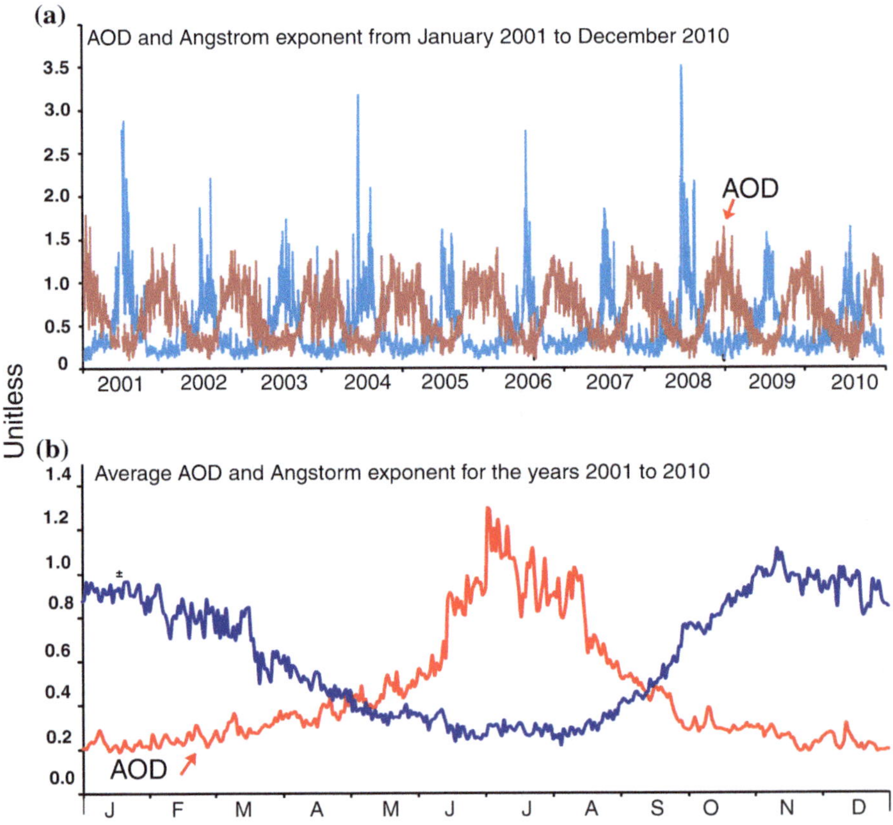

Fig. 7.2 Aerosol optical depth at 550 nm (*red line*) from MODIS (average of Aqua and Terra) satellites between 1 January 2001 and 31 December 2010 (**a**) for the northern Arabian Sea (Lat 18–25 and Long 58–72.5). *Blue line* indicates the Angstrom exponent (α) computed using 550 and 865 nm. The Angstrom exponent provides information on the particle size-larger the exponent, the smaller the particle size. Average AOD and α values (**b**) for the period 2001–2010

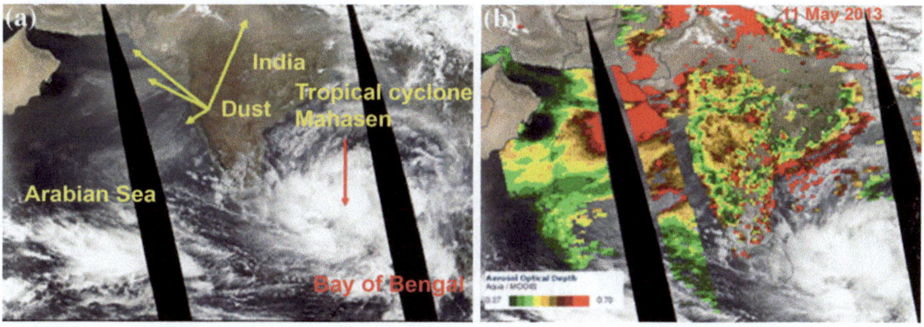

Fig. 7.3 Identification and confirmation of tropical cyclones and dust plumes. Storms, depressions and tropical cyclones were visually identified and then confirmed based on cyclone warnings and tracks published by the Indian Meteorological Department (http://imd.gov.in). Dust plumes were initially identified visually in satellite images (**a**) as they have a *yellowish tinge* in true color images. They were confirmed as dust plumes if the AOD values (**b**) exceed >0.5 units and have angstrom exponent values (α) less than 0.4. The image shows the effect of cyclonic storm Mahasen in the Bay of Bengal on drawing dust plumes from the Saudi Arabian region over the Indian Peninsula

the period 1 January 2001–31 December 2010. The above area was chosen as it is located close to the dust source regions and has low cloud cover. AOD values over water are more reliable than those over dust source regions as the latter have high reflectivity pattern similar to dust plumes.

Dust plumes were identified visually in satellite images as they have a *yellowish tinge* in true color images (Figs. 7.3, 7.4, 7.5, 7.6, 7.7 and 7.8). AOD values and Angstorm exponent values for the same images were used from confirmation of

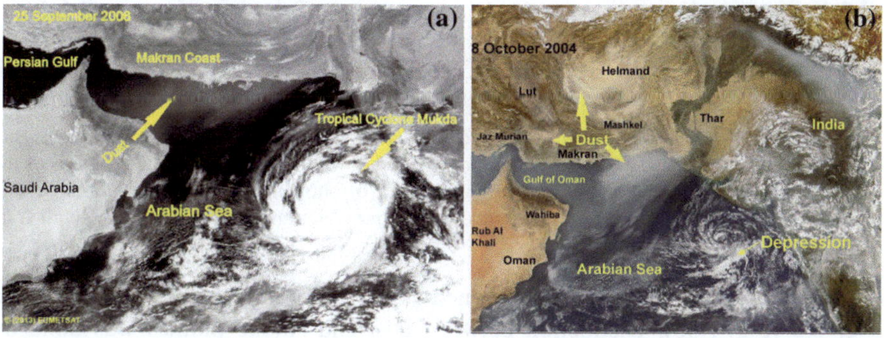

Fig. 7.4 Direct effects of tropical cyclone in the Arabian sea on dust entrainment and transport. Eumetsat image of 25th September 2006 (**a**), showing tropical cyclone "Mukda" in the north-eastern Arabian Sea and a major dust storm over the SW Asia. The counter clockwise winds transport dust from the Makaran coast towards Oman. Sea-Wifs image of 8 October 2004 (**b**) showing a minor depression in the NE Arabian Sea with the winds transporting dust from eastern Makaran Coast towards Oman. Fine-mode aerosols from the Indus plains are also being carried towards the western Arabian Sea

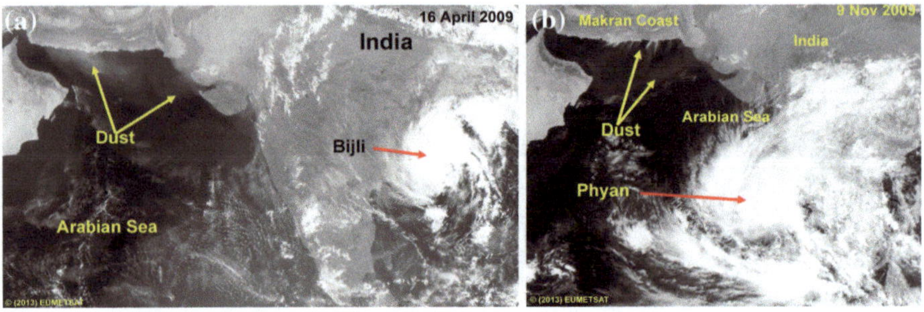

Fig. 7.5 Eumetsat image of 16 April 2009 (**a**) showing effects of tropical cyclone "Bijli" in the Bay of Bengal on dust transport. Dust plume from a frontal system in the Makaran coast is being transported towards the east due to the presence of a tropical cyclone in the Bay of Bengal. Dust streaks can be seen from Makaran coast in Iran up to peninsular India. Eumetsat image of 9th November 2009 of tropical cyclone "Phyan" in the eastern Arabian Sea (**b**). Dust from Makaran coast can be seen being transported by counter clockwise cyclonic winds towards western Arabian Sea. A mixture of dust and fine-mode aerosols from the Indus plains is also being transported towards the central Arabian Sea

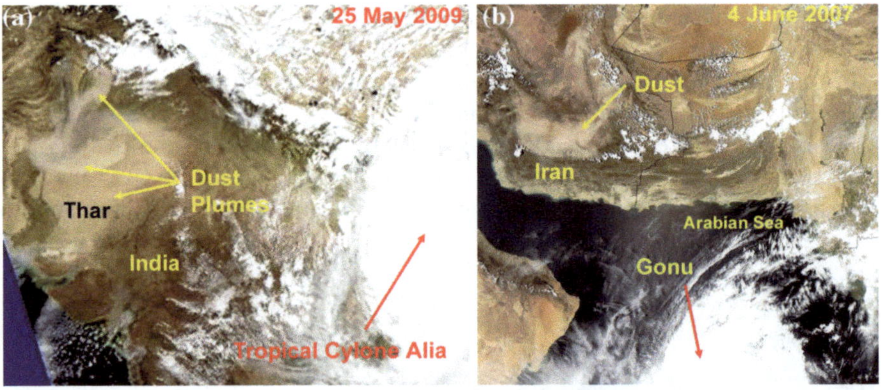

Fig. 7.6 MODIS Aqua image of Tropical cyclone "Aila" 25th May 2009 (**a**) in the northern Bay of Bengal. Dust from the Thar Desert being transported down the highly populated regions of the Gangetic plains. Eumetsat image of 4th June 2007 (**b**) showing super-cyclone Gonu in the western Arabian Sea and a large dust events in the Djazz Murian and Lut Basins in south Iran

dust plumes. The plumes were confirmed as dust plumes if the AOD values were over 0.5 (Fig. 7.3b). Distinction between dust and smoke plumes was done by examining the Angstrom coefficient (α) values which are inversely related to the size of the particles. Dust plumes have relatively large particle size and low α values (<0.4) while smoke particles are generally sub-micron size particles having α values much higher than 0.4 (Fig. 7.2).

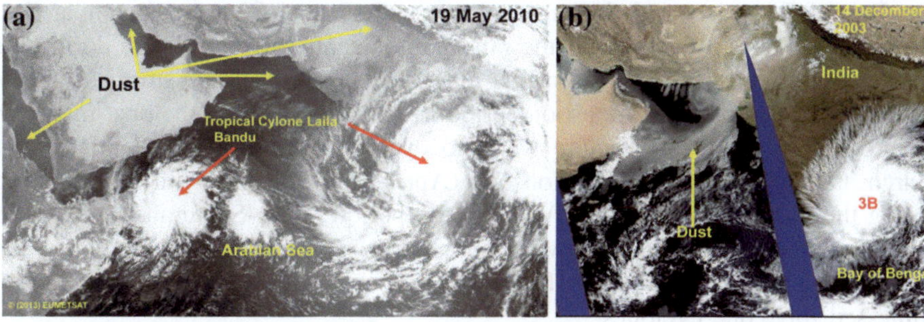

Fig. 7.7 Eumetsat image of 19th May 2010 (**a**) showing two cyclones in the northern Indian Ocean. Tropical cyclone Bandu is the western Arabian Sea and tropical cyclone Laila in the Bay of Bengal. High dust can be seen in the Red Sea, northern Persian Gulf and northern Arabian Sea. MODIS Aqua Image of 14 December 2003 (**b**). Tropical cyclone 3B in the central Bay of Bengal in Dec 2003. A cold front created a large dust storm in the Persian gulf region. A thick dust plume can be seen being transported towards the low pressure region of the SE Arabian Sea

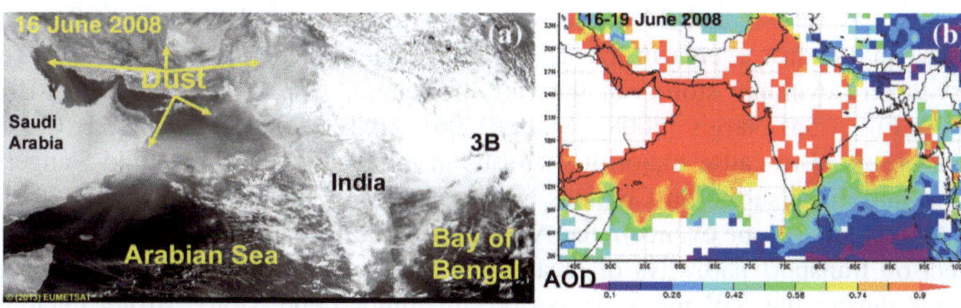

Fig. 7.8 Eumetsat image of 16th June 2008 (**a**) showing cyclone 3B in the northern Bay of Bengal pulling dust from the Saudi Arabia over the Arabian Sea, India and up to the eastern Bay of Bengal. Confirmation of long-range dust plumes is from MODIS (Aqua), average AOD values (unit less) for the period 16th–19th June 2008 (**b**). AOD values above 0.7 over the sea indicate major dust events. High dust can be seen over the Persian Gulf, Red Sea, Arabian Sea and northern Bay of Bengal up to the Andaman Sea

7.3 Results

7.3.1 Storms and Tropical Cyclone Between 2001 and 2010

Ninety five cyclone events (tropical cyclones, storms and depressions) were recorded in the northern Indian Ocean between January 2001 and December 2010 (Table 7.2). The effect of each cyclone on aerosols were studied from a number of satellite images. A brief description of the effects of each storm on aerosol transport is given in Table 7.2. While tropical cyclones in the northern Indian Ocean can occur any time between April and January the maximum number of strong events are noticed during May and between October and December.

The Bay of Bengal recorded more cyclones and storms (69 events) compared to the Arabian Sea (26 events). Of the ninety five events, fifty five were during the post-monsoon period. However, of the nine very severe cyclonic storms (VSCS), seven were during the pre-monsoon period and only two during the post-monsoon period. During 2001–2010 there were no cyclone events during February and March. Of the fifteen pre-SW Monsoon (April–May) cyclones thirteen occurred during the month of May. The SW monsoon period (June–September) witnessed thirty nine events; most of them were minor depressions in the northern part of the Bay of Bengal and Arabian Sea. The post-SW monsoon period (October–January) witnessed forty one cyclones.

During April–May the entire South Asia and Middle-East region has low soil moisture levels with local convective winds creating a number of dust events. The summer Shamal winds start by mid-May. All the fifteen tropical cyclones during the pre-monsoon months had a major effect on dust transport by directly raising dust events or altering the transport pathways (Table 7.2). Notable among them were the Arabian Sea Cyclones like Phet (2010, Fig. 7.7a), Bandu (2010, Fig. 7.7a), 1A (2004), 1A (2001) which were the cause of major dust events. Cyclone 1A (2004) and 3A (2004) recurved towards the NE Arabian Sea and caused major dust events in the Thar and Afghanistan regions (Table 7.2). The very severe cyclonic storms in the Bay of Bengal, Aila, (2009, Fig. 7.6a) Bijli (2009, Fig. 7.5a) Nargis (2008), Mala (2006), 2B (2004), 1B (2003), 2B (2002) were observed to cause major dust storms in the Thar Desert and Indo-Gangetic plains and disperse the dust plumes eastward towards the Bay of Bengal.

Almost all cyclone events in the Arabian Sea and Bay of Bengal had significant effect on transport pathways of dust raised by the Shamal Winds. Though most storms during the SW monsoon period are minor depressions, strong cyclonic storms can occur in beginning of June and end of September. Satellite images show that strong to severe cyclones during the SW monsoon period such as Gonu (2007, Fig. 7.5b), Yemyin (2007), Pyaar (2005) 3A (2001) had severe impact on dust entrainment and transport over the entire region.

Between October and January, dust events are less numerous because of gentler winds and enhanced soil moisture levels. Occurrence of tropical cyclones however, is at a maximum during this period. The frontal systems passing over the region north of the Arabian Sea create dust events which are dispersed by the prevailing weak NE trade winds. Whenever passage of the frontal system coincides with a tropical cyclone a major deviation in dust transport pathway occurs. Out of forty one cyclones between October and January, ten cyclonic storm events coincided with dust storms caused by passage of frontal events (Table 7.2). In most cases the Arabian Sea cyclones transported the dust towards the Arabian Peninsula as the principal dust sources are to the north of the cyclone where the winds blows towards the west or south-west. The northern Bay of Bengal cyclones on the other hand dispersed the aerosols and dust mostly in the eastward direction as the winds south of the tropical cyclone blows mainly in the SE or eastward direction. Post-SW monsoon cyclones also had major effect on fine-mode anthropogenic aerosols. Since fine-mode aerosols levels are high between October and March almost all post-monsoon tropical cyclones play a major role in dispersing them. Out of the forty one cyclones, thirty five had a significant effect on anthropogenic aerosol dispersion from the Indian subcontinent to over the adjacent Arabian Sea and Bay of Bengal (Table 7.2).

7.3.2 Aerosol Optical Depth and Angstrom Exponent Variations

MODIS satellite Aerosol optical depth (AOD) values (average of Terra and Aqua) from the northern Arabian Sea between 1st January 2001 and 31 December 2010 are shown in Fig. 7.2. AOD values are generally high (>0.5) between May and September and comparatively low (<0.5) between October and April. High AOD values indicate high particle concentration in the atmospheric column. The Angstrom exponent (α) is inversely related to the average size of the particles; larger the particles, smaller the α values. α values less than 0.4 units are found during the summer monsoons indicating particles dominated by coarse mineral-dust particles. During October–April large α values indicate dominance of sub-micron size particles derived mostly from burning of fossil fuels and biomass (Gautam et al. 2009; George et al. 2008; Prasad and Singh 2007). Combining the AOD and α values we can generalize that coarse mode mineral-dust particles dominate during May–September and fine mode particles dominate between October and April. The large dust inputs between end of May and September is due to the persistent summer Shamal Winds. During the low dust period (October–April) occasional high AOD values (>0.7) are seen due to short-term dust events due to winter Shamals associated with passage of cold fronts or as will be discussed below, due to tropical cyclones.

7.4 Discussion

7.4.1 Direct Effect of Cyclonic Events Which Migrate Close to Dust Source Areas

The wind speeds associated with different categories of tropical cyclonic events in the northern Indian Ocean range from 8.8 m s^{-1} to over 61 m s^{-1} (Table 7.2). The threshold velocity for initiation of dust events are in the range of 5–20 m s^{-1} which means that most cyclonic storms impinging on dust source regions should be able to initiate dust storms. Severe and very severe cyclonic storms in the Arabian Sea, if close to dust source areas, should be able to initiate major dust events over large areas. Between 2001 and 2010 fifteen tropical cyclones and storms Phet (2010) Bandu (2010), O3B (2008), O5A Gonu 2007; Mukda 2007;

Agni O4A, 10th Oct, O3A (2004) O2A (2003) O1A (2002); TS3, TS2, TS1 (2001) made landfall or passed close to dust source areas. Satellite images show that strong anti-clockwise winds associated with these cyclones are able to entrain dust and transport it mostly in a SW direction (Fig. 7.4a, b). However, there is no direct relation between intensity of the tropical cyclone and amount of dust raised, because other factors like soil moisture, crust formation etc. can raise the threshold for dust entrainment. Large dust plumes were observed when the cyclonic storms move towards the north or northwestern Arabian Sea because of proximity to dust source areas along the Makaran Coast and the endhoreic basins north of the coast. The Makaran coast and the endhoreic basisn of Helmand (Seistan), Jaz Murian and Mashkel are dust prone because of the presence of fine dry loose soils derived from uplifted recent marine sediment and dry lake beds. Moreover, these sediments do not have much large pebbles or boulders to form desert pavements which can inhibit dust storms. The northern coast of the Persian Gulf (Iran) do not have large-scale dust events because of the presence of numerous salt domes in the region which contribute to salt-crust formation. The main dust sources in the Rub Al Khali desert or Mesopotamia are located far away from the Arabian Sea and are less responsive due to drop in wind speed because of greater friction over land. Dust events in the above regions are seen only during very severe cyclonic storms. The southern Oman or Yemen coast highland regions receive high rainfall during the summer monsoon and do not have major dust source areas. However, cyclones moving close to the Gulf of Aden are observed to raise large dust plumes along both sides of the Red Sea coast (Fig. 7.7a) and especially off the Tokar Delta.

7.4.2 Indirect Effects by Tropical Cyclones Located Further Away from Dust Source Regions

Tropical cyclone located further away from the source region can affect dust entrainment by intensification of the pressure gradient leading to greater wind speeds over dust source regions (Badrinath et al. 2009, 2010). Also, if dust raised by Shamal or local winds are present then the transport direction of the dust plumes can be significantly altered. Tropical cyclones during the SW monsoon period, though rare, have a major effect on the dust transport as significant amount of dust is present in the western and northern Arabian Sea due to the summer Shamal Winds. A example of this is tropical cyclone 2A located in the central Arabian Sea on 6th June 2008. From satellite images, dust plumes from Oman and Saudi Arabian Peninsula could be traced up to the western Arabian Sea but not eastward of the cyclone. Ten days later a tropical cyclone (3B 16–18 June 2008) struck the northern Bay of Bengal (Fig. 7.8). Huge dust plumes could be traced from the Middle-East right across the Arabian Sea, Peninsular India and up to the eastern Bay of Bengal (Fig. 7.8b). The highest AOD values recorded in the Arabian Sea during the 10 year study period was during this period (Fig. 7.2).

The most important effect of tropical cyclones on dust transport is deflection of dust plumes away from their normal pathways. Mineral dust is normally transported by the relatively steady trade winds. As the dust plumes comes under the influence of cyclonic winds its transport direction is sharply altered because of the counter clockwise winds. In case of the Arabian Sea cyclones, the dust sources lies towards the north or east of the cyclone hence the plumes are generally deflected in a southwesterly direction. The Bay of Bengal cyclones lie much further eastwards of dust sources and it is observed that the dust plumes are generally deflected towards the east or southeasterly direction. An illustration of this is tropical cyclone Bijli (Fig. 7.5a) which formed in the central Bay of Bengal and migrated northward towards the Bangladesh coast between 14 and 17 April 2009. Dust plumes from the Makaran coast of Iran can be seen being pulled in a southeasterly direction towards India. A similar case is a large tropical cyclone Aila in May 2009 (Fig. 7.6a) where east flowing winds south of the cyclone transported dust from the Thar Desert down the Gangetic plains towards the Bay of Bengal. Another example of the effect of tropical cyclone on deflection of dust plumes is tropical cyclone Phyan (9–11 Nov 2009) in the north-western Arabian Sea (Fig. 7.5b). Dust from Makaran is blown by the counter-clockwise winds in a southwesterly direction towards the Arabian Coast.

After the retreat of the SW monsoon in September, wind speeds are generally low and no dust storms are expected in the Thar Desert region between October and December. However, major cyclones in the Bay of Bengal or north-eastern Arabian Sea during the post SW monsoon period have led to major dust storms in the region. For example, tropical cyclone Sidr in October 2007 created a major dust storm over the Thar Desert and the plume was traceable for more than a 1,000 km up to the northern Bay of Bengal (Badrinath et al. 2009). Another example is the tropical cyclone Agni (28 Nov–2 Dec 2004) in the western Arabian Sea which caused a major dust storm and plume to be blown from the Indus-Thar region into the northern Arabian Sea (Table 7.2).

Tropical cyclones also have a major effect on deflection of dust plumes created by the winter Shamal dust events. The dust raised by these north westerly winds transport dust from the Mesopotamian region towards the Persian Gulf. The winter frontal systems also affect the endorheic basins of SW Asia (Iran, Afghanistan, and Baluchistan) and the Makaran coast. The dust from these regions is normally transported by the prevailing northeasterly trade winds towards Saudi Arabia and North Africa. Winter Shamals are short lived events occurring between October and April and lasting between 3 and 5 days. The post SW-Monsoon tropical cyclones in the northern India Ocean last for 2–8 days and the probability of these two events coinciding are high between September and December. If these events coincide then the dust transport pathway can be significantly deviate from its normal path. In December 2003 a massive dust plume generated in Iran and Saudi Arabian Peninsula was affected by the tropical cyclone 3B in the central Bay of Bengal (Fig. 7.7b). The dust plume could be traced up to the southeastern Arabian Sea. Another example of combination of events is the 10th October 2004 event where a cold front caused a major dust storm in the Afghanistan, Thar, and

Makaran region. A minor cyclonic depression in the northern Arabian Sea deflected the plumes towards the north-western Arabian Sea (Fig. 7.4b).

7.4.3 Tropical Cyclones and Dispersal of Anthropogenic Aerosols

During October to end of March, the Indian subcontinent has high aerosol levels dominated by fine mode particles from mostly low temperature biomass burning, especially over the highly populated Indo-Gangetic plains (Ramanathan et al. 2001b). High fine-mode aerosols levels can persist near source regions for long periods as prevailing wind speeds are low. Enhanced wind circulation during tropical cyclones can disperse there fine-mode particles to long distances. Tropical cyclones have a major effect on dispersing these fine-mode particles as they are the major cause of high wind-speed events during winter. Even minor storms can help in dispersion of the fine mode particles because, unlike dust, no energy is required for entraining them. The Himalayan mountain chain with a height of over 4.5 km forms a formidable barrier and prevents dispersal of fine mode particles from south Asia to the north. However the fine mode aerosols are easily dispersed towards the south and adjoining Arabian Sea and Bay of Bengal. Satellite images show plumes of grey aerosol clouds trailing out into the Arabian Sea and Bay of Bengal during tropical cyclone events (Fig. 7.6). These plumes can be traced even south of the equator and westwards to the Middle-East. The fine-mode particle plumes can be easily identified by their ash grey color and high angstrom exponent (α) values.

Post SW monsoon cyclones disperse fine-mode and mineral dust particles (Fig. 7.4b) from different sources. The two size-classes of particles can collide leading to formation of complex aggregates. Fine mode particles have more sulphate ions which when dissolved in water form acidic waters which can enhance the solubility and release of trace-nutrients from mineral dust particles (Ooki and Uematsu 2005). Dispersal of dust and fine-mode particles by tropical cyclones over the oceans can therefore contribute significantly to increase in ocean productivity and capacity of oceans to absorb atmospheric carbon dioxide.

7.5 Summary

Study of ninety five cyclonic storms between 2001 and 2010 show that they have a major effect on dust entrainment and transport. Tropical cyclones which move close to dust source regions in the north and northwestern Arabian Sea raise dust but transport them mainly westwards towards the western Arabian Sea and Arabian Peninsula. Tropical cyclones located further away in the SE Arabian Sea and Bay of Bengal alter the transport pathway of dust and transport them towards the

Indian Peninsula. Severe tropical cyclones in the northern Bay of Bengal aid advection of dust plumes from southwest Asia and Thar Desert over highly populated regions of the Indo-Gangetic plains. Tropical cyclones also disperse fine-mode aerosols over South Asia leading to formation of complex aerosol-dust mixtures.

Acknowledgments I thank Danling Tang for discussions and encouraging me to write this paper. I am grateful to the MODIS science data support team for providing access to images (http://rapidfire.sci.gsfc.nasa.gov; http://earthdata.nasa.gov/) and data via the Giovanni website (http://giovanni.gsfc.nasa.gov/). Satellite images and data were also obtained from the IRS-P4 Oceansat-II, and as well as the KALPANA, SeaWifs and Eumetsat websites. This is NIO Contribution No. 5453

References

Ackerman, S.A., Cox, S.K.: Dust outbreaks associated with the southwest monsoon region. Meteorol. Atmos. Phys. **41**(1), 19–34 (1988)

Ackerman, S.A., Cox, S.K.: Surface weather observations of atmospheric dust over the southwest summer monsoon region. Meteorol. Atmos. Phys. **41**, 19–34 (1989)

Anon: Tropical cyclone operational plan for the Bay of Bengal and Arabian Sea. 2012 edn. World Meteorological Organization Technical Document No. 84. http://www.imd.gov.in/section/nhac/dynamic/TCP-21_2012.pdf (2012)

Arimoto, R.: Eolian dust and climate: relationship to sources, tropospheric chemistry, transport and deposition. Earth-Sci. Rev. **54**, 29–42 (2001)

Badarinath, K.V.S., Kharol, S.K., Sharma, A.R., Ramaswamy, V., Kaskaoutis, D.G., Kambezidis, H.D.: Investigations of an intense aerosol loading during 2007 cyclone SIDR–A study using satellite data and ground measurements over Indian region. Atmos. Environ. **43**, 3706–3708 (2009)

Badarinath, K.V.S., Kharol, S.K., Kaskaoutis, D.G., Sharma, A.R., Ramaswamy, V., Kambezidis, H.D.: Long-range transport of dust aerosols over the Arabian Sea and Indian region—a case study using satellite data and ground-based measurements. Global Planet. Change **72**, 164–181 (2010)

Das, P.K.: The Monsoon, p. 162. National Book Trust, New Delhi (1968)

Duce, R.A., et al.: The atmospheric input of trace species to the world ocean. Glob. Biogeochem. Cycles **5**(3), 193–259 (1991)

Gautam, R., Zhaoyan, L., Singh, R.P., Hsu, N.C.: Two contrasting dust-dominant periods over India observed from MODIS and CALIPSO data. Geophys. Res. Lett. **36**, L06813 (2009)

George, S.K., Nair, P.R., Parameswaran, K., Jacob, S., Abraham, A.: Seasonal trends in chemical composition of aerosols at a tropical coastal site of India. J. Geophys. Res. **113**, D16209 (2008)

Givehchi, R., Arhami, M., Tajrishy, M.: Contribution of the Middle Eastern dust source areas to PM10 levels in urban receptors: case study of Tehran, Iran. Atmos. Environ. **75**, 287–295 (2013)

Goudie, A.S., Middleton, N.J.: Desert Dust in the Global System, pp. 117–133. Springer, Berlin (2006)

Jickells, T.D., An, Z.S., Andersen, K.K., Baker, A.R., et al.: Global iron connections between desert dust, ocean biogeochemistry, and climate. Science **308**, 67–71 (2005)

McTanish, G.H., Strong, C.S.: The role of aeolian dust in ecosystems. Geomorphology **89**, 39–54 (2007)

Middleton, N.J.: A geography of dust storms in southwest Asia. J. Climatol. **6**, 183–197 (1986)
Muni Krishna, K.: Intensifying tropical cyclones over the North Indian Ocean during summer monsoon—Global warming. Global Planet. Change **65**, 12–16 (2009)
Ooki, A., Uematsu, M.: Chemical interactions between mineral dust particles and acid gases during Asian dust events. J. Geophys. Res. **110**, D03201 (2005)
Prasad, A.K., Singh, R.P.: Changes in aerosol parameters during major dust storm events (2001–2005) over the Indo-Gangetic Plains using AERONET and MODIS data. J. Geophys. Res. **112**, D09208 (2007)
Prospero, J.M.: Long-range transport of mineral dust in the global atmosphere: impact of African dust on the environment of the southeastern United States. Proc. Natl. Acad. Sci. U.S.A. **96**, 3396–3403 (1999)
Prospero, J.M., Ginoux, P., Torres, O., Nicholson, S.E., Gill, T.E.: Environmental characterization of global sources of atmospheric soil dust identified with the nimbus 7 total ozone mapping spectrometer (TOMS) absorbing aerosol product. Rev. Geophys. **40**(1), 1002 (2002)
Pye, K.: Aeolian Dust and Dust Deposits, p. 334. Academic Press, London (1987)
Ramanathan, V., Crutzen, P.J., Kiehl, J.T., Rosenfeld, D.: Aerosols, climate and the hydrological cycle. Science **294**, 2119–2124 (2001a)
Ramanathan, V., Crutzen, P.J., elieveld, J., Mitra, A.P., et al.: Indian Ocean experiment: an integrated analysis of climate forcing and effects of the great Indo-Asian haze. J. Geophys. Res. **106**, 28371–28398 (2001b)
Rastogi, N., Sarin, M.M.: Long-term characterization of ionic species in aerosols from urban and high-altitude sites in western India: Role of mineral dust and anthropogenic sources. Atmos. Environ. **39**, 5541–5554 (2005)
Tegen, I., Fung, I.: Contribution to the atmospheric mineral aerosol load from land surface modification. J. Geophys. Res. **100**, 18707–18726 (1995)
Tegen, I., Lacis, A.A., Fung, I.: The influence on climate forcing of mineral dust disturbed soils. Nature **380**, 419–422 (1996)

Chapter 8
Rainfall Prediction for Landfalling Tropical Cyclones: Perspectives of Mitigation

Kevin K. W. Cheung, Lisa T.-C. Chang and Yubin Li

Abstract The torrential rainfall associated with landfalling tropical cyclones (TCs) often represents the major impact to coastal regions, but at the same time an enormous challenge to meteorologists and forecasts. This chapter first discusses the complex dynamical processes involved in TC landfalls, which are related to the increased surface roughness and reduced surface moisture fluxes of land. The result is often certain patterns of convection and rainfall asymmetry in the landfalling TCs, but these patterns are not well explained by current theories or conceptual models. With emphasis of development of rainfall prediction techniques according to the needs of mitigation, the requirements on the skill of rainfall forecasts from the perspectives of mitigation are reviewed. Then, the operation and performance of several statistical TC rainfall models are discussed including the rainfall climatology-persistence model (R-CLIPER) for the Taiwan area. A topographic component is developed for R-CLIPER through multiple regression analyses, which improves the model's performance in reproducing the local extreme rain that is lacking in the original model. Finally, the importance of utilizing remote-sensing data in TC rainfall prediction is discussed, and how TC rainfall statistical models can be applied to risk analyses under the consideration of global changes.

K. K. W. Cheung (✉) · Y. Li
Department of Environment and Geography, Macquarie University,
Sydney, NSW 2109, Australia
e-mail: kevin.cheung@mq.edu.au

L. T.-C. Chang
Department of Environmental Management, Tungnan University, New Taipei City, Taiwan

L. T.-C. Chang
Risk Frontiers Natural Hazards Research Centre, Macquarie University, Macquarie, Australia

Y. Li
School of Energy and Environment, City University of Hong Kong, Kowloon, Hong Kong

D. L. Tang and G. J. Sui (eds.), *Typhoon Impact and Crisis Management*,
Advances in Natural and Technological Hazards Research 40,
DOI: 10.1007/978-3-642-40695-9_8,

Keywords Tropical cyclone rainfall · Landfall processes · Climatology-persistence model · Topographic effect · Orographic rain · Mitigation

Prepared for "Remote Sensing of Typhoon Impact and Crisis Management".

May 2011

8.1 Introduction

Tropical cyclones (TCs) are severe weather systems that develop over most of the ocean basins, and thus their impacts are worldwide that include the Gulf and East coast of the U.S., South East Asia, India, the tropical Australian regions as well as African countries located by the Southwest Indian Ocean. It is well known that the hazards associated with TCs are multiple in nature and mutually interconnected. Direct damages to infrastructure may mostly due to the destructive winds within the core circulation of TCs, however, very often it is the torrential rainfall accompanying most TCs that bring huge economic loss and even that of human lives. Moreover, the coastal areas are exposed to the risk of storm surges that could be disastrous and may have more extensive areas of impact than the TC landfall location. Given this wide range of sources of hazards, however, their relative impacts in various geographic locations highly vary. This is partially due to the fact that there are different characteristics of TCs that developed in different ocean basins. For example, it is known that TCs in the western North Pacific have a larger average size compared with those in other ocean basins, and thus possibly have a larger radius of gale-force winds when the extent of impact is considered. On the other hand, the local geography, persistent synoptic situations and topography often determine the nature of hazards that would occur in a particular region.

Since there is no weather station and rain gauge to measure rainfall over the oceans, estimation of rainfall from TCs before they make landfall relies on satellite observations, which are only available in the recent few decades. Different algorithms for estimating rainfall exist depending on which part of the electromagnetic spectrum that the satellite observations are from. For example, there are algorithms that estimate rain rate from just the brightness temperature of the infrared-channel satellite imageries (Ebert et al. 2007). Due to the capability of penetrating into clouds and that microwave emission and scattering are more directly related to precipitation, passive microwave data is considered superior compared with IR algorithms in rain rate estimation. An example is the satellite that was launched in the Tropical Rainfall Measuring Mission (TRMM) by the National Aeronautics and Space Administration (NASA) in 1997, which included the TRMM Microwave Imager (TMI). Moreover, the TRMM satellite platform was the first to have the precipitation radar installed, which was often taken as the reference for calibrating IR and microwave algorithms. Based on TMI rain rate estimates, Lonfat et al. (2004) examined the rainfall distribution within TCs over the global oceans. It was found that for azimuthally averaged rain rate, the radial profile (or variation) is similar to that of wind speed with a maximum near the eyewall. The maximum rain rate of about 12 mm h^{-1} was found for the most intense TC category of Cat 3-5 according to the Saffir-Simpson hurricane scale (Simpson 1971; Saffir 1977). When TCs intensify, the average radius with maximum rain rate decreases from about 50 km for tropical storms (systems with sustained near-surface wind speeds larger than 34 kt or 17.5 m s^{-1}) to about 35 km for Cat 3-5 TCs. On the other hand, the asymmetric component of rainfall distribution in TCs was found to concentrate at the front quadrants with respect to the motion vector, which is related to convergence between the TC circulation and the environmental flow (Fig. 8.1). For tropical storms in the Northern Hemisphere, the maximum rain

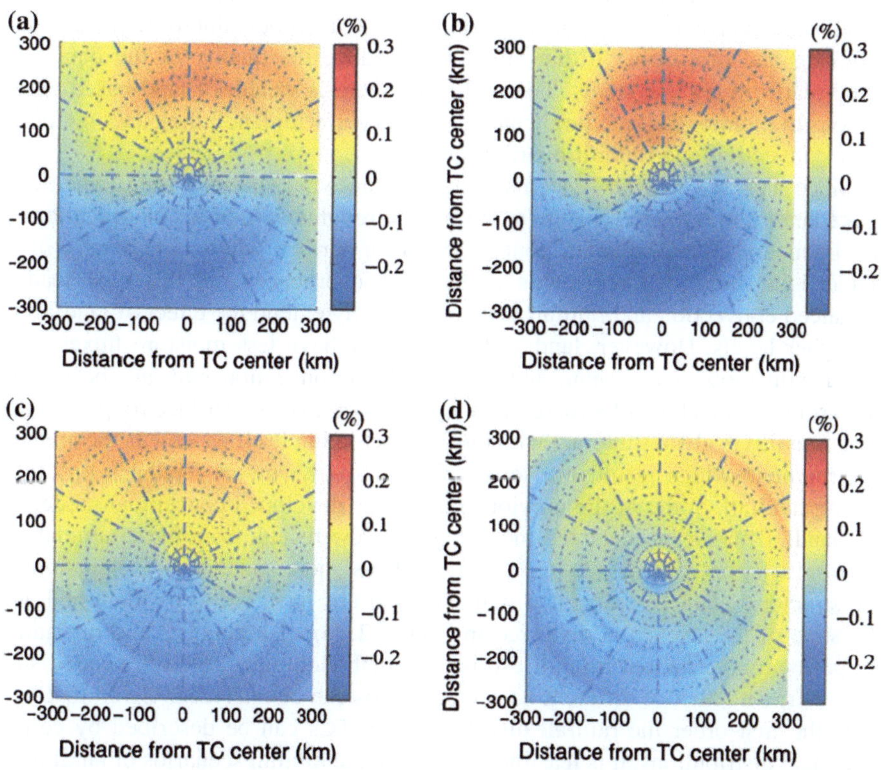

Fig. 8.1 Rainfall asymmetry calculated in 10 km rings around the storm center, as a function of storm intensity: **a** 2121 TC observations from TCs globally, **b** TS, **c** CAT12, **d** CAT35. The storm motion vector is aligned with the positive y axis. The color scale indicates the amplitude of the normalized asymmetry. The *red* corresponds to the maximum positive anomaly and the *blue* the minimum rainfall within the storm [adapted from Londfat et al. (2004)]

rate is likely in the front-left quadrant, and then shifts to the front-right quadrant when intensify to the Cat 3-5 more intense TCs.

When TCs make landfall, rain gauges installed over land provide a valuable source of in situ observations of rain rate within the TCs. However, rain-gauge networks usually do not have enough spatial resolution to resolve the detailed rainfall distributions within TCs and identify extreme rain rates. For these purposes, land-based Doppler radar systems serve much better to monitor torrential rainfall because the spatial resolution of the reflectivity data from most radar systems is of the order of kilometers, and rain estimates based on radar reflectivity can have resolution about 2–3 km or even better (e.g., Chen et al. 2007). Although there are issues such as topographic effect to radar-based rain estimation, a lot of algorithms have been developed to suit rainfall systems with different origins such as those from the tropics and midlatitudes (e.g., Krajewski et al. 2010). Advanced radar techniques such as measurement of the raindrop size distributions in clouds largely increase the capability of more accurate rain estimate based on radar data. Moreover, there are mobile radar systems that have been developed to observe the local-scale convective systems especially those in mountainous regions where blocking and orographic lifting effect may reduce the quality of the data from fixed-platform radar systems. That is to say, there is wealth of land-based observations to monitor the structural changes associated with landfalling TCs including their convection and rainfall distributions.

Since TCs are weather systems with extreme wind speeds within, when they make landfall the sudden change in the surface roughness imposes modifications in the boundary-layer winds. In general, the near-surface winds over land will be reduced in speed that represents low-level convergence, and this may lead to enhanced convection and rainfall that subsequently extend the effect of landfall up to higher levels. However, land surfaces usually have less moisture fluxes compared with those over ocean, and the effects to convection may not be simple. Nevertheless, this kind of structural changes associated with surface roughness and moisture fluxes always interact with the TC circulation itself as well as the ambient environment, which result in a series of complex dynamical feedback processes during landfall. Some of the previous researches developed conceptual models of TC landfall based on idealized modeling, but few results from these simulations have actually been verified by observations. Thus, more conclusive explanations of the structural changes associated with TC landfalls have yet to be identified, and the state of the science is reviewed in Sect. 8.2. For risk analysis and mitigation purposes, more efficient rainfall estimation methodologies are often needed such that a large number of different hazard scenarios can be analyzed. Given that to at least the first order the rainfall distributions in TCs can be described by certain statistical distributions, it is feasible to estimate the rainfall scenarios of landfalling TCs based on some kinds of statistical models that take information from TC track forecasts, rainfall climatology and other possible effects such as topography. Some recent research in this aspect is discussed in Sects. 8.3 and 8.4. After that, Sect. 8.5 briefly discusses the potential impacts to trends of TC rainfall from natural and anthropogenic global changes, and a summary is given in Sect. 8.6.

8.2 Dynamical Processes that Govern the Structure and Rainfall Distribution of Tropical Cyclones During Landfall

8.2.1 Dissipation Over Land

The intensity of a TC is defined as either the maximum surface wind or the minimum central surface pressure (Wang and Wu 2004). Specifically, maximum surface wind is directly associated with damages and impacts caused by TCs, and when it is used to represent the intensity the wind speed at height of 10 m averaged over a period of 1 min (for U.S. agencies) or 10 min (most of the other regions) is referred to. When TCs develop over the oceans, influences on intensity include inner-core characteristic, large-scale atmospheric flow especially the vertical wind shear, ocean processes and interactions between air and sea (e.g., Kaplan et al. 2010). When TCs make landfall, ample observational and theoretical evidence show that TCs decay rapidly and finally dissipate as they move over land, because they are primarily maintained by the latent heat release of water vapor extracted from the oceans (Wong et al. 2008 and references therein). However, a lot of factors affect how the intensities of TCs vary before and after landfall. For TCs that made landfall at the U.S. Gulf coast, statistical results show that on average the weakest TCs strengthen at most before landfall, and the strongest hurricanes weaken the most, with the threshold of intensity separating the two groups about 85–100 kt (1 kt $= 0.514$ m s^{-1}; Rappaport et al. 2010). Previous studies show that the decay rate of TCs is related to their respective central pressures at landfall, and the wind speed decays to a small but nonzero background wind speed (Schwerdt et al. 1979; Ho et al. 1987; Kaplan and DeMaria 1995; Wong et al. 2008; Ramsay and Leslie 2008). Based on the data of U.S. landfalling TCs during the period 1967–1993, Kaplan and DeMaria (1995) developed an empirical model for predicting the maximum wind of landfalling TCs. In the model, the wind speed is determined from a simple two-parameter exponential decay function. An optional correction can also be added that considers storms that move inland slowly or rapidly. This model from Kaplan and DeMaria can explain 91 % (93 % if the motion speed correction is made) of the variance in intensity change after landfall. On the other hand, Wong et al. (2008) developed a similar model for South China coast landfalling TCs based on a dataset during 1975–2005. Besides the landfall intensity and landward speed, it was found that excess of 850 hPa moist static energy has significant influence on the intensity decay rates.

8.2.2 Complex Interactions of the Dynamical Processes

The structure of a TC is characterized by its wind, pressure, temperature fields and water distribution. Many factors determine TC structure such as the size of the low-level cyclonic circulation as measured by the outermost closed isobar or the radius of sustained gale-force winds, the radius of maximum wind (RMW), the convection distribution including that at the eyewall, and both the inner and outer spiral rainbands (Chan et al. 2004; Dastoor and Krishnamurti 1991; Wang and Wu 2004). Numerous studies have been conducted to understand how land surface affects the structure of TCs (Chang 1982; Bender et al. 1985, 1987; Dastoor and Krishnamurti 1991; Chang et al. 1993; Zehnder 1993; Yeh and Elsberry 1993a, b; Lin et al. 1999; Wu and Kuo 1999; Wu 2001; Farfan and Zehnder 2001; Wu et al. 2002, 2009; Ramsay and Leslie 2008). The results from these studies generally show that when a TC is getting close to the coast, the outer circulation of a TC interacts with the land region and have significant influence on the TC's structure. Firstly, the surface wind speed over land is significantly reduced owing to the increased surface roughness, while the turbulent kinetic energy over land is obviously enhanced for the same reason (e.g., Ramsay and Leslie 2008). Secondly, the moisture supply is different between the on-shore and off-shore winds (e.g., Tuleya and Kurihara 1978). Besides, the moisture supply is also affected by mesoscale systems introduced by land-sea difference and influences from monsoon (e.g., Gao et al. 2009). Thirdly, topography near the coastal region imposes additional mechanisms that affect TC structure during landfall. The orographic lifting effect highly enhances rainfall at the windward side by forcing convection, but the resulted rainfall distribution is sensitive to the location of landfall and movement speed of the TC (Lin et al. 2002; Wu et al. 2002). On the leeward side, down-slope winds can be generated, and large turbulent eddies enhanced by complex terrains may transport strong winds to the surface (e.g., Ramsay and Leslie 2008).

These effects are organically connected together, and will further affect the TC structure by interacting with each other and with the ambient environment, which makes the processes associated with structure changes of landfalling TCs even more complicated (Chan 2010; Fig. 8.2). For example, the changed surface wind and moisture field over land will further affect the upper-lever atmosphere through momentum and moisture exchange. The consequent may be changes in the vertical wind shear, which is another critical factor to determine the development of convection within the TC. Through the tangential and vertical winds of TC circulation, the aforementioned effects can be advected back to the offshore areas over the sea and/or upper levels, which leads to the question of whether there are guidelines in estimating the asymmetry of rainfall during TC landfall?

8.2.3 Asymmetry in Convection

Accurately forecasting the precipitation pattern of landfalling TCs is of significant value, as the torrential rainfall accompanying a TC cause major threats to the coastal areas. The TC rainfall distribution is determined by many factors, including, but not limited to, environmental factors such as vertical wind shear, sea surface temperature, moisture distribution, as well as the TC-specific factors such as

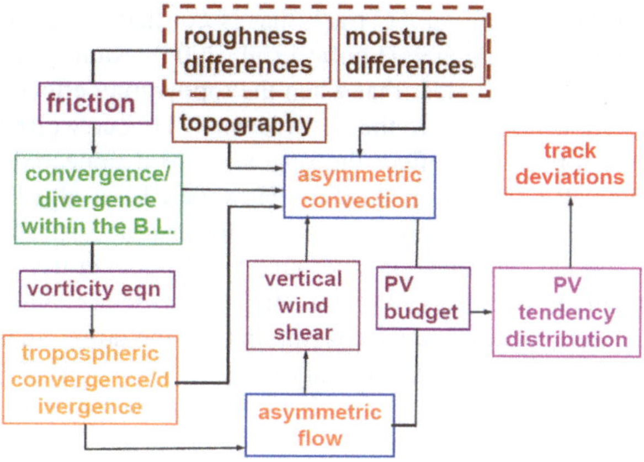

Fig. 8.2 Schematics to show the feedback processes between the land-sea roughness difference, moisture difference, topography, asymmetric flow and convection, and the resulted TC track deviations [adapted from Chan (2010)]

intensity, location, and translation speed (Lonfat et al. 2004). The average axi-symmetric distribution of rainfall within a TC is an intense-rainfall ring that is associated with the eyewall convection and the rainbands that extend from the storm center. However, in reality TC rainfall distributions may have large deviations from this average distribution, which sometimes are caused by environmental factors (e.g., monsoonal effects) or mesoscale systems embedded within TCs.

When a TC is approaching land, the friction-induced convergence in the boundary layer will affect its precipitation pattern (e.g., Marks 1985). Early studies such as Koteswaram and Gaspar (1956), Dunn and Miller (1960), and Miller (1964) already observed this asymmetry in rainfall pattern and explained that this frictional convergence is induced to the right side of a landfalling TC, which follows the on-shore flow, in the Northern Hemisphere and strengthen the rainfall there. Later observational and numerical studies also claimed that coastal rainfall inclines to the right of the track of Northern Hemisphere TCs (Tuleya and Kurihara 1978; Jones 1987; Powell 1987). However, other studies have also found convective activity to be more pronounced to the left of the track (e.g., Parrish et al. 1982; Blackwell 2000; Chan and Liang 2003; Chan et al. 2004). Examination of several TC landfall cases over the South China Sea in Chan et al. (2004) showed that enhanced convection occurs to the west of the TC (i.e., to the left of track) in the mid- to lower troposphere and was then advected to the southward side (i.e., off shore) of the upper troposphere by cyclonic flow and rising motion. For this type of disagreement on the location of enhanced rainfall, Powell (1982) explained that the frictional convergence reasoning cannot be the only determining factor of the coastal rainfall maximum, but other parameters such as local topography can also be in action. The idealized simulations in Chan and Liang (2003) further this point in that even without topography, the less moisture fluxes from land can possibly increase the convective instability and enhance rainfall. Then the area of enhanced convection is advected by the TC circulation to regions that are different from the on-shore location where the strongest frictional convergence is found. Meanwhile in the Southern Hemisphere, observational studies show that the rainfall is stronger in the left side of TC track that is consistent with the location of enhanced frictional convergence (Lonfat et al. 2004). Ramsay and Leslie's (2008) numerical study made it clear that when the topography effect was removed, the rainfall was obviously moved to the left side when TC Larry (2006) was landfalling, while with the topography effect the opposite result appeared, which is in agreement with Powell's (1982) explanation. However, Li et al. (2011; 2013) recently examined several TC landfall cases at northwestern Australia where topography effect is minimal (Fig. 8.3). It was found that low-level frictional convergence cannot fully explain the observed asymmetry in convection, while physical processes as in the aforementioned idealized simulation including feedbacks to the ambient environment may play important roles.

8.3 Statistical Models to Estimate Tropical Cyclone Rainfall During Landfall

8.3.1 Consideration of Requirements from Mitigation

There have been substantial improvements in the skill of numerical weather prediction (NWP) models for simulating rainfall in the recent decade, which partially attribute to fundamental improvements in the physical process representations in

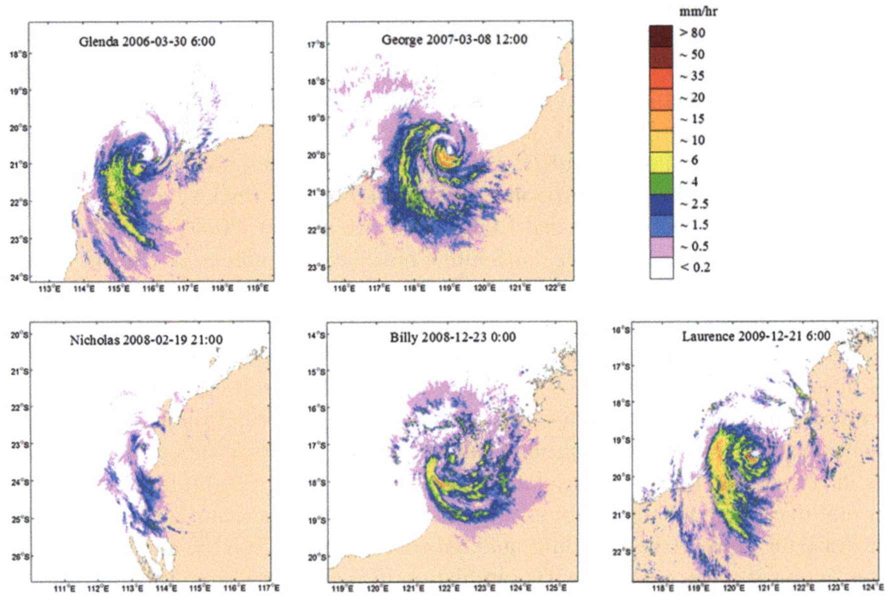

Fig. 8.3 Rain rate estimates from radar reflectivity prepared by the Bureau of Meteorology for five landfalling TCs at northwestern Australia showing their asymmetry in convection distributions [adapted from Li et al. (2011, 2013)]

the numerical models. In addition, the availability of high-performance computing resources allows explicit simulations of the rainfall systems such as TCs rather than relying on the so-called cumulus parameterization schemes that are sources of large uncertainties in rainfall simulations. Explicit moisture or cloud-resolving simulation is a trend of development in global general circulation models as well.

Given these advances in NWP models, their applications in rainfall forecasts for mitigation purposes do not always benefit because sometimes the models' performances vary largely with the timescales concerned. The multiple nature of hazards associated with TCs, however, need accurate rainfall forecasts within a wide range of timescales. For example, flash flood potential estimation requires that the extreme rain rate within hours be well captured (e.g., Chen et al. 2008). For estimation of the extent of flood events, good estimates of the accumulated rainfall over a longer period of time are necessary. Similarly for landslide and debris flow potential estimation because the avalanche processes depend much on the absorbed water in the slopes within days, and thus the accumulated rainfall at particular locations is of most concern. However, very often NWP models are tuned to perform best at a specific timescale only, through either settings in the physical process representations or post processing. This is the reason why different techniques of rainfall estimation other than NWP models are utilized in mitigation. For nowcasting, rainfall estimation algorithms based on radar reflectivity data are widely applied (e.g., Chen et al. 2007). For accumulated rainfall estimation from timescales of hours to days, statistical models of rainfall are common practices because they have the additional advantage of able to generate a large number of different rainfall scenarios with much shorter time compared with running an ensemble of dynamical models. These statistical rainfall models, which are usually developed for specific weather systems, have the limitations of low ability to reproduce extreme rainfall events and that they are not simulating the realistic environment such as topographic effect. The following sections describe several TC-related statistical rainfall models and their recent developments.

8.3.2 The Rainfall Climatology-Persistence Model for the Taiwan Area

Cheung et al. (2008) examined the TC rainfall characteristics in Taiwan and evaluated the The rainfall climatology-persistence (R-CLIPER) model. The following discussions are based on the Cheung et al. study. The spatial characteristics of TC rainfall is studied by first setting up climatology based on the 62 TC cases that passed through the domain 118°–126°E, 18°–27°N during 1989–2002 (Fig. 8.4). When one of the 62 TC tracks passes through one of the grids in the domain, the rainfall data of the 371 rain gauges in Taiwan are recorded. Thus after examining all 62 TC cases, statistics on the average, maximum and minimum rainfall, standard deviation and the number of TCs ever passed through each grid

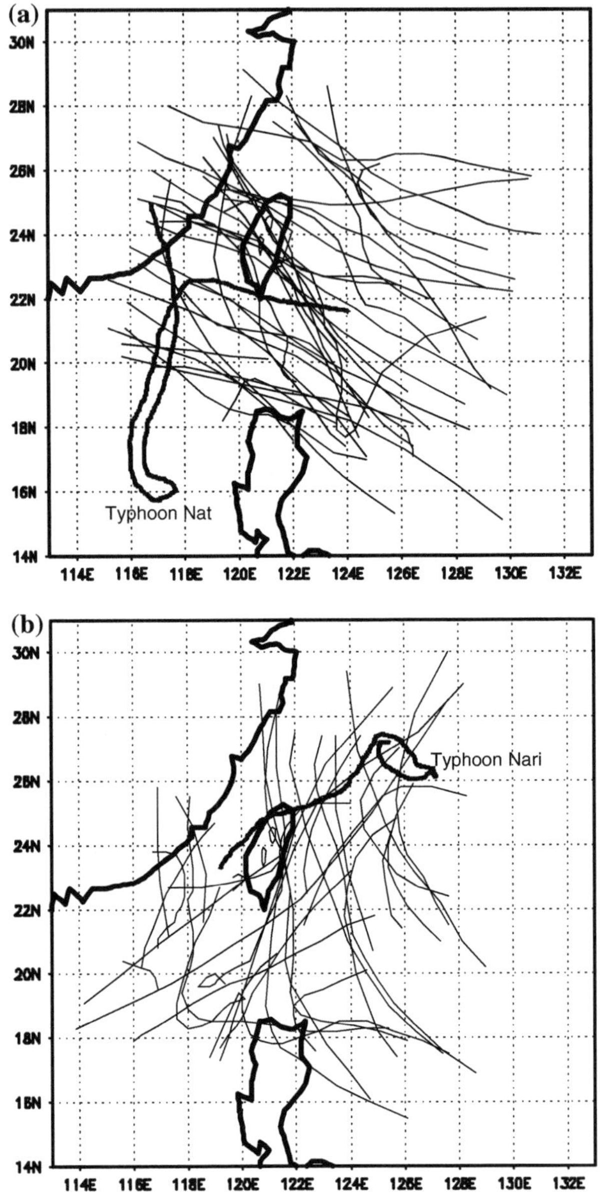

Fig. 8.4 The (**a**) 34 East–West oriented and (**b**) 27 south-north oriented TC tracks during 1989–2002 in the Taiwan area. Note the special track types of Typhoons Nat (1991) and Nari (2001) [adapted from Cheung et al. (2008)]

box and for each rain gauge are obtained. In other words, for each rain gauge there is a map of rainfall describing the climatology of that particular station when a TC is situated at different position in the domain (Fig. 8.5). For example, Fig. 8.5 indicates that Taipei station (46692) will be receiving more rainfall when the TC center is located northeast of Taiwan or over land west of the Central Mountain Range (CMR; Fig. 8.6).

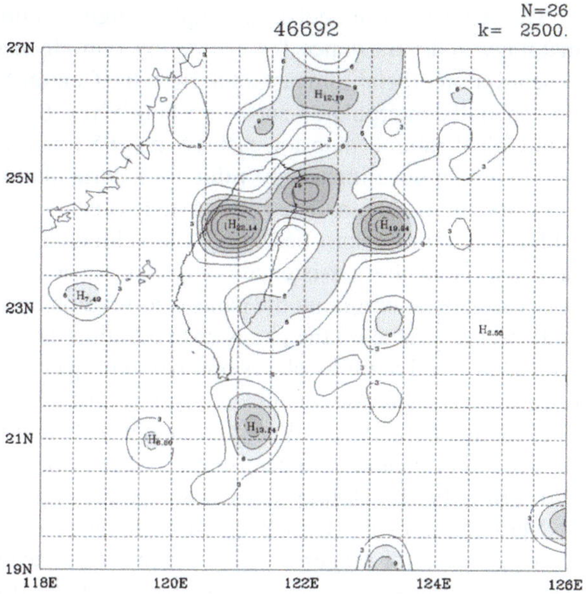

Fig. 8.5 Computation domain and the 0.5° × 0.5° latitude/longitude grid boxes used in computing the statistics of TC rainfall during 1989–2002. The rainfall climatology map shown is for the Taipei station (46692) after a Barnes objective analysis to a fine 0.1° × 0.1° latitude/longitude grid [adapted from Cheung et al. (2008)]

Fig. 8.6 Topography of Taiwan showing the Central Mountain Range with contour levels of 500, 1,500 and 3,000 m [adapted from Chang et al. (2011)]

The R-CLIPER model is a simple weighted combination of TC rainfall climatology and persistence for estimating rainfall in the future (Lee et al. 2006). Whenever an official forecast track from the Central Weather Bureau is issued, it is first spatially interpolated such that hourly positions are available. Then the climatological rainfall map for each of the 365 main-island stations is looked up in turn to obtain the climatology value for a particular TC center position. When this rain amount is accumulated along the forecast TC track, the rain distribution within a certain forecast period (i.e., the climatology component of R-CLIPER) is obtained. Next the persistence component of R-CLIPER is determined. Various tests based on maximizing the pattern correlation between R-CLIPER forecast rain and observed value on the persistence duration indicate that a 3-h period, which is typical convective timescale, is optimal to project a reasonable short-range rainfall estimate. For determining the relative contributions from climatology and persistence in the final rainfall prediction, optimal pattern correlation is realized when the ratio of climatology to persistence is 4/6 (7/3) in the 0–3-h (3–6-h) time period, and then only climatology is used after 6 h.

Verification of R-CLIPER based on independent TC cases finds that the root-mean-square error ranges from 11.4 mm for 3-h accumulated rain to 86.6 mm for 24-h accumulated rain, indicating that the R-CLIPER does provide reasonable estimates of rainfall amount in the short range. Equitable threat score (ETS), which measures the number of forecasts (locations) that match the observed threshold amount (Tuleya et al. 2007; see definition in Cheung et al. 2008) is used to further evaluate the rain prediction skill. For 24-h forecasts, the ETS values of 0.22, 0.13, 0.08 and 0.04 are obtained for the thresholds of 50, 130, 200 and 350 mm, respectively. These ETS values are similar to those of the R-CLIPER for Atlantic hurricane rainfall (Marchok et al. 2007). In Cheung et al. (2008) more verification are performed with ETS values for shorter forecast periods and for individual stations such that the geographic distributions of the model's performance can be examined. Several TC cases are also used to illustrate the applications of R-CLIPER. The readers are referred to that chapter for details.

8.3.3 Recent Developments

Since the R-CLIPER uses the rainfall climatology (average of historical TC cases in the model database) as the major component in the model, it often underestimates the observed rainfall especially for the extreme rainfall in mountainous regions (Cheung et al. 2008). Lee et al. (2008) improved the original R-CLIPER and developed a version 2.0 of the model. The major aspects of improvement are stratifications of the rainfall estimation according to river basins, TC track types, intensity and size. For example, rainfall climatology usually accounts for only about 30–40 % of the observed value, but this percentage varies for different river basins. Thus, it is found that for some river basins the rainfall estimation can be improved by calculating the average of the higher rainfall in the historical database and replacing the climatology component in the model, which is referred to as 'threshold adjustment' in Lee et al. (2008).

The above-mentioned adjustment procedure is certainly case dependent. Through examination of many typical TC cases in the Taiwan area, Lee et al. (2008) found that for the westward-moving TCs that make landfall to Taiwan, correlation between rainfall and intensity is higher than other parameters because the severe convection in the intense TCs is bringing direct impact to the landfall location together with torrential rainfall. On the other hand, for the TC tracks that pass through the Taiwan area but are quite distant from the island, the correlation between rainfall and TC size is higher because the more extensive rain bands usually accompanying the larger TCs have better chance of bringing heavy rainfall to the island even though TC centers do not make landfall at all.

The essential philosophy of the R-CLIPER version 2.0 is its flexibility to adjust the climatology component in different river basins, i.e., what percentile of the historical extreme rainfall to retain. The adjustment procedure is a semi-subjective process of identifying some historical analogues (usually TC cases with similar tracks) of the real-time forecast case, and then assigning appropriate 'thresholds' to the river basins according to the intensity and size of the current TC under forecast.

By way of example from Lee et al. is the case of Typhoon Krosa (2007) that moved westward and then northwestward and made landfall at northeastern Taiwan (Fig. 8.7). The original R-CLIPER estimates heavy rainfall only at the landfall location and near the core of the Typhoon, but actually the heaviest rainfall occurred in central to southwestern Taiwan. By adjusting the climatology component in individual river basins according to some historical analogues and the fact that Typhoon Krosa had a quite large radius of gale-force winds ($\sim$300 km), the estimation of accumulated rainfall distribution from the new R-CLIPER is much improved over the original version.

8.4 Topographic Effect

The Taiwan R-CLIPER discussed in the last section does not consider topographic effect explicitly, although the effect was embedded in the rainfall climatology. Similar situation applies to the Atlantic version of R-CLIPER that utilizes rain estimates from the TMI and derives storm-centered mean rain rate distribution stratified for different TC intensities (Marks et al. 2002). This mean rain rate distribution is azimuthally symmetrical, and when accumulated rain amount is computed along the forecast TC track, the rainfall distribution is also symmetrical with respect to the track, which is not realistic. This is why Lonfat et al. (2007)

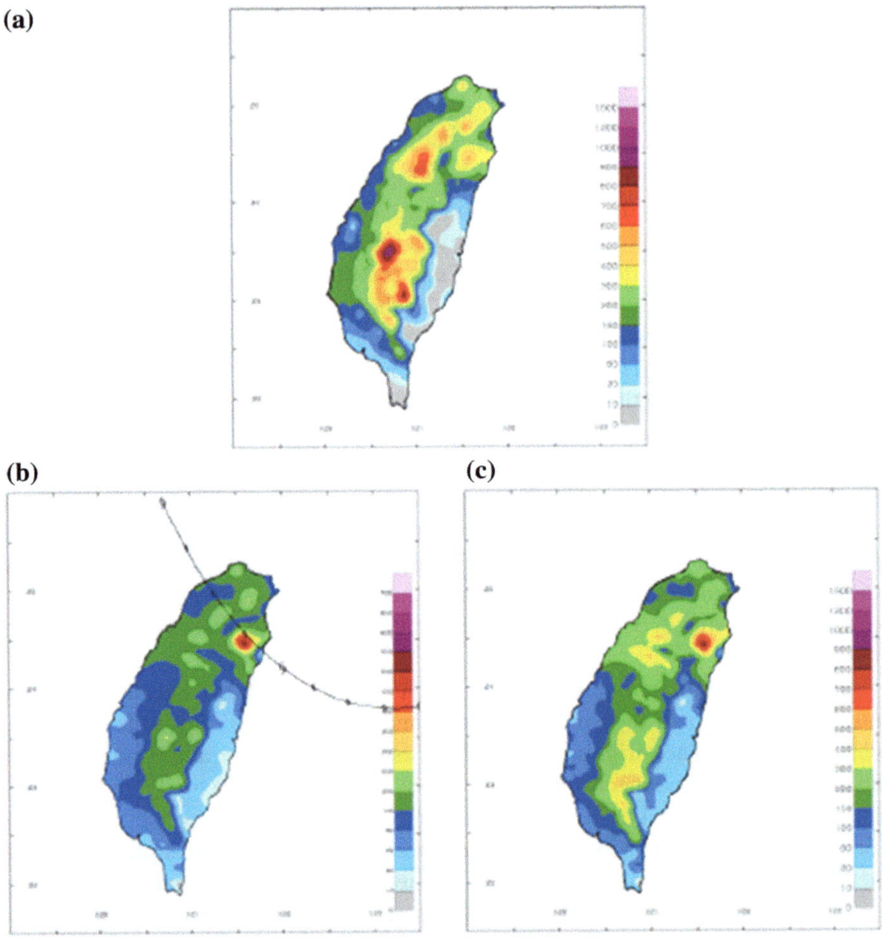

Fig. 8.7 **a** Observed, **b** estimate from the original R-CLIPER, and **c** estimate from the R-CLIPER version 2.0 of accumulated rainfall distribution for Typhoon Krosa (2007) during the period when the Typhoon was in the model domain [adapted from Lee et al. (2008)]

developed the Parametric Hurricane Rainfall Model (PHRaM) that adds a vertical wind shear term and a topography term to the original R-CLIPER model equation. The effect of topography is modeled by evaluating changes in elevation of flow parcels within the storm circulation and correcting the rainfall field in proportion to those changes (Fig. 8.8). Chang et al. (2011) carried out similar effort to make the Taiwan version of R-CLIPER considers topographic effect explicitly in order to improve the model's ability to estimate extreme rainfall. The following discussions are based on the Chang et al. study.

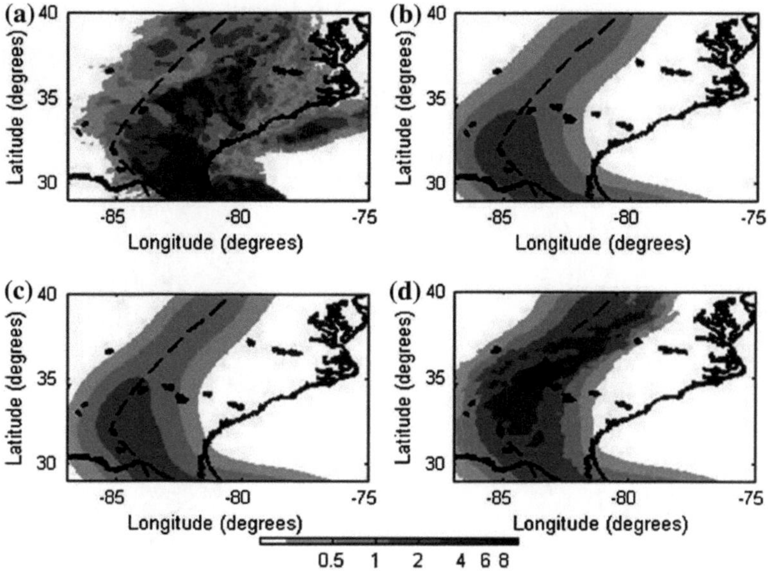

Fig. 8.8 Rainfall accumulation (in.) in Hurricane Frances (2004) for its second U.S. landfall: **a** stage-IV observations, **b** R-CLIPER, **c** PHRaM-NoTopo, and **d** PHRaM. The *dashed line* indicates the best track for Frances [adapted from Lonfat et al. (2007)]

8.4.1 Correlation Between Orographic and Observed Rainfall

The 33 river basins of Taiwan are divided into 5 groups (Fig. 8.9): three to the West (named W1 to W3) and two to the East (E1 and E2). The advantage of analyzing according to the river basins is that if there is heavy rainfall at the catchment area, subsequent flood likely occurs at the lower-level plain area. Thus, additional statistical analyses are also performed for rain stations with high altitudes (>200 m) in order to isolate the effect from topography to the large slopes at the CMR.

The individual relationships between observed rainfall, climatology and orographic rain are examined. For this purpose, linear regression equation in the form of $y = ax + b$, where x and y are the variates, are fitted to observed 24-h accumulated rainfall from the rain stations in Taiwan and either the climatology values or orographic rain estimates. The same flux model as in the preliminary investigation in Cheung and McAneney (2008) is used to calculate the orographic rain. The model considers orographic lifting of horizontal winds and assumes that all lifted moisture is transformed to rainfall (Lin et al. 2002; Chang et al. 2008). Namely,

$$\mathrm{ORO} = (d_a/d_w) \times E \times (\mathbf{V} \cdot \nabla H) \times q \times (L_s/C_s)$$

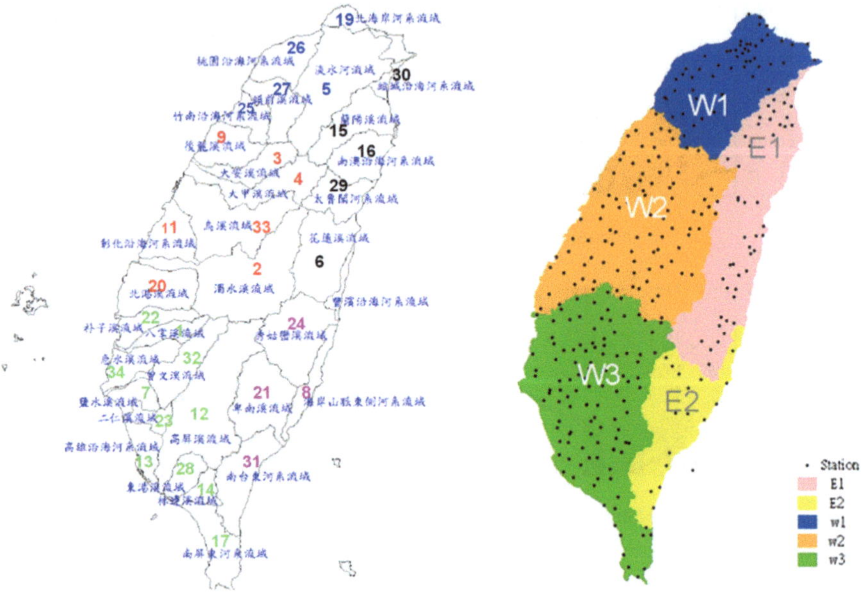

Fig. 8.9 The 33 river basins in Taiwan (*left panel*), which are separated into five groups E1–E2, W1–W3 with locations of the rain stations (*right panel*) [adapted from Chang et al. (2011)]

where d_a and d_w are the density of air and water respectively, E rainfall efficiency assumed to be 1.0, **V** horizontal wind vector, ∇H terrain gradient, and q the mixing ratio or specific humidity. (L_s/C_s) measures the length scale and speed of a mesoscale convective system that passes through a point of interest and has the unit of time, and is assigned 24 h for our estimation of accumulated rainfall.

When the explained variances (R^2) in the linear regression are examined, it is found that climatology explains from nearly 50 % to about 66 % of observed rainfall (Table 8.1), which are consistent with the results from similar analyses for

Table 8.1 Explained variance R^2 of observed rainfall by climatology and orographic rain estimate, respectively for all stations or only those with altitude above 200 m

	R^2 (Obs-Clim) all stations	R^2 (Obs-Clim) stations >200 m	R^2 (Obs-Oro) all stations	R^2 (Obs-Oro) stations >200 m
W1	0.486	0.412	0.166	0.193
W2	0.562	0.507	0.115	0.191
W3	0.573	0.569	0.103	0.148
W123	0.542	0.502	0.110	0.148
E1	0.597	0.611	0.119	0.186
E2	0.662	0.633	0.174	0.264
E12	0.604	0.620	0.140	0.229

W123 represents the combined western regions, and E12 the combined Eastern regions of Taiwan

R-CLIPER in Cheung et al. (2008). The R^2 values for the Eastern regions are all higher than those for the western region, which indicates that more complex processes are affecting rainfall in western Taiwan, such as TC structures changed by the CMR and monsoon effects. Thus, rainfall estimation based on simple climatology or R-CLIPER is more reliable at Eastern Taiwan than in the western region. As expected, the orographic rain estimate explains less percentage of the observed rainfall, which ranges from 10 to 20 %. The increasing trend in R^2 from West to East is not clear because it is a single mechanism of topographic effect under consideration in this calculation, and the East–West contrast is not the dominating factor.

The explained variances for rain stations with altitude higher than 200 m are also examined. For these rain stations on the slopes, the R^2 values between observed rain and climatology change only little compared with all stations. The slight decreases in R^2 likely attribute to smaller number of stations in calculating the statistics. However, it can be seen that orographic rain can explain more variance of observed rainfall for those rain stations on the mountains with an increase from a few percents to nearly 10 %. Thus, the mechanism of orographic lifting does contribute more to the total rainfall at the slopes compared with plain areas, which has implications for further TC rainfall model development especially with respect to the consideration of regional dependence.

8.4.2 Multiple Regression Analyses

In order to explicitly take into account the contribution from topographic effect in TC rainfall estimation, a multiple regression model is established for 24-h accumulated rainfall. The model takes the form

$$\mathrm{OBS} = \mathrm{A1} \times \mathrm{ORO} + \mathrm{A2} \times \mathrm{CLIM} + \mathrm{B},$$

where OBS, ORO and CLIM are observed accumulated rainfall, orographic rain as calculated with the above formula, and the climatology value, respectively. A1, A2 and B are the regression coefficients. The regression is performed for all the 62 TC cases as in R-CLIPER, for each of the regions (W1–W3, E1–E2) as well as the combined regions of W123 (i.e., all western stations) and E12 (i.e., all Eastern stations). There are a total of 37,752 pieces of data that includes all TC cases and stations over Taiwan, and then these data are partitioned into different regions. Due to the different number of TCs that passed through each of the sub-regions, the available number of data that is put into the regression varies from region to region. For example, the western regions of W123 has 20,416 pieces of data and the Eastern regions E12 has 3,504 (columns termed No. of data in Table 8.2). These numbers of data dropped to 9,223 and 1,292 respectively when only stations higher than 200 m are considered (Table 8.3).

Table 8.2 Regression coefficients and explained variance R^2 for the five separate regions and combined regions

	All stations					
	No. of data	A1	A2	B	R^2	Significance of model
W1	5,319	0.014 (0.000)	0.834 (0.000)	−5.896 (0.000)	0.491	0.000
W2	7,231	0.001 (0.153)	1.049 (0.000)	−4.063 (0.000)	0.563	0.000
W3	7,866	0.001 (0.886)	1.100 (0.000)	−9.089 (0.000)	0.573	0.000
W123	20,416	0.005 (0.000)	0.973 (0.000)	−6.273 (0.000)	0.543	0.000
E1	2,127	0.001 (0.950)	1.120 (0.000)	−30.174 (0.000)	0.597	0.000
E2	1,377	0.014 (0.000)	1.050 (0.000)	−18.949 (0.000)	0.627	0.000
E12	3,504	0.004 (0.086)	1.080 (0.000)	−22.912 (0.000)	0.604	0.000

The numbers in parentheses are the statistical significances for individual coefficients, and the number of data input to each regression analysis is also shown

Table 8.3 As in Table 8.2 except for rain stations with altitudes higher than 200 m

	Stations >200 m					
	No. of data	A1	A2	B	R^2	Significance of model
W1	2,443	0.033 (0.000)	0.695 (0.000)	−15.142 (0.000)	0.440	0.000
W2	4,084	0.011 (0.000)	0.918 (0.000)	−10.961 (0.000)	0.514	0.000
W3	2,696	0.002 (0.346)	1.110 (0.000)	−13.867 (0.000)	0.569	0.000
W123	9,223	0.016 (0.000)	0.877 (0.000)	−15.622 (0.000)	0.511	0.000
E1	597	0.018 (0.011)	1.087 (0.000)	−60.946 (0.000)	0.615	
E2	695	0.020 (0.000)	1.030 (0.000)	−29.086 (0.000)	0.642	
E12	1,292	0.015 (0.000)	1.039 (0.000)	−36.028 (0.000)	0.624	

Comparison between Tables 8.1 and 8.2 indicates that after adding the orographic term to climatology, the explained variance (R^2) increases slightly when all rain stations are considered. However, when only the stations above 200 m are considered (Table 8.3), the increase in explained variances are more than marginal, and this applies to almost all sub-regions. Note that when interpreting the regression coefficients, their magnitudes do not reflect the relative importance of the variables. For example, the coefficients for orographic term are always much lower than those for climatology because the rain estimates from the orographic flux formula are much higher than reasonable rainfall amount (Cheung and McAneney 2008), and thus the coefficients serve to scale down to near-climatology values.

A noteworthy point is that the coefficients for region W3 (southwestern Taiwan) and to certain degree W2 (western Taiwan) are always very small compared with other regions. That is, the contribution from topographic effect is minimal in these two regions as shown by our regression analyses. However, extreme rainfall events associated with TCs did occur in southwestern Taiwan [e.g., Typhoon Mindulle (2004) that is discussed in Cheung et al. (2008) and the recent Typhoon Morakot (2009)] and as a matter of fact that region is the most disastrous area that suffers from heavy rainfall. It is believed that other mechanisms other than topographic effect are enhancing the rainfall in that region.

The above conclusion in regard to the southwestern regions in Taiwan is also confirmed by statistical significance tests for the regression models. The values of statistical significance given in Tables 8.2 and 8.3 represent the probability that the predicted values from the regression model or the individual regression coefficients are from a random distribution. It can be seen from Tables 8.2 and 8.3 that no matter when all stations or only those with high altitudes are considered, the regression models are statistical significant for all regions. This certainly makes sense since it is known that at least the rainfall climatology in the regression model accounts for a large portion of rainfall variance. However, the confidence level associated with the coefficient A1 for region W3 (and for E1 and W2 in the all-station model), which scales the contribution from orographic rain, is quite low. This recapitulates the early point that for southwestern Taiwan, adding the orographic lifting term to the regression model does not help much because they are not very different from adding a random variate. One effect not considered in this model comes from the summer southwesterly monsoon that coincides much with the peak typhoon season such that this kind of effects is most easily manifested. For region E1 (northeastern Taiwan), the non-effectiveness of adding the orographic term in the regression model may be due to effects from the winter northeasterly monsoon, which have been discussed in Cheung et al. (2008) and Wu et al. (2009). However, there are always fewer typhoon occurrences in the winter season and thus the number of late-season typhoons with rainfall distribution deviated much from climatology and simple orographic estimation is relatively small. Nevertheless, the statistical significance of the orographic term in the high-altitude regression model is valid at confidence level of nearly 99 %, which indicates that consideration of topographic effect in this region is still critical.

Besides a quantitative understanding of the relative contribution from climatology and topographic effect to TC rainfall, the established multiple regression models can be applied as an improved forecast tool to R-CLIPER. In particular, since regionally specific models are available, improvements in terms of the more localized topographic effects can be expected when the W1–W3 and E1–E2 models are combined into one for the entire Taiwan area. These have been illustrated in Chang et al. (2011) with several TC cases.

8.4.3 Importance of Remote-Sensing Techniques and Data

The development of the Taiwan R-CLIPER and the subsequent multiple regression analyses benefit from the high-resolution network of automatic rain station. However, for regions with less density of rain stations such as in China, the ability of station-based rainfall climatology to resolve the rainfall distributions from TCs will be much reduced. Therefore, a major effort in research recently is to develop satellite-based algorithms for rain rate estimation that take into account the effects of land. Two recent studies are briefly summarized here. Jiang et al. (2008) investigated whether the 'rainfall potential' derived from the NASA TRMM multisatellite precipitation analysis is capable of predicting TC rainfall overland. The 'rainfall potential' is defined by using the satellite-derived rain rate, the satellite-derived storm size, and the storm translation speed. A total of 37 TCs that impacted the U.S. were examined and rainfall distributions for overland and over ocean are compared. It is found that the TC rainfall over ocean has a stronger relationship with TC intensity. Nevertheless, high correlations between rain potentials before landfall, in particular on the day prior to landfall, and the maximum storm total rain overland are identified. A TC overland rainfall index is introduced based on the satellite-derived rain potential. This technique will be useful when R-CLIPER-type model is extended to other regions without access to in situ rain gauge observations needed to develop the rainfall climatology or the rain gauge density is small.

Yu et al. (2009) compared the precipitation retrievals from three satellite datasets: the TRMM algorithm 3B42, the Climate Prediction Center Morphed (CMORPH) product, and a methodology based on the Geostationary Meteorological Satellite-5 infrared brightness temperature data (GMS5-TBB) developed by the Shanghai Typhoon Institute. The verification of the three datasets was based on rain gauge measurement associated with 50 TCs that affected China during 2003–2006, with 25 of them actually made landfall. It is found that in general the three datasets are able to provide reasonable rainfall distribution over land, but the skill of rain estimate decreases with latitude and rainfall amount. Because the TRMM-3B42 algorithm applies rain gauge adjustment, it has slightly better skill than the CMORPH dataset. However, both the TRMM-3B42 and CMORPH underestimate moderate to heavy rainfall but overestimate light precipitation. The GMS5-TBB method first analyzes the spatial structure of infrared brightness temperature data to identify the convective core and associated rain rate. For non-convective area, stratiform rain rate is also estimated. For rainfall over land associated with landfalling TCs, the method also adjusts the satellite estimates by rain gauge data. The result is that the GMS5-TBB method overestimates light

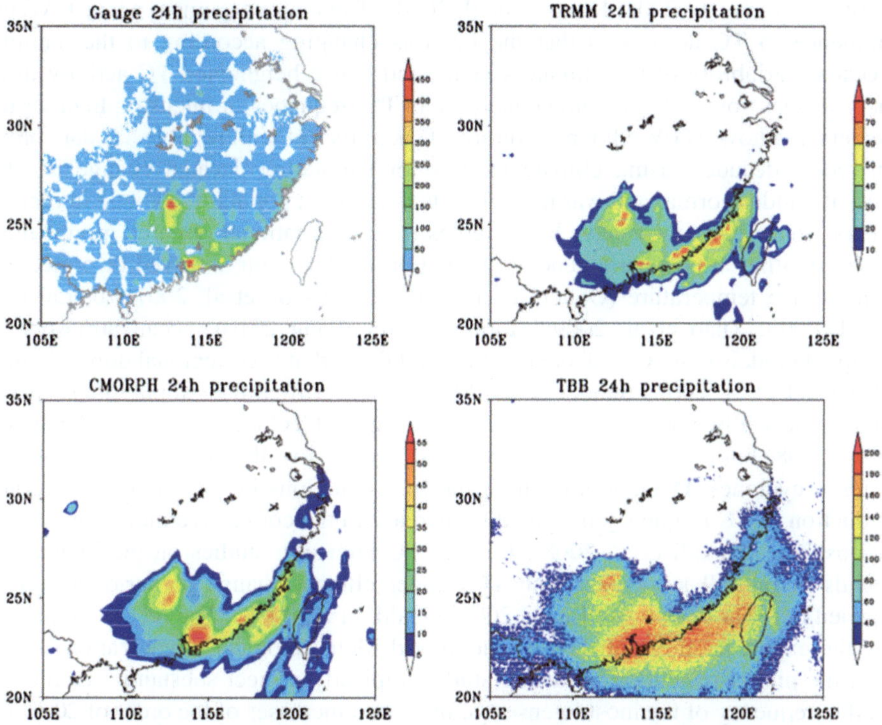

Fig. 8.10 Comparison of the 24-h accumulated rainfall associated with Typhoon Bilis during 14–15 Jul 2006 from the four datasets of rain gauge, TRMM-3B42, CMORPH, and GMS5-TBB, respectively [adapted from Yu et al. (2009)]

precipitation more than the TRMM-3B42 and CMORPH, but performs much better than the other two algorithms in estimating heavy rainfall events. An example of Typhoon Bilis (2006) was given in Yu et al. (Fig. 8.10), which showed that the maximum accumulated rainfall estimate of 200–300 mm from GMS5-TBB at South China is much closer to the rain gauge measurement of about 400 mm than the other two methods.

8.5 Impacts from Global Changes

The future impacts from TCs under global changes depend on a lot of factors that include changes in the characteristics of TC activity, vulnerability of the coastal areas and resilience of our community to TC-related hazards. In recent years there have been debates on whether anthropogenic global warming (AGW) is modifying global TC activities, and in which aspects such as frequency and intensity are mostly affected (e.g., Emanuel 2005; Webster et al. 2005; Landsea 2005; Chan

2006; Vecchi et al. 2008; Knutson et al. 2010). The major argument against AGW influence to TC activity is that the latter is changing according to the natural decadal variability of the climate system, and hence heightened TC activity in a decade does not imply monotonic increase in TC frequency or intensity. In order to understand how AGW influences future TC activity but with reliable TC data only in recent decades, using climate models for simulating decadal change in TC activity under normal and warming climate becomes a popular technique in recent years. Since 2006, there have been studies from international research groups that utilized either atmospheric general circulation models with observed or projected sea surface temperature (Oouchi et al. 2006; Bengtsson et al. 2007a, b; LaRow et al. 2008; Zhao et al. 2009; Murakami and Wang 2010), ocean–atmosphere coupled models (e.g., Gualdi et al. 2008), and those that used regional downscaling (Emanuel et al. 2008; Bender et al. 2010). For warm-climate simulations, these studies basically agree on a reducing trend of global TC frequency (e.g., Knutson et al. 2008) but deviations exist on regional trends as well as the frequency of the intense cyclones. Diagnoses of these model simulations indicate that the overall reduction in TC frequency is related to increase in the convective inhibition of the atmosphere (Gualdi et al. 2008). Besides TC intensity, studies on the long-term trends in rainfall associated with TCs under climate change scenarios are also carried out (Hasegawa and Emori 2005; Gualdi et al. 2008). The recent summary by the expert team established by the World Meteorological Organization states that the higher-resolution modeling studies typically project substantial increases in the frequency of the most intense cyclones, and increases of the order of 20 % in the precipitation rate within 100 km of the storm center (Knutson et al. 2010). That is, more extreme convection episodes will occur in future TCs. Given this likely increase in TC rainfall under global changes, what are the roles played by the statistical rainfall models? Although these statistical models, such as those discussed herein, generally are not able to reproduce all of the extreme rainfall in TCs, they are capable for applications in estimating future TC-related rainfall scenarios. For example, it is convenient in execution of the statistical models such that a large number of simulations can be carried out for estimating future rainfall footprints in a particular region if there are changes in the TC tracks as identified in Murakami and Wang (2010). When more sophisticated statistical rainfall models that consider environment influences such as vertical wind shear, topography and synoptic effects like monsoonal flows, the models will continue to serve the forecast community for real-time mitigation, risk analysis as well as analyzing impacts of global change.

8.6 Summary

The torrential rainfall associated with landfalling TCs often represents the major impact to coastal regions, but at the same time an enormous challenge to meteorologists and forecasts. This chapter first describes the dissipation process of TCs over land, and then discusses the complex dynamical processes involved in TC landfalls. These dynamical processes are introduced by the increased surface roughness and reduced surface moisture fluxes of land. Whereas each of these effects has quite straightforward impacts to the TC dynamical and convective structure, when they combine the resultant behavior of the TC is much less predictable. On one hand the effects from the land-sea contrast are interacting with the cyclonic circulation of the TC, on the other hand continuous feedback processes between the TC circulation and the ambient environment add substantial complexity to the problem. Although there have been a large number of observational and numerical studies on TC landfall in the last few decades, there is no single theory or conceptual model that can explain the asymmetry in convection and rainfall distribution of landfalling TCs in all ocean basins well.

The development of rainfall prediction techniques according to the needs of mitigation is emphasized in this chapter. Thus, after reviewing the requirements on the skill of rainfall forecasts from the perspectives of mitigation, the operation and performance of several statistical TC rainfall models are discussed. These include the PHRaM for predicting Atlantic hurricane rainfall, the R-CLIPER for the Taiwan area and the effort to introduce a component in R-CLIPER to consider topographic effect. While the R-CLIPER can usually explain 50–60 % of rainfall variance, the additional topographic component improves the model's performance in reproducing the local extreme rain that is lacking in the original model. Whereas NWP models are improving rapidly in terms of rainfall prediction, this kind of statistical models such as PHRaM and R-CLIPER has the advantage of efficiency in generating a large number of different rainfall scenarios such that they are not only applicable in real-time mitigation, but for risk analyses under the consideration of global changes as well.

Acknowledgments The first two authors (KKWC and LTCC) would like to thank Swiss Re for supporting their work on the rainfall climatology-persistence model. The continuous encouragement, support and comments from Prof. John McAneney of the Risk Frontiers Natural Hazards Research Centre of Macquarie University are much appreciated. The third author (YL) is supported by the Higher Degree Research project support fund of Macquarie University. The third author (YL) was supported by a Macquarie University Research Excellence Scholarship (MQRES) and Higher Degree project support fund. YL is currently supported by a postdoctoral fellowship from the City University of Hong Kong.

References

Bender, M.A., Tuleya, R.E., Kurihara, Y.: A numerical study of the effect of a mountain range on a landfalling tropical cyclone. Mon. Wea. Rev. **113**, 567–583 (1985)

Bender, M.A., Tuleya, R.E., Kurihara, Y.: A numerical study of the effect of island terrain on tropical cyclones. Mon. Wea. Rev. **115**, 130–155 (1987)

Bender, M.A., Knutson, T.R., Tuleya, R.E., Sirutis, J.J., Vecchi, G.A., Garner, S.T., Held, I.M.: Modeled impact of anthropogenic warming on the frequency of intense atlantic hurricanes. Science **327**, 454–458 (2010)

Bengtsson, L., Hodges, K.I., Esch, M.: Tropical cyclones in a T159 resolution global climate model: comparison with observations and reanalyses. Tellus **59A**, 396–416 (2007a)

Bengtsson, L., Hodges, K.I., Esch, M., Keenlyside, N., Kornblueh, L., Luo, J–.J., Yamagata, T.: How may tropical cyclones change in a warmer climate? Tellus **59A**, 539–561 (2007b)

Blackwell, K.G.: The evolution of Hurricane Danny (1997) at landfall: Doppler-observed eyewall replacement, vortex contraction/intensification, and low-level wind maxima. Mon. Wea. Rev. **128**, 4002–4016 (2000)

Chan, J.C.L.: Comment on changes in tropical cyclone number, duration and intensity in a warming environment. Nature **311**, 1713 (2006)

Chan, J.C.L.: Changes in track and structure of tropical cyclones near landfall. In: Extended Abstract, 29th Conference on Hurricanes and Tropical Meteorology, Tuscon, Arizona, U.S., American Meteorological Society (2010)

Chan, J.C.L., Liang, X.: Convective asymmetries associated with tropical cyclone landfall. Part I: f -plane simulations. J. Atmos. Sci. **60**, 1560–1567 (2003)

Chan, J.C.L., Liu, K.S., Ching, S.E., Lai, E.S.T.: Asymmetric distribution of convection associated with tropical cyclones making landfall along the south China coast. Mon. Wea. Rev. **132**, 2410–2420 (2004)

Chang, C.P., Yeh, T.-C., Chen, J.-M.: Effects of terrain on the surface structure of typhoons over Taiwan. Mon. Wea. Rev. **121**, 734–752 (1993)

Chang, L.T.-C., Chen, G.T.-J., Cheung, K.K.W.: Mesoscale simulation and moisture budget analyses of a heavy rain event over southern Taiwan in the Meiyu season. Meteor. Atmos. Phys. **101**, 43–63 (2008)

Chang, L.T.-C., Cheung, K.K.W., McAneney, J.: A statistical analysis of the topographic effect on rainfall enhancement for tropical cyclones in the Taiwan area, p. 26, Research Report for Swiss Re, Risk Frontiers Natural Hazard Research Centre, Macquarie Univeristy (2011)

Chang, S.: The orographic effects induced by an island mountain range on propagating tropical cyclones. Mon. Wea. Rev. **110**, 1255–1270 (1982)

Chen, C.-Y., Lin, L.-Y., Yu, F.-C., Lee, C.-S., Tseng, C.-C., Wang, A.-X., Cheung, K.K.W.: Improving debris flow monitoring in Taiwan by using high-resolution rainfall products from QPESUMS. Nat. Hazards **40**, 447–461 (2007). doi:10.1007/s11069-006-9004-2

Chen, C.-Y., Chen, L.-K., Yu, F.-C., Lin, S.-C., Lin, Y.-C., Lee, C.-L., Wang, Y.-T., Cheung, K.W.: Characteristics analysis for the flash flood-induced debris flow. Nat. Hazards **47**, 245–261 (2008). doi:10.1007/s11069-008-9217-7

Cheung, K.K.W., McAneney, J.: Development of a statistical model for tropical cyclone rainfall in Taiwan for reinsurance loss analyses: preliminary results, p. 29, Research Report for Swiss Re, Risk Frontiers Natural Hazard Research Centre, Macquarie Univeristy (2008)

Cheung, K.K.W., Huang, L.-R., Lee, C.-S.: Characteristics of rainfall during tropical cyclone periods in Taiwan. Nat. Hazards Earth Syst. Sci. **8**, 1463–1474 (2008)

Dastoor, A., Krisiinamurti, T.N.: The landfall and structure of a tropical cyclone: the sensitivetity of model predictions to soil moisture parameterizations. Boundary-Layer Meteorol. **55**, 345–380 (1991)

Dunn, G.E., Miller, B.I.: Atlantic Hurricanes, p. 377. Louisiana State University Press, Louisiana (1960)

Ebert, E.E., Janowiak, J.E., Kidd, C.: Comparison of near-real-time precipitation estimates from satellite observations and numerical models. Bull. Amer. Meteor. Soc. **88**, 47–64 (2007)

Emanuel, K.: Increasing destructiveness of tropical cyclones over the past 30 years. Science **436**, 686–688 (2005)

Emanuel, K., Sundararajan, R., Williams, J.: Hurricanes and global warming: results from downscaling IPCC AR4 simulations. Bull. Amer. Meteor. Soc. **89**, 347–367 (2008)

Farfan, L.M., Zehnder, J.A.: An analysis of the landfall of Hurricane Nora (1997). Mon. Wea. Rev. **129**, 2073–2088 (2001)

Gao, S., Meng, Z., Zhang, F., Bosart, L.F.: Observational analysis of heavy rainfall mechanisms associated with severe tropical storm Bilis (2006) after its landfall. Mon. Wea. Rev. **137**, 1881–1897 (2009)

Gualdi, S., Scoccimarro, E., Navarra, A.: Changes in tropical cyclone activity due to global warming: results from a high-resolution coupled general circulation model. J. Climate **21**, 5204–5228 (2008)

Hasegawa, A., Emori, S.: Tropical cyclones and associated precipitation over the western North Pacific: T106 atmospheric GCM simulation for present-day and doubled CO_2 climates. SOLA **1**, 145–148 (2005)

Ho, F.P., Su, J.C., Hanevich, K.L., Smith, R.J., Richards, F.P.: Hurricane climatology for the Atlantic and Gulf Coasts of the United States, p. 195, NOAA Technical Report NWS 38 (1987)

Jiang, H., Halverson, J.B., Simpson, J., Zipser, E.J.: Hurricane "rainfall potential" derived from satellite observations aids overland rainfall prediction. J. Appl. Meteor. Climatol. **47**, 944–959 (2008)

Jones, R.W.: A simulation of hurricane landfall with a numerical model featuring latent heating by the resolvable scales. Mon. Wea. Rev. **115**, 2279–2297 (1987)

Kaplan, J., DeMaria, M.: A simple empirical model for predicting the decay of tropical cyclone winds after landfall. J. Appl. Meteor. **34**, 2499–2512 (1995)

Kaplan, J., DeMaria, M., Knaff, J.A.: A revised tropical cyclone rapid intensification index for the Atlantic and east Pacific basins. Weather Forecast. **25**, 220–241 (2010)

Knutson, T., Sirutis, J., Garner, S., Wecchi, G., Held, I.: Simulated reduction in Atlantic hurricane frequency under twenty-first-century warming condition. Nat. Geosci. p. 22. (2008). doi:10.1038/ngeo202

Knutson, T.R., McBride, J.L., Chan, J., Emanuel, K., Holland, G., Landsea, C., Held, I., Kossin, J.P., Srivastave, A.K., Sugi, M.: Tropical cyclones and climate change. Nature Geosci., **3**. doi:10.1038/ngeo779 (2010)

Koteswaram, P., Gaspar, S.: The surface structure of tropical cyclones in the Indian area. Ind. J. Meteor. Geophys. **7**, 339–352 (1956)

Krajewski, W.F., Gabriele, V., Smith, J.A.: RADAR-rainfall uncertainties. Bull. Amer. Meteor. Soc. **91**, 87–94 (2010). doi:10.1175/2009BAMS2747.1

Landsea, C.W.: Hurricanes and global warming. Science **438**, E11–E12 (2005)

LaRow, T., Lim, Y.-K., Shin, D., Chassignet, E., Cocke, S.: Atlantic basin seasonal hurricane simulations. J. Climate **21**, 3191–3206 (2008)

Lee, C.-S., Huang, L.-R., Shen, H.-S., Wang, S.-T.: A climatological model for forecasting typhoon rainfall in Taiwan. Nat. Hazards **37**, 87–105 (2006)

Lee, C.-S., et al.: Improvements to a river basin and catchment-based quantitative precipitation forecast technique during meiyu and typhoon periods (I), p. 248, Research report for the Water Resources Agency, Taiwan, National Taiwan University (2008) (In Chinese, English abstract available)

Li, Y., Cheung, K.K.W., Chan, J.C.L.: An observational study on tropical cyclone landfall processes in the northwestern Australian region. Presentation at the AMOS and MetSoc NZ Joint Conference: Extreme Weather, Te Papa, Wellington, New Zealand, Australian Meteorological and Oceanographic Society and the Meteorological Society of New Zealand (2011)

Li, Y., Cheung, K.K.W., Chan, J.C.L., Tokuno, M.: Rainfall distribution of five landfalling tropical cyclones in the northwestern Australian region. Aust. Met. Oceanog. J. **63**, 325–338 (2013)

Lin, Y.-L., Han, J.G., Hamilton, D.W., Huang, C.-Y.: Orographic influence on a drifting cyclone. J. Atmos. Sci. **56**, 534–562 (1999)

Lin, Y.-L., Ensley, D.B., Chiao, S., Huang, C.-Y.: Orographic influences on rainfall and track deflection associated with the passage of a tropical cyclone. Mon. Wea. Rev. **130**, 2929–2950 (2002)

Lonfat, M., Marks Jr, F.D., Chen, S.: Precipitation distribution in tropical cyclones using the Tropical Rainfall Measuring Mission (TRMM) microwave imager: a global perspective. Mon. Wea. Rev. **132**, 1645–1660 (2004)

Lonfat, M., Rogers, R., Marchok, T., Marks Jr, F.D.: A parametric model for predicting hurricane rainfall. Mon. Wea. Rev. **135**, 3086–3097 (2007)

Marchok, T., Rogers, R., Tuleya, R.: Validation schemes for tropical cyclone quantitative precipitation forecasts: evaluation of operational models for U.S. landfalling cases. Weather Forecast. **22**, 726–746 (2007)

Marks Jr, F.D.: Evolution of the structure of precipitation in Hurricane Allen (1980). Mon. Wea. Rev. **113**, 909–930 (1985)

Marks, F.D., Kappler, G., DeMaria, M.: Development of a tropical cyclone rainfall climatology and persistence (R-CLIPER) model. In: Preprints, 25th Conference Hurricane and Tropical Meteorology, pp. 327–328, San Diego, American. Meteorological Society (2002)

Miller, B.L.: A study of the filling of Hurricane Donna (1960) over land. Mon. Wea. Rev. **92**, 389–406 (1964)

Murakami, H., Wang, B.: Future change of north Atlantic tropical cyclone tracks: projection by a 20 km-mesh global atmospheric model. J. Climate **23**, 2699–2721 (2010)

Oouchi, K., Yoshimura, J., Yoshimura, H., Mizuta, R., Kusunoki, S., Noda, A.: Tropical cyclone climatology in a global-warming climate as simulated in a 20 km-mesh global atmospheric model: Frequency and wind intensity analyses. J. Meteor. Soc. Japan **84**, 259–276 (2006)

Parrish, J.R., Burpee, R.W., Marks Jr, F.D., Grebe, R.: Rain patterns observed by digitized radar during the landfall of Hurricane Frederic (1979). Mon. Wea. Rev. **110**, 1933–1944 (1982)

Powell, M.D.: The transition of the Hurricane Frederic boundary-layer wind field from the open Gulf of Mexico to landfall. Mon. Wea. Rev. **110**, 1912–1932 (1982)

Powell, M.D.: Changes in the low-level kinematic and thermodynamic structure of Hurricane Alicia (1983) at landfall. Mon. Wea. Rev. **115**, 75–99 (1987)

Ramsay, H.A., Leslie, L.M.: The effects of complex terrain on severe landfalling tropical cyclone Larry (2006) over Northeast Australia. Mon. Wea. Rev. **136**, 4334–4354 (2008)

Rappaport, E.N., Franklin, J.L., Schumacher, A.B., DeMaria, M., Shay, L.K., Gibney, E.J.: Tropical cyclone intensity change before U.S. Gulf coast landfall. Weather Forecast. **25**, 1380–1396 (2010)

Saffir, H.S.: Design and construction requirements for hurricane resistant construction, p. 20. American Society of Civil Engineers, Preprint Number 2830 (1977)

Schwerdt, R.W., Ho, F.P., Watkinds, R.R.: Meteorological criteria for standard project hurricane and probable maximum hurricane windfields, Gulf and East Coasts of the United States, p. 317, NOAA Technical Report NWS 23 (1979)

Simpson, R.H.: A proposed scale for ranking hurricanes by intensity. In: Minutes of the 8th NOAA, NWS Hurricane Conference, Miami, Florida, National Oceanic and Atmospheric Administration (1971)

Tuleya, R.E., Kurihara, Y.: A numerical simulation of the landfall of tropical cyclones. J. Atmos. Sci. **35**, 242–257 (1978)

Tuleya, R.E., DeMaria, M., Kuligowski, R.J.: Evaluation of GFDL and simple statistical model rainfall forecasts for US landfalling tropical storms. Weather Forecast. **22**, 56–70 (2007)

Vecchi, G.A., Swanson, K.L., Soden, B.L.: Whither hurricane activity? Science **322**, 687–689 (2008)

Wang, Y., Wu, C–.C.: Current understanding of tropical cyclone structure and intensity changes: a review. Meteorol. Atmos. Phys. **87**, 257–278 (2004)

Webster, P.J., Holland, G.J., Curry, J.A., Chang, H.-R.: Changes in tropical cyclone number, duration and intensity in a warming environment. Nature **309**, 1844–1846 (2005)

Wong, M.L.M., Chan, J.C.L., Zhou, W.: A simple empirical model for estimating the intensity change of tropical cyclones after landfall along the South China coast. J. Appl. Meteor. Climatol. **47**, 326–338 (2008)

Wu, C–.C., Kuo, Y.-H.: Typhoons affecting Taiwan: current understanding and future challenges. Bull. Amer. Meteor. Soc. **80**, 67–80 (1999)

Wu, C–.C.: Numerical simulation of Typhoon Gladys (1994) and its interaction with Taiwan terrain using the GFDL hurricane model. Mon. Wea. Rev. **129**, 1533–1549 (2001)

Wu, C–.C., Yen, T.-H., Kuo, Y.-H., Wang, W.: Rainfall simulation associated with Typhoon Herb (1996) near Taiwan. Part I: the topographic effect. Weather Forecast. **17**, 1001–1015 (2002)

Wu, C.-C., Cheung, K.K.W., Lo, Y.-Y.: Numerical study of the heavy rainfall event due to the interaction of Typhoon Babs (1998) and the northeasterly monsoon. Mon. Wea. Rev. **137**, 2049–2064 (2009)

Yeh, T.-C., Elsberry, R.L.: Interaction of typhoons with the Taiwan orography. Part I: upstream track deflections. Mon. Wea. Rev. **121**, 3193–3212 (1993a)

Yeh, T.-C., Elsberry, R.L.: Interaction of typhoons with the Taiwan orography. Part II: continuous and discontinuous tracks across the island. Mon. Wea. Rev. **121**, 3213–3233 (1993b)

Yu, Z., Yu, H., Chen, P., Qian, C., Yue, C.: Verification of tropical cyclone-related satellite precipitation estimates in mainland China. J. Appl. Meteor. Climatol. **48**, 2227–2241 (2009)

Zehnder, J.A.: The influence of large-scale topography on barotropic vortex motion. J. Atmos. Sci. **50**, 2519–2532 (1993)

Zhao, M., Held, I.M., Lin, S.-J., Vecchi, G.A.: Simulations of global hurricane climatology, interannual variability, and response to global warming using a 50 km resolution GCM. J. Climate **22**, 6653–6678 (2009)

Part III
Typhoon Impacts on Marine Ecosystems

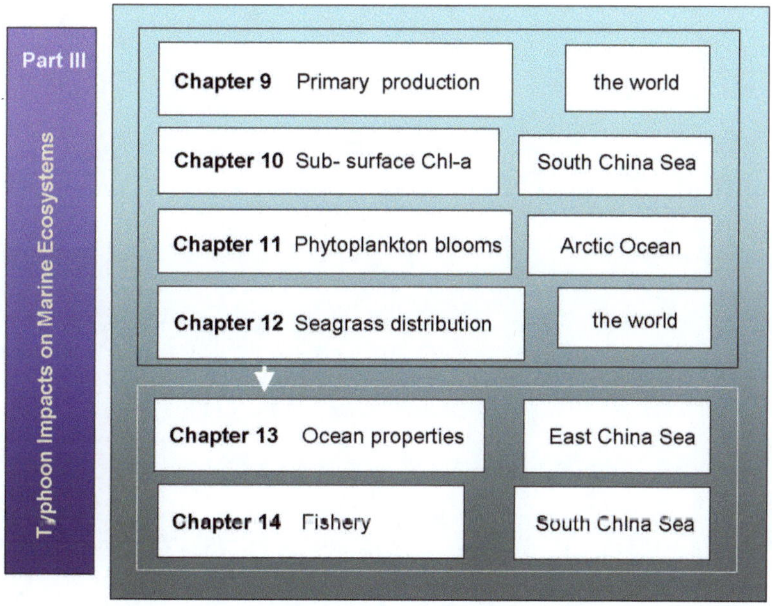

Chapter 9
Typhoon Impacts on the Surface Phytoplankton and Primary Production in the Oceans

SuFen Wang, DanLing Tang, Yi Sui, Gang Pan and Dmitry Pozdnyakov

Abstract This chapter introduces recent research about typhoon impacts on marine phytoplankton and related oceanic ecological environment in the world. Typhoon can induce the increase of Chlorophyll *a* (Chl *a*) concentration in the up water, and introduce phytoplankton blooms in the different regions of the oceans in world. Typhoons can also have important contribution to the marine primary production. Many studies suggested that typhoons indeed contribute to local primary production and carbon fixation through nutrient pumping to the surface. Phytoplankton blooms in response to typhoons are closely related to their speeds and intensities, which must be considered when estimating primary production. Different classification of typhoons may have different influence on the marine ecology. The mechanisms of typhoon impacts on phytoplankton blooms and primary production were summarized at the end of the chapter.

Keywords Typhoon · Phytoplankton bloom · Marine primary production · Remote sensing

S. F. Wang · D. L. Tang (✉) · G. Pan
Research Center for Remote Sensing and Marine Ecology and Environment, State Key Laboratory of Tropical Oceanography, South China Sea Institute of Oceanology, Chinese Academy of Sciences, Guangzhou, China
e-mail: lingzistdl@126.com
URL: http://lingzis.51.net

Y. Sui
Department of Oceanography, University of Dalhousie, Bible Hill, Canada

Y. Sui
Department of Physics and Physical Oceanography, Memorial University, St. John's, Newfoundland, Canada

D. Pozdnyakov
Nansen International Environmental and Remote Sensing Centre, St. Petersburg, Petersburg, Russia

D. L. Tang and G. J. Sui (eds.), *Typhoon Impact and Crisis Management*, Advances in Natural and Technological Hazards Research 40,
DOI: 10.1007/978-3-642-40695-9_9,

9.1 Introduction

Typhoon is an extremely strong wind event, which can have dramatic effects on the upper ocean. Typhoon induced winds can cause entrainment, strong vertical mixing, and upwelling, as well as the cooling of near surface water on the right-hand side of the storm track (Emanuel 1999). Marine ecosystem is sensitive to environmental factors, including typhoon/tropical cyclone. In the past several decades, typhoon activities have been strengthening in both intensity and spatial coverage, along with global changes; however, our knowledge about the impact of typhoons upon the marine ecosystem is very scarce.

Ocean primary production plays significant roles in the earth environmental system. Besides being the base of the ocean food chain, half of the world's oxygen is produced by phytoplankton. Like all plants, phytoplankton uses carbon dioxide, sunlight and nutrients to photosynthesize and multiply. Thus, primary production also affects the uptaking of carbon dioxide, an important greenhouse gas and a major cause of natural and man-made climate changes (Eppley and Peterson 1979). Typhoons also play an important role in phytoplankton blooms and primary productivity in typhoon-dominated oligotrophic oceanic waters (Lin et al. 2003; Babin et al. 2004; Walker et al. 2005; Zheng and Tang 2007).

To understand how could typhoon impact on marine ecosystems, scientists conducted series studies about phytoplankton concentration, sea surface temperature (SST) and related factors before, during, and after typhoon by satellite remote sensing and in situ observation data; those have been reviewed and analyzed in this chapter.

9.2 Typhoons Induce Phytoplankton Blooms on Ocean Surface Water

9.2.1 Phytoplankton Blooms in Coastal Water

Chl *a* concentrations can increase in the coasal water after typhoon. In September 2005, Typhoon Damrey (Category 2), the strongest typhoon over Hainan Island (HNI) in 32 year, and south China in 10 year (according to Xinhua News Agency), visited the northern South China Sea (SCS) and led to Chl *a* increases in two large areas. Two large area of phytoplankton blooms with increases of Chl *a* concentrations appeared in the SCS in September 2005 after typhoon Damrey, one was offshore bloom, another wass nearshore bloom. The offshore bloom was due to nutrient increase from mixing and upwelling, and the nearshore feature were resulted from rainwater discharge. By these two mechanisms, both typhoon winds and rain can enhance production of marine phytoplankton (Zheng and Tang 2007) (Fig. 9.1).

Another study investigated two phytoplankton blooms near or off the Pearl River Estuary (PRE) triggered by Category 2 Typhoon Nuri with moderate wind

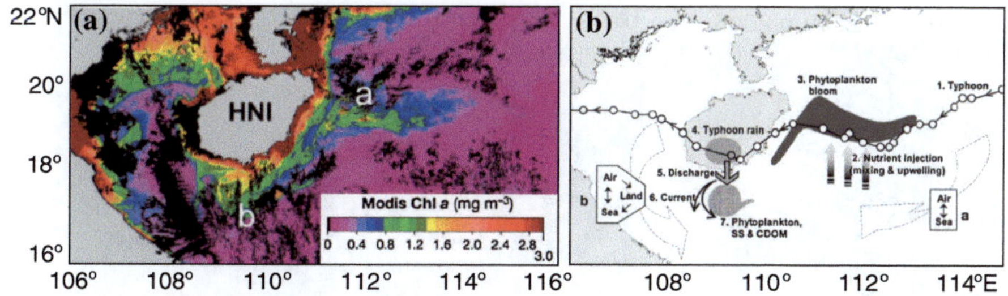

Fig. 9.1 **a** Appearance of Chl *a* anomalies in 2005 after the Typhoon, shown by comparing weekly (28 September–4 October) mean Chl *a* concentrations. *a* Offshore Chl *a* increase in the shape of a tongue; *b* a nearshore Chl *a* increase in an eddy pattern after the passage. **b** The oceanic mechanisms of two blooms (from Zheng and Tang 2007)

intensity in the SCS using both remotely sensed and in situ data sets (Zhao et al. 2009). The nearshore bloom was probably due to the increased discharge from the PRE and favorable current, as well as mixing, upwelling, and near-inertial resonance driven by strong typhoon wind. The offshore bloom may be triggered by upwelling and entrainment induced by strong typhoon wind and the topography of islands and islets. This study suggests that one typhoon may nourish phytoplankton biomass by inducing transport of nutrient-rich water from both the Pearl River Estuary to offshore and the sublayer to surface (Zhao et al. 2009) (Fig. 9.2).

Typhoon also showed effect on phytoplankton bloom in the different regions of the world ocean. Shi and Wang (2007) describe a Hurricane Katrina (in the August of 2005) induced phytoplankton bloom in the Gulf of Mexico using dataset from various satellite sensors and model analyses. A notable phytoplankton bloom centered at (24.4°N, 84°W) in the Gulf of Mexico is observed four days after Katrina's passing. There was no evidence of sediment re-suspension and transport from nearby locations into the bloom region. The phytoplankton bloom is attributed to the enhanced nutrient supply brought up by the wind-driven upwelling and vertical mixing (Fig. 9.3)

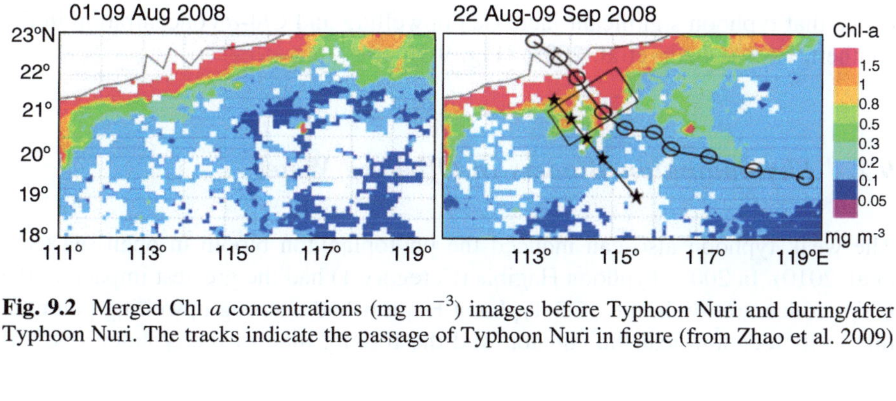

Fig. 9.2 Merged Chl *a* concentrations (mg m^{-3}) images before Typhoon Nuri and during/after Typhoon Nuri. The tracks indicate the passage of Typhoon Nuri in figure (from Zhao et al. 2009)

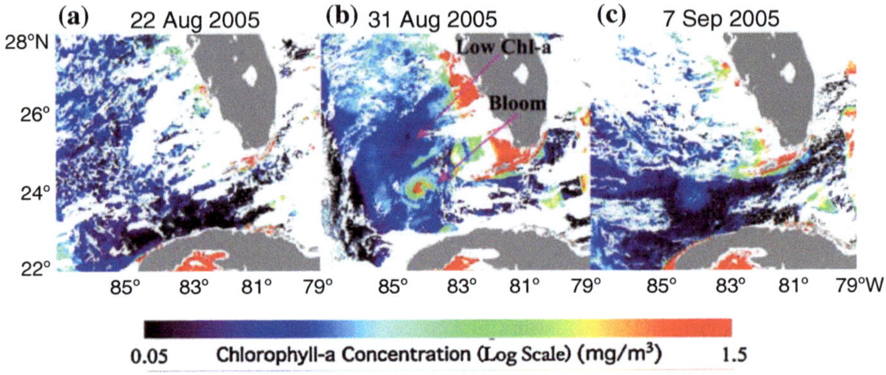

Fig. 9.3 Hurricane Katrina-induced ocean biological responses as indicated by MODIS-derived ocean color products for Chl-*a* concentration. The images were acquired on August 22 (before Katrina), August 31 (4-days after Katrina), and September 7 (11-days after Katrina) of 2005, respectively (from Shi and Wang 2007)

Clear skies, subsequent to Hurricane Ivan's passage across the Gulf of Mexico in September 2004, provided a unique opportunity to investigate upper ocean responses to a major hurricane. Oceanic cyclonic circulation was rapidly intensified by the hurricane's wind field (59–62 m s^{-1}), maximizing upwelling and surface cooling (3–7 °C) in two large areas along Ivan's track. Upward isothermal displacements of 50–65 m, computed from wind stress and sea surface height changes, caused rapid ventilation of thermoclines and nutriclines, leading to phytoplankton blooms with peak concentrations 3–4 days later (Walker et al. 2005) (Fig. 9.4).

The serial remote sensing based imageries clearly revealed large scale of upwelling within large regional enhancement of Chl *a* concentration in the southern East China Sea (ECS) after the passage of super typhoon Hai-Tang in July 2005. The large increased upwelling persisted for more than a week. Ocean color images also reveled that high Chl-*a* concentration of >3.0 mg m^{-3} appeared in the shelf region, where the high Chl-*a* pattern matched the upwelling in terms of location and time. The result of research provided clear and high resolution evidence that typhoon significant increased upwelling and Chl-*a* concentration in the southern ECS (Chang et al. 2008) (Fig. 9.5).

9.2.2 Phytoplankton Blooms in Offshore Water

The weak typhoon also can induced the phytoplankton bloom in open sea (Sun et al. 2010). In 2007, Typhoon Hagibis (Category 1) had the greatest impact on the Chl *a* concentration in the SCS, typhoon Hagibis had a strong upwelling potential due to its location near the equator, and the forcing time of the typhoon (>82 h)

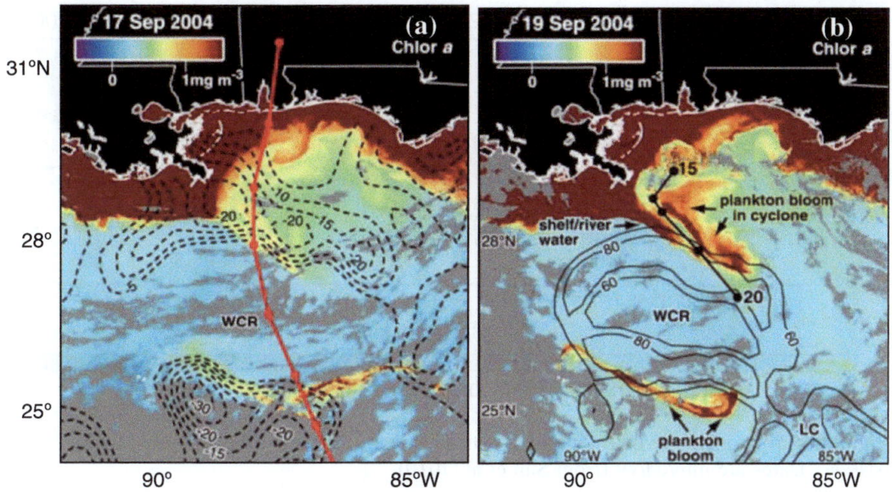

Fig. 9.4 Chl *a* (mg m^{-3}) computed from mean values on 17 Sep and 19 Sep (from Walker et al. 2005)

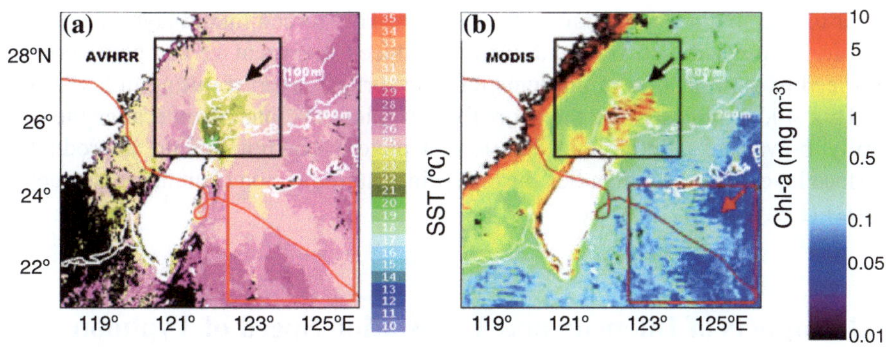

Fig. 9.5 **a** AVHRR SST images on 22 July after typhoon; **b** weekly composites of MODIS a with unitary color scale on 22–29 July, 1 week after typhoon (*arrows* indicate the decreased SST and increased Chl-*a*) (from Chang et al. 2008)

was much longer than the geostrophic adjustment time (~63 h). The higher upwelling velocity and the longer forcing time increased the depth of the mixed-layer, which consequently induced a strong phytoplankton bloom that accounted for about 30 % of the total annual Chl *a* concentration in the middle of the SCS. Induction of significant upper ocean responses can be expected if the forcing time of a typhoon is long enough to establish strong upwelling (Fig. 9.6).

The effect of a tropical cyclone on the variation of phytoplankton biomass in terms of surface Chl *a* is brought out based on satellite observations and mixed layer model simulations in the Arabian Sea during 21 May–3 June 2001. Along the

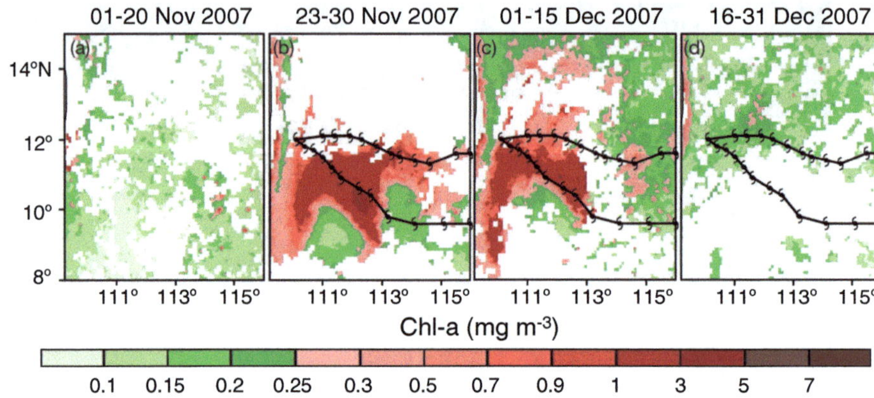

Fig. 9.6 Strong incidents of chlorophyll a enhancement (lasting >3 week) during the typhoon's passage. Dates are mm/dd/yy (from Sun et al. 2010)

cyclone's passage, Chl *a* was high with extreme values (5–8 mg m^{-3}) in the blooms of phytoplankton. The model simulations indicate deepening of mixed layer on the southeastern edge of the cyclone. This forced mixed layer deepening, due to intense wind stirring and cyclone-induced divergent geostrophic currents, has lead to the injection of nutrients into the surface layer, resulting in higher Chl *a*. This study suggests that the short-lived tropical cyclones would alter the generally prevailing oligotrophic (nutrient depleted) conditions into a productive surface layer in the Arabian Sea during spring intermonsoon (Subrahmanyam et al. 2002) (Fig. 9.7).

9.3 Impacts of Intensity and Translation Speed of Typhoon on Marine Condition

Does the intensity and translation speed of typhoon showed the same effect on the phytoplankton bloom? Zhao et al. (2008) conducted a research to compare two typhoon's impacts in the SCS, their results show that the intensity and translation speed of typhoon showed different effect on the phytoplankton bloom.

Two phytoplankton blooms triggered by two typhoons (Ling-Ling 2001; Kai-Tak 2005) were compared with different intensities and translation speeds in SCS. The data showed that the slow-moving typhoon induced phytoplankton blooms of higher Chl-*a*, the strong typhoon induced phytoplankton blooms of a large area. Typhoon Ling-Ling in 2001 was strong, with a maximum sustained surface wind speed of 59 m s^{-1}, and fast-moving with a mean translation speed of 4.52 m s^{-1}. Typhoon Kai-Tak in 2005 was weak with a maximum sustained surface wind speed of 46 m s^{-1}, and slow-moving with a mean translation speed

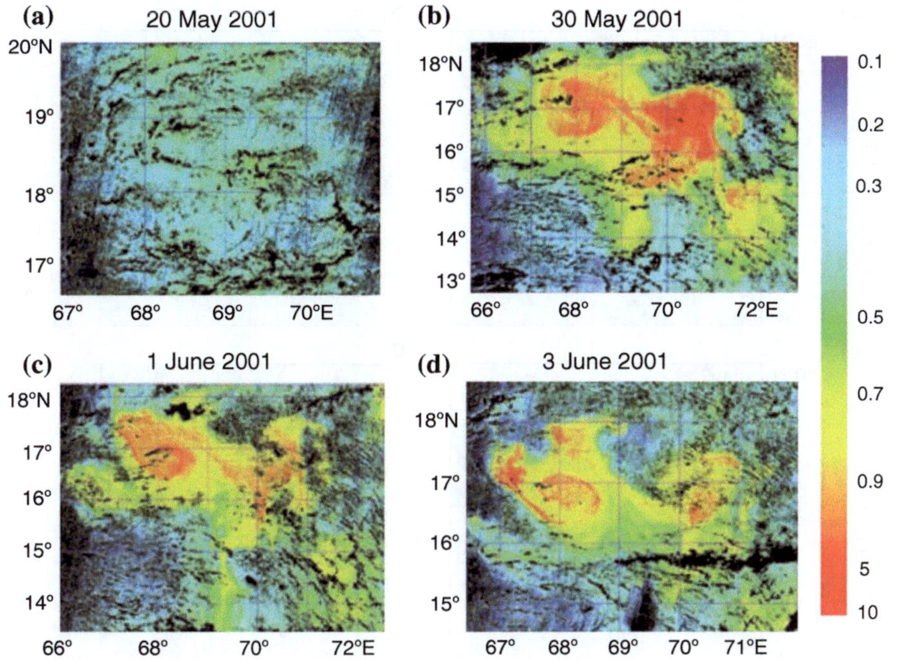

Fig. 9.7 Chl-*a* concentration (mg m^{-3}) from IRS-P4(OCEANSAT-1) ocean color monitor (*OCM*) at various stages of cyclone. **a** Prior to its formation on May 20 and, **b–d** after it intensified and weakened towards north on May 30, June 1, June 3, 2001 in the Arabian Sea (from Subrahmanyam et al. 2002)

of 2.87 m s^{-1}. The weak, slow-moving typhoon Kai-Tak induced phytoplankton blooms with higher Chl *a* concentrations, while the strong, fast-moving typhoon Ling-Ling induced blooms over a larger area (Zhao et al. 2008) (Fig. 9.8).

9.4 Eddy and Eddy-Shape Blooms Induced by Typhoons

Two eddies were observed 4 days after a cyclone, then two eddy-shape phytoplankton blooms appeared at the same location after a while. The analysis indicated that cyclone may have induced two cold eddies, and the cold eddies then produced nutrients to phytoplankton blooms by upwelling. This observation may help better understand the mechanism of tropical cyclone impact on phytoplankton (Fig. 9.9).

This is the first time to report moving eddies and eddy-shape phytoplankton blooms associated with tropical cyclone, the relationship among tropical cyclone, cold eddy upwelling and eddy-shape phytoplankton bloom may give some viewpoint on the tropical cyclone's affection on the mesoscale circulation.

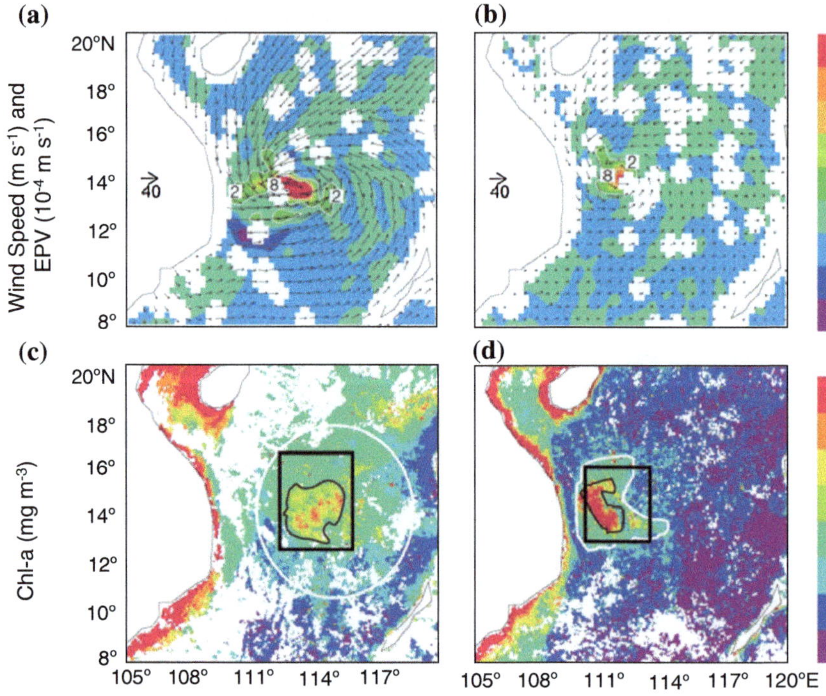

Fig. 9.8 Typhoon Ling-Ling (L-L) and Typhoon Kai-Tak (K-T). **a** and **b** Surface wind vectors (*arrows* represent wind speeds in m s^{-1} and their directions) and Ekman pumping velocity (*EPV*) (*color shaded* in 10–4 m s^{-1}); **c** and **d** Chl *a* (*color shaded*, mg m^{-3}). *Black boxes* in **c** and **d** highlight areas of high, typhoon-induced primary production (from Zhao et al. 2008)

9.5 Typhoons Contribute to Marine Primary Production

Typhoon can induce phytoplankton bloom by providing nutrients to the ocean surface, and that may increase ocean primary production. New evidence based on recent satellite data is presented to provide a rare opportunity in quantifying the long-speculated contribution of tropical cyclones to enhance ocean primary production. In July 2000, moderate cyclone Kai-Tak passed over the SCS. During its short 3-day stay, Kai-Tak triggered an average 30-fold increase in surface Chl *a* concentration. The estimated carbon fixation resulting from this event alone is 0.8 Mt, or 2–4 % of SCS's annual new production. Given an average of 14 cyclones passing over the SCS annually, we suggest the long-neglected contribution of tropical cyclones to SCS's annual new production may be as much as 20–30 % (Lin et al. 2003) (Fig. 9.10).

A large patch of enhanced Chl *a* concentration, lower sea surface temperature (SST), and lower sea surface height (SSH) was revealed in the central SCS in November 2001 after the passage of typhoon Lingling. A background level of

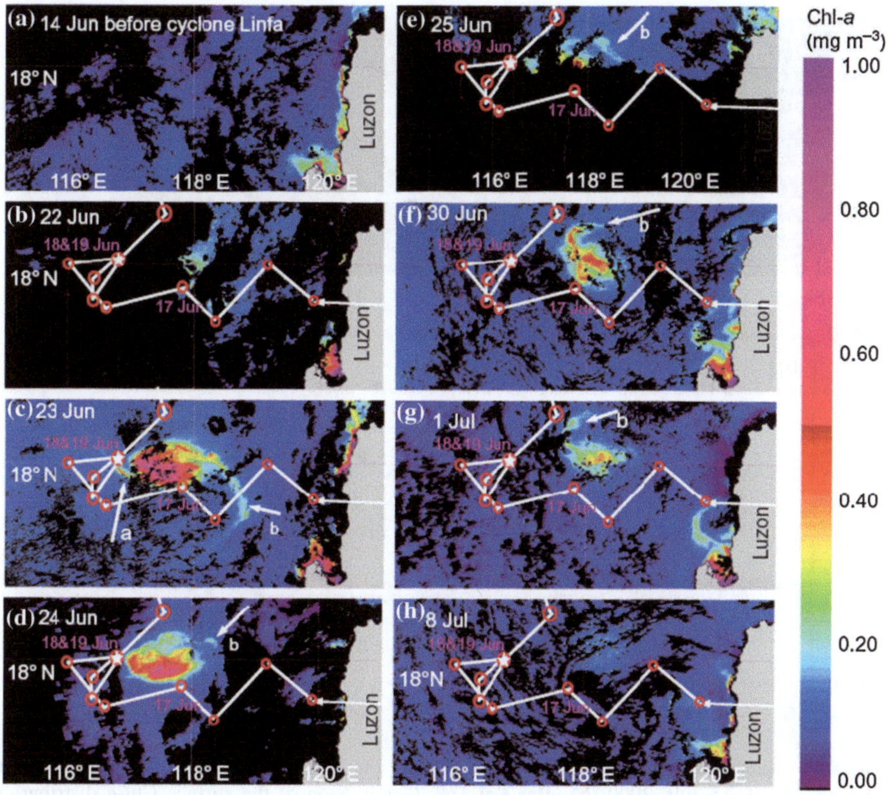

Fig. 9.9 MODIS-derived sea surface Chl-*a* concentration. **a** Before cyclone Linfa; **b–g** after cyclone Linfa. The *white arrows* show the dynamics of the arms around the bloom over the time (from Chen and Tang 2012)

0.08 mg m^{-3}, average Chl *a* within the area of 12.60–16.49°N, 112.17–117.05°E increased to 0.14 mg m^{-3} on 11/12 and then to 0.37 mg m^{-3} on 11/14. The area under Lingling's impact covered ca. 3° latitude and 4° longitude, which is much greater than the two summer cases previously observed in the northern SCS. This event lasted for ca. 15 days, and resulted in carbon fixation in the order of 0.4 Mt. Such a drastic response was attributed to the coupling of typhoon-induced nutrient pumping with the pre-established cyclonic gyre in the central SCS driven by the prevailing northeast monsoon.

Hence, these unambiguous results, together with earlier reports for the northern SCS, suggest that typhoons indeed contributed to local primary production and carbon fixation through nutrient pumping to the surface. Because numerous hurricanes/typhoons occur every year in the global ocean, their significance to global primary production deserves more consideration, and further investigation using the most updated bio-optical models is advocated in order to better quantify their

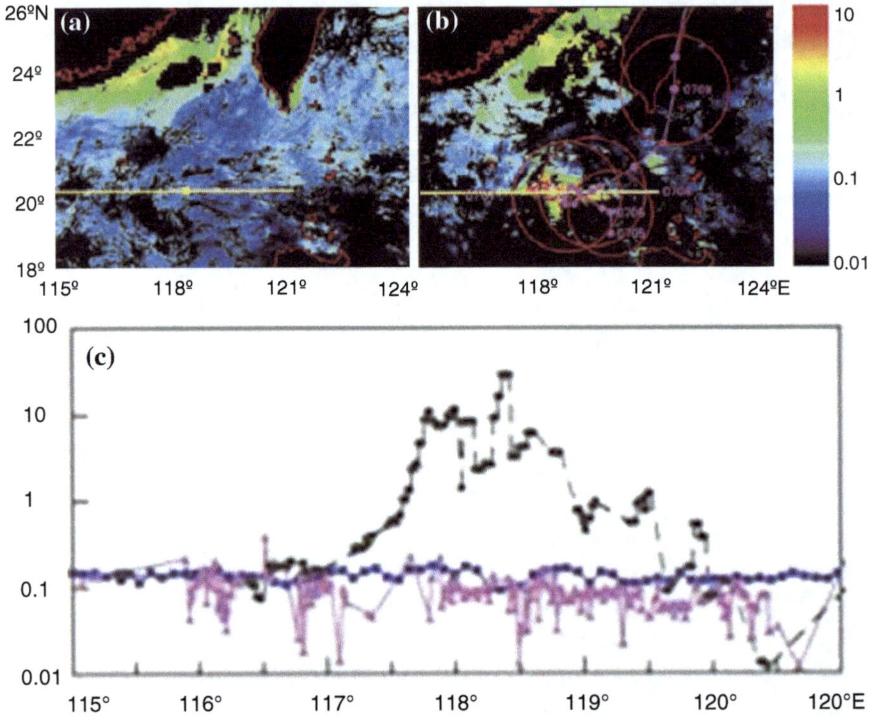

Fig. 9.10 SeaWIFS surface Chl-*a* image composite on 27 June–4 July 2000, before Kai-Tak (**a**) and 12, 14, 15 July 2000 (**b**) after Kai-Tak. Compassion of the surface Chl-*a* distribution along tr1, *pink* before (from Fig. 9.10a), *Green* after, (from Fig. 9.10b), *Blue* The 3-year (1998, 1999, 2001) climatological average of surface Chl-*a* concentration for July (from Lin et al. 2003)

contribution and to reduce uncertainty in the global carbon budget (Shang et al. 2008) (Fig. 9.11).

The enhancement of primary productivity in the southern East China Sea (ECS) following 16 typhoon passages was investigated using ocean color data and a primary productivity model. PPenh tended to be higher when typhoons traversed slowly with trajectories that allowed strong southerly winds to prevail over Yonaguni Island. Such long-lasting southerly winds were believed to push the Kuroshio current axis shelfward, enhancing the upwelling of nutrients, hence promoting new productivity (NP). The importance of long-lasting southerly winds as a proxy for physical perturbations underlying PPenh was expressed by an empirical equation by which 88 % of PPenh variation could be explained. Applying this equation, we assessed that typhoon passages accounted for a minimum of 0.6–11.8 % of the ECS summer–fall NP, suggesting that typhoon passage over the southern ECS is an important phenomenon supporting NP in the ECS (Siswanto et al. 2009) (Fig. 9.12).

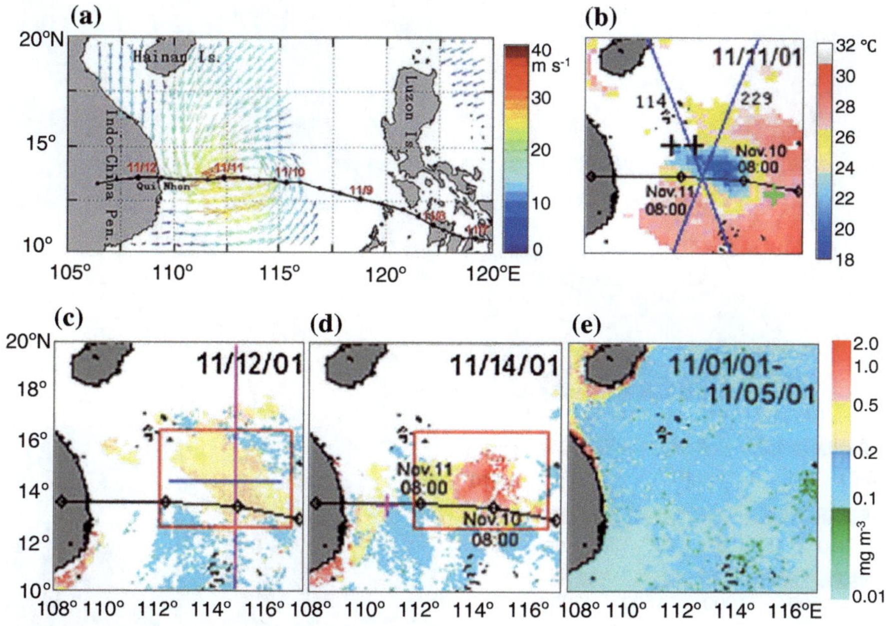

Fig. 9.11 **a** QuikSCAT wind image on 11/11/2001 showing Lingling and the South China Sea. Lingling's 6-h positions are shown as *black dots*, while the time of each annotated data is 08:00 local time. **b** TMI SST on 11/11. *Black line* shows the track of Lingling. **c** SeaWIFS OC4-Chl *a* on 11/12. *Blue line* 14.4°N; *Purple line* 115.2°E; *Red square* area for estimates of the mean Chl *a*. **d** SeaWIFS OC4-Chl *a* on 11/14. **e** SeaWIFS OC4-Chl *a* of 11/1-11/5 (before Lingling) (from Shang et al. 2008)

9.6 Summary

Typhoons have important impacts on marine phytoplankton and related ecological environments. Typhoon can induce increase of Chl-*a* concentration in the upper water and introduce phytoplankton blooms in the different region of the world ocean, and also have important contribution to the marine primary production.

Typhoons indeed contribute to local primary production and carbon fixation through nutrient pumping to the surface. Because numerous hurricanes/typhoons occur every year in the global oceans, their significance to global primary production deserves more consideration, and further investigation using the most updated bio-optical models is advocated in order to better quantify their contribution. With the increasing of primary production induced by typhoon, the fishery resources increased (Yu et al. 2013).

The main reason of typhoon introducing phytoplankton bloom is that typhoons can produce strong vertical mixing, eddy and upwelling, those can bring up cold and nutrient-rich water to the euphotic zone. The typhoon brings large rainfall also can introduce offshore phytoplankton bloom (Table 9.1).

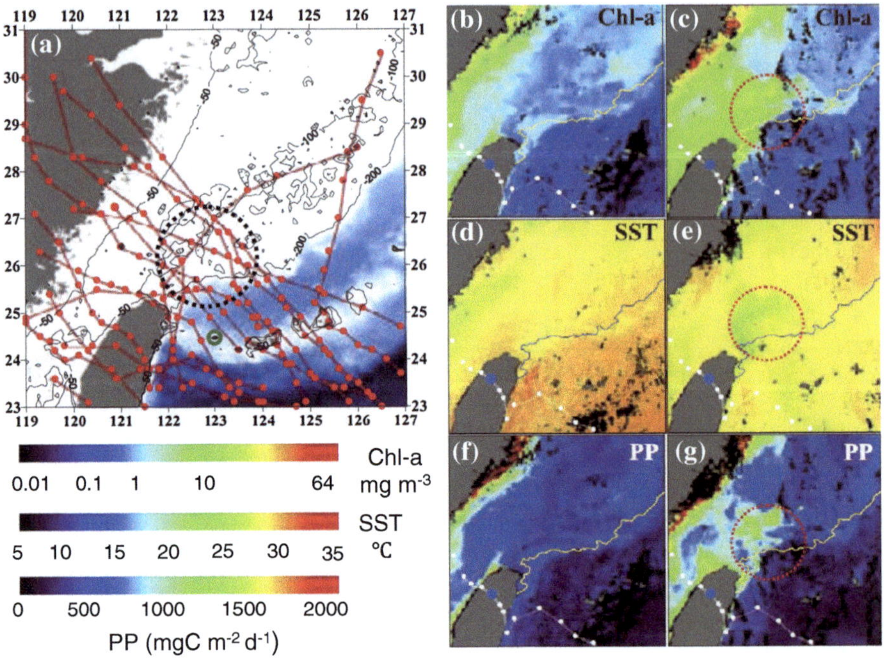

Fig. 9.12 **a** The best tracks of 16 typhoons (*red lines with dots*) investigated in this study overlaid on the ECS bathymetric map. *Green circles* indicated the location of Yonaguni Island. (**b**, **d**, and **f**) Pre-Hai-Tang (12–16 July) and (**c**, **e**, and **g**) post-Hai-Tang (22–24 July) composite images Chl-*a*, MODIS SST, and estimated PP before (16 July) and after (21 July) Hai-Tang, respectively. *Dotted black circle* in **a** and *dotted red circles* in **c**, **e**, and **g** denote the region over which PPenh was computed, and is considered to have minimum suspended sediment interference on satellite Chl-*a* retrieval. *Yellow* (or *blue* in SST images) *contour line* in Chl-*a* and PP images indicates 200-m isobath. *White line with dots* in **b–g** is Hai-Tang's best track. Hai-Tang position on 18 July 12:00 is marked with a *blue dot* over Taiwan (from Siswanto et al. 2009)

Table 9.1 Typhoon impacts on phytoplankton blooms, primary production

Typhoon effects	Phtoplankton blooms situation	References
Upwelling	Off shore or coastal water	Chang et al. (2008)
Subsurface upwelling	Off shore water	Yu et al. (2013)
Vertical mixing	Open seas or coastal water	Zhao et al. (2009); Sun et al. (2010)
Cold eddy	Open seas	Chen and Tang (2012); Yang and Tang (2010)
Rainwater	Coastal water	Zheng and Tang (2007)
River runoff after typhoon	Estuary	Zhao et al. (2008); Yu et al. (2013)

Acknowledgments The present research was supported by the following grants awarded to D. L. Tang: (1) National Natural Science Foundation of China (41006068, 31061160190, 40976091, 40811140533) and Guangdong Natural Science Foundation, China (8351030101000002,); (2) South China Sea Institute of Oceanology, CAS (SQ201004); (3). Innovation Group Program of State Key Laboratory of Tropical Oceanography (LTOZZ1201); In addition, SKLEC201204, LMEB201106, and National Natural Science Foundation of China (41006068, 41106105) also jointly supported this research.

References

Babin, S. M., Carton, J. A., Dickey, T.D., Wiggert, J.D.: Satellite evidence of hurricane-induced phytoplankton blooms in an oceanic desert. J. Geophys. Res. 109, C03043 (2004). doi:10.1029/2003JC001938

Chang, Y., Liao, H.T., Lee, M.A., Chan, J.W., Shieh, W.J., Lee, K.T., Wang, G.H., Lan, Y.C.: Multisatellite observation on upwelling after the passage of typhoon Hai-Tang in the southern east China sea. Geophys. Res. Lett. **35**, L03612 (2008). doi:10.1029/2007GL03285

Chen, Y.Q., Tang, D.L.: Eddy-feature phytoplankton bloom induced by tropical cyclone in the South China Sea. Int. J. Remote Sens. **33**(23), 7444–7457 (2012)

Emanuel, K.: Thermodynamic control of hurricane intensity. Nature 401, 665–669, (1999)

Eppley, R., Peterson, B.J.: Particulate organic matter flux and planktonic new production in the deep ocean. Nature **282**, 677–680 (1979)

Lin, I., Liu, W.T., Wu, C.C., Wong, G.T.F., Hu, C., Chen, Z., Liang, W.D., Yang, Y., Liu, K.K.: New evidence for enhanced ocean primary production triggered by tropical cyclone. Geophys. Res. Lett. **30**(13), 1718 (2003). doi:10.1029/2003GL017141

Shang, S., Li, L., Sun, F., Wu, J., Hu, C., Chen, D., Ning, X., Qiu, Y., Zhang, C., Shang, S.: Changes of temperature and bio-optical properties in the South China Sea in response to typhoon lingling, 2001. Geophys. Res. Lett. **35**, L10602 (2008). doi:10.1029/2008GL033502

Shi, W., Wang, M.: Observations of a hurricane Katrina induced phytoplankton bloom in the Gulf of Mexico. Geophys. Res. Lett. **34**, L11607 (2007). doi:10.1029/2007GL029724

Siswanto, E., Morimoto, A., Kojima, S.: Enhancement of phytoplankton primary productivity in the southern East China Sea following episodic typhoon passage. Geophys. Res. Lett. **36**(11), (2009). doi: 10.1029/2009GL037883

Subrahmanyam, B., Rao, K.H., Rao, N.S., Murty, V.S.N., Sharp, R.J.: Influence of a tropical cyclone on chlorophyll-a concentration in the Arabian Sea. Geophys. Res. Lett. **29**(22), 2065 (2002). doi:10.1029/2002GL015892

Sun, L., Yang, Y.J., Xian, T., Lu, Z.M., Fu, Y.F.: Strong enhancement of chlorophyll a concentration by a weak typhoon. Mar. Ecol. Prog. Ser. **404**, 39–50 (2010)

Tang, D.L., Zhao, H., Satyanarayana, B., Zheng, G.M., Singh, R.P., LV, J.H.: Enhancement of chlorophyll-a in the northeastern Indian Ocean After the 2004 south Asian tsunami. Int. J. Remote Sens. **30**(17), 4553–4565 (2009)

Walker, N.D., Leben, R.R., Balasubramanian, S.: Hurricane-forced upwelling and chlorophyll a enhancement within cold-core cyclones in the Gulf of Mexico. Geophys. Res. Lett. **3**, L18610 (2005). doi:10.1029/2005GL023716

Yang, X.X., Tang, D.L.: Location of sea surface temperature cooling induced by typhoon in the South China Sea. J. Trop. Oceanogr **29**, 26–31 (2010). (Chinese with English abstract)

Yu, J., Tang, D.L., Li, Y.Z., Huang, Z.R., Chen, G.B.: Increase in fish abundance during two typhoons in the South China Sea. Adv. Space Res. **51**, 1734–1749 (2013)

Ye, H.J., Sui, Y., Tang, D.L., Afanasyev, Y.D.: A subsurface chlorophyll a bloom induced by typhoon in the South China Sea. J. Mar. Syst. (2013, in press)

Zhao, H., Tang, D.L., Wang, D.X.: Phytoplankton blooms near the pearl river estuary induced by typhoon nuri. J. Geophys. Res. (Oceans) **114**, C12027 (2009). doi:10.1029/2009JC005384

Zhao, H., Tang, D.L., Wang, Y.: Comparison of phytoplankton blooms triggered by two typhoons with different intensities and translation speeds in the South China Sea. Mar. Ecol. Prog. Ser. **365**, 57–65 (2008)

Zheng, G.M., Tang, D.L.: Offshore and nearshore chlorophyll increases induced by typhoon winds and subsequent terrestrial rainwater runoff. Mar. Ecol. Prog. Ser. **333**, 61–72 (2007)

Chapter 10
Typhoon Impacts on Subsurface Marine Ecosystems

DanLing Tang, HaiJun Ye, Yi Sui, Y. D. Afanasyev and SuFen Wang

Abstract Typhoons usually can induce phytoplankton blooms in the surface waters. This chapter, using satellite and cruise survey data, introduces recent observations that Chlorophyll a (Chl-a) blooms occur not only on the surface, but also in the interior, just above the thermocline after the passage of a typhoon. The analysis of the physical and biological characteristics in the South China Sea after the passage of the typhoon Nuri in August of 2008 show that a subsurface (20–100 m depth) Chl-a bloom (1.31 mg m^{-3}) occurred and lasted for 3 weeks; stronger and longer than the surface Chl-a bloom (0.48 mg m^{-3}). The maximum value of the Chl-a at 2.10 mg m^{-3} was detected at a 50 m depth. This value was approximately 4–5 times higher than the background value of 0.48 mg m^{-3} measured at non-blooming areas at the same time and about 7.5 times higher than the mean Chl-a value of 0.28 mg m^{-3} measured over a period of 5 years. The mixed layer-depth and the thickness of the Chl-a bloom increased after the typhoon. This study clearly shows that a subsurface upwelling caused by the passage of the typhoon, transported nutrients to the euphotic zone and supported the Chl-a bloom. This study identified three phases for a typhoon-induced Chl-a bloom: The normal or pre-typhoon phase; The early post-typhoon phase and; The late-post typhoon phase. These observations provide some insights on the effect of typhoons on marine ecosystems, especially as related to the Integrated Primary Production.

D. L. Tang (✉) · H. J. Ye · S. F. Wang
Research Center for Remote Sensing and Marine Ecology and Environment,
State Key Laboratory of Tropical Oceanography, South China Sea Institute of Oceanology,
Chinese Academy of Sciences, Guangzhou, China
e-mail: Lingzistdl@126.com

D. L. Tang · H. J. Ye
Graduate University of the Chinese Academy of Sciences, Beijing, China

Y. Sui · Y. D. Afanasyev
Department of Physics and Physical Oceanography, Memorial University,
Newfoundland, Canada

D. L. Tang · H. J. Ye · Y. Sui · Y. D. Afanasyev · S. F. Wang
Nansen International Environmental and Remote Sensing Centre, St. Petersburg, Russia

D. L. Tang and G. J. Sui (eds.), *Typhoon Impact and Crisis Management*,
Advances in Natural and Technological Hazards Research 40,
DOI: 10.1007/978-3-642-40695-9_10,

At the end of this chapter a summary of the impacts of typhoons on marine ecosystems, including surface, subsurface, and deep water systems, and their related physical processes is presented.

10.1 Introduction

Nutrients can be pumped from the deep waters of the ocean to the surface and subsurface where they initiate a sequence of biological processes and, as a result, physical processes occur on different time scales (Price 1981; Tang et al. 2004a, b; Wang et al. 2010). Recently, subsurface chlorophyll has received a lot of attention in modeling studies as well as in primary production studies because of its significant contribution to ecological well-being (Sarma and Aswanikumar 1991; Fennel and Boss 2003; Perry et al. 2008; Lu et al. 2010). In the open ocean, the subsurface chlorophyll maximum depth remains relatively stable. In the near shore waters, the subsurface chlorophyll maximum related to shoals weakens when the surface layer is affected by river discharges (Lu et al. 2010). In the Northern part of the South China Sea (SCS), the depth where the maximum value of chlorophyll a (Chl-a) concentration, is observed is lowest (20–30 m) in the winter, reaches intermediate levels in May and is the deepest (50–90 m) in other months (Chen et al. 2006).

Recent studies show that typhoons, hurricanes or tropical cyclones can cause significant increases of Chl-a concentration in surface waters. The tropical cyclone "Kai-Tak", triggered a 30-fold increase in surface Chl-a concentrations in the SCS in the year 2000 (Lin 2003). Other examples where the increase of Chl-a concentrations were observed offshore include typhoons "Damrey", in 2005 (Zheng and Tang 2007), and Nuri, in 2008 (Zhao et al. 2009). In the year 2007, typhoon "Hagibis" caused a notable effect on Chl-a concentrations exhibiting a forcing time much longer than the typical geostrophic adjustment time in the middle of the SCS (Sun et al. 2010). Tropical cyclone "Linfa", in 2009, triggered an eddy-like Chl-a bloom (Walker et al. 2005). Typhoon "Morakot" changed the diatom abundance and species composition of samples due to strong winds and, as well, caused nutrient entrainment from the upwelling and nutrient-enriched floodwaters (Chung et al. 2012) that resulted. In the above studies the surface Chl-a blooms were detected using satellite data designed for tracking surface Chl-a.

Typhoons also affect particulate organic carbon flux. In subsurface waters (Chen et al. 2009; Hung et al. 2010) detected suspended matter concentrations, re-suspension and terrestrial runoff, entrainment of riverine-mixing, and Integrated Primary Production (IPP) (Shiah et al. 2000; Kundu et al. 2001; Gautam et al. 2005; Chen et al. 2009; Hung et al. 2010) resulting in fish abundance (Yu et al. 2013). In contrast, however, Shiah et al. (2000) mentioned that the spatial and temporal variation patterns of Chl-a were quite heterogeneous and that a detailed collection of information on Chl-a variations in the subsurface was lacking. There clearly remains certain gaps in the collection of information on subsurface phytoplankton; on the occurrence of blooms in subsurface waters; and on the mechanisms of blooming. In August 2008, typhoon Nuri passed over the northern part of the SCS and provided an opportunity to study biological and physical changes in the subsurface waters. Several days after the passage of Nuri, a research cruise obtained samples of in situ data in this region. The results documented a subsurface Chl-a bloom that lasted for a long time, even after the surface Chl-a bloom had disappeared.

This chapter focuses on the sub-surface phytoplanktons induced by the typhoon, and summarizes the possible responses and the oceanic physical process mechanisms that followed.

10.2 Data and Methods

10.2.1 Study Areas and In-Situ Data

The northwest Pacific Ocean is the region with the most numerous and violent typhoons in the world (Fumin et al. 2002), and the SCS is the largest, semi-enclosed, marginal sea in the Northwest Pacific and has an area of about 3.5×10^6 km^2 (Fig. 10.1). On average, more than 10 typhoons pass through this region annually, and the typhoon tracks are mostly in the Westward direction over the northern SCS (Elsner and Liu 2003; Wu et al. 2005; Zheng and Tang 2007). Compared with other ocean ecosystems, the Summer SCS is notable for its shallow "Mixed Layer Depth" (MLD) which is typically less than 50 m in depth (Karl and Lukas 1996; Chen et al. 2006). During the Summer season, the Northern SCS exhibits generally stratified and oligotrophic conditions (Chen et al. 2003).

In-situ data of Chl-a, taking into consideration temperature and salinity, were measured at 14 stations (Fig. 10.1) during a Research Cruise of the South China Sea by the Institute of Oceanology, Chinese Academy of Sciences, from 18 August to 6 September, 2008 following typhoon Nuri (Table 10.1). The 1–2 L water samples drawn from five depths (0, 25, 50, 75, 100 m) were collected at each station in order to measure Chl-a concentrations. Sea water was filtered through a GF/F filter (Whatman, 25 mm) and the filter paper was wrapped in aluminum foil and stored at temperatures of −20 °C pending analysis. Chl-a concentrations were determined via fluorescence with a Turner Design 10 fluorometer which was used following the equations given by Parsons et al. (1984). Temperature and salinity data were obtained by a Conductivity-Temperature-Depth (CTD) probe. The MLD was determined based on 0.125 unit potential density criterion (Levitus 1982). In order to obtain information about mean MLD in previous years the study used CTD data obtained during cruise studies conducted in 2004, 2005, 2006, and 2007 (Table 10.2). Vertical profiles of Chl-a, temperatures and salinity were obtained along the Transect North (TN) (Stations 2–9) and Transect South (TS) routes

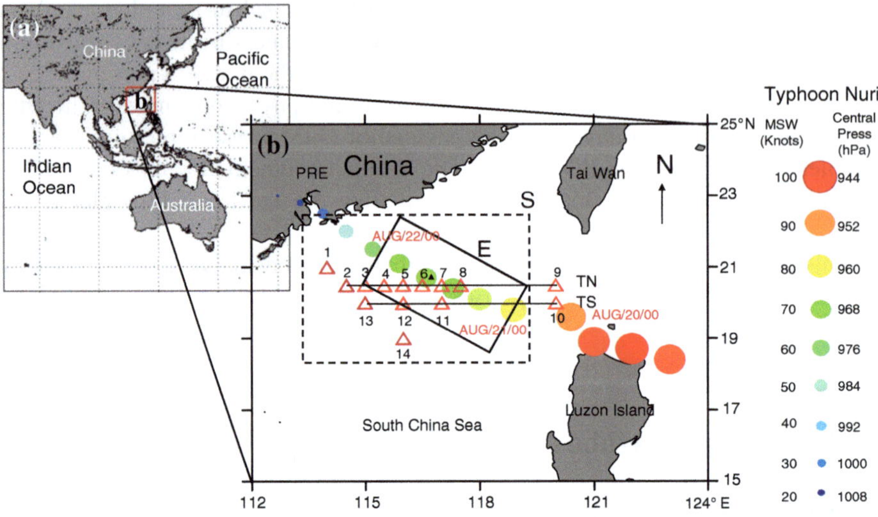

Fig. 10.1 **a** Location of the South China Sea. **b** Study area (Box S) and the track of typhoon Nuri. The typhoon *center* positions for every 6 h are indicated by the colored *circles*. The *circle color* and *size* represent the central pressure (hPa) and the maximum sustained wind speeds (MSW, knots, 1 knot = 0.514 m s^{-1}), respectively. Box E is typhoon effect area (1.46 × 10^5 km^2; clockwise from the *top*, 116°W–22.5°N, 119.5°W–20.5°N, 118.5°W–18.5°N and 115°W–20.5°N); *TN* Transect North (Stations 2–9); *TS* Transect South (Stations 10–13); Pearl River Estuary *(PRE)*; the *black triangle*: Dongsha Island. All times are in Coordinated Universal Time (*UTC*)

Table 10.1 In situ measurements at different stations in the South China Sea during 2008

Station	Station locations	Date/time
1	114.00°E, 21.00°N	0248 Sep 6
2	114.50°E, 20.50°N	2041 Sep 5
3	115.00°E, 20.50°N	2045 Sep 5
4	115.50°E, 20.50°N	1623 Sep 5
5	116.00°E, 20.50°N	1255 Sep 5
6	116.50°E, 20.50°N	0830 Sep 5
7	117.00°E, 20.50°N	0615 Sep 5
8	117.50°E, 20.50°N	0215 Sep 5
9	120.00°E, 20.50°N	2033 Aug 18
10	120.00°E, 20.00°N	0033 Aug 19
11	117.00°E, 20.00°N	1519 Sep 4
12	116.00°E, 20.00°N	1326 Sep 3
13	115.00°E, 20.00°N	0541 Sep 3
14	116.00°E, 19.00°N	0305 Sep 2

(Stations 10–13) (Fig. 10.1). Relevant published information on Chl-a concentrations in the SCS were used in this study and are summarized in Table 10.3 (Ye et al. 2013).

Table 10.2 The mixed layer depth (MLD) in our study area at different sampling years

(Aug–Sep) year	Number of observations	MLD (m)
2004	10	30.5 ± 7.58
2005	6	36.0 ± 7.59
2006	7	41.4 ± 17.54
2007	4	33.5 ± 3.11
2008	10	63.0 ± 9.49

Values were presented as mean ± standard deviation

Table 10.3 The number of Chl-a data in our study area and data sources

Date	Number of observations	Data was obtained within 15 days after typhoon passed	Sources
12 Jun-7 Jul, 1998	2	No	Ning et al. (2005)
Sep, 1998	1	No	Chen et al. (2006)
14–18 Sep, 1998	3	No	Liu et al. (2002)
20–24 Jul 2004	5	Yes (6–9 days)	Lu et al. (2010)
18–20 Sep, 2004	6	No	Present study
6–7 Sep, 2005	2	No	He et al. (2009)
14 Aug, 2007	1	No	Liu et al. (2009)
2–6 Sep, 2008	5	Yes (12–16 days)	Present study
16–17 Sep, 2010	4	No	Present study

10.2.2 Typhoon and Satellite Data

Typhoon Nuri's track, based on the best track data issued by the Joint Typhoon Warning Center (JTWC), was sourced from Unisys Weather Web (http://weather.unisys.com/hurricane/w_pacific/index.html). Information such as typhoon center locations, central pressure and Maximum Sustained Winds (MSW) were gathered every 6 h. The translation speed of the typhoon was estimated using positions pinpointing its center that were recorded every 6 h. The MSW and the central-pressure of the typhoon are shown in Fig. 10.1.

The daily, merged 8 day Chl-a concentrations and sea surface temperatures (SST), established at data Level 3, with a spatial resolution of 4 km, were derived from the ocean color sensor *Moderate Resolution Imaging Spectroradiometer* (MODIS) and were obtained from the Distributed Active Archive Center (DAAC) of NASA, the US National Aeronautics and Space Administration (NASA) data sources (http://oceancolor.gsfc.nasa.gov/cgi/l3). Sea level anomaly (SLA) data were gathered from the TOPEX/Poseidon and JASON altimeters (http://www.aviso.oceanobs.com/) equipment and were used to analyze surface circulations.

The microwave scatter-ometer SeaWinds (QuikScat) data (http://www.remss.com) was used to obtain wind vectors at sea surface levels for periods before, during and after typhoon Nuri. The wind-field data allowed researchers to estimate wind stress levels ($\vec{\tau}$) on the ocean surface using the conventional formula:

$$\vec{\tau} = \rho_a C_D \vec{U}_{10} \vec{U}_{10} \tag{10.1}$$

where ρ_a is the density of air and $\vec{U}_{10}$ is wind vector, and the drag coefficient C_D (Jarosz et al. 2007), and is expressed as:

$$C_D = (2.229 + 0.2983\, U_{10} \;-\; 0.00468\, U_{10}^2) \;\times\; 10^{-3}. \tag{10.2}$$

In order to estimate the rate of exchange between the deeper layers and the top layers researchers calculated the Ekman pumping velocity W_E as:

$$W_E = \mathrm{curl}(\vec{\tau}/\rho f), \tag{10.3}$$

where ρ is the density of seawater and f is the Coriolis parameter. Thermocline displacement $\Delta\eta$ due to a typhoon can be estimated using the formula given by

$$\Delta\eta = \tau/(\rho f U_t), \tag{10.4}$$

(Price et al. 1994) where U_t is the translation speed of the typhoon.

Alternately, the thermocline displacement can be estimated from the SLA data. Assuming a purely baroclinic response such that the thermocline displacement compensates for the surface displacement and is of the opposite sign, thus a simple relation of (Shay et al. 2000)

$$\Delta\eta g' = -g\Delta h \tag{10.5}$$

where $g' = g\Delta\rho/\rho$ is the reduced gravity, $\Delta\rho$ is the density difference between the upper and lower layers and Δh is the SLA. The thermocline displacement is then

$$\Delta\eta = -\Delta h\rho/\Delta\rho \tag{10.6}$$

10.3 Results

10.3.1 Surface Chl-a Bloom during Typhoon Nuri in the SCS

Nuri (2008) was a Category 2 (based on the Saffir-Simpson scale) typhoon that impacted a wide area from the Luzon Strait to the Pearl River Estuary. It initially developed as a tropical depression in the northern Pacific, strengthening to typhoon status on 18 August 2008. The typhoon migrated towards the northwest and, as its center passed through the DongSha Island area on the 21st of August, 2008, it then moved northwest towards PRE (Fig. 10.1).

Before the typhoon formed, from 9 to 16 August, 2008, winds in the northern SCS was weak ($\sim$7 m s^{-1}) (Fig. 10.2a-a). During the passage of Nuri the winds grew strong, in excess of 30 m s^{-1} (Fig. 10.2a-b). After the typhoon, the wind speed returned to a low value of approximately 5 m s^{-1} (Fig. 10.2a-c). According to the JTWC, the maximum sustained winds reached 51 m s^{-1} on 20 August in the Luzon Straits.

In the study area, average wind speeds were 35 m s^{-1} from 1,800 UTC on 20 August to 0000 UTC on 22 August. In the same period the translation speed of the typhoon was $U_t = 4.2$ m s^{-1}. Although Nuri was initially weak, it became a fast-moving typhoon, and heavy rains brought by Nuri led to a significant direct economic impact of about ¥400 million (http://finance.sina.com.cn/roll/20080827/09172394980.shtml).

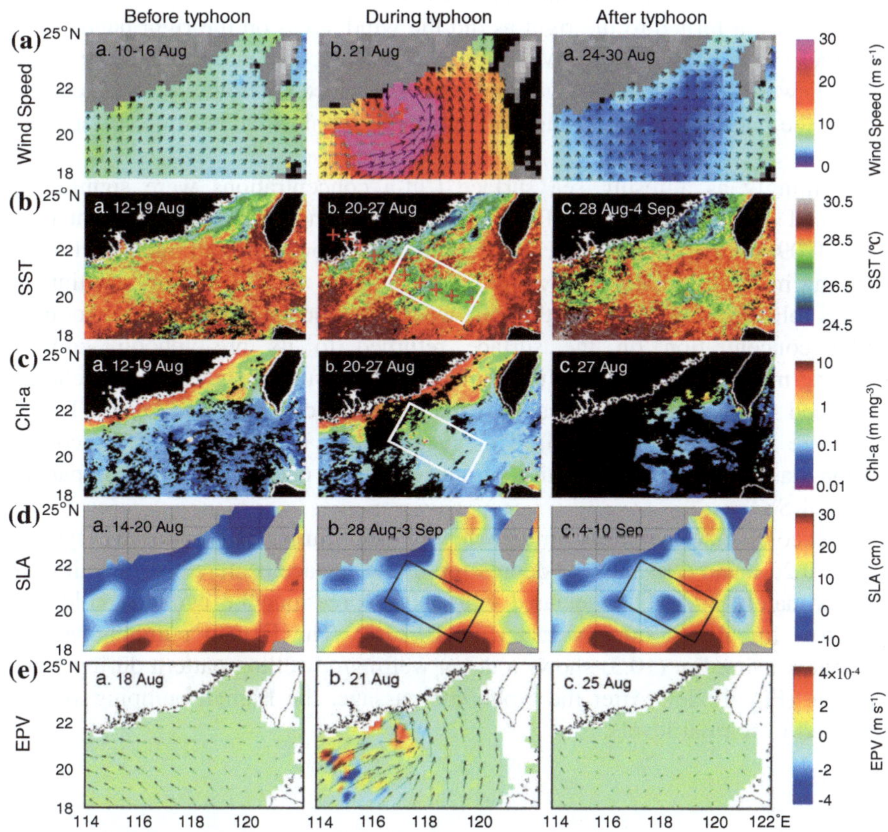

Fig. 10.2 Optical and biological response of the ocean to the passage of typhoon Nuri. **a** QuikScat wind-vector fields. **b** Sea surface temperature. **c** MODIS-derived ocean color products for Chl-a concentration. **d** Sea Level anomaly from AVISO. **e** Ekman pumping velocity. *Rectangles* in Fig. 10.2**a**-*b*, 2**c**-*b* 2**d**-*b* and 2**d**-*c* indicate the study area. *Red crosses* show the track of typhoon Nuri in Fig. 10.2**b**-*b*

During the period of 12–19 August, before Nuri's passage, the map of SST shows that the study area was dominated by warm waters with SST exceeding 29 °C (Fig. 10.2b-a). The SST map obtained during the passage of Nuri demonstrates that a significant cooling occurred along the track of the typhoon with an $\Delta SST = -2.0$ °C, approximately (Table 10.4). The lateral extent of the low SST patch was approximately 100 km. The low temperature patch persisted for about two weeks after the passage of Nuri (Fig. 10.2b-c) and it is interesting to note a clear rightward bias of in ΔSST distribution. The low SST patch was located mainly to the right of the typhoon track so that its centerline shifted by approximately 50 km from the projected track. The rightward bias was observed previously by a number of researchers (Fedorov et al. 1979; Sanford et al. 1987) and was discussed by Price (1981), Greatbatch (1983) and Price et al. (1994). The bias seemed to occur based on the effective coupling of the wind stress of the moving typhoon and the upper ocean inertial currents.

The current turned to the right as did the wind stress on the right hand side of the track. This caused a positive feedback related to the current. As a result the currents were stronger to the right of the track; surface horizontal divergences and, hence, the pumping of cold water to the surface in this area, became stronger as a result.

Seemingly as a result, sea surface Chl-a concentrations were significantly enhanced after Nuri's passage. Pre-typhoon Chl-a concentrations were in the range from 0.09 to 0.10 mg m^{-3} (Fig. 10.2c-a). The concentration increased up to 0.24 mg m^{-3} within the following 2 days and up to 0.50 mg m^{-3} within the 4 days following. However, within one week after the passage of the typhoon, the Chl-a concentration on the surface returned to its pre-typhoon value of 0.11 mg m^{-3} (Fig. 10.2c-c). It is interesting to further note that patches of enhanced Chl-a concentration approximately coincided with the low SST patch (compare Fig. 10.2b-b and c-b). This suggests that the enhanced Chl-a concentration was due to the introduction of nutrients from deeper waters accompanying the cool SST.

The comparison of SLA images before and immediately after the typhoon (Fig. 10.2d-a and d-b) shows a depression of the surface centered at the latitude of approximately 20.5°N. Associated with this depression, the cyclonic cold eddy persisted for almost three weeks (Fig. 10.2d-c). The Ekman pumping velocity estimated using Eq. (10.3) shows a down welling/upwelling pattern during Nuri's passage (Fig. 10.2e). After the typhoon's passage, the Ekman pumping velocity declined to normal levels.

10.3.2 Variations of Hydrographic Conditions in the SCS

The changes of physical hydrography in the upper ocean were examined using temperature and salinity measures at 14 stations in the area during the 12–16 day period following the typhoon (Fig. 10.3). The results showed that the highest

Table 10.4 Comparisons of environmental variables between surface and subsurface and the potential mechanism induced by Nuri

Variables	Nuri induced Chl-a				Climatology Chl-a		Tem change (°C)	
	Maximum	Mean	Lasted time	Bloom size (10^5 km^2)	Maximum	Mean	Maximum	Mean
Surface	1.10[R]	0.48 ± 0.23[R]	1 week	0.60[R,S] (Chl-a ≥0.50 mg m^{-3})	0.20[I]	0.10 ± 0.07[I]	3.40[R]	1.70[R]
Subsurface	2.10[I]	1.31 ± 0.47[I]	~3 weeks	1.00[I,S] (Chl-a ≥1.00 mg m^{-3})	1.00[I]	0.28 ± 0.13[I]	3.50[I]	2.00[I]
Mechanism	Short-lasting vertical mixing and long-lasting subsurface upwelling							

[R] Remote sensing data; [I] in situ data; [S] bloom size calculated from Figs. 10.2c-b and 10.4c

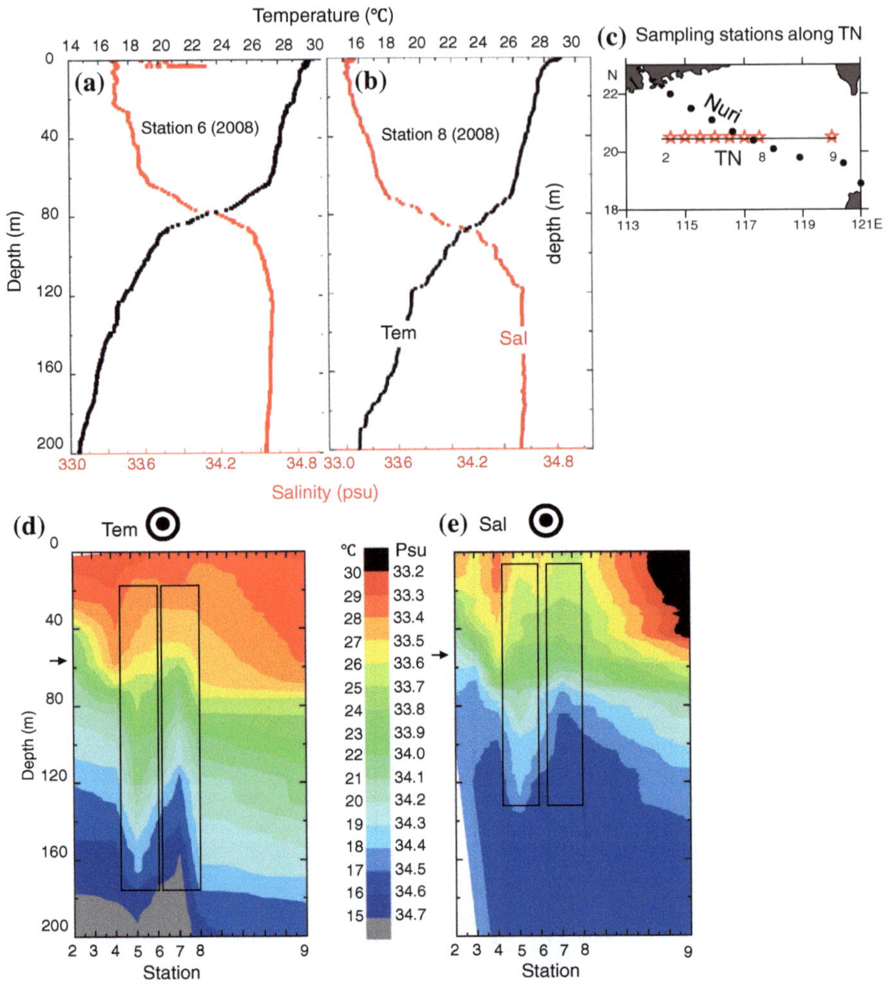

Fig. 10.3 Temperature (*black*) and salinity (*red*) profiles after Nuri's passage at station 6 (**a**) and 8 (**b**). *Vertical* distribution of temperature (**d**) and salinity (**e**) in the upper 200 m at the TN Transect North, with contour intervals of 1 °C and 0.10 Psu, respectively. The *circles* on the *top* of (**d**) and (**e**) indicate Nuri's track. The *black rectangles* show the area where decrease in temperature or increase in salinity is observed. **c** Sampling locations along the transect TN

temperatures and the lowest salinity levels occurring at sea-surface levels were as one might expect; A peak in isolines of temperature and salinity centered around station 7, indicating a region of upwelling. Compared with surrounding stations, data associates this region with a sharp drop in temperature of 3–4 °C and an increase in salinity of approximately 0.2 psu between 60 and 120 m in depth. A region of equally strong down welling was observed at station 5. The physical mechanisms of this phenomenon are the focus of this area and appear in Sect. 10.4.

The greatest temperature and salinity changes occurred at depths of 60 m. The sharp drop of temperature and sharp rise of salinity indicated that the typhoon induced a subsurface upwelling. The upwelling lifted deeper waters, from approximately 140 m upwards to 20 m in depth and were sustained for at least 16 days before returning to normal climatology levels.

The MLD in this region is typically about 35 m at this time of year, as data from previous years indicates (Table 10.2). Station 2, located to the extreme left of the typhoon track where the influence of typhoon was minimal, shows a similar value of approximately 40 m. However, in the area affected by the mixing of elements induced by the typhoon, the mean MLD increased significantly to approximately 70 m.

10.3.3 Subsurface Chl-a Blooming following Typhoon Nuri

The Chl-a concentrations at the surface level of the sea returned to their normal climatology levels after about two weeks following Typhoon Nuri's passage (Fig. 10.4a and b). However, high subsurface Chl-a concentrations with average values of 1.31 ± 0.47 mg m^{-3} at 50 m depth were observed in the vicinity of Nuri's track (Fig. 10.4c, d, e and Table 10.4). The maximum Chl-a value of

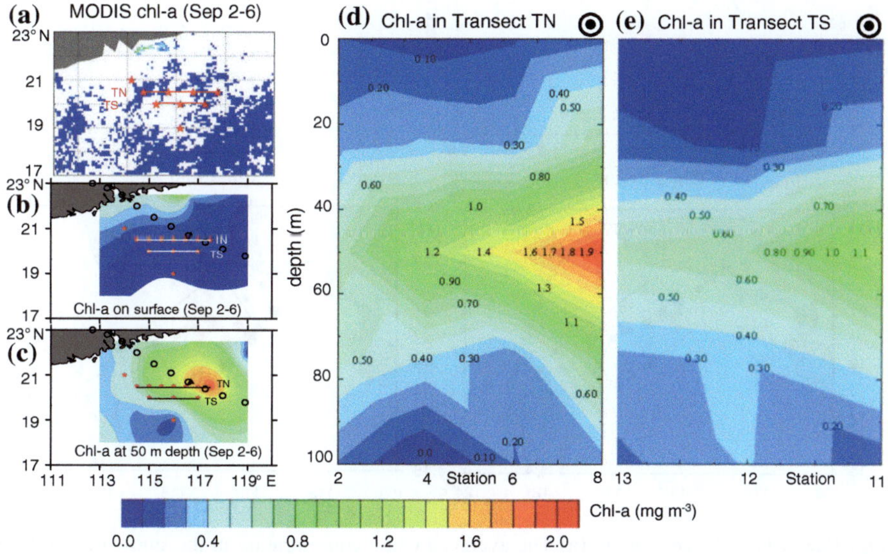

Fig. 10.4 The Chl-a contours at the surface (**a, b**) and at 50 m (**c**) depth after Nuri's passage. The vertical distribution of Chl-a at the transect TN (Stations 2, 4, 6 and 8) (**d**) and the transect TS (Stations 11–13) (**e**) after Nuri's passage. The *black circles* in *top* of (**d**) and (**e**) indicate the locations of typhoon Nuri. The *black triangle* indicates the Dongsha Island

2.10 mg m^{-3} (Fig. 10.4c, d and Table 10.4) was observed at a depth of 50 m (Station 8). The maximum Chl-a values measured at stations away from the typhoon track (Stations 1, 2, 13 and 14) were less than 0.60 mg m^{-3}. Vertically, the Chl-a values decreased sharply in both upward and downward directions when moving away from the subsurface chlorophyll maximum layer of 50 m (Fig. 10.4d and e). The enhanced subsurface Chl-a concentrations were observed 12–16 days after Nuri's passage, though its duration may have been longer.

Figure 10.5a shows typical climatological profiles of Chl-a concentrations in the open sea. The profiles were calculated using data provided by Liu et al. (2002), Ning et al. (2005), Chen et al. (2006), He et al. (2009), Liu et al. (2009), Lu et al. (2010) and from data developed via the present study. The concentrations are typically low at the surface and increase slowly with depth until reaching a maximum between 50 and 75 m in depth. The concentrations then decrease towards the bottom of the euphotic layer (Fig. 10.5a and b show data only for depths <100 m). The maximum value of Chl-a concentrations after Typhoon Nuri were enhanced to approximately 7.5 times the climatological values (0.28 ± 0.13 mg m^{-3}) (Table 10.4).

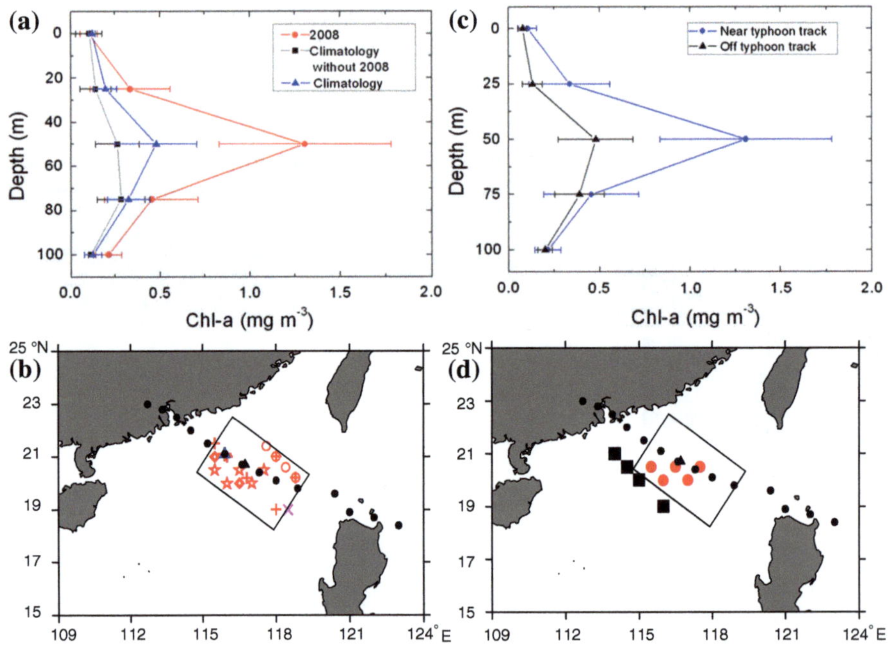

Fig. 10.5 Chl-a Concentration. (**a**) The averaged Chl-a concentration in the study area in 2008 after the typhoon Nuri (*red dots*). Chl-a climatology (*squares*) averaged over 1998, 2004, 2005, 2007, 2008 and 2010; Chl-a climatology (*triangles*) where 2008 data is excluded. (**c**) The averaged Chl-a concentration (*blue dots*) in typhoon affected area compared with the data from outside the Nuri's track. In (**b**) *red symbols* show sample sites, and (**d**) *red dots* and *black squares* show sample sites. In (**b**) and (**d**), the *black dots* show Nuri's track

10.4 Discussion

10.4.1 Typhoon Induced High-Salinity and Low-Temperature Patches

It is useful to estimate the principle physical parameters of flows generated by typhoons in order to understand the observed patterns of temperature and salinity. A linear solution describing the response of the ocean to a moving storm was provided by Geisler (1970). Nonlinear behavior of the flow was further investigated using numerical models developed by Chang and Anthes (1978), Price (1981), Greatbatch (1983) and Price et al. (1994). Geisler (1970), in particular, showed that the baroclinic response of a two-layer ocean is more important than the barotropic one and includes an inertia-gravity wave wake in the tail of the storm. The wake pattern of a storm translating at 5 m s^{-1} over an ocean with MLD of 50 m is given in Fig. 10.1 in Greatbatch (1983). The pattern consists of alternating regions of horizontal divergence and convergence. These regions correspond to upwelling and down-welling respectively. This effect was called "inertial pumping" (Price et al. 1994).

The regions of upwelling and down-welling shown in Fig. 10.3 of this study correspond to those stated in the theoretical solution. Note that the transect TN (stations 2–9) is not perpendicular to the typhoon track but rather at an angle of about 30°, thus we are able to observe the alternating patches of upwelling/down-welling along the typhoon track. Equations (10.1) and (10.3) give the estimate for the mean wind stress $\vec{\tau} = 3.5$ N m^{-2}. Equation (10.4) then allows us to calculate the thermocline displacement to be $\Delta\eta = 21$ m which is in reasonable agreement with the observed displacement shown in Fig. 10.3. We can also estimate the thermocline displacement using the altimetry data. Figure 10.2d-b shows that a surface depression of approximately $\Delta h = -5$ cm occurred after the passage of Nuri. Equation (10.6) then gives $\Delta\eta = 20.5$ m. Here we take the value of the density difference between the upper and lower layers to be $\Delta\rho = 2.5$ kg m^{-3}. This value was calculated using temperature and salinity data measured at station 2 where these characteristics were almost undisturbed by the typhoon and corresponded closely to the climatology data (Table 10.2). The thermocline displacement calculated using SLA is surprisingly in close agreement with that calculated using the wind stress and with the observed displacement.

An important control parameter for the flow induced by a moving storm is the ratio of the translation speed of the storm to the velocity of the baroclinic gravity waves, $c = (g'H)^{1/2}$, where H is the depth of the upper (mixed) layer (before the typhoon). Taking $\Delta\rho = 2.5$ kg m^{-3} and $H = 35$ m, we obtain $c = 0.9$ m s^{-1}. Thus, the ratio U_t/c which is in fact the Mach number and is quite high in this case, $U_t/c = 4.7$. This indicates that typhoon Nuri was a fast-moving storm. Coincidently, most of the control parameters in the current case are very close to those derived by Chang and Anthes (1978) and Greatbatch (1983) in their numerical simulations. Thus, Fig. 10.1 in Greatbatch (1983) gives a good picture of the flow under examination in the present case. In particular, it shows that maximum vertical velocity was approximately 3×10^{-3} m s^{-1}. According to Greatbatch (1983) the along-track wavelength of the trailing wave pattern can be estimated using the inertial wavelength $\lambda = 2\pi\ U_t/f = 530$ km. We did not observe another maximum/minimum further down the track of the typhoon because of the proximity of the coast.

In order to get some understanding of the vertical exchange in the SCS, we calculated the Ekman pumping velocity W_E using Eq. (10.3). The wind stress curl was calculated using the QuikScat wind data before and after the typhoon (Fig. 10.2a-a and 2a-c). The typical values of velocity were quite low, $W_E = 0.5 \times 10^{-4}$ m s^{-1} which is not surprising for the region with an oligothrophic upper layer. The estimate of W_E for typhoon winds (Fig. 10.2a-b) gave an order of magnitude higher value of $W_E = 5 \times 10^{-4}$ m s^{-1} near the center of the typhoon. This enhanced vertical velocity that occurred during the passage of the typhoon can result in a significant displacement of the isolines of temperature and salinity (Sun et al. 2010).

The effects of pumping (the Ekman pumping during the typhoon and the inertial pumping in the wake of the typhoon) together with an intense turbulent mixing of the upper layer caused by the typhoon result in the occurrence of the low-temperature and high-salinity patch that was observed. Measures taken during our study showed that the patch lasted for the period of about three weeks (Figs. 10.2b and 10.3). The duration of the patch is consistent with the previous observations in the SCS by Zheng and Tang (2007) and Zhao et al. (2009).

10.4.2 Short-Term Surface Chl-a Blooms and Long-Term Subsurface Chl-a Blooms

Nutrients are vital for growth and propagation of Chl-a in the open sea. The ocean's primary production takes place in the euphotic zone where there is enough sunlight for photosynthesis (Hader et al. 1995; Kumar et al. 2004). Nutrients are, however, often deficient at the upper layer but increase in quantity with depth (Levitus and Boyer 1994). Typhoons with higher wind speeds and a slow translation speed typically induce longer-lasting Chl-a blooms in surface water (Zhao et al. 2008). In the present study, the typhoon was one with a fast translation speed and relatively weak wind speed inducing a short term Chl-a bloom in surface water (Table 10.4). The enhanced surface Chl-a concentrations can last for 2–3 weeks (Babin et al. 2004; Zheng and Tang 2007), and are a result of upward entrainment of phytoplankton to the surface from the subsurface Chl-a maximum. Another result may be a new production of Chl-a because of enhanced nutrient concentration in the upper layer caused by the mixing that occurred during the passage of a typhoon. Although it is difficult to establish with certainty, it is likely that both of the above processes contributed to the observed surface Chl-a bloom in the present case. Walker et al. (2005) reached a similar conclusion for hurricane Ivan in the Gulf of Mexico.

Although the surface Chl-a bloom disappeared (Fig. 10.4a and b) 1 week after Nuri's passage, the subsurface Chl-a concentration showed extremely high values (1.31 ± 0.47 mg m^{-3}) (Fig. 10.4c, d, e and Table 10.4) before returning to normal after three weeks had passed. The long-term subsurface Chl-a bloom resulted from the introduction of nutrients from the deep water by subsurface upwelling. A description of the nutrients concentration might help to explain the subsurface Chl-a bloom, but there was no supporting data available related to the study area at that time.

It is important to discuss the enhancement of the subsurface Chl-a for the IPP. Whether typhoons, hurricanes and tropical storms have a significant integrated impact on the IPP in the ocean is still an open question (Lin 2003; Sun et al. 2010). The IPP is routinely estimated via the remotely sensed Chl-a data using one of the available mathematical models. Such a model, and one that is widely used, was developed by Behrenfeld and Falkowski (1997). This Vertically Generalized Production Model (VGPM) is used to calculate the primary production in depth-integrated euphotic zones using surface Chl-a concentration data. In VGPM model, the climatology (August) Chl-a ratio (subsurface to surface) is 3 (0.28 mg m^{-3}/ 0.10 mg m^{-3}), where the Chl-a ratio after Nuri was 12 (1.31 mg m^{-3}/0.11 mg m^{-3}) (Fig. 10.5a and Table 10.4) which shows an increase of 3 times. If, by using this model, which is based on surface Chl-a, to estimate the integrated impact of typhoons, we may underestimate the IPP due to missing the contribution of the enhanced subsurface Chl-a induced by typhoon, the part that is underestimated as IPP can be a significant amount in the SCS because about 10 typhoons occur every year.

During 20–24 July 2004, a research cruise was carried out along transect A (Fig. 10.6a) (Lu et al. 2010), the same area as our present study. In-situ observations from the data of the above cruise indicated a subsurface bloom (Fig. 10.6b), and satellite data analysis proved that an increase of surface Chl-a near the Station A3 was present. The maximum Chl-a concentration at the Station A3 was approximately 2.80 mg m^{-3} at 50 m depth (Fig. 10.6b), and the thickness (40 m) of the subsurface Chl-a bloom reached the depth of 40–80 m (Fig. 10.6b and c). However, simulated data showed that the thickness of Chl-a bloom was only 10 m and it appeared at 30–40 m in depth (Fig. 10.6d), with a Chl-a concentration of 1.80 mg m^{-3}. Therefore, it was very clear that the in situ observations differed when compared to the simulated data.

In the view of the results of the present study, we paid carefully attention to the reasons for the above situations occurring in 2004. According to the information provided by the JTWC, tropical storm "Kompasu" (MSW 21.6 m s^{-1}, translation speeds 5.5 m s^{-1}) passed through the northern SCS near transect A (Fig. 10.6a) approximately 1 week before the research cruise in July 2004. The findings from that cruise matched with the results of this study and provided additional supporting evidence that typhoons could induce subsurface Chl-a blooms. Further, these blooms have longer durations and higher concentrations of Chl-a than the surface blooms.

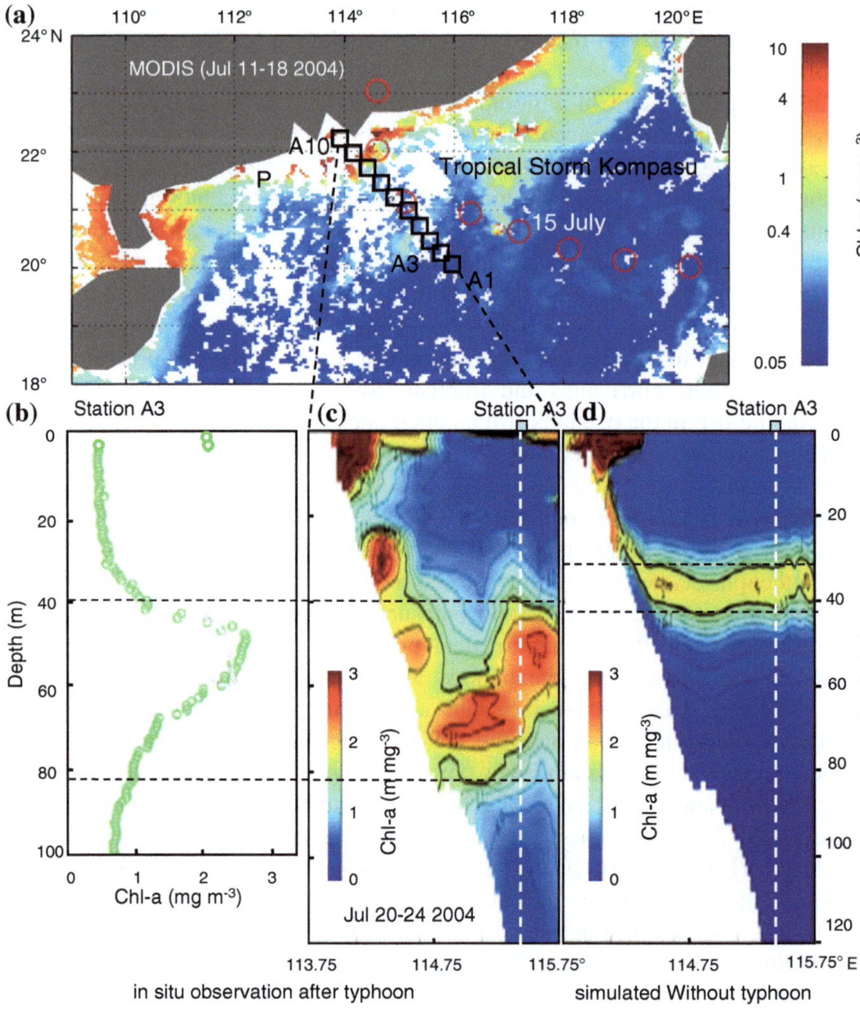

Fig. 10.6 Comparison of Chl-a values between in situ observation after tropical storm "Kompasu" (**b**, **c**) and simulated data **d**. **a** Track of tropical storm "Kompasu" (*red circles*), MODIS Chl-a (Jul 11–18), and sample stations (*black squares*) along the transect A; (**b**) Profiles of observed Chl-a concentration at Station A3 (20.5N, 115.5E) after tropical storm "Kompasu"; (**c**) *vertical* distribution of observed (Jul 20–24 2004) long the transect A after tropical storm "Kompasu"; (**d**) simulated vertical distribution of Chl-a concentration long transect A. (**b**, **c** and **d** are modified from Lu et al. (2010)

10.5 Three Phases of the Chl-a Bloom and Related Physical Processes

The normal, or pre-typhoon, phase is shown in Fig. 10.7a. Nutrients are low at the surface but increase rapidly at the nitricline which is approximately 60–80 m deep (Chen et al. 2006; Wang et al. 2010). The MLD is approximately 35 m. The Chl-a concentration is low at the surface and reaches a maximum value of approximately 0.30 mg m^{-3} at 50 m depth. The early post-typhoon phase (Fig. 10.7b) includes a period of about 1 week after the passage of the typhoon. MLD is increased to 40–50 m because of intense mixing generated by the typhoon. As a result, the level of nutrients in the mixed layer is increased which stimulates the growth of Chl-a both at the surface and in the subsurface layer. The late post-typhoon phase (Fig. 10.7c) starts approximately two weeks after the passage of the typhoon. When the surface nutrients are depleted, the Chl-a bloom at the surface decays. With less Chl-a at the sea surface level, more radiation is available for photosynthesis in the subsurface layer. This, together with the subsurface upwelling, results in further enhancement of the subsurface Chl-a bloom. The concentration of nutrients falls to lower levels while MLD increases to approximately 60 m. After about three weeks following the passage of the typhoon, upwelling decreases and the re-supply of nutrients becomes slower. In addition, the grazing pressure of zooplankton filter feeders becomes significant. It should be noted that the intensive grazing pressure was the main cause of the termination of the diatom bloom induced by the typhoon "Morakot" (Chung et al. 2012). As a result, Chl-a concentrations return to normal, or may fall even lower than normal levels.

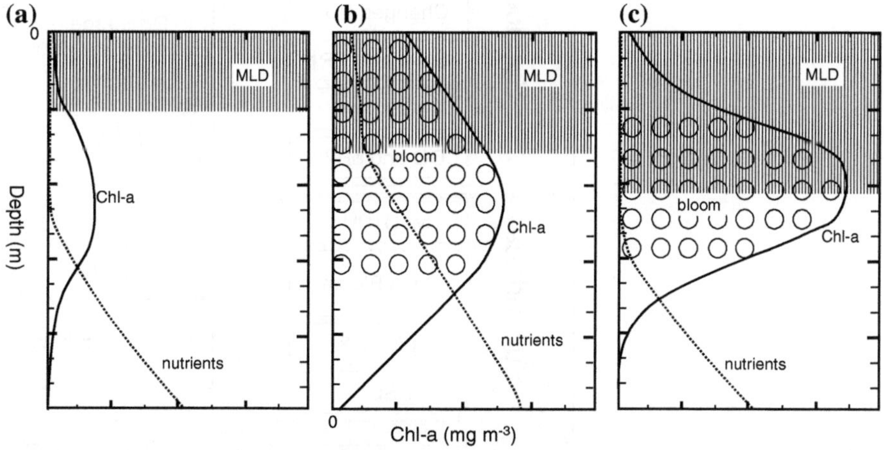

Fig. 10.7 Concept model showing four phases of Chl-a blooms induced by typhoon: (**a**): normal phase; (**b**): early post-typhoon phase; (**c**): late post-typhoon phase. The *solid thick lines* indicate Chl-a concentration; the *dashed lines* indicate nutrient; the *solid thin lines* indicate the MLD; the *open circles* in Fig. 10.7b and c indicate Chl-a bloom (surface $\geq$0.50 mg m^{-3}, subsurface $\geq$1.00 mg m^{-3})

Previous studies reported that typhoons could enhance the subsurface Chl-a and IPP in continental ecosystems (Chen et al. 2009; Hung et al. 2010). In addition, Hung et al. (2010) mentioned that subsurface phytoplankton blooming occurred offshore one week after the typhoon. Nevertheless, the duration of the subsurface bloom was not known. Depending on the magnitude of the typhoon's MSW, translation speeds, pre-existed conditions, and early and late post-typhoon phases can be sustained for different periods of time. The typhoons with strong MSW, low translation speeds can induce longer early and late post-typhoon phases than in our case where they lasted for one, and three weeks, respectively.

As mentioned, typhoons have a great impact on the oceans. On the ocean surface, typhoons, hurricanes or tropical cyclones can enhance the Chl-a, IPP, ρCO_2, decrease the ρCO_2, SST, SLA and change the phytoplankton size classes (PSC) due to typhoon induced mixing and upwelling, discharge, and current and eddy (Fig. 10.8). However, a typhoon's impact on the dissolved oxygen and pH is unclear. Further research is needed to perfect our understanding of a typhoon's ecological effect, to quantiativly analysis a typhoon's impact on surface elements and to establish the relationships between the typhoon's impact and the typhoon's MSW, translation speed, size, transmit time and the pre-typhoon ocean environment (Yang and Tang 2010; Lin 2012).

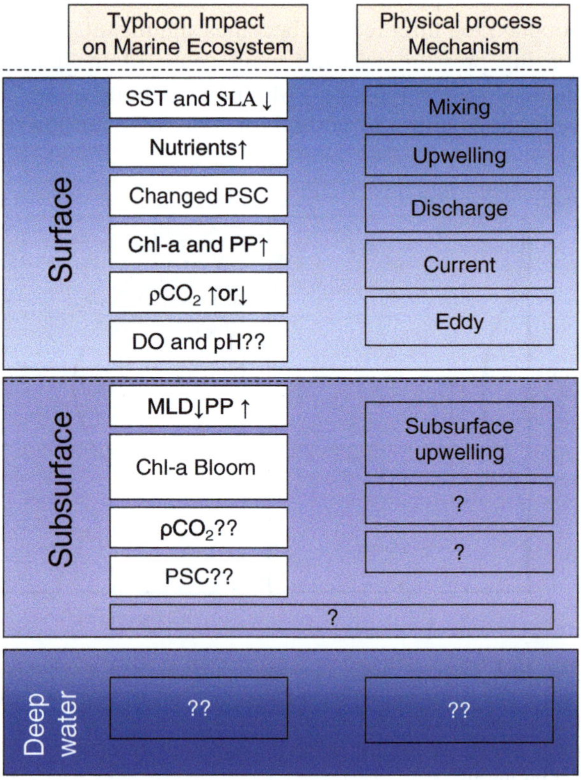

Fig. 10.8 The summary of the typhoon impacts on the marine system

Few papers have studied the impacts on subsurface water of typhoons (Chen et al. 2009; Hung et al. 2010). In the present study, typhoon induced subsurface upwelling deepened the MLD and made subsurface Chl-a bloom lasted longer and to be stronger than surface Chl-a bloom. But we still don't know what had happened to the subsurface PSC, ρCO_2 and so on. In addition, we must ask; Can typhoons have impacts on water deeper than the euphotic depth? Further research is needed to analyze the impact of typhoon on deep and subsurface water and to reveal the related mechanisms. Further perspectives on the replenishing mechanics of a typhoon's ecological effect also show great potential for further study.

10.6 Conclusions

In this chapter we documented the typhoon-induced subsurface Chl-a bloom showing that it was stronger and lasted for longer periods of time than the surface Chl-a blooms. The surface Chl-a concentrations increased suddenly on the 2nd day after the passage of typhoon Nuri and lasted for about one week. The subsurface Chl-a bloom lasted for at least three weeks; Chl-a concentrations were approximately 5 times higher than normal levels. Large increases of the MLD indicated that Nuri induced significant vertical mixing. The occurrence of low SLA observed by satellite altimeter as well as measured distributions of temperature and salinity demonstrated that Nuri induced an upwelling. As a result, the typhoon induced subsurface Chl-a blooms lasted for a relatively long period of time.

The impacts of the typhoon on both surface and subsurface Chl-a could be divided into three phases due to the magnitude of the MSW and the translation speeds of the typhoon: a, The normal or pre-typhoon phase; b, The early post-typhoon phase; and c, The late post-typhoon phase. The stronger the MSW joined with lower translation speeds of the typhoon and greater SLA, the longer lasting are the early and late post-typhoon phases.

We conclude that the impact of a typhoon on the Integrated Primary Production (IPP) may be greater than originally estimated due to the occurrence of strong and long-lasting subsurface Chl-a blooms. Accurate numerical modeling is required in order to obtain a more accurate estimate of the impacts of typhoons on IPP.

Acknowledgments This study was supported by the following research projects awarded to DL Tang: (1) National Natural Sciences Foundation of China (31061160190, 40976091, 41006068, and NSFC-RFBR Project-41211120181 of DL Tang and D. Pozdnyakov); (2) Guangdong Sciences Foundation (2010B031900041, 8351030101000002); (3) Innovation Group Program of State Key Laboratory of Tropical Oceanography (LTOZZ1201). YD Afanasyev has been supported by the Natural Sciences and Engineering Research Council of Canada. The authors thank the South China Sea Open Cruise, R/V *Shiyan* 3 (SCSIO, CAS); The authors are grateful to David Lai of NorthWest Research Associates for providing valuable comments to the manuscript, to J. D. C. Craig for thorough proofreading of the manuscript and to all members of the Research Center for Remote Sensing and Marine Ecology and Environment (RSMEE).

References

Babin, S.M., Carton, J.A., Dickey, T.D., Wiggert, J.D.: Satellite evidence of hurricane-induced phytoplankton blooms in an oceanic desert. J. Geophys. Res. **109**(C03043), 1–21 (2004)

Behrenfeld, M.J., Falkowski, P.G.: Photosynthetic rates derived from satellite-based chlorophyll concentration. Limnol. Oceanogr. **42**, 1–20 (1997)

Chang, S.W., Anthes, R.A.: Numerical simulations of the ocean's nonlinear, baroclinic response to translating hurricanes. J. Phys. Oceanogr. **8**, 468–480 (1978)

Chen, Y.L., Chen, H.Y., Lin, Y.H.: Distribution and downward flux of Trichodesmium in the South China Sea as influenced by the transport from the Kuroshio Current. Mar. Ecol. Prog. Ser. **259**, 47–57 (2003)

Chen, C., Shiah, F., Chung, S., Liu, K.: Winter phytoplankton blooms in the shallow mixed layer of the South China Sea enhanced by upwelling. J. Marine. Syst. **59**, 97–110 (2006)

Chen, Y.L., Chen, H.Y., Jan, S., Tuo, S.: Phytoplankton productivity enhancement and assemblage change in the upstream Kuroshio after typhoons. Mar. Ecol. Prog. Ser. **385**, 111–126 (2009)

Chung, C.C., Gong, G.C., Hung, C.C.: Effect of Typhoon Morakot on microphytoplankton population dynamics in the subtropical Northwest Pacific. Mar. Ecol. Prog. Ser. **448**, 39–49 (2012)

Elsner, J.B., Liu, K.: Examining the ENSO-typhoon hypothesis. Clim. Res. **25**, 43–54 (2003)

Fedorov, K., Varfolomeev, A., Ginzburg, A., Zatsepin, A., Krasnopevtsev, A.Y., Ostrovsky, A., Skylarov, V.: Thermal reaction of the ocean on the passage of Hurricane Ella. Okeanologiya **19**, 992–1001 (1979)

Fennel, K., Boss, E.: Subsurface maxima of phytoplankton and chlorophyll: steady-state solutions from a simple model. Limnol. Oceanogr. **48**, 1521–1534 (2003)

Fumin, R., Gleason, B., Easterling, D.: Typhoon impacts on china's precipitation during 1957–1996. Adv. Atmos. Sci. **19**, 943–952 (2002)

Gautam, R., Singh, R.P., Kafatos, M.: Changes in ocean properties associated with Hurrican Isabel. Int. J. Remote Sens. **26**, 643–649 (2005)

Geisler, J.E.: Linear theory of the response of a two layer ocean to a moving hurricane. Geophys. Astro. Fluid. **1**, 249–272 (1970)

Greatbatch, R.J.: On the response of the ocean to a moving storm: the nonlinear dynamics. J. Phys. Oceanogr. **13**, 357–367 (1983)

Hader, D.P., Worrest, R.C., Kumar, H.D., Smith, R.C.: Effects of increased solor ultraviolet radiation on aquatic ecosystems. Ambio **24**, 174–180 (1995)

He, L., Yin, K., Yuan, X., Li, D., Zhang, D., Harrison, P.J.: Spatial distribution of viruses, bacteria and chlorophyll in the northern South China Sea. Aquat. Microb. Ecol. **54**, 153–162 (2009)

Hung, C.C., Gong, G.C., Chou, W.C., Chung, C.C., Lee, M.A., Chang, Y., Chen, H.Y., Huang, S.J., Yang, Y., Yang, W.R., Chung, W.C., Li, S.L., Laws, E.: The effect of typhoon on particulate organic carbon flux in the southern East China Sea. Biogeosciences **7**, 3007–3018 (2010)

Jarosz, E., Mitchell, D.A., Wang, D.W., Teague, W.J.: Bottom-up determination of air-sea momentum exchange under a major tropical cyclone. Science **315**, 1707–1709 (2007)

Karl, D., Lukas, R.: The Hawaii Ocean time-series (HOT) program: background, rationale and field implementation. Deep-Sea Res. II **43**, 129–156 (1996)

Kumar, S.P., Nuncio, M., Narvekar, J., Kumar, A., Sardesai, S., de Souza, S.N., Gauns, M., Ramaiah, N., Madhupratap, M.: Are eddies nature's trigger to enhance biological productivity in the Bay of Bengal? Geophys. Res. Lett. **31**(L07309), 1–5 (2004)

Kundu, S.N., Sahoo, A.K., Mohapatra, S., Singh, R.P.: Change analysis using IRS-P4 OCM data after the Orissa super cyclone. Int. J. Remote Sens. **22**, 1383–1389 (2001)

Levitus, S., Boyer, T.P.: World Ocean Atlas 1994, vol. 2, Oxygen. National Environmental Satellite, Data, and Information Service, Washington, DC (United States) (1994)

Levitus, S.: Climatological atlas of the world ocean. United States Government Printing (1982)
Lin, I.: New evidence for enhanced ocean primary production triggered by tropical cyclone. Geophys. Res. Lett. **30**, 1–5 (2003)
Lin, I.: Typhoon-induced phytoplankton blooms and primary productivity increase in the western North Pacific subtropical ocean. J. Geophys. Res. **117**(C03039), 1–15 (2012)
Liu, K.K., Chao, S.Y., Shaw, P.T., Gong, G.C., Chen, C.C., Tang, T.: Monsoon-forced chlorophyll distribution and primary production in the South China Sea: observations and a numerical study. Deep-Sea Res. I **49**, 1387–1412 (2002)
Liu, H.X., Song, X.Y., Huang, L.M., Tan, Y.H., Zhong, Y., Zhou, L.B., Liang, X.P.: Phytoplankton biomass and primary production in the northern South China Sea during summer, 2007. In: Proceedings of the Northern South China Sea Open Cruise 2004–2009, pp. 515–521 (2009)
Lu, Z., Gan, J., Dai, M., Cheung, A.Y.Y.: The influence of coastal upwelling and a river plume on the subsurface chlorophyll maximum over the shelf of the northeastern South China Sea. J. Marine. Syst. **82**, 35–46 (2010)
Ning, X., Chai, F., Xue, H., Cai, Y., Liu, C., Zhu, G., Shi, J.: Physical-biological oceanographic coupling influencing phytoplankton and primary production in the South China Sea. J. Geophys. Res. **109**(C10005), 1–20 (2005)
Parsons, T.R., Maita, Y., Lalli, C.M.: Manual of Chemical and Biological Methods for Seawater Analysis. Pergamon, Oxford (1984)
Perry, M.J., Sackmann, B.S., Eriksen, C.C., Lee, C.M.: Seaglider observations of blooms and subsurface chlorophyll maxima off the Washington coast. Limnol. Oceanogr. **53**, 2169–2179 (2008)
Price, J.F.: Upper ocean response to a Hurricane. J. phys. Oceanogr. **11**, 153–175 (1981)
Price, J.F., Sanford, T.B., Forristall, G.Z.: Forced stage response to a moving hurricane. J. Phys. Oceanogr. **24**, 233–260 (1994)
Sanford, T.B., Black, P.G., Haustein, J.R., Feeney, J.W., Forristall, G.Z., Price, J.F.: Ocean response to a hurricane. Part I: Observations. J. Phys. Oceanogr. **17**, 2065–2083 (1987)
Sarma, V.V., Aswanikumar, V.: Subsurface chlorophyll maxima in the north-western Bay of Bengal. J. Plankton Res. **13**, 339–352 (1991)
Shay, L.K., Goni, G.J., Black, P.G.: Effects of a warm oceanic feature on Hurricane Opal. Mon. Weather Rev. **128**, 1366–1383 (2000)
Shiah, F.K., Chung, S.W., Kao, S.J., Gong, G.C., Liu, K.K.: Biological and hydrographical responses to tropical cyclones (typhoons) in the continental shelf of the Taiwan Strait. Cont. Shelf Res. **20**, 2029–2044 (2000)
Sun, L., Yang, Y., Xian, T., Lu, Z., Fu, Y.: Strong enhancement of chlorophyll a concentration by a weak typhoon. Mar. Ecol. Prog. Ser. **404**, 39–50 (2010)
Tang, D.L., Kawamura, H., Doan-Nhu, H., Takahashi, W.: Remote sensing oceanography of a harmful algal bloom off the coast of southeastern Vietnam. J. Geophys. Res. **109**(C03014), 1–5 (2004a)
Tang, D.L., Kawamura, H., Van Dien, T., Lee, M.A.: Offshore phytoplankton biomass increase and its oceanographic causes in the South China Sea. Mar. Ecol. Prog. Ser. **268**, 31–41 (2004b)
Walker, N.D., Leben, R.R., Balasubramanian, S.: Hurricane-forced upwelling and chlorophyll a enhancement within cold-core cyclones in the Gulf of Mexico. Geophys. Res. Lett. **32**, 1–5 (2005)
Wang, J.J., Tang, D.L., Sui, Y.: Winter phytoplankton bloom induced by subsurface upwelling and mixed layer entrainment southwest of Luzon Strait. J. Marine. Syst. **83**, 141–149 (2010)
Wu, L., Wang, B., Geng, S.: Growing typhoon influence on east Asia. Geophys. Res. Lett. **32**(L18703), 1–4 (2005)
Yang, X.X., Tang, D.L.: Location of sea surface temperature cooling induced by typhoon in the South China Sea. J. Trop. Ocea. **29**, 26–31 (2010). (Chinese with English abstract)
Ye, H.J., Sui, Y., Tang, D.L., Afanasyev Y.D.: A subsurface chlorophyll a bloom induced by Typhoon in the South China Sea. J. Mar. Syst. (2013, in press)

Yu, J., Tang, D.L., Li, Y., Huang, Z., Chen, G.: Increase in fish abundance during two typhoons in the South China Sea. Adv. Space Res. **51**, 1734–1749 (2013)

Zhao, H., Tang, D.L., Wang, Y.: Comparison of phytoplankton blooms triggered by two typhoons with different intensities and translation speeds in the South China Sea. Mar. Ecol. Prog. Ser. **365**, 57–65 (2008)

Zhao, H., Tang, D.L., Wang, D.X.: Phytoplankton blooms near the Pearl River Estuary induced by Typhoon Nuri. J. Geophys. Res. **114**(C12027), 1–9 (2009)

Zheng, G.M., Tang, D.L.: Offshore and nearshore chlorophyll increases induced by typhoon winds and subsequent terrestrial rainwater runoff. Mar. Ecol. Prog. Ser. **333**, 61–74 (2007)

Chapter 11
A Pilot Satellite-Based Investigation of the Impact of a Deep Polar Cyclone Propagation on the Phytoplankton Chlorophyll Spatial and Temporal Dynamics in the Arctic Ocean

Dmitry Pozdnyakov, DanLing Tang, Leonid Bobylev, Pavel Golubkin, Elizaveta Zabolotskikh, Dmitry Petrenko and Evgeny Morozov

Abstract A pilot satellite-based investigation of modulations exerted upon mixed-layer phytoplankton fields by deep cyclones is performed for the first time across the northern hemisphere polar region, viz. the Arctic Ocean. Resorting to a synergistic approach, polar cyclones were first identified from NCEP/NCAR data for the summer time period during 2002–2005, and their propagation throughout the Barents Sea was further traced down. The above water wind force was retrieved from QuikSCAT data. These data were further accompanied by ocean colour data from SeaWiFS, and MODIS to examine the spatial and temporal distributions of surficial phytoplankton chlorophyll concentration dynamics along the trajectory of the cyclone's footprint across the sea. When the wind speed, bathymetric features and cloud conditions proved conjointly favorable, appreciable increases in phytoplankton chlorophyll concentration (for basically oligotrophic waters of the Arctic Ocean) have been observed following the cyclone passage with a time lag of about 5 days. This implies that with the ongoing amplification of climate warming at high northern latitudes, the increase in chlorophyll discussed above is potentially capable of boosting the primary production in the Arctic Ocean. However, further studies are certainly required to

---

D. Pozdnyakov (✉) · L. Bobylev · P. Golubkin · E. Zabolotskikh · D. Petrenko · E. Morozov
Nansen International Environmental and Remote Sensing Centre,
St. Petersburg, Russia
e-mail: dmitry.pozdnyakov@niersc.spb.ru

D. Pozdnyakov · L. Bobylev
Nansen Environmental and Remote Sensing Centre, Bergen, Norway

D. L. Tang
State Key Laboratory of Tropical Oceanography, South China
Institute of Oceanology, Guangzhou, China

P. Golubkin · E. Zabolotskikh · D. Petrenko
Russian State Hydrometeorological University, St. Petersburg, Russia

D. L. Tang and G. J. Sui (eds.), *Typhoon Impact and Crisis Management*,
Advances in Natural and Technological Hazards Research 40,
DOI: 10.1007/978-3-642-40695-9_11,

extend the observational data up to 2012 and further on in order to statistically and phenomenologically underpin and further our understanding of the actual mechanisms of changes in the Arctic Ocean ecosystem functioning.

11.1 Introduction

Among the variety of environmental effects produced by ongoing climate change, significant variations in phytoplankton primary productivity and time- and area-integrated production are becoming increasingly evident (Greene and Pershoing 2007). Undoubtedly, the shifts in primary production (PP) amply documented are a reflection of serious alterations occurring to the ecosystems of the world's oceans. Revealed in a wide variety of marine/oceanic regions, the alterations are driven by a host of climate change-related mechanisms that are yet insufficiently understood (Hanshaw et al. 2008).

Deep baric formations in the atmosphere are shown to be able to strongly affect PP variations across oceanic/marine tracts (Chang et al. 1996; Lin et al. 2003; Davis and Yan 2004; Tang et al. 2004a, b, 2006; Walker et al. 2005; Zhao and Tang 2006; Rao et al. 2006; Zheng and Tang 2007; Zhao et al. 2008, 2009; Toratani2008; Byju and Kumar 2011; Sarangi 2011; Lin 2012; Chung et al. 2012; Chen and Tang 2012 and more). This is evidenced by quite a number of satellite-based studies employing synergistic approaches. These studies seemingly indicate that the phytoplankton biomass increase provoked by cyclones might arise from a variety in-water processes (Le Fouest et al. 2011). Nearly invariably they were conducted over low-latitude waters in the northern hemisphere: the Western North Pacific (reportedly, the area of the highest incidence of tropical cyclones), and more specifically, South China Sea, as well as in the Indian Ocean (Bay of Bengal and Eastern Arabian Sea), and northern Atlantic Ocean (Gulf of Mexico and the 24°N–38°N latitudinal belt).

There are reasons to expect that the effect of deep cyclones on the primary production in the Arctic Ocean can also be appreciable/consequential (Le Fouest et al. 2011) regardless of the fact that it is a low production region of the world's oceans (Arrigo and van Dijken 2011). We are unaware, however, of any satellite-based investigations of this phenomenon at these latitudes.

Although the Arctic Ocean ecosystems are subject to atmospheric and hydrodynamic forcing of different nature (Bobylev et al. 2003), identification of the role of deep cyclones in PP variations and quantification of the ensuing consequences is an attainable task due to the specific spatial and temporal scales inherent in this driving mechanism. Indeed, when studying a single Nordic Sea, the relevant scales are of a few hundred kilometers and a few days, which are not characteristic of many, but not all, other options of external forcing (Benzeman 2009).

Although attainable, the above task must be challenging because of the high incidence of heavy cloud conditions over ice-free tracts of the Arctic

(Chernokulsky and Mokhov 2012). It necessitates the application of spatio-temporal averaging of data in the visible, which, inevitably, will affect the spatio-temporal resolution of spaceborne images, and, thus, complicate the quantitative assessment of the effect sought for. Obviously, a reasonable trade-off should be found to overcome the above impediment.

Since, in comparison with low-latitude productive waters, the PP levels and variations in the Arctic are rather low, the retrieval error can be significant Petrenko et al. 2013), and this imposes strict requirements to the inference of the desired information. Statistical substantiation of the data is a prerequisite in this case.

To increase the analytical capacity of the research, a synergistic approach is required in order to consider not solely the ocean color data (from which phytoplankton chlorophyll concentration and PP are retrievable), but also sea surface temperature (SST) and near-surface winds, as well as currents, frontal zones/zones of convergence and divergence, the bathymetry of the target region, and meteorological data on the baric fields.

Below we present the results of our pilot study aimed at revealing and quantitatively assessing the impact of deep cyclones on phytoplankton chlorophyll (*chl*) spatial and temporal variations in the Arctic. At this stage of research, the target region was confined to the Barents Sea (BS), and the time period was limited to 2002–2004. In the results analysis presented here not all the abovementioned auxiliary data were yet employed in depth. This will be done at later stages of the research implementation.

11.2 Barents Sea General Description

Bordered by the shelf edge towards the Norwegian Sea in the west, the Islands of Svalbard in the northwest, and the Islands of Franz Josef Land and Novaya Zemlya in the northeast and east, the BS (Fig. 11.1) is characterized by a relatively shallow shelf and fairly complex hydrography. A seasonably variable ice-cover with its edge retreating far to the north in summer, an inflow of warm and saline Atlantic waters (t > 3 °C, salinity > 35 psu) with the Norway Current and their blending with cold and less saline Arctic waters (t < 0 °C, salinity < 35 psu) makes this sea very sensitive to atmospheric, hydrodynamic and, ultimately climate change forcing (Reigstad et al. 2002). The average depth is ~300 m plunging to a maximum of 600 m in the major Bear Island Trench.

The interaction between warm Atlantic and cold Arctic waters occurs mainly in the Polar Front Zone (PFZ). In summer the PFZ is located at 78–80°N extending amid the BS between Svalbard and the southern coast of Novaya Zemlya.

In spring the PFZ structure is frequently affected by mesoscale eddies with characteristic sizes of 25–40 km and the water vertical velocity reaching up to 20 m day-1. The tidal/ebbing water motions are significant in the Barents Sea with

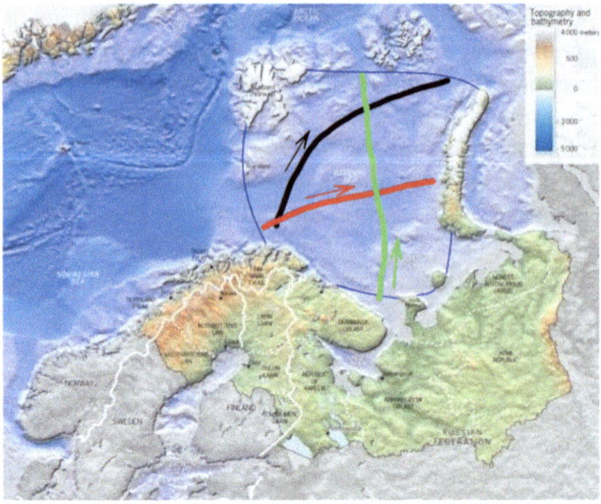

Fig. 11.1 The bathymetry and limits (*blue line*) of the BS. *Black*, *red* and *green lines* and *arrows* indicate tracts and direction of propagation of the cyclones passed over the sea on 15 May 2003, 24–25 2003, 23–25, 2004, and respectively

the tidal amplitude and current direction varying greatly. The atmospheric cyclonic activity over the Barents Sea is very pronounced throughout the year.

As the PFZ enhances the vertical and horizontal mixing in the region, and hence channels the nutrients up to the euphotic zone, it is a site of high biological activity. It largely explains that, compared to other marine waters of similar latitude, the BS is a relatively productive high-latitude marine ecosystem. However, it should not be overlooked that in addition to the PFZ, all the aforementioned atmospheric and hydrodynamic factors can also efficiently modulate and spur up the BS productivity.

11.3 Sources of Data

The occurrence of cyclones moving across the BS has been traced down using the reanalysis data on the geopotential of the 1,000 mbar isobaric surface (http://www.esrl.noaa.gov/psd) from the National Center for Environmental Prediction/National Center for Atmospheric Research (NCEP/NCAR) (Kalnay et al. 1996).

Satellite data on wind speed and direction over the ocean employed in the study are from the National Aeronautics and Space Administration (NASA) QUIcK SCATterometer (QuikSCAT) (http://winds.jpl.nasa.gov/missions/quikscat/index.cfm). The spatial resolution of the QuikSCAT and NCEP/NCAR data are, respectively, 25 km and 2.5° (~275 km) with the revisiting frequency of about twice and four times per day, respectively.

Ocean colour data were downloaded from three satellite sensors: Sea-viewing Wide Field-of-view Sensor (SeaWiFS), and Moderate Resolution Imaging Spectroradiometer (MODIS) [on Aqua and Terra] (http://oceancolor.gsfc.nasa.gov)

at a spatial resolution of 1.0 and 1.1 km, respectively, and the revisiting frequency of 2–3 and 4–6 times per day for SeaWiFS and MODIS, respectively. SeaWiFS and MODIS (conjointly on Aqua and Terra) ocean colour data were blended daily to obtain a composite image for each specific day. In addition to ocean colour data, MODIS sensors also provide SST data at the space and time resolution equaling that for *chl* fields.

11.4 Methodology

11.4.1 Algorithms

The concentration of *chl* was retrieved with the NASA standard algorithms OC4 for SeaWiFS (O'Reilly et al. 1998), and OC3 for MODIS Aqua and Terra (O'Reilly et al. 2000). These are modified cubic polynomial functions based on the band-ratio paradigm and employing remote sensing spectral reflectance in the visible and near infrared channels as an input parameter. The NASA data reprocessed only recently were employed in the present study: MODIS-Aqua was reprocessed by reprocessing R2012.0; MODIS-Terra and SeaWiFS were reprocessed by reprocessing R2010 (http://oceancolor.gsfc.nasa.gov/WIKI/OCReproc.html).

To determine SST, "window-split" retrieval algorithms were employed. They are based on the difference between the satellite-observed water surface *apparent* (*brightness*) temperature, T_i determined in two several spectral channels centered at 11 μm (T_{11}) and 12 μm (T_{12}). The NASA algorithm is a four-term expression with proportionality coefficients c_1-c_4: $\text{SST} = c_1 + c_2T_{11} + c_3(T_{11} - T_{12}) + c_4[(\sec ə - 1)(T_{11} - T_{12})$, where ə is the satellite zenith angle (http://yyy.rsmas.miami.edu/groups/rrsl/pathfinder/Algorithm/algo_index.html#algo-pathsst and http://modis.gsfc.nasa.gov/data/atbd/atbd_mod25.pdf) (Robinson 1994).

The principal of wind speed measurements rests in the backscatter of QuikSCAT transmitted microwave pulses by the water surface. The backscattered microwave signal is modified by the wind-roughened surface in a certain (logarithmic) proportion to the wind speed and direction. Application of inverse modeling allows retrieving the sought for wind parameters. The system measures winds between 3 and 30 ms^{-1} with an accuracy better than 2 ms^{-1} or 10 % in speed and 20° in direction with a spatial resolution of 25 km (Callahan and Lungu 2006).

11.4.2 Detection of Cyclone Impact Approach

Analyzing the *NCEP/NCAR information,* a metadata base was compiled to reflect all cases of cyclones passing through the Barents during the vegetation period, i.e. the time of arrival, passage through and leaving off the BS.

For each identified case of cyclone passage, ocean colour images were acquired for the periods of 5 days prior to the cyclone arrival to the confines of the BS, and about 10 days after the definitive cyclone departure from the target marine zone.

Using the NCEP/NCAR data on the cyclone track, a relatively narrow strip of about 200 km width centered at the cyclone's "eye" was selected from the colour image(s) of the BS, and processed to yield the *chl* field within the strip.

To mitigate the screening effect due to cloudiness, the retrieved *chl* fields were blended from the three ocean colour sensors, and a daily to a few days averaging was performed; the exact time period of averaging was prompted in each case by the specific features of the cloud coverage.

The strength of the cyclone (the baric pressure in the "eye's" zone) and the speed of wind above the ocean were identified for each case and used for further analysis.

To assess the response of the phytoplankton surficial abundance to a cyclone passing through the sea, the *chl* average level for the period preceding the cyclone passage was compared to the counter value obtained 3–5 days after the cyclone definitive departure.

The same approach has been exploited for SST field characterization to be further collated with the respective *chl* data.

In the case of a cyclone tandem coming to the Barents Sea, the changes triggered by both were assessed in a similar way, but the period of averaging depended on the time-interval between the passages of two baric formations.

11.5 Results and Dicussion

Here we are presenting our results for three events of cyclones that passed across the BS on May 15, 2003 and 23–25 June and 24–25 July, 2004. As an example, Fig. 11.1 shows the trajectories of all three cyclones.

On May 15, 2003 the maximum value of the geopotential height of the isobaric surface of 1,000 mbar was –86 m^2 s^{-2}. Figure 11.2 illustrates the field of surface winds over the marine area for the moment when the cyclone started moving across the BS.

Figure 11.3 illustrates, for the 15 May cyclone, the time and space averaged *chl* concentration fields (encompassing the footprint area of this cyclone as it moved from the mainland into the BS) for two time periods, viz. 4 days prior to (*a*) and 3–5 days after (*b*) the cyclone left the sounded area of the water surface. Unfortunately, the cloud conditions did not permit to extend spaceborne data on *chl* over the entire cyclone footprint area. Even a cursory comparison of data in Fig. 11.3a, b reveals a notable increase on the *chl* concentration after the cyclone passage.

Figure 11.4 presents schematically the temporal variations of sea surface temperature (SST) and *chl* concentration in the study area. As seen from Fig. 11.4, the cyclone passage through the target area resulted in a rather substantial increase in *chl* concentration (from ~0.5 to 1.1 μg/l): it occurred on the fifth day after the

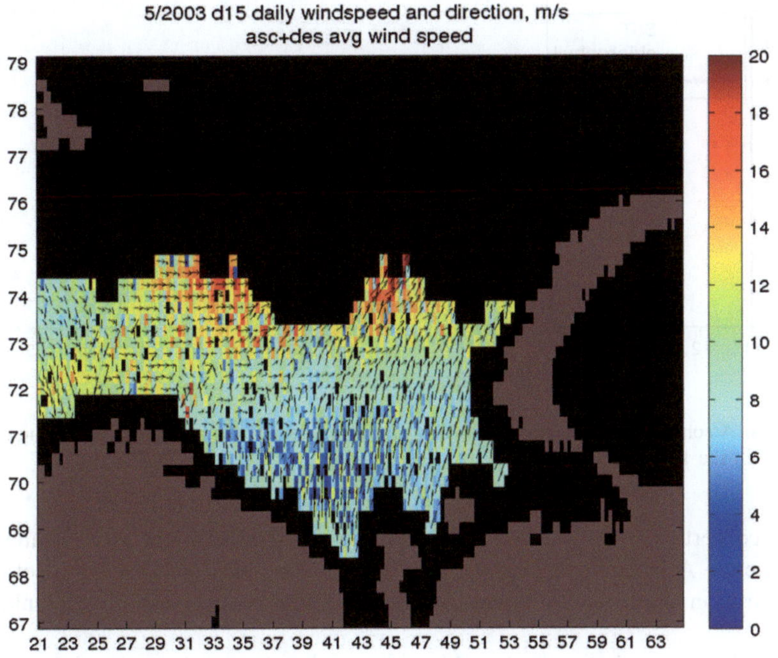

Fig. 11.2 The SST field on May 15, 2003 in the area of the cyclone coming out in the BS from the mainland. The *colour bar* is in m s^{-1}. *Numbers* on the horizontal and vertical axes are degrees of latitude and longitude, respectively

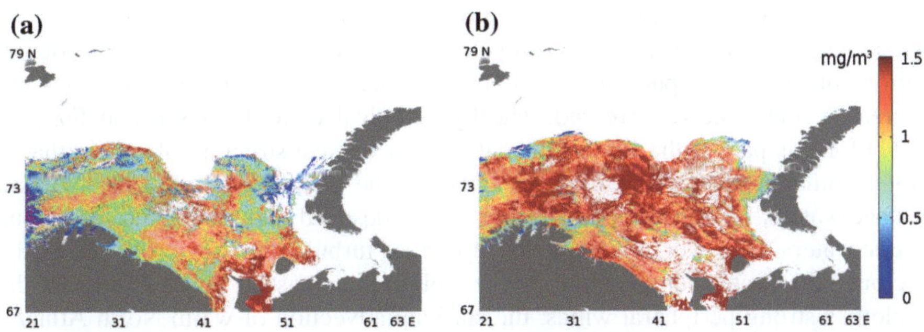

Fig. 11.3 The *chl* (mgm^{-3}) field on May 15, 2003 in the area of the cyclone coming out in the BS from the mainland. *Left panel* 4 days prior to and *right panel* 5 days after the cyclone passage. *Numbers* on the horizontal and vertical axes are degrees of latitude and longitude, respectively

cyclone passage and lasted for about 2 days. Later on May 22, the *chl* concentration returned to its pre-cyclone values.

It can be assumed that the mechanism of the *chl* concentration increase documented in this case is the same as the one reported by other workers: cyclone

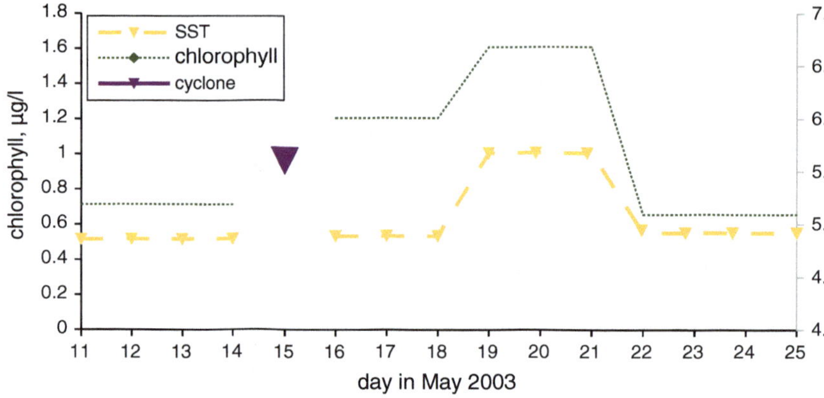

Fig. 11.4 A schematic presentation of SST and *chl* variation patterns for the case of the cyclone that passed the BS on 15 May, 2003

provoked vertical water column mixing bringing into the upper layers waters rich in nutrients. A three day delay between the cyclone passage and the onset of *chl* concentration increase is dictated by the time required for the phytoplankton to develop.

The differences between the *chl* concentration increase on the right and left sides of the cyclone footprint proved to be very small and were within the retrieval error (O'Reighly et al. 1998).

The pattern of SST variations in Fig. 11.4 formally resembles (but does not duplicate) that of *chl*: 3 days after the passage, SST increased from 0.7 to 1.2 °C. However, unlike *chl*, during the next 3 days, SST increased further up to 1.6 °C. After that, SST returned to its pre-cyclone event value. Thus, unlike the observations of cyclone impact on aquatic environment at low latitudes, SST has not decreased, but instead increased. The hydrological data (Jakobsen and Ozhigin 2011) for this part of the BS are indicative of a vertical structure of waters that is largely influenced by Atlantic water. Turning the Kolar Peninsular, the Atlantic surface waters, being warmer and more saline, descend under cold and less saline Arctic waters. However, due to intensive vertical turbulent mixing provoked by the cyclone, the water temperature in the column increased. Further, due to the cyclone's strong peripheral winds, the surficial advection of warm North Atlantic Waters resulted in an additional temporally longer increase in SST: with the propagation of the cyclone further northeast, this mechanism subsided, and SST started dropping to eventually reach the pre-cyclone event level.

The data relating to two other summer time cyclones indicate that the consequences of cyclone passage across the BS are very similar with respect to the sequence of variations in *chl*: the increase in *chl* is observed on the fifth day after the cyclone passage (although the increase proved to be less than that described for the 15 May cyclone) and practically disappears on the third day after the cyclone departure from the target area of observations. SST temporal variations resembled those presented in Fig. 11.4, but are significantly less in amplitude. This is easily explainable as both summer cyclones proved to be shallower (the minimum value of the geopotential was 4 and 19 $m^2 s^{-2}$ for the cyclones in June and July, respectively), and hence, it had less potential to affect strongly the aquatic environment they were passing over. In both cases, the cyclone's impact on the phytoplankton production increase becomes pronounced on the fifth day after the cyclone passage.

11.6 Concluding Remarks

For the first time the impact of cyclones on marine primary production was revealed and documented for the BS in the Arctic Ocean. It was found that (1) the cyclone-driven increase in *chl* is far less significant compared to low latitude marine environments, and (2) the typical lag between the cyclone passage over the marine environment and the increase in *chl* is about five days, i.e. longer than that most frequently reported for low-latitude seas. Both specific features can be explained by less favorable conditions for the phytoplankton growth: lower levels of incident light and rather low water temperatures. This leads to lower productivity rates and longer times required for the enhancement of *chl* levels observable from space.

The pattern of SST variations associated with the phenomenon of cyclone passage is unlike that observed in low-latitude seas. Instead of SST lowering with the cyclone passage, we observed an increase in SST. We assume that this increase resulted, at first, from the ascension of submerged warmer and more saline Atlantic waters due to cyclone-driven vertical mixing and further, at a later stage, by the horizontal advection to the target site of North Atlantic surficial waters (also warmer than the Arctic waters) due to the cyclone peripheral winds.

These pilot studies indicate that cyclonic activity in the Arctic, like that at lower latitudes, is able to appreciably modulate the marine primary productivity and hence alter the marine ecosystem state. These changes are powered by vertical water mixing processes and obviously result in short term alterations of the structure of surface waters. However, unlike the cyclone-driven SST changes at low latitudes, in the BS, due to a very specific interaction of the Arctic and Atlantic waters, the cyclone passage is conducive to a short-term increase in SST.

The present study is in its initial stage. Further research is required through extending the number of cases of cyclone passages across the BS throughout 2003–2012, and establishing the actual meteorological and hydrodynamic mechanisms responsible for the ensuing alterations of primary productivity in the Arctic Seas.

Acknowledgments The present study has been performed under the joint Russian-Chinese Project 12-05-91170- ГФ УР_a, no 01201255442 "Investigation of factors driving changes in phytoplankton surficial fields as an aftermath of passage of hurricanes in tropical and polar regions" (01.01.2012–12.31.2013). The authors express their gratitude to the Russian Foundation for Basic Research (RFBS) and the National Natural Science Foundation of China (NSFC) for the provided financial support.

References

Arrigo, K., van Dijken, G., Secular trends in arctic ocean net primary production. J. Geophys. Res. **116**(C09011), 1–15. doi:10.1029/2011JC007151 (2011)

Benzeman, V.Y.: Space-and Time Variability of Hydrophysical Fields in the Ocean, p. 320. Russian Ministry of Natural Resources, St. Petersburg (2009)

Bobylev, L.P., Kondratyev, K.Y., Johannessen, O.M.: Arctic Environment Variability in the Context of Global Change, p. 470. Springer-Praxis, Chichester (2003)

Byju, P., Kumar, S.P.: Physical and biological response of the Arabian sea to tropical cyclone Phyan and its implications. Mar. Environ. Res. **71**(5), 325–330 (2011)

Callahan, P.S., Lungu, T.: QuikSCAT Science Data Product User's Manual (v3.0), p. 91. Jet Propulsion Laboratory, Pasadena (2006)

Chernokulsky, A., Mokhov, I.I.: Climatology of total cloudiness in the Arctic: an interpretation of observations and reanalysis. Adv. Meteorol. (2012). doi: 10.1155/2012/542093

Chang, J., Chung, C.-C., Gong, G.-C.: Influences of cyclones on chlorophyll a concentration and synechococcus abundance in a subtropical western Pacific coastal ecosystem. Mar. Ecol. Prog. Ser. **140**, 199–205 (1996)

Chen, Y., Tang, D.L.: Eddy-feature phytoplankton bloom induced by a tropical cyclone in the south China sea. Int. J. Remote Sens. **33**(23), 7444–7457 (2012)

Chung, C.-C., Gong, G.-C., Hung, C.-C.: Effect of typhoon morakot on microphytoplankton population dynamics in the subtropical northwest Pacific. Mar. Ecol. Prog. Ser. **448**, 39–49 (2012)

Davis, A., Yan, X.-H.: Hurricane forcing on chlorophyll-a concentration off the northeast coast of the U.S. Geophys. Res. Lett. **31**, L17304 (2004)

Le Fouest, V., Postlethwaite, C., Marqueda, M.A.M., Belanger, S., Babin, M.: On the role of tides and strong wind events in promoting primary production in the Barents sea. Cont. Shelf Res. **31**, 1869–1879 (2011)

Greene, C., Pershoing, A.: Climate drives sea change. Science **315**(5815), 1084–1085 (2007)

Hanshaw, M.N., Lozier, M.S., Palter, J.B.: Integrated impact of tropical cyclones on sea surface chlorophyll in the North Atlantic. Geophys. Res. Lett. **35**(1), L01601 (2008)

Jakobsen, T., Ozhigin, V.: The Barents Sea: Ecosystem, Resources, Management: Half a Century of Russian–Norwegian Co-operation, p. 670. Tapir Academic Press, Trondheim (2011)

Kalnay, E., et al.: The NCEP/NCAR 40-year reanalysis project. Bull. Am. Meteorol. Soc. **77**, 437–471 (1996)

Lin, I., Liu, W.T., Wu, C.-C., Wong, G.T.F., Hu, C., Chen, Z., Liang, W.-D., Yang, Y., Liu, K.-K.: New evidence for enhanced ocean primary production triggered by tropical cyclone. Geophys. Res. Lett. **30**(13), 1718 (2003)

Lin, I.-I.: Typhoon-induced phytoplankton blooms and primary productivity increase in the western North Pacific subtropical ocean. J. Geophys. Res. **117**, C03039 (2012)

O'Reilly, J.E., Maritorena, S., Mitchell, B.G., Siegel, D.A., Carder, K.L., Garver, S.A., Kahru, M., McClain, C.: Ocean color chlorophyll algorithms for SeaWiFS. J. Geophys. Res. **103**(C11), 24937–24953 (1998)

O'Reilly, J.E., et al.: SeaWiFS postlaunch calibration and validation analyses, part 3, NASA Tech. Memo. In: Hooker S.B. Firestone E.R. (eds.) vol. 206892(11), pp. 49 NASA Goddard Space Flight Center, Maryland (2000)

Petrenko, D., Pozdnyakov, D., Johannessen, J., Counillon, F., Sychov, V.: Satellite derived multi-year trend in primary production in the Arctic Ocean. Int. J. Remote Sens. 34(11), 3903–3937 (2013)

Rao, K.H., Smitha, A., Ali, M.M.: A study on cyclone induced productivity in south-western Bay of Bengal during November–December 2000 using MODIS (SST and chlorophyll-a) and altimeter sea surface height observations. Indian J. Mar. Sci. **35**(2), 153–160 (2006)

Sarangi, R.K.: Impact of cyclones on the Bay of Bengal chlorophyll variability using remote sensing satellites. Indian J. Geo-Mar. Sci. **40**(6), 794–801 (2011)

Reigstad, M., Wassmann, P., Riser, C.W., Oygarden, S., Rey, F.: Variations in hydrography, nutrients and chlorophyll a in the marginal ice-zone and central Barents sea. J. Mar. Syst. **38**, 9–29 (2002)

Robinson, I.S.: Satellite Oceanography: An Introduction for Oceanographers and Remote-Sensing Scientists, p. 455 Praxis Publishing, Chichester (1994)

Tang, D.L., Kawamura, H., Dien, T.V., Lee, M.A.: Offshore phytoplankton biomass increase and its oceanographic causes in the south China sea. Mar. Ecol. Prog. Ser. **268**, 31–41 (2004a)

Tang, D.L., Kawamura, H., Doan-Nhu, H., Takahashi, W.: Remote sensing oceanography of a harmful algal bloom off the coast of southeastern Vietnam. J. Geophys. Res. **109**, C03014 (2004b)

Tang, D.L., Kawamura, H., Shi, P., Takahashi, W., Guan, L., Shimada, T., Sakaida, F., Isoguchi, O.: Seasonal phytoplankton blooms associated with monsoonal influences and coastal environments in the sea areas either side of the Indochina Peninsula. J. Geophys. Res. **111**, G01010 (2006)

Toratani, M.: Primary production enhancement by typhoon ketsana in 2003 in western north Pacific. In: Frouin R.J., Andrefouet S., Kawamura H., Lynch M. J., Pan D., Platt T. (eds.) Remote Sensing of Inland, Coastal, and Oceanic Waters, Proceedings of SPIE, pp. 715013–715013 (2008)

Walker, N.D., Leben, R.R., Balasubramanian, S.: Hurricane-forced upwelling and chlorophyll a enhancement within cold-core cyclones in the Gulf of Mexico. Geophys. Res. Lett. **32**, L18610 (2005)

Zhao, H., Tang, D.L.: The spatial distribution of chlorophyll-a and its responses to oceanographic environments in the south China sea. Adv. Geosci. **5**, 7–14 (2006)

Zhao, H., Tang, D.L., Wang, Y.: Comparison of phytoplankton blooms triggered by two typhoons with different intensities and translation speeds in the south China sea. Mar. Ecol. Prog. Ser. **365**, 57–65 (2008)

Zhao, H., Tang, D.L., Wang, D.: Phytoplankton blooms near the pearl river estuary induced by typhoon Nuri. J. Geophys. Res. **114**, C12027 (2009)

Zheng, G.M., Tang, D.L.: Offshore and nearshore chlorophyll increases induced by typhoon winds and subsequent terrestrial rainwater runoff. Mar. Ecol. Prog. Ser. **333**, 67–74 (2007)

Chapter 12
Effects of Typhoon on Seagrass Distribution

DingTian Yang and ChaoYu Yang

Abstract Typhoons inflicted great press on coastal environment. Seagrass, one of the most important constituents of coastal ecosystem, have been also greatly affected by typhoon. This chapter reviews effects of typhoons, hurricane and cyclone on seagrass distribution variation in the world, with much attention to China. The 5 major issues have been investigated : (1) Increased turbulence generated by cyclone and hurricane can result in direct physical damage; (2) Sediments resuspended by huge waves and coastal land use change decreased water transparency correspondingly decreased the efficiency of photosynthesis; (3) Hurricane, cyclone and typhoon increased the rainfall, which decreased nearshore and river mouth water salinity, correspondingly changed seagrass growth environment; (4) New species brought in by Hurricane, cyclone and typhoon, changed the composition of seagrass. (5) Satellite remote sensing is a good method for detecting seagrass distribution change.

Keywords Seagrass · Typhoon · Remote sensing

D. T. Yang (✉)
State Key Laboratory of Oceanography in the Tropics, South China Sea Institute of Oceanology, Chinese Academy of Sciences, Guangzhou 510301, China
e-mail: dtyang@scsio.ac.cn

C. Y. Yang
South China Sea Branch, State Oceanic Administration, Guangzhou 510301, China

D. T. Yang
State Key Laboratory of Lake Science and Environment, Nanjing 210008, China

C. Y. Yang
Département de Biologie/Pavillon Alexandre-Vachon, Université Laval Québec (Québec), G1V 0A6 Québec, Canada

D. L. Tang and G. J. Sui (eds.), *Typhoon Impact and Crisis Management*,
Advances in Natural and Technological Hazards Research 40,
DOI: 10.1007/978-3-642-40695-9_12,

12.1 Introduction

Seagrasses are marine vascular plants that are widely distributed around the margins of continents and islands. Calcium-carbonate rich sediments in many tropical and subtropical locations contain seagrass beds, the epiphytes of which provide a major portion of the calcium carbonate that accumulates on shallow water banks. These large expanses of vegetation provide the major biological habitat in many shallow-water carbonate sediments. Seagrass beds are an ecologically significant marine habitat providing food and shelter for bottom dwelling animals, which in turn provide food for fish, crustaceans and shorebird communities. Loss of seagrass habitat has been reported frequently in recent years. Many factors have been implicated, including cyclones and storms. Hurricane affects the distribution of seagrass by reducing transparence of water, eroding bottom mud, and bringing new species which can change the population structure and environment of the seagrass. The effects of hurricanes on seagrass meadows have not been widely studied. Hurricanes can affect seagrass beds in several ways. Increased turbulence associated with waves and currents generated by hurricane-force winds can result in physical tearing, stripping, and breakage of plant leaves and shoots. Scouring of bottom sediments and removal of the associated seagrasses and other organisms can occur, especially near reefs, shorelines, or other elevational features. Sediments in the water column can be redistributed by the storm and then deposited on top of seagrass beds, resulting in partial or complete burial. Additional sediments eroded from adjacent beaches and other landforms can be washed into the seagrass meadow and cause burial and changes in substrate quality, as shown in Chandeleur Sound, Louisiana, by Michot et al. (1999). And lastly, increased overland precipitation associated with hurricanes caused flooding and increased runoff, resulting in the deposition of huge amounts of riverine sediments into bays, estuaries, and adjacent marine systems. Following Hurricane Gilbert (September 1988), Marba and others (1994) studied its effects on the Thalassia meadows of the Caribbean coast of Yucatan, Mexico. They found differences in internodal lengths (plastochrone interval, or PI) that were attributable to the storm's impact on the seagrass beds in terms of increased sediment deposition. Cabello-Pasini et al. (2002) found that Zostera marina (eelgrass) meadows on the Pacific coast of Mexico were significantly reduced by light limitation that resulted from storm-induced resuspension of bottom sediments. Hurricane impacts to seagrass beds have also been shown to have occurred in Guadeloupe and Puerto Rico (Bouchon et al. 1991; Rodriguez et al. 1994; Salazar-Vallejo 2002). By contrast, Posey and Alphin (2002) found that sedimentation from three hurricanes had little effect on offshore benthic communities on the Atlantic coast of North Carolina's Outer Banks (Fig. 12.1).

However, the influence of typhoon on the distribution of seagrass in Southeast China is not clear. Southeast China often is attacked by typhoon. The 90 % typhoon lands in Southeast China (China Typhoon Web). The ecological environment along the shore suffers huge influences, and typhoon and cyclone have increased in

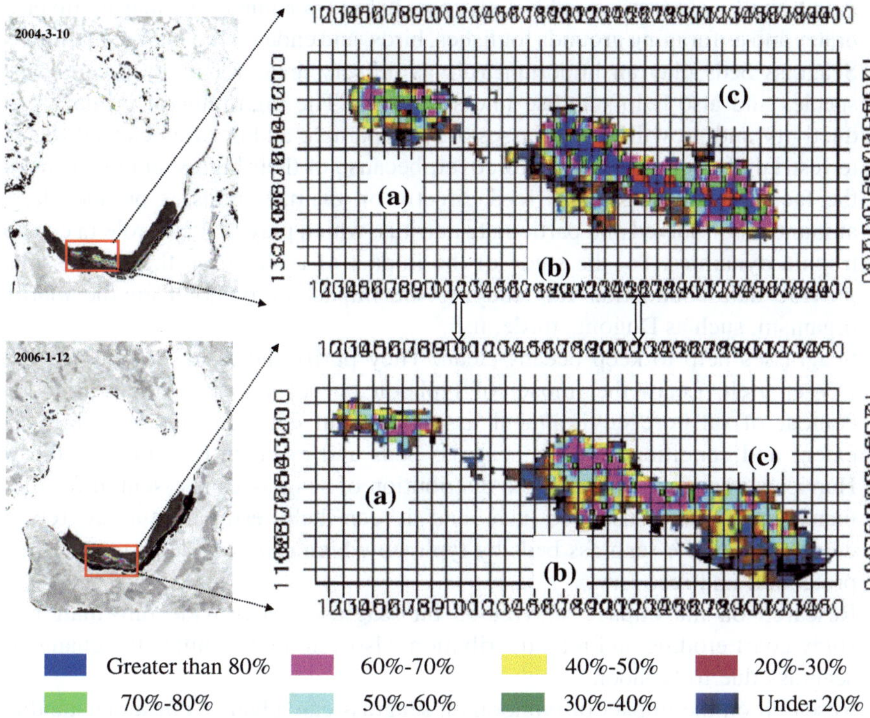

Fig. 12.1 Seagrass distribution change affected by typhoon Tianying (2005-7-29) and Dawei (2005-9-21) (with 10 × 10 pixels, 20 × 20 m every pixel. The *upper* of the figure is seagrass distribution in March 10, 2004; the lower is seagrass distribution in January 12, 2006. The *left* of image shows seagrass location in Xincun bay) (Yang and Huang 2011)

frequency and intensity with the appearance of El Nino phenomenon. In order to understand the impact of typhoon on the living conditions of seagrass, to protect seagrass and find out the condition of the recovery, the study of the impact of typhoon on the distribution of the seagrass is significant, as follows:

1. Seagrass can protect bank by reducing the impact of storm surge, waves and erosion of the shore, but seagrass suffer seriously damage while seagrass protect the bank. Successful management and restoration of coastal vegetation requires a quantitative process-based understanding of the impact of typhoon on seagrass. We can learn the process of recovery and relapse by evaluating the effects of typhoon on the living conditions of seagrass which provide the basis for protection and recovery of the seagrass.
2. Seagrass, as one of the most important aquatic biotopes, influence coastal processes and are of great importance to humans (Vicente 1992; Connolly 1994; Heck et al. 1997; Michot 1997; Arrivillaga and Baltz 1999a; Michot 2000). They significantly modify the physical, chemical, and sediment properties of coastal areas and provide nursery habitats that sustain coastal fishery resources.

Additionally, seagrass meadows serve as habitat for numerous marine invertebrates and as foraging grounds for fishes, birds, and endangered marine mammals. Seagrass beds play an important role as nursery areas for many juvenile fish species and also help stabilise fine sediments. The high rates of primary productivity and growth of seagrasses are comparable and/or often exceed that of terrestrial agricultural crops. Moreover, because of their high standing crop and the fact that few organisms feed directly on them, seagrasses produce large amounts of dissolved and particulate detritus, which plays a major role in coastal trophic dynamics. So, the study of the impact of typhoon on the seagrass can provide data foundation for studying the impact of typhoon on the marine organism, such as Dugong, turtle, fish.

3. Seagrasses help to keep oceans clean. They purify the water and ensure it is oxygen rich. Learn the impact of typhoon on seagrass can help enrich the content of offshore environment assessment. In addition, understanding the extent and condition of these seagrass beds and evaluate how the landfall of Hurricane may have affected the distribution of seagrasses is essential for their management and sustained use, and it can help estimate the degree of destruction of the seagrass beds by typhoon which can provide instructions for protection and recovery.
4. Research on the impact of typhoon on seagrass can provide information to study coast eroding and mud distribution, also can help simulate the change of seagrass due to typhoon.
5. Research on the impact of typhoon on seagrass can advance our understanding of the destruction of offshore biology by typhoon.

There is a need for protecting and preserving these seagrass ecosystems, thus requiring a quantitative method for monitoring them. Such an approach can be provided by remote sensing and image analysis. The traditional technique of studying the distribution and living conditions of seagrass mainly are in situ measurements. Seagrasses are highly productive flowering plants that often form vast meadows over shallow, unconsolidated sediments in coastal areas. Seagrass usually grow in the shallow water, especially in mud. People and boat can not be there easily, so remote sensing is significant for studying seagrass. Remote sensing can be used as an effective method for assessing and monitoring seagrass. Analysis using multispectral sensors such as Landsat and IKONOS are available (Purkis and Pasterkamp 2004; Zhang et al. 2003; Mishra et al. 2005), but effectiveness using this data has been limited due to the low spectral information of these instruments (i.e. relatively broad wavelength bands or channels), which are insufficient for a detailed benthic classification. Multispectral information stands short when there is need to know a detailed assessment of the components in each pixel. That is one of the most important features that can be extracted when high spectral resolution is available. Hyperspectral technology allows us to extract multiple layer of information from a complex optical signal can be extracted providing the necessary information for identifying the fractional contribution of classes or endmembers that relate to actual components in the study area in both a spatial and temporal way. As a simple example, assuming that seagrass and other benthic habitats can be subdivided into live coral, seagrass and sand, an image can be classified in order to obtain a spatial indication of variations in these three main reef components (Goodman and Ustin 2004). It is important to mention that combination of other endmembers is possible. Temporal changes may be obtained through the observation of different classification results over a sequence of images over time. For example, changes of classifications from the same area may indicate a loss of seagrass, coral and an increase of algae, or vice versa. Some works combine two existing models into a single algorithm for estimating the optical properties of the water, bathymetry, bottom albedo and benthic habitat classification from hyperspectral imagery.

12.2 The Progress of Related Research in the World

To a first approximation, a hurricane is a symmetric vortex circulation (Ooyama 1982). However, stationary and propagating asymmetric features are known to be ubiquitous features that have an impact on the hurricane (Willoughby et al. 1984). The precise manner in which asymmetries, either in the environment or in the near-core region, influence of the hurricane circulation is not yet well understood.

The influence of the hurricane, typhoon, and cyclone on seagrass can be mainly classified as: (1) the huge swell can damage directly seagrass, and wind-driven oceanic waves and currents can break seagrass leaves or uproot whole plants. In addition, they may destroy seagrass regenerative capacity, often resulting in the loss of plants, species and meadows. If the speed of the swell is 90 cm/s, it can break the leaves of Thalassia; when the speed of the swell is 150 cm/s, mud can be moved, and the rhizomes of Thalassia will be eroded (Scoffin 1970). (2) Typhoon, hurricane, and cyclone can re-suspend the mud which heighten the turbid of the water, lower the transmittance of light; when the suspended solids subside and are deposited, they will recover the seagrass leaves which affect the effectiveness of photosynthesis. (3) Typhoon, cyclone, and hurricane can increase rainfall which leads to lower the offshore salinity, and change the ecological environment. (4) Because people overly develop land, the surface water loss and soil erosion is serious. Typhoon, cyclone, and hurricane cause heavy rainfall. In addition, suspended solid and a lot of nutrimental materials are brought to offshore water which increase the turbidity of the water and phytoplankton. So, the transmittance of the water is further reduced (Listun 1990). (5) Typhoon, cyclone, and hurricane can bring new species which will invade the old ecological environment and change the structure of the species population. Some researches point out that hurricane is the main reason for the change of seagrass (Connell 1978).

Hurricanes and tropical storms may cause destruction of complete seagrass beds (Poiner et al. 1989; Preen et al. 1995; Rodriguez et al. 1994; USGS 2005) or cause “blowouts”, which are patches devoid of vegetation in continuous seagrass beds (Clarde and Kirkman 1989; Patriquin 1975; Rodriguez et al. 1994). But frequently, the impacts of these major meteorological events on the seagrass beds are less obvious, causing loss of above- or below-ground plant parts without destroying the beds completely (see Cruz-palacios and Tussenbroek 2005). Hurricanes or storms often cause erosion or deposition of sediments (Fourqurean and Rutten 2004; Rodriguez et al. 1994; Van Tussenbroek 1994) resulting in changes in sediment levels over seagrass beds. Cruz-Palacios and Van Tussenbroek (2005) simulated hurricane-like effects by artificially changing sediment levels in situ at the Puerto Morelos reef lagoon in the Mexican Caribbean. The experimental treatments caused selective removal of some species or groups of species, resulting in changes in the community composition. The area of interest was surveyed just before the passage of the category 4 hurricane Wilma (October 21–23, 2005). The duration of hurricane-force winds was 60 h (Pasch et al. 2006), and the average wave height in the back-reef lagoon was 3 m (Hydrological and meteorological service, unidad acade mica puerto morelos, unpublished data). The coastline of the Puerto Morelos reef lagoon was highly impacted and a 10–60 m wide nearshore fringe of the seagrass community was destroyed completely, but further into the lagoon and near the reefs the seagrass beds persisted (Tussenbroek unpublished data). These persistent beds covered 95 % of the reef lagoon, and although they were not destroyed, they may have been impacted by the hurricane in other ways. The experimental sites of Cruz-Palacios and Van Tussenbroek (2005) were located within such persistent beds, and we returned to these sites within 2 months of hurricane Wilma to verify whether the effects of the hurricane on the vegetation were in accordance with the results of the experiments by latter authors. Seagrasses are resilient to many hurricanes; however, very strong storms like Ivan can produce severe seagrass loss, as was observed in Santa Rosa Sound, FL after the passage of another major storm, Hurricane Opal (Heck et al. 1996). Thus, recent landfall of Hurricane Ivan in Baldwin County may have negatively impacted the abundance of seagrasses in Alabama coastal waters by uprooting or burying them. The Mobile Bay NEP funded a survey of submerged aquatic vegetation (SAV) in coastal Alabama in Summer and Fall 2002, and this work included photo interpreted aerial imagery of SAV that was ground truthed (Vittor and Associates Inc. 2003). To evaluate how the landfall of Hurricane Ivan may have affected the distribution of seagrasses in coastal Alabama, Heck carried out a rapid field assessment of the seagrass resources identified by Vittor and Associates Inc. (2003).

The hurricane impacts are affected by the speed and orientation of the hurricane and cyclone, especially in the damage to seagrass bed. Some hurricane and cyclone are too huge to form swell which destroy seriously seagrass. Poiner et al. (1989) reported that the hurricane Sandy with speed of 220 km/h generated a swell with height 12 m which caused 70 % seagrass death in Carpinteria Bay of Australia. Preen et al. (1995) published that the 1,000 km^2 seagrass bed in Hervey Bay of Australia was damaged by the swell, flood, and increasing turbidity due to the cyclone and hurricane. Fourqurean and Rutten (2004) studied the influence of Hurricane Georges on the seagrass bed in Florida. Based on his research, they found that 3 % of *T. testudinum* and 19 % of *S. filiforme* were damaged. Sometimes the speed of the hurricane and cyclone is fast, but the influence on the seagrass is comparatively less. Ball et al. (1967) studied the category 4 hurricane. Based on the research on the hurricane Donna which passed the seagrass bed in Florida in 1960, he found that the influence of hurricane on the seagrass bed and ooze offshore is smaller. The category 4 hurricane Andrew which destroyed the mangrove and biological ecosystem in Florida in 1992 had a little influence on seagrass bed (Thomas et al. 1961; Smith et al. 1994; Armentano et al. 1995). It demonstrates that the speed of ocean current is the main reason for the damage of seagrass beds.

The seagrass beds in the area where is comparatively far from the center of the hurricane and cyclone only got several local blowouts. Hurricane Hugo struck the Commonwealth of Puerto Rico on the September 18, 1989, with winds greater than 240 km/h (150 mph). Based on the research, Rodriguez et al. (1994) found that over 100,000 m^2 of sand form the Escollo de Arenas was spread out over adjacent seagrass beds. Hugo caused severe to moderate erosion of the beaches in the San Juan area. Hugo created new backshore scarps temporarily increasing the vulnerability of the coast to future high-energy (greater than 3 m swell) storms. Rodriguez and Webb (1990) discussed the immediate impact of the storm on the coastal resources. Aerial observations right after the storm suggested that a major portion of the Escollo de Arenas had been dispersed over the surrounding seagrass beds and that beach erosion occurred along several sections of the northeast coast. Bush (1991) noted how the steep, rocky nature of the coastal zone, combined with low maximum storm surges, helped keep property damage at a minimum. However, crowding of building into coastal lowlands and poor construction quality worked to increase damage in places. In addition, Cambridge (1975) stated that one of the most obvious features of these seagrass meadows is the abundant blowouts. These scars, in the integral Posidonia meadows, were probably formed by catastrophic storms. In the blowouts, other seagrass species grow and changes in cover appear to be occurring. It was obvious from the results of other studies that colonisers, such as Heterozostera tasmanica (Martens ex Aschers) den Hartog and Halophila ovalis (R. Br.) Hook were growing rapidly during summer but that the larger, dominant species such as *P. sinuosa* Cambridge and Kuo and the Amphibolis species were not extending quickly into sandy areas (Kirkman 1985, 1990).

Hurricane and cyclone can also cause indirect damage to seagrass beds. Salinity will be lower due to rainfall into the river which was caused by hurricane and cyclone. Following major flooding in the Fitzroy River catchment (Brodie and Mitchell 1992; Preker 1992), low salinity plume water was observed offshore for a period of 3 weeks (O'Neill et al. 1992). Low salinity water (down to 8) caused significant coral and seagrass mortality (Van Woesik 1991) to the fringing coral reefs around the Keppel Islands.

All these works can help improve the research works on the change in the seagrass distribution. However, if the destruction of typhoon on seagrass in South China has a same or similar model is needed to further testify and study.

Besides detection of the change in seagrass distribution due to destruction, how seagrass recovery is the most import work. After destruction, seagrass beds recovery process is completely different, and it varies with the damaged area and extent of damage. Usually, the seagrass beds can recover within several weeks or months if the blowout area is small. But, when the destruction area is larger, the seagrass beds need 2–4 years or more than 5 years to recover. Of course, it needs the other environment index keep same. If the turbidity increases, some areas need more than decade years to recover. In addition, once the water quality is worse, the seagrass in these areas hardly recover (Erftemeijer and Lewis 2006). The loss and recovery of intertidal seagrass meadows were assessed by Campbell and McKenzie (2004) following the flood related catastrophic loss of seagrass meadows in February 1999 in the Sandy Strait, Queensland. Meadow-scale assessments of seagrass loss and recovery focused on two existing Zostera capricorni monitoring meadows in the region. Mapping surveys showed that approximately 90 % of intertidal seagrasses in the northern Great Sandy Strait disappeared after the February 1999 flooding of the Mary River. Based on their observation, they found that full recovery of all seagrass meadows took 3 years. At the two study sites (Urangan and Creek) the onset of Z. capricorni germination following the loss of seagrass occurred 14 months post-flood at Wanggoolba Creek, and at Urangan it took 20 months for germination to occur. By February 2001 (24 months post-flood) seagrass abundance at Wanggoolba Creek sites was comparable to pre-flood abundance levels and full recovery at Urangan sites was complete in August 2001 (31 months post-flood). Reduced water quality characterised by 2–3 fold increases in turbidity and nutrient concentrations during the 6 months following the flood was followed by a 95 % loss of seagrass meadows in the region. They thought that reductions in available light due to increased flood associated turbidity in February 1999 were the likely cause of seagrass loss in the Great Sandy Strait region, southern Queensland.

The recovery speed of seagrass is different when the sediments covered the seagrass leaves are different. When Fourqurean studied the influence of the hurricane Georges, he found that these seagrass without sediment cover can recover soon; and, for those area covered with less sediment, seagrass also recover comparatively soon. However, when the thickness of the sediments is more than 50 cm, the area only grows several seagrass after 3 years. For the area where the soft mud was flashed away by the huge swell, the recover speed is most slow.

The recovery and tolerance ability of the different species of seagrass is also different. This is directly correlated with morphologic and physiological condition. For the seagrass which the size is small and grow fast, such as Halophila ovalis or Halodule wrightii, has less tolerance to typhoon, but can recover comparatively fast (Clarke 1987). Those seagrass which the size is large, such as *Thalassia* or *Posidonia* sp., has more tolerance (Chesire et al. 2002). But, they need more time to recover, once they are destroyed.

The remote sensing techniques and aerial photography are the main study methods to detect the seagrass distribution. Bastish (2002) used Lansat TM data to study the impact of land use on the benthic habitats and seagrass. He noted that the hurricane and heavy rainfall is main reason for the soil runoff causing the transparency lower and changing the seagrass distribution. Hashim et al. (2001) analyzed the spectral characteristics of seagrass with Land TM in northern Sabah coastline. Ralph and Short (2002) researched the impact of the wasting disease pathogen, Labyrinthyla zosterae on the photobiology of eelgrass, Zostera marina. Ward et al. (1997) observed the distribution and studied the stability of eelgrass beds at Izenmbek lagoon. The results reflected that the seagrass distribution in the lagoon is comparatively stable. Bouvet et al. (2003) evaluated the large-scale unsupervised classification of Caledonia reef ecosystems using landsat 7 ETM+ imagery. They tested the capacity of the Landsat 7 Enhanced Thematic Mapper Plus sensor to classify the shallow benthic ecosystems of New Caledonia (South Pacific) using a novel unsupervised classification method. The classes were defined by using a set of multiple spectral decision rules based on the image spectral bands. Firstly, a general model was applied to the entire Southwest lagoon (5,500 km^2) and tested on three representative sites: a section of the barrier reef, a cay reef flat rich in corals, and a cay reef flat rich in algae and seagrass beds. The classification results were compared with a locally optimized model, with aerial color photographs and extensive ground-truthed observations. Results showed that a reconnaissance of the main benthic habitats in shallow areas (<5 m depth) was possible, at a geomorphological scale for coral reef structure and at a habitat scale for seagrass beds. However, results directly issued from the model must be cautiously interpreted according to empirical spatial rules, especially to avoid confusion between coral slopes and shallow dense seagrass.

With the development of remote sending applications, satellite data has become the uppermost data source to monitor large-scale seagrass condition and detect seagrass distribution, such as SPOT, QuickBirds, and IKONOS. Hyperspectral remote sensing images are now widely available. They are characterized by hundreds of spectral bands. For the detection and classification task, the increased dimensionality of the data increases the capability to detect various classes with a better accuracy. Pasqualini (2005) use SPOT data which provides multispectral imagery with a spatial resolution of 10 m and fused imagery with a spatial resolution of 2.5 m for mapping seagrass. These types of satellite imagery were used for mapping beds of *Posidonia oceanica* in the Mediterranean Sea, where it is a dominant species forming monospecific beds in a structurally simple environment (four classes: sand, photophilous algae on rock, patchy seagrass beds and continuous seagrass beds). Supervised classifications by depth range were made of both types of image. Through the direct comparison of overall accuracy between SPOT 2.5 m and SPOT 10 m, they concluded that this tool provided accurate mapping in both cases (between 73 and 96 % accuracy). Although SPOT 2.5 m provides lower overall accuracy than SPOT 10 m, it is a very useful tool for the mapping of *P. oceanica*, as it allows the patchiness of the formations to be better taken into account. The opportunity to use a reliability scale, which takes into account the effects of extrinsic factors on the processing of the images, confirmed the usefulness of the option of using a reduced pixel size in order to obtain an improved match between the results from mapping and field observations. Andréfouët et al. (2001) also studied seagrass in Tuamotu archipelago (French Polynesia) by using SPOT-HRV and IKONOS data. But at the same time, classical classification techniques are facing the problem of statistical estimation in high dimensional spaces. Due to the high number of features and small number of training samples, reliable estimation of statistical parameters is difficult (Lee and Landgrebe 1993).

12.3 The Recent Research Advances in China

Recently, the technique of remote sensing is applied widely in China. And, the study on bio-optics of ocean ecosystems also got development, for example, the methods of water spectra measurement and analysis (Tang et al. 1998, 2004; Lee 2002), the inversing model of water transparency using the SeaWiFS data (Pan et al. 2004), and the application of satellite remote sensing on seagrass in China (Yang and Yang 2009; Yang et al. 2010, 2011). The remotely sensed information of island intertidal zone and wetland was extracted, and Dongsha Island was taken as an example (Lee et al. 2005). All these are foundation for water column correction and bio-optical model of seagrass which is the necessary for seagrass mapping. However, there are still many problems on the seagrass detection before and after typhoon landfall. How can we distinguish the difference of the influence between human activity and typhoon? In addition, offshore water quality often varies widely, chlorophyll-a, CDOM, and suspend solid can affect the accuracy of the bio-optical model. Besides, water column influence on extracting the seagrass substrates information is need to research further.

Typhoon often attack the southeast of China. There are averagely 27 typhoon generated in Pacific Northwest Ocean, while the landfall typhoon are averagely 3.2, and the biggest number is 9 (Xu and Gao 2005). Based on the data of recent 50 years, changing characteristics of typhoons in the northwest Pacific Ocean and landing at China are analyzed. At the same time, a quantitative method of assessment to annual disaster caused by typhoon is developed over rural region on the basis of social economic data and losses. The storm tide and swell due to typhoon wash away loose sediments and seagrasses that are loosely rooted in sediments and the benthic organism living at the sediment surface. In addition seagrass may be buried which affect the seagrass growth. However, the records on the influence of typhoon on seagrass are comparatively less. Only one paper recorded the destruction of seagrass in GuangXi due to the typhoon in September, 2002 (Huang et al. 2006). But, the detail research on the influence of typhoon on seagrass can not be found.

Acknowledgments The National Basic Research Program of China (973 Program) under grant No. 2010CB951203 and 2013CB956503; the National Natural Sciences Foundation of China under grant No. 41176161; State Key Laboratory of Lake Science and Environment (2012SKL004); Foundation of Guangdong Province and Chinese Academy of Sciences cooperation under grant No. 2011B090300059; Foundation of Science and technology of Guangdong Province under grant No. 2012A031100008.

References

Andréfouët, S., Claereboudt, M., Matsakis, P.: Typology of atolls rims in Tuamotu Archipelago (French Polynesia) at landscape scale using SPOT-HRV images. Int. J. Remote Sens. **22**(6), 987–1004 (2001)

Armentano, T.V., Doren, R.F., Platt, W.J., Mullins, T.: Effect of Hurricane Andrew on coastal and interior forests of Southern Florida: overview and synthesis. J. Coast. Res. **21**, 111–144 (1995)

Arrivillaga, A., Baltz, D.: Comparison of fishes and macroinvertebrates on seagrass and bare-sand sites on Guatemala's Atlantic coast. Bull. Mar. Sci. **65**, 301–319 (1999)

Ball, M.M., Shinn, E.A., Stockman, K.W.: The geologic effects of hurricane Donna in South Florida. J. Geol. **75**, 583–597 (1967)

Batish, S.: Demonstration of the impact of land use on the benthic habitats of the U.S. Virgin islands with LandSat satellite imagery. Littoral, the Changing Coast, pp. 203–204 (2002)

Brodie, J.E., Mitchell, A.W.: Nutrient composition of the January 1991 Fitzroy River flood plume. In: Byron, G.T. (ed.) Workshop on the Impacts of Flooding. Workshop Series No. 17. Great Barrier Reef Marine Park Authority, pp. 56–74, Townsville (1992)

Bouchon, C., Bouchon-Navaro, Y., Imbert, D., Louis, M.: The effect of hurricane Hugo on the coastal environment of Guadeloupe Island (FWI). In: Annales de l'Institut Oceanographique, vol. 67, pp. 5–33. Paris, Nouveau Series (1991)

Bouvet, G., Ferraris, J., Andrefouet, S.: Evaluation of large-scale unsupervised classification of new Caledonia reef ecosystems using Landsat 7 ETM+ imagery. Oceanol. Acta **26**, 281–290 (2003)

Bush, D.M.: Impact of Hurricane Hugo on the rocky coast of Puerto Rico. J. Coastal Res. **8**, 49–69 (1991)

Cabello-Pasini, A., Lara-Turrent, C., Zimmerman, R.C.: Effect of storms on photosynthesis, carbohydrate content and survival of eelgrass populations from a coastal lagoon and the adjacent open ocean. Aquat. Bot. **74**, 149–164 (2002)

Cambridge, M.L.: Seagrasses in south-western Australia with special reference to the ecology of *Posidonia australis* Hook *f.* in a polluted environment. Aquat. Bot. **1**, 149–162 (1975)

Campbell, S.J., McKenzie, L.J.: Flood related loss and recovery of intertidal seagrass meadows in southern Queensland, Australia. Estuar. Coast. Shelf Sci. **60**, 477–490 (2004)

Chesire, A.C., Miller, D.J., Murray-Jones, S., Scriven, L., Sandercock, R.: The Section Bank: ecological communities and strategies for the minimization of dredging impacts. A report to the Office for Coast and Marine National Parks and Wildlife, South Australia, Department for Environment and Heritage. SARDI Aquatic Sciences, West Beach. ISBN:0730852857 (2002)

Clarke, S.M.: Sediment-seagrass dynamics in Holdfast Bay: summary. Safish **11**, 4–10 (1987)

Clarke, S.M., Kirkman, H.: Seagrass dynamics, in Biology of Seagrasses. In: Larkum, A.W.D.,Mccomb A.J., Shepherd, S.A. (eds.). A Treatise on the Biology of Seagrasses with Special Reference to the Australian Region, pp. 304–345. Elsevier, The Netherlands (1989)

Connell, J.H.: Diversity in tropical rainforests and coral reefs. Science **199**, 1302–1310 (1978)

Connolly, R.M.: Removal of seagrass canopy: effects on small fish and their prey. J. Exp. Mar. Biol. Ecol. **184**, 99–110 (1994)

Cruz-Palacios, V., Van Tussenbroek, B.I.: Simulation of hurricane-like disturbances on a Caribbean seagrass beds. J. Exp. Mar. Biol. Ecol. **324**(1), 44–60 (2005)

Erftemeijer, P.L.A., Robin Lewis, R.R.: Environmental impacts of dredging on Seagrass: a review. Mar. Pollut. Bull. **52**(12), 1553–1572 (2006)

Fourqurean, J.W., Rutten, L.M.: The impact of hurricane George on soft bottom, back reef communities: site- and species-specific effects in South Florida seagrass beds. Bull. Mar. Sci. **75**(2), 239–257. In: Lee M.J. (ed.) Back Reef Habitats: Ecological Analysis and Characterization (Special Issue). Stocking Island, Bahamas, University of Miami: Rosenstiel School of Marine and Atmospheric Science. ISSN 0007-4977 (2004)

Goodman, J.A., Ustin, S.L.: Hyperspectral remote sensing of coral reefs: deriving bathymetry, aquatic optical properties and a benthic spectral unmixing classification using AVIRIS data in the Hawaiian Islands, pp. 74–135. Hydrologic Sciences Department, Ph.D. Dissertation, University of California, Davis (2004)

Hashim, M, Rahman R. A., Muhammad M.: Spectral characteristics of seagrass with Landsat TM in northern Sabah coastline, Malasia. Paper presented at the 22nd Asian conference on remote sensing, Singapore (2001)

Heck Jr, K.L., Sullivan, M.J., Zande, J.M., Moncreiff, C.A.: An ecological analysis of Seagrass meadows of the Gulf Islands National Seashore. Final Report to the National Park Service, Gulf Islands National Seashore (1996)

Heck, K.L., Nadeau, D.A., Thomas, R.: The nursery role of seagrass beds. Gulf Mex. Sci. **1**, 50–54 (1997)

http://www.nhc.noaa.gov/pdf/TCR-AL252005_Wilma.pdf

http://www.nwrc.usgs.gov/hurricane/katrina-post-hurricane-flights.htm

Huang, X., Huang, L., Lee, Y.: Seagrass in south China under stress and reflects changes in biomass 57(S1), 114–119 (2006)

Kirkman, H.: Community structure in seagrasses in south Western Australia. Aquat. Bot. **21**, 363–375 (1985)

Kirkman, H.: Seagrass distribution and mapping. In: Phillips, R.C. (ed.) Seagrass Research Methods. McRoy, C.P. (ed.) Monographs on Oceanographic Methodology 9, pp. 19–26. UNESCO, Paris (1990)

Kirkman, H., Kuo, J.: Pattern and process in southern Western Australian seagrasses. Aquat. Bot. 37, 367–382 (1990)

Lee, C., Landgrebe, D.A.: Analysing high dimensional multispectral data. IEEE Trans. Geosci. Remote Sens. **31**, 792–800 (1993)

Lee, Z.-P., Carder, K.L.: Effect of spectral band numbers on the retrieval of water column and bottom properties from ocean color data. Appl. Opt. **41**(12), 2191–2201 (2002)

Lee, Z.-P., Darecki, M., Carder, K.L., Davis, C.O., Stramski, D., Rhea, W.J.: Diffuse attenuation coefficient of downwelling irradiance: An evaluation of remote sensing methods. J. Geophys. Res. 110(C02017), doi:10.1029/2004JC002573 (2005)

Listun, P.W.: Spatial variability and covariability of chlorophyll and zooplankton on the Great Barrier Reef, p. 204. Ph.D. Thesis, James Cook University, Townsville (1990)

Michot, T.C., Hoese, H.D., Valentine, J.M.: Impacts of Hurricanes Camille and Georges, and other tropical cyclones, on biological resources of Chandeleur Islands, Louisiana, and Chandleur Sound. In: Symposium Proceedings of the Impact of Hurricane Camille: A Storm Impact Symposium to Mark the 30th Anniversary. University of New Orleans, Louisiana, 17–18 Aug 1999, Program and Abstracts (unpaginated) (1999)

Michot, T.C.: Carrying capacity of seagrass beds predicted for redheads wintering in Chandeleur Sound, Louisiana, USA. In: Goss-Custard, J., Rufino, R., Luis, A. (eds.) Effect of habitat loss and change on waterbirds, vol. 42, pp. 93–102. Institute of Terrestrial Ecology Symposium No. 30, Wetlands International Publication. Louisiana, USA (1997)

Michot, T.C.: Comparison of wintering redhead populations in four Gulf of Mexico seagrass beds. In: Comin, F.A., Herrera, J.A., Ramirez, J. (eds.) Limnology and Aquatic Birds: Monitoring, Modelling, and Management, pp. 243–260. Universidad Autonoma de Yucatan, Merida, Mexico (2000)

Mishra, D.R., Narumalani, S., Rundquist, D., Lawson, M.: High-resolution ocean color remote sensing of benthic habitats: a case study at the Roatan Island, Honduras. IEEE Trans. Geosci. Remote Sens. **43**(7), 1592–1604 (2005)

O'Neill, J.P., Byron, G.T., Wright, S.C.: Some physical characteristics and movement of 1991 Fitzroy River flood plume. In: Great Barrier Reef Marine Park Authority Workshop Series: G.T., GBRMPA, Townsville, vol. 17, pp. 36–51 (1992)

Ooyama, K.V.: Conceptual evolution on the theory and modeling of the tropical cyclone. J. Meteor. Soc. Japan **60**, 369–379 (1982)

Pasch, R.J., Blake, E.S., Cobb, H.D., Roberts, D.P.: Available from National Hurricane Center (2006)

Pasqualini, V., Pergent-Martini, C., Pergen, G.: Use of SPOT 5 for mapping seagrasses: an application to Posidonia oceanica. Remote Sens. Environ. **94**, 39–45 (2005)

Patriquin, D.G.: Migration' of blowouts in seagrass beds at Barbados and Carricou, West Indies, and its ecological and geological implications. Aquat. Bot. **1**(1), 163–189 (1975)

Poiner, I.R., Walker, D.I., Coles, R.G.: Regional-studies seagrasses of tropical Australia. In: Larkum, A.W.D., Mccomb, A.J., Shepherd, S.A. (eds.) Biology of Seagrasses, a Treatise on the Biology of Seagrasses with Special Reference to the Australian Region, pp. 279–303. Elsevier, The Netherlands (1989)

Posey, M., Alphin, T.: Resilience and stability in an offshore benthic community: responses to sediment borrow activities and hurricane disturbance. J. Coastal Res. **18**, 685–697 (2002)

Preker, M.: The effects of the 1991 central Queensland floodwaters around Heron Island, Great Barrier Reef. In: Byron, G.T. (ed.) Workshop on the Impacts of Flooding. Workshop Series No.17. Great Barrier Reef Marine Park Authority, p. 75. Townsville (1992)

Preen, A.R., Long, W.J., Coles, R.G.: Flood and cyclone related loss, and partial recovery, of more than 1,000 km^2 of seagrasses in Harvey Bay. Queensland, Australia, Aquatic Botany **52**(1), 3–17 (1995)

Purkis, S.J., Pasterkamp, R.: Integration in situ reef-top reflectance spectra with Landsat TM imagery to aid shallow-tropical benthic habitat mapping. Coral Reefs **23**, 5–20 (2004)

Ralph, P.J., Short, F.: Impact of the wasting disease pathogen, Labyrinthula zosterae, on the photobiology of eelgrass, Zostera marina, vol. 226, pp. 265–271. Marine Ecology Progress Series (2002)

Rodriguez, R.W., Webb, R.M.T.: Impact of hurricane Hugo on coastal resources of Puerto Rico(abstract). American association of petroleum geologists annual convention, program with abstracts California, 157, San Francisco (1990)

Rodriguez, R.W., Webb, R.M.T., Bush, D.M.: Another look at the impact of hurricane Hugo on the shelf and coastal resources of Puerto Rico, USA. J. Coast. Res. **10**(2), 278–296 (1994)

Salazar-Vallejo, S.I.: Huracanes y biodiversidad costera tropical. Revista de Biología Tropical **50**, 415–428 (2002)

Scoffin, T.P.: The trapping and binding of subtidal carbonate sediments by marine vegetation in Bimini Lagoon, Bahamas. J. Sediment. Petrol. **40**, 249–273 (1970)

Smith III, T.J., Robblee, M.B., Wanless, H.R., Doyle, T.W.: Mangroves, hurricanes and lightning strikes. BioScience **44**, 256–262 (1994)

Tang, J.W., Chen, O.L., Tan, S.X., Dong, Q., Miao, W., Feng, L.: Methods of oceanic spectral data measurement and analysis. Mar. Sci. Bull. **17**(1), 71–79 (1998)

Tang, J.W., Tian, G.L., Wang, X.Y., Wang, X.M., Song, Q.J.: The methods of water spectra measurement and analysis I: above water method. J. Remote Sens. **8**(1), 37–44 (2004)

Thomas, L.P., Moore, D.R.: Effects of Hurricane Donna on the turtle grass beds of Biscayne Bay, Florida. Bull. Mar. Sci. Gulf and Carib. **11**, 191–197 (1961)

USGS: Post Hurricane Katrina Flights Over Louisiana's Barrier Islands. Available (20 Sept 2005)

Van Tussenbroek, B.I.: The impact of hurricane Gilbert on the vegetative development of Thalassia testudinum in Puerto Morelos coral reef lagoon, Mexico: a retrospective study. Bot. Mar. **37**(5), 421–428 (1994)

Van Woesik, R.: Immediate Impact of the January 1991 floods on the coral assemblages of the Keppel Islands, p. 30. Great Barrier Reef Marine Park Authority Research Publication (GBRMPA), Townsville (1991)

Vicente, V.P.: A summary of ecological information on the seagrass beds of Puerto Rico. In: Seeliger, U. (ed.) Coastal plant communities of Latin America: San Diego, pp. 123–133. Academic Press, Inc, California (1992)

Vittor, Associates, Inc.: Mobile Bay Submerged Aquatic Vegetation. Final Report to Mobile Bay National Estuary Program (2003)

Ward, D.H., Markon, C.J., Douglas, D.C.: Distribution and stability of eelgrass beds at Izembek lagoon, Alaska. Aquat. Bot. **58**, 229–240 (1997)

Willoughby, H.E., Marks Jr, F.D., Feinberg, R.J.: Stationary and moving convective bands in hurricanes. J. Atmos. Sci. **41**, 3189–3211 (1984)

Xianqiang, H., Delu, P., Zhihua, M., Zhu, Q.: The study on the inversing model of water transparency using the SeaWiFS data. Acta Oceanol. Sinica **26**(5), 55–62 (2004)

Xu, L.Y., Gao, G.: Features of typhoon in recent 50 years and annual disaster assessment. Meteorol. Mon. **31**(3), 41–44 (2005)

Yang, C.Y., Yang, D.T., Cao, W.X., Zhao, J., Wang, G.F., Sun, Z.H., Xu, Z.T., Kumar, M.S.R.: Analysis of seagrass reflectivity by using a water column correction algorithm. Int. J. Remote Sens. **31**(17–18), 4595–4608. ISSN:0143-1161 (2010)

Yang, D.T., Huang, D.J.: Impacts of Typhoons Tianying and Dawei on seagrass distribution in Xincun Bay, Hainan Province, China. *Acta Oceanologica Sinica.* ISSN:0253-505X. **30**(1), 32–39 (2011)

Yang, D.T., Yang Y., Yang C., Zhao, J., Sun, Z.: Detection of seagrass in opticalshallow water with Quickbird in Xincun Bay, Hainan province, China. *IET Image Processing.* ISSN:1751-9667 **5**(5), 363–368 (2011)

Yang, D., Yang, C.: Detection of Seagrass Distribution Changes from 1991 to 2006 in Xincun Bay, Hainan, with Satellite Remote Sensing. Sensors **9**, 830–844 (2009)

Zhang, Y., Pulliainen, J.T., Koponene, S.S., Hallikainen, M.T.: Water Quality Retrievals From Combined Landsat TM Data and ERS-2 SAR Data in the Gulf of Finland. IEEE Trans. Geosci. Remote Sens. **42**(3), 622–629 (2003)

Chapter 13
Impacts of Typhoons on Nutrient Supply and Potential Fish Production in the Southern East China Sea

Chin-Chang Hung, Gwo-Ching Gong, Ming-An Lee, Cheng-Hsin Liao, Yi Chang, Yung-Yen Shih, Kuo-Shu Chen, Meng-Hsien Chen and Peter H. Santschi

Abstract Each year, typhoons (also called tropical cyclones or hurricanes) cause billions in property damage and great human toll. Besides, many typhoons occurring in tropical regions of the open ocean and their collective effects upon the marine environment and fishery remain undefined because of sampling difficulties. Recent satellite observations have shown that phytoplankton biomass can be enhanced for several days after typhoons, while in situ hydrographic observations, and data on biogeochemical properties and potential fish production induced by typhoons are limited. In this chapter, field observations are used to evaluate

C.-C. Hung (✉) · Y.-Y. Shih · K.-S. Chen
Institute of Marine Geology and Chemistry, and Asia–Pacific Ocean Research Center, National Sun Yat-sen University, Kaohsiung 80424, Taiwan
e-mail: cchung@mail.nsysu.edu.tw

C.-C. Hung · G.-C. Gong
Taiwan Ocean Research Institute, National Applied Research Laboratories, Kaohsiung 85243, Taiwan

G.-C. Gong
Institute of Marine Environmental Chemistry and Ecology, and Center of Excellence for Marine Bioenvironment and Biotechnology, National Taiwan Ocean University, Keelung 20424, Taiwan

M.-A. Lee · C.-H. Liao
Department of Environmental Biology and Fisheries Science, National Taiwan Ocean University, Keelung, Taiwan

Y. Chang
Institute of Ocean Technology and Marine Affairs, National Cheng Kung University, Tainan, Taiwan

M.-H. Chen
Department of Marine Biotechnology and Resources, National Sun Yat-sen University, Kaohsiung, Taiwan

P. H. Santschi
Laboratory for Oceanography and Environmental Research, Texas A and M University at Galveston, 200 Seawolf Parkway, Galveston, TX 77553, USA

D. L. Tang and G. J. Sui (eds.), *Typhoon Impact and Crisis Management*, Advances in Natural and Technological Hazards Research 40,
DOI: 10.1007/978-3-642-40695-9_13,

possible impacts of different typhoons on nutrient supply and potential fish production in the Southern East China Sea (SECS). The results provide evidence that typhoons add significantly to the nutrient supply and biological productivity in the SECS. The specific highlights are: (1) the typhoon-induced nitrate supplies after typhoons Fungwong (8.2×10^9 g N d^{-1}) and Morakot (1.3×10^{10} g N d^{-1}) are roughly 10-fold higher than that (1×10^9 g N d^{-1}) during non-typhoon periods, and (2) the field observations demonstrate that typhoons can have a profound influence on nutrient supply and potential fish food sources in marginal seas.

13.1 Introduction

The East China Sea (ECS) is one of the larger and more productive marginal seas of the world (Fig. 13.1). It extends from the Cheju Island (~33°N) in the north to the northern coast (~25°N) of Taiwan in the south. It is bounded to the east by the Kuroshio and to the west by continental China. Extensive exchanges occur between the ECS and the Kuroshio. The warm, saline, and nutrient poor Kuroshio surface water finds its way to the shelf through frontal processes. The cold and nutrient-rich subsurface Kuroshio water can reach the shelf through topographically induced upwelling in the Southern ECS (SECS) near the northern tip of Taiwan (Chern and Wang 1990; Liu et al. 1992; Gong et al. 1995; Wu et al. 2008; Jan et al. 2011), resulting in forming a major ECS fishing ground.

Mackerel and swordtip squid are two of the most important fishery resources in the SECS, with a production season from approximately spring to early winter (Sassa et al. 2008; Wang et al. 2008). Elevated phytoplankton biomass on the SECS shelf break has been documented (Gong et al. 1995, 2000; Chen et al. 2001) and most likely provides an important food source for both adult and larval mackerels (Sassa et al. 2008). However, some reports suggest that phytoplankton production in the ECS is limited in the summer because of nutrient deficiency (Chen et al. 2001; Liu et al. 2010). Moreover, satellite-derived average sea surface temperature (SST) in the SECS is about 28 °C from July to September (Wang et al. 2008), suggesting that SECS surface water is likely nitrate-limited, based on the relationship between nitrate and temperature (Gong et al. 1995). Furthermore, Hung and Gong (2011) suggested that nutrient supply may not support a high biological productivity in the SECS in summer.

Many extreme weather events (including dust storms, floods and typhoons, which are also termed hurricanes or tropical cyclones) can increase nutrient supply (Shiah et al. 2000; Hung et al. 2005, 2009; Zheng and Tang 2007; Zhao et al. 2008; Gong et al. 2011), affect marine ecosystems functioning (Karl 1999; Mallin et al. 1999), and enhance marine biological activity and geochemical cycling (Chang et al. 1996; Chen et al. 2003; McKinnon et al. 2003; Chen et al. 2009; Siswanto et al. 2009; Hung et al. 2010, 2012; Chung et al. 2012) and possible fishery production (Qiu et al. 2010; Gong et al. 2011; Hung and Gong 2011).

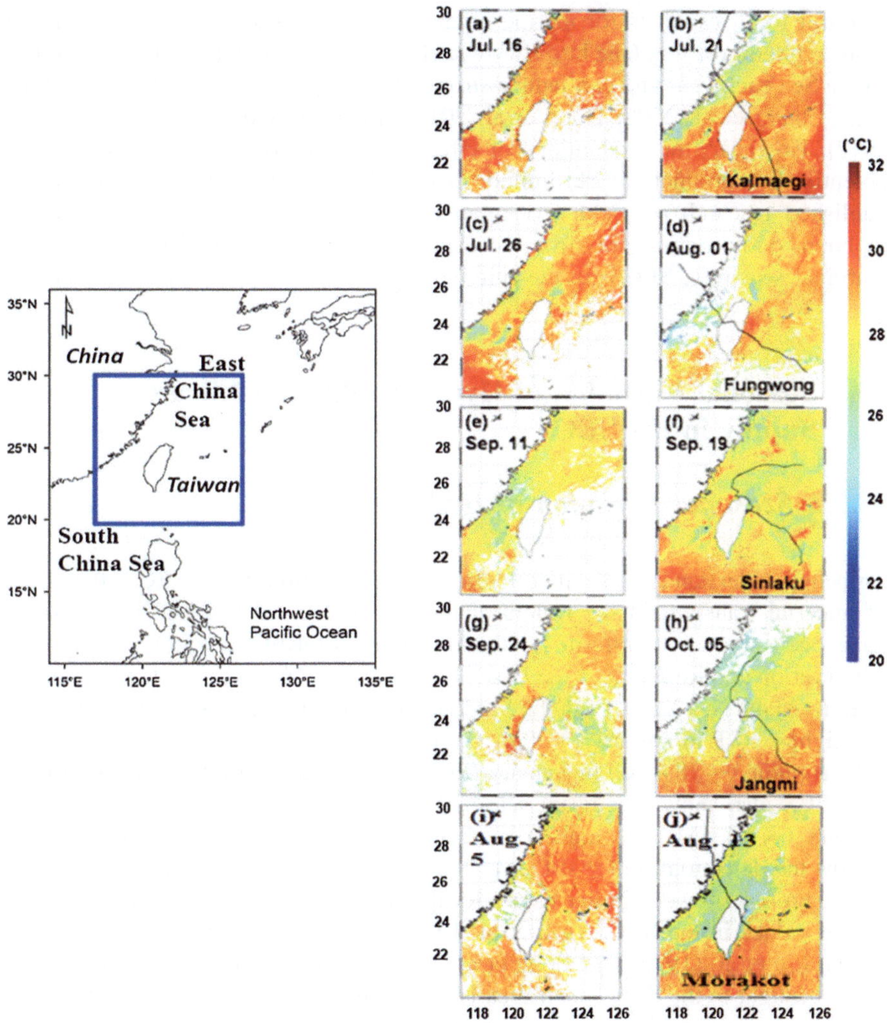

Fig. 13.1 Sea surface temperature (*SST*) before (*left* panels) and after (*right* panels) typhoons (Kalmaegi (**a**, **b**), Fungwong (**c**, **d**), Sinlaku (**e**, **f**), Jangmi (**g**, **h**) and Morakot (**i**, **j**) in the southern East China Sea). *Solid lines* on panels **b**, **d**, **f**, **h** and **j** represent typhoon tracks in the northwestern Pacific Ocean in 2008

Numerous satellite-image investigations have documented the effects of typhoons on phytoplankton biomass (e.g. chlorophyll, Chl-*a*), primary productivity (PP) and phytoplankton ecology in marginal seas, but most of these studies have not been conducted via sea-going observations. Recently, Qiu et al. (2010) suggested that typhoons are a strong forcing factor in the mobilization of nutrients, through increased water circulation, wind mixing and upwelling. In other words, typhoons may increase phytoplankton biomass and PP, and possibly provide plentiful food resources for marine organisms (Hung et al. 2010). Based on previous typhoon records (see http://www.cwb.gov.tw), many typhoons pass the East China Sea each year and can induce strong upwelling in the SECS (Figs. 13.1 and 13.2, Chang et al. 2008; Hung et al. 2010; Hung and Gong 2011; Chung et al. 2012). However, the evaluations of effects of typhoons on nutrient supply and potential fish production are limited. In this chapter, we review past and recent published research on field investigations to better understand the impacts of typhoons on nutrient supply from oceanic sources and possible fish production on marginal seas (e.g. SECS) in the summer following the passage of typhoons.

13.2 Seasonal Variation of Hydrography and Nitrate Supply in the SECS

13.2.1 Seasonal Variation of Hydrography

To better understand the detailed hydrographic variation in the SECS, the effects of material transport in the waters of the SECS (located at 25.40°N, 122.45°E) have been studied long-term by the integrated LORECS program (Long-term Observation and Research of the East China Sea). SST (sea surface temperature) at station (25.40°N, 122.45°E) in the SECS shows remarkable seasonal variation with the highest SST in August ($\sim$29.0 °C), followed by autumn and spring, and the coldest value ($\sim$18.4 °C) in January (Fig. 13.3). Surface nitrate concentrations in the SECS also show significant seasonal variations, with the maximum value in January ($\sim$4.1 μM), the highest median value in spring, and the lowest value (below the detection limit, $\leq$0.1 μM) in the summer (Fig. 13.3). Previous monthly investigations showed similar values for the seasonal variation of SST and surface nitrate values in the SECS (Liu et al. 1992; Gong et al. 1995, 2003; Chen et al. 2001). Liu et al. (1992) and Gong et al. (1995) reported that the cold, nitrate-rich water brought to the SECS shelf area is mainly from bottom intrusion of Kuroshio water during autumn, winter, and spring. Despite the persistent southwest monsoon during summer over the Taiwan Strait (June to September), nitrate-rich upwelled water at the surface has seldom appeared in the SECS (Liu et al. 1992; Gong et al. 1995). In other words, surface nitrate concentration was below detection levels in the SECS during the summer months when no typhoons occurred (Fig. 13.3).

Generally, the integrated nitrate inventory (I-NO_3, integrated from the surface to 75 m) in the SECS water column in March 2007, December 2008, and February 2009 were 0.236, 0.323, and 0.444 mol N m^{-2}, respectively. Conversely, the I-NO_3 (0.05–0.293 mol N m^{-2}, Fig. 13.3) in the SECS in the summer is lower than that in other seasons. These observations suggest that nitrate supply in the whole water column in the SECS in the summer is indeed lower than that in other seasons. According to Gong et al. (1995), the presence of the upwelled Kuroshio

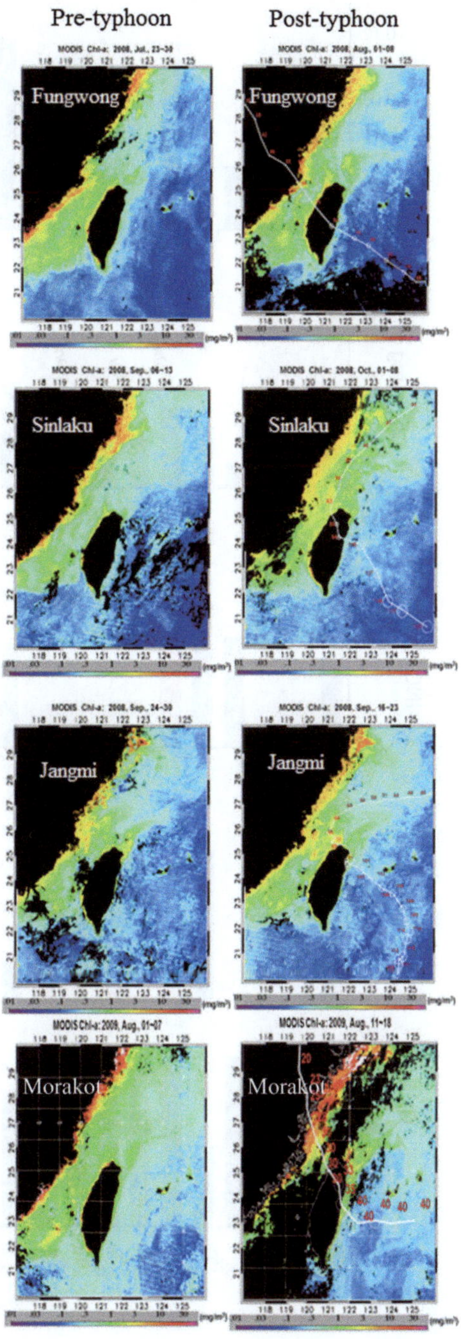

Fig. 13.2 The derived surface Chl-*a* concentrations by MODIS *before* and *after* typhoons Fungwong, Sinlaku, Jangmi and Morakot, respectively. A *white line* represents the track of each typhoon

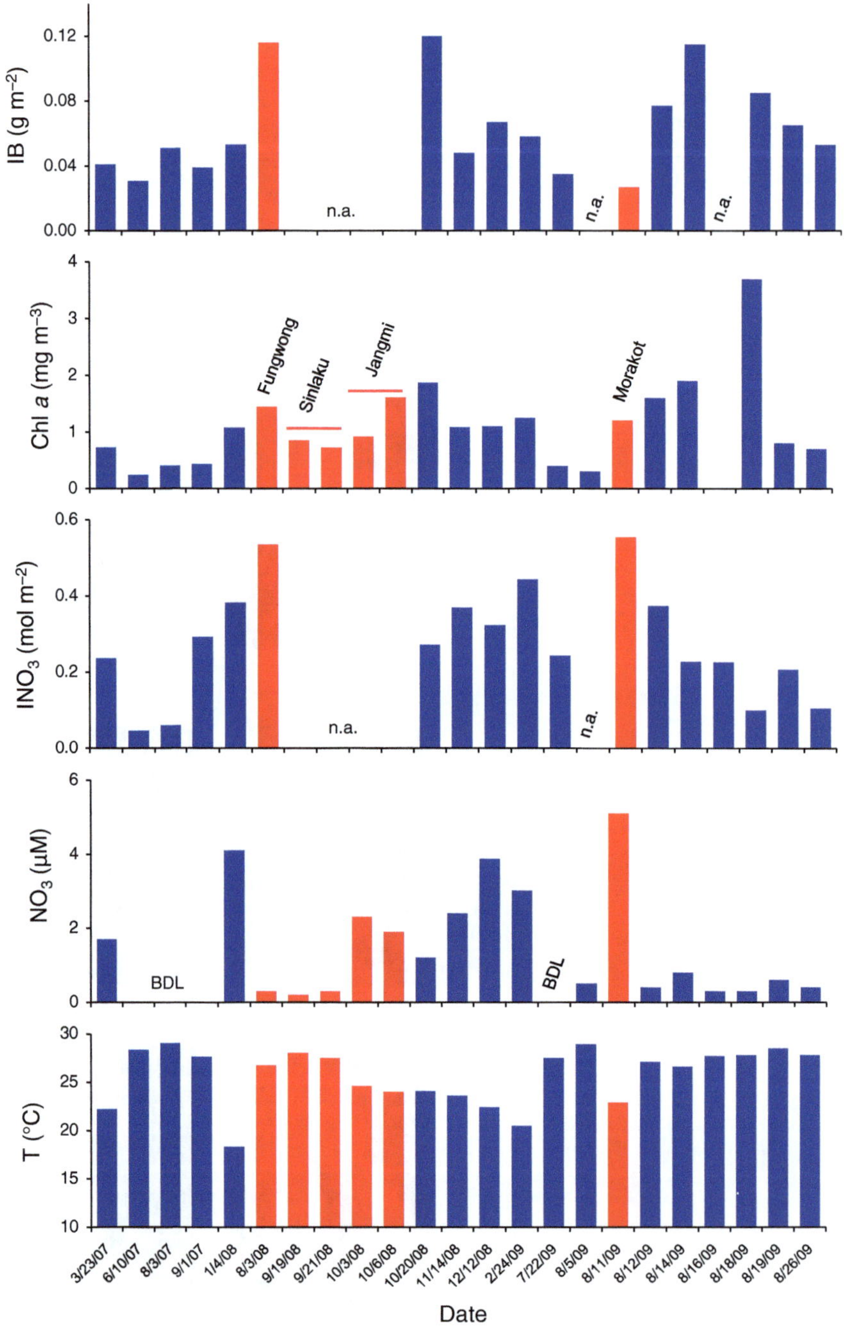

Fig. 13.3 Field-measured integrated chlorophyll (IB, 0–75 m), surface chlorophyll *a* concentration (Chl *a*), integrated nitrate (INO_3), surface nitrate concentration (NO_3), sea surface temperature (T) in the southern East China Sea under non-typhoon conditions (*blue bars*) and after typhoon conditions (*red bars*) from 2007 to 2009. *Note* There are no surface nitrate concentrations during non-typhoon periods available because they fell below the detection limit (BDL, <0.1 μM). *n.a* data not available

subsurface water is a major nutrient source to the SECS shelf. The riverine input of nutrients, which is revealed by the less-saline coastal water with a salinity <34, is seldom observed in the mid and outer shelf. Moreover, Liu et al. (1992) reported that benthic regeneration of nutrients is unimportant in the shelf region, where the sediments are composed of relic sand and the organic carbon content (<0.3 %) is very low. Additionally, atmospheric nitrogen (nitrate and ammonium) deposition ($\sim$1.4 mg N m^{-2} d^{-1}, Chen and Chen 2008) should be a minor source as compared to the nitrogen flux ($\sim$32 mg N m^{-2} d^{-1}) measured in the euphotic zone (Hung et al. 2009). Based on sea-going surveys, our current understanding of the summer nutrient supply in the SECS is that it cannot support the high average PP ($\sim$1,000 mg C m^{-2} d^{-1} in summer) reported by Hung et al. (2000) and Chen et al. (2001). These concentrations imply that any combination of regular nitrate upwelled under non-typhoon conditions, atmospheric deposition, and riverine input cannot account for the observed high nitrogen export flux out of the euphotic zone. Other nitrogen sources need to be considered to balance the nitrogen budget.

13.2.2 Nitrate Supply in the Summer During Non-typhoon Conditions

As mentioned above, nutrient supply is most likely limited in the southern East China Sea during summer. Therefore, we need to understand how much the nutrient supply in the SECS in the summer is during non-typhoon conditions. According to intensive field investigations conducted previously (Liu et al. 1992; Gong et al. 1995), the estimated transport of upwelled water in the SECS in summer (e.g. July to September) is about 0.2 Sv (1 Sv = 10^6 m^3 s^{-1}) and the average upwelling area (here defined as SST < 27 °C) is approximately 2,900 km^2 (Liu et al. 1992) under good weather conditions. In other words, the estimated average upwelling speed in the SECS is approximately 5.4 m d^{-1} (Liu et al. 1992).

The average nitrate concentration in the water column (0–75 m) in the SECS in summer under non-typhoon conditions is about 2.67 μM based on field observations (see data in Hung and Gong 2011). The level of nitrogen associated with POC concentration in the water column must be added to this nitrate budget. The average POC concentration in the water column in the SECS in summer under non-typhoon conditions is 0.04 g C m^3 (data from Table 13.1 in Hung and Gong 2011). Assuming Redfield stoichiometry (Redfield et al. 1963), the POC would be associated with 0.6 μM N (as nitrate). Thus the net average nitrate concentration in the upper 75 m of the water column is 3.27 μM N in summer. Additionally, the removal by the sinking nitrogen flux needs to be considered. The average POC flux in the SECS is approximately 175 ± 31 mg C m^{-2} d^{-1} in the summer under non-typhoon conditions (Hung and Gong 2011) and would be associated with 2.2 mmol N m^{-2} d^{-1}. The daily nitrogen transport through sinking particles is therefore 0.03 μM N in the summer. The estimated summer daily nitrate transport

Table 13.1 Wind speed, sea surface temperature (SST), nitrate concentration, and chlorophyll *a* (Chl *a*) concentration before (B) and after (A) typhoons from 2008 to 2009

Typhoon	Wind speed (ms^{-1})		Landfall (mm/dd/yy)	Affecting time (days)	Cruise (mm/dd)	SST (°C)		Nitrate (μM)		Chl *a* ($mg\ m^{-3}$)	
	Max	Near ECS				B	A	B	A	B	A
Fungwong	43	37	07/28/08	3	07/21–07/25	28.7		<0.1		n.a.	
Fungwong					08/03		26.7		0.3		1.48
Sinlaku	51	41	09/13/08	5	09/08–09/11	28.0		<0.1		0.39	
Sinlaku					09/19		28.0		0.2		0.85
Sinlaku					09/21		27.5		0.3		0.73
Jangmi	53	50	09/28/08	3	09/25–09/26	27.5		< 0.1		n.a.	
Jangmi					10/03		24.6		2.3		0.92
Jangmi					10/06		24.0		1.9		1.61
Morakot	43	37	08/7/09	2	07/22, 08/5	28.91	22.9	0.5		0.65	
Morakot					08/11–08/26				5.1		3.70

Note SST and Chl *a* data for 25.2°N–25.7°N, 122.1°E–122.6°E before typhoon passage were derived using AVHRR infrared sensors and MODIS ocean color, respectively. Nitrate concentrations in 2008 before the typhoons were derived according to Hung et al. (2010). SST, nitrate, and Chl *a* data were measured in the field shortly after typhoon passage at the station located at 25.40°N, 122.45°E. *ECS* the East China Sea to the top 75 m (the euphotic zone about 30–50 m) will be 8.5×10^8 g N d^{-1} (=0.2 Sv $\times$ 3.3 mmol m^{-3}) under non-typhoon conditions.

Alternatively, taking a specific example, the average nitrate concentration in the water column (0–75 m) in the SECS on July 22, 2009 (e.g. pre-typhoon period) was 2.89 μM before Typhoon Morakot. Due to lacking POC data before Morakot, we use the same POC values (~0.6 μM N) as nitrate source under non-typhoon conditions. As a result, the average nitrate concentration in the upper 75 m of the water column is 3.5 μM. The estimated daily water transport in the upwelling region (1,865 km^2) was 0.11 Sv based on the average upwelling speed (5.4 m d^{-1}) in the SECS. The estimated daily nitrate transport to the top 75 m will be 5.4×10^8 g N d^{-1} before Typhoon Morakot. The model estimated nitrate transport in the SECS is approximately 2×10^9 g N d^{-1} based on sea-going monthly investigations (Liu et al. 1992). If we consider the uncertainty of both methods, the estimated nitrate transport value (~1×10^9 g N d^{-1}) in this study is slightly lower than the previously reported data by Liu et al. (1992) if monthly variation of nitrate transport is considered. Of course, nitrate transport in the summer in the SECS is much lower than in the cold seasons (fall to spring) (Gong et al. 1995; Hung and Gong 2011); therefore, our estimated daily nitrate transport in the summer under non-typhoon conditions is reasonable.

13.3 Nitrate Supply in the SECS After Typhoons

The nutrient dynamics, phytoplankton species changes, and particulate organic carbon fluxes in the SECS before and after typhoons have reported by Hung et al. (2010), Hung and Gong (2011), Chung et al. (2012). Based the field investigations, typhoons clearly provide a significant source of nutrients, but how important is nutrient supply induced by typhoons responsible for phytoplankton growth in the SECS in summer? Here we use two cases (typhoons Fungwong and Morakot) to demonstrate the nitrate supply brought to the euphotic zone of the SECS by typhoon-caused upwelling.

13.3.1 The Case 1 (Typhoon Fungwong in 2008)

The average nitrate inventory in summer in 2007 was 0.052 mol m^{-2} during non-typhoon conditions due to low nitrate concentrations in the water column before the passage of Fungwong. When averaging over the upper 75 m of the water column, the increase in NO_3 concentration (Fig. 13.3) associated with this upwelling was 6.41 μM [= (0.534 − 0.052 mol m^{-2})/75 m = 0.0064 mol m^{-3}]. Again, the level of nitrogen associated with the increase in POC concentration must be added to this figure. The change in POC concentration was 2.39 μM POC (see data in Hung et al. 2010) and would be associated with an increase of 0.36 (= 2.39/6.625) μM in nitrogen assuming Redfield ratios. Additionally, the removal by the sinking nitrogen flux after Typhoon Fungwong needs to be considered. The POC flux in the SECS 5 days after Fungwong was 265 ± 14 mg C m^{-2} d^{-1} (see data in Hung and Gong 2011) and will be associated with 1.1 mmol N m^{-2} d^{-1}. Thus the net increase in nitrogen in the upper 75 m of the water column averaged 6.79 μM. Because the field observation of hydrographic settings was conducted 5 days after the typhoon passed the ECS, it is difficult to estimate the upwelling speed of cold water patch. As shown in Table 13.1, Typhoon Fungwong was similar to Morakot in terms of typhoon strength and pathway. So, we assumed that the same average upwelling speed after Typhoon Fungwong is 8.6 m d^{-1} (see description below). The upwelling area [e.g. a cold (<27 °C) water patch] ranged from ~5,000 to 14,400 km^2 (on average ~10,000 km^2, see data in Hung et al. 2010) after Typhoon Fungwong. The upwelling velocity yielded a total volume transport of 8.6×10^{10} m^3 d^{-1} to the top 75 m in the upwelling area of 10,000 km^2. The nitrate transport would be 8.2×10^9 g N d^{-1} to the top from 75 m in the upwelling area of 10,000 km^2 if an average nitrate concentration of 6.79 μM and a time period of 2 days were adopted. If we use a similar two end-member approach (see procedure in case two), the net nitrate transport from the Kuroshio subsurface water is 3.7×10^9 g N after considering the vertical mixing effect and heat correction. The enhanced nitrate supply in the SECS after the passage of Typhoon Fungwong resulted in considerably higher biological production (see detailed results in Hung et al. 2010).

13.3.2 The Case 2 (Typhoon Morakot in 2009)

According to Hung et al. (2013), the cold water patches due to upwelling after Typhoon Morakot increased to approximately 32,587 km^2 on 11 August from 1,800 km^2 on August 3–5, 2009. As the average nitrate concentration in the water column (0–75 m) in the SECS 2 days (11 August) after Typhoon Morakot was 7.54 μM. Therefore, we use 7.5 μM to represent the average nitrate concentration in the SECS after typhoons and do not consider the POC-nitrogen contribution because particles in fresh upwelling water are not as important as aged upwelling water. We conservatively assume that the average upwelling area after a typhoon is approximately 10,000 km^2. We did not have a vertical transport velocity of upwelling water after Typhoon Morakot. The upwelling velocity after Typhoon Morakot should be significantly higher than 5.4 m d^{-1} (Liu et al. 1992) because temperature and nitrate concentration in the surface water cannot change rapidly in just 2 days (water only move about 11 m) after the passage of Typhoon Morakot. Here we use two ways to estimate an upwelling speed in the cold water patches. First, we used temperate as a proxy to estimate upwelling speed. The temperature at 75 m in the study area during 21–22 July ranged from 20.4 to 21.8 °C. The temperature at 70 m in the study area on 5 August was about 20–21 °C. We assume that the center of upwelling water did not mix with surrounding water while it first came out from the Kuroshio subsurface water. Based on the temperature data, the water sources should then be from about 60 m. Second, the surface nitrate concentrations in the center of upwelling water 2 days after Typhoon Morakot was 5.1 μM (Fig. 13.3). The nitrate concentrations at 50 m and 70 m in the center of upwelling water on July 22 were 3.3 and 7.1 μM, respectively. Remarkably, the source of nitrate should be from water deeper than 50 and 60 m should be a reasonable depth. The estimated upwelling velocity would be 24 m d^{-1} (= 60 m/2.5 days) if a time period of 2.5 days (9–11 August) is considered. In fact, the upwelling velocity at the center of upwelling should be a maximum, if the upwelling velocity decreases along the center to the edge of the upwelling area. The average upwelling velocity in the whole cold water patch should be 12 m d^{-1}. The upwelling velocity yielded a total volume transport of 1.2×10^{11} m^3 d^{-1} to the top 75 m in the upwelling area of 10,000 km^2. The nitrate transport would be 1.26×10^{10} g N d^{-1} to the top from 75 m in the upwelling area of 10,000 km^2 if average nitrate concentration of 7.5 μM is used. The net nitrate transport from the oceanic source after Typhoon Morakot would then be 2.5×10^{10} g N to the top of water column if a time period of 2 days is adopted.

However, one has to consider that the cooling phenomenon is also caused by vertical mixing rather than upwelling only. We did not have vertical velocity data to demonstrate the effect of vertical mixing. Here we refer the one-dimensional turbulence model of Mellor and Yamada (1982) to estimate the effect of the passage of a typhoon on the variation mixed layer. The wind speed reached 44 m s^{-1} at the study area at 9 a.m. on 8 August (local time) by in situ measurement (Jan et al. 2013). Therefore, the wind speed is given as 40 m s^{-1}, the simulation time is 36 h, and the surface mixed layer deepened from <20 m to deeper than 70 m within 36 h. The simulated SST decreases from 29 to about 26.7 °C ($\Delta C = 2.3$ °C) after 36 h. The SST was 23 °C 2 days after the typhoon. If we assume 1 °C due to heating loss, the estimated cooling caused by upwelling should account for 45 % of that (= 2.7/6 °C). Thus, we estimate that the net nitrate transport from an oceanic source is 1.1×10^{10} g N after vertical mixing effect and heat correction are considered. The nitrate supply in the SECS provided by Morakot is about twofold higher than that by Fungwong, but they are of the same magnitude. The reasons caused the difference could be complicated because the tracks of two typhoons, transition speed of typhoons and pre hydrographic conditions may be different. More recently, Hung et al. (2013) used a numerical model provided by Tsai et al. (2013) to calculate nitrogen supply in the same area of the SECS and the estimated total nitrogen supply after Typhoon Morakot was 2.5×10^{10} g N, which is of the same magnitude as compared to the estimated value (1.1×10^{10} g N) using a simple method. Overall, the strong nitrate supply in the SECS after the passage of Typhoon Morakot results in higher biological production and zooplankton production as well (see detailed results in Chung et al. 2012; Hung et al. 2013).

13.4 Possible Impact on Fish Production in the SECS by Typhoons

The nutrient transport to the euphotic zone in the SECS after typhoons Fungwong and Morakot were 8.2×10^9 and 1.3×10^{10} g N d^{-1}, respectively. Remarkably, the nitrate supplies after two typhoons are significantly higher than that (1×10^9 g N d^{-1}) during non-typhoon periods (Fig. 13.4). If all nitrate would have been taken up by phytoplankton, the fixed carbon after typhoons Fungwong and Morakot are 21×10^3 and 32×10^3 t C d^{-1} using a Redfield ratio (C:N = 6.6:1), respectively. In comparison, the typhoon-induced potential carbon uptake by phytoplankton is similar to the Yangtze River flood-induced phytoplankton bloom (147×10^3 t C d^{-1}) in the ECS (Gong et al. 2011).

As mentioned in the introduction, mackerel and sword tip squid (mean tropic level of 3) are two of the most important fishery resources in the SECS. According to the assessed nitrate supplies after typhoons and mean tropic level of 3, the potential fish production in the ECS after typhoons Fungwong and Morakot can be estimated (equation from Pauly and Christensen 1995) to be 1.9×10^3 and

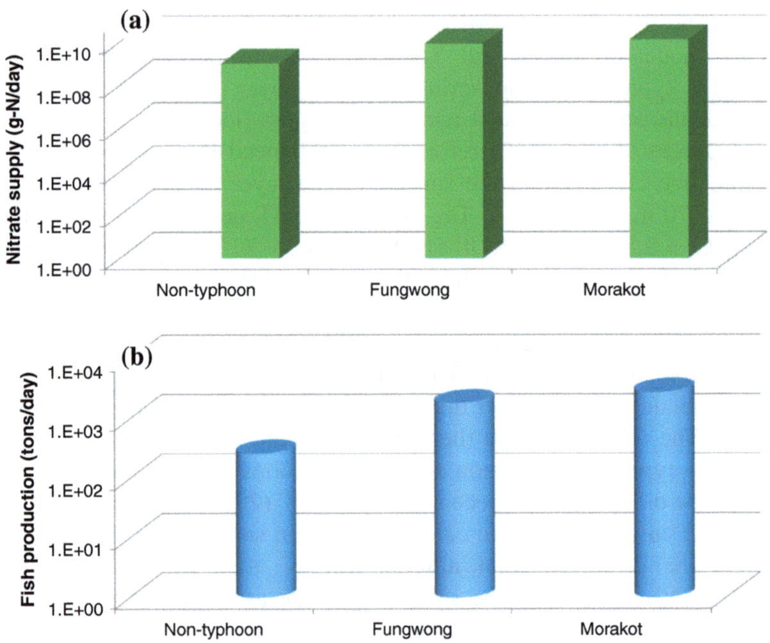

Fig. 13.4 Estimated nitrate supply (g N d^{-1}) and potential fish production during non-typhoon conditions and after typhoons Fungwong and Morakot in the southern East China Sea. Potential fish production was estimated by the following equation: [Primary production required] = [Catches/9] $\times$ $10^{(TL-1)}$, where TL is Trophic Level, which is assumed to be 3 (Pauly and Christensen 1995)

2.9×10^3 tons-fish d^{-1} (mean value = 2.4×10^3 tons-fish d^{-1}, Fig. 13.4), respectively. Conversely, the estimated potential fish production in the ECS during non-typhoon conditions is only 1.2×10^1 tons-fish d^{-1}. If we further consider that there are 2 typhoons passing through the ECS in the summer each year and the cold water induced by typhoons is sustained over 2–3 days (on average 2.5 days), the estimated nitrate-induced fish production in the ECS will be 4.6×10^4 tons-fish in summer (assuming summer time 90 days). On the other hand, the estimated nitrate-induced fish production in the ECS will be 1.1×10^3 tons-fish (= $90 \times 1.2 \times 10^1$) in the summer under non-typhoon conditions. In fact, the mean monthly and annual mackerel catches in the ECS are approximately 5×10^3 and 6×10^4 tons, respectively (Hung and Gong 2011). The potential fish production induced by typhoons is closed to annual mackerel catches in the ECS, suggesting that typhoon events indeed play an important role for biological productivity. These results strongly suggest that normal nutrient supply may not support the high biological productivity in the SECS in summer and that extra nutrient and food resources (caused by typhoons) are needs to support both adult and larval fish.

13.5 Conclusions and Recommendations

Based on sea-going observations before and shortly after the passage of typhoons, our results provide evidence that typhoons significantly affect nutrient supply and potential fish production over a continental shelf break region. Most importantly, nutrient supply after the passage of typhoons results in enhanced carbon uptake rate and an increase in potential fish production.

Specific conclusions are: (1) the typhoon-induced nitrate supply after typhoons Fungwong (8.2×10^9 g N d^{-1}) and Morakot (1.3×10^{10} g N d^{-1}) are significantly higher than that (1×10^9 g N d^{-1}) during non-typhoon periods; and (2) typhoons greatly enhance biological productivity and potential fish production as well.

Although intensive sea-going observations provide clear evidence, we still need to monitor or directly measure the upwelling speed for the cold water patches after the passage of typhoons, rather than estimating it. Moreover, the hydrographic settings in the water column are another important factor affecting the calculation of nitrate supply after the passage of typhoons. However, it is difficult to conduct sea-going experiments before and during a typhoon event due to dangerous conditions. Therefore, we recommend that a moored system with ADCP and biogeochemical sensors would be a good approach to obtain such valuable information during the passage of a typhoon.

Acknowledgments We acknowledge the assistance of the crew of R/V *Ocean Researcher* 2, G.-S. Hsieh, J.-M. Wu, L.-C. Wei, C.-C. Chung, and C.W. Tseng. This research was supported by the National Science Council of Taiwan (NSC101-2116-M-110-001, NSC101-2611-M-110-015-MY3).

References

Chang, J., Chung, C.-C., Gong, G.-C.: Influences of cyclones on chlorophyll *a* concentration and *Synechococcus* abundance in a subtropical western Pacific coastal ecosystem. Mar. Ecol. Prog. Ser. **140**, 199–205 (1996)

Chang, Y., Liao, H.-T., Lee, M.-A., Chan, J.-W., Shieh, W.-J., Lee, K.-T., Wang, G.-H., Lan, Y.-C.: Multisatellite observation on upwelling after the passage of Typhoon Hai-Tang in the southern East China Sea. Geophys. Res. Lett. **35**, L03612 (2008)

Chen, H.-Y., Chen, L.-D.: Importance of anthropogenic inputs and continental-derived dust for the distribution and flux of water-soluble nitrogen and phosphorus species in aerosol within the atmosphere over the East China Sea. J. Geophys. Res. **113**, D11303 (2008)

Chen, Y.L.L., Chen, H.-Y., Lee, W.-H., Hung, C.-C., Wong, G.T.F., Kanda, J.: New production in the East China Sea, comparison between well-mixed winter and stratified summer conditions. Cont. Shelf Res. **21**, 751–764 (2001)

Chen, C.T.A., Liu, C.-T., Chuang, W.S., Yang, Y.J., Shiah, F.-K., Tang, T.Y., Chung, S.W.: Enhanced buoyancy and hence upwelling of subsurface Kuroshio waters after a typhoon in the southern East China Sea. J. Mar. Syst. **42**, 65–79 (2003)

Chen, Y.L.L., Chen, H.-Y., Jan, S., Tuo, S.-H.: Phytoplankton productivity enhancement and assemblage change in the upstream Kuroshio after typhoons. Mar. Ecol. Prog. Ser. **385**, 111–126 (2009)

Chern, C.-S., Wang, J.: The exchange of Kuroshio and East China Sea shelf water. J. Geophys. Res. **95**, 16017–16023 (1990)

Chung, C.-C., Gong, G.-C., Hung, C.-C.: Effect of Typhoon Morakot on microphytoplankton population dynamics in the subtropical Northwest Pacific. Mar. Ecol. Prog. Ser. **448**, 39–49 (2012)

Gong, G.-C., Liu, K.-K., Pai, S.-C.: Prediction of nitrate concentration from two end member mixing in the southern East China Sea. Cont. Shelf Res. **15**, 827–842 (1995)

Gong, G.-C., Shiah, F.-K., Liu, K.-K., Wen, Y.-H., Liang, M.-H.: Spatial and temporal variation of chlorophyll *a*, primary productivity and chemical hydrography in the southern East China Sea. Cont. Shelf Res. **20**, 411–436 (2000)

Gong, G.-C., Wen, Y.-H., Wang, B.-W., Liu, G.-J.: Seasonal variation of chlorophyll *a* concentration, primary production and environmental conditions in the subtropical East China Sea. Deep-Sea Res. II **50**, 1219–1236 (2003)

Gong, G.-C., Liu, K.-K., Chiang, K.-P., Hsiung, T.-M., Chang, J., Chen, C.-C., Hung, C.-C., Chou, W.-C., Chung, C.-C., Chen, H.-Y., Shiah, F.-K., Tsai, A.-Y., Hsieh, C.-H., Shiao, J.-C., Tseng, C.-M., Hsu, S.-C., Lee, H.-J., Lee, M.-A., Lin, I.-I., Tsai, F.: Yangtze River floods enhance coastal ocean phytoplankton biomass and potential fish production. Geophys. Res. Lett. **38**, L13603 (2011)

Hung, C.-C., Gong, G.-C.: Biogeochemical responses in the southern East China Sea after typhoons. Oceanography **24**, 42–51 (2011)

Hung, C.-C., Wong, G.T.F., Liu, K.-K., Shiah, F.-K., Gong, G.-C.: The effects of light and nitrate levels on the relationship between nitrate reductase activity and $^{15}NO_3^-$ uptake: field observations in the East China Sea. Limnol. Oceanogr. **45**, 836–848 (2000)

Hung, C.-C., Warnken, K.W., Santschi, P.H.: A seasonal survey of carbohydrates and uronic acids in the Trinity River, Texas. Org. Geochem. **36**, 463–474 (2005)

Hung, C.-C., Gong, G.-C., Chung, W.-C., Kuo, W.-T., Lin, F.-C.: Enhancement of particulate organic carbon export flux induced by atmospheric forcing in the subtropical oligotrophic northwest Pacific Ocean. Mar. Chem. **113**, 19–24 (2009)

Hung, C.-C., Gong, G.-C., Chou, W.-C., Chung, C.-C., Lee, M.-A., Chang, Y., Chen, H.-Y., Huang, S.-J., Yang, Y., Yang, W.-R., Chung, W.-C., Li, S.-L., Laws, E.: The effect of typhoon on particulate organic carbon flux in the southern East China Sea. Biogeosciences **7**, 3007–3018 (2010)

Hung, C.-C., Gong, G.-C., Santschi, P.H.: ^{234}Th in different size classes of sediment trap collected particles from the Northwestern Pacific Ocean. Geochim. Cosmochim. Acta **91**, 60–74 (2012)

Hung, C.-C., Chung, C.C.C., Gong, G.-C., Jan, S., Tsai, Y., Chen, K.-S., Chou, W.-C., Lee, M.-A., Chang, Y., Chen, M.-H., Yang, W.-R., Tseng, C.-J., Gawarkiewicz, G.: Nutrient supply in the southern East China Sea after Typhoon Morakot. J. Mar. Res. **71**, 133–150 (2013)

Jan, S., Chen, C.-C., Tsai, Y.-L., Yang, Y.J., Wang, J., Chern, C.-S., Gawarkiewicz, G., Lien, R.-C., Centurioni, L., Kuo, J.-Y.: Mean structure and variability of the cold dome northeast of Taiwan. Oceanography **24**, 98–107 (2011)

Jan, S., Wang, J., Yang, Y.J., Hung, C.-C., Chern, C.-S., Gawarkiewicz, G., Lien, R.-C., Centurioni, L., Kuo, J.-Y., Wang, B.B.: Observation of a freshwater pulse induced by Typhoon Morakot off the northern coast of Taiwan in August 2009. J. Mar. Res. **71**, 19–46 (2013)

Karl, D.M.: A sea of change: biogeochemical variability in the North Pacific subtropical gyre. Ecosystems **2**, 181–214 (1999)

Liu, K.-K., Gong, G.-C., Lin, S., Yang, C.-Y., Wei, C.-L., Pai, S.-C., Wu, C.-K.: The year-round upwelling at the shelf break near the northern tip of Taiwan as evidenced by chemical hydrography. Terr. Atmos. Oceanic Sci. **3**, 243–276 (1992)

Liu, H.-C., Gong, G.-C., Chang, J.: Lateral water exchange between shelf-margin upwelling and Kuroshio waters influences phosphorus stress in microphytoplankton. Mar. Ecol. Prog. Ser. **409**, 121–130 (2010)

Mallin, M.A., Posey, M.H., Shank, G.C., McIver, M.R., Ensign, S.H., Alphin, T.D.: Hurricane effects on water quality and benthos in the Cape Fear watershed: natural and anthropogenic impacts. Ecol. Appl. **9**, 350–362 (1999)

McKinnon, A.D., Meekan, M.G., Carleton, J.H., Furnas, M.J., Duggan, S., Skirving, W.: Rapid changes in shelf waters and pelagic communities on the southern Northwest Shelf, Australia, following a tropical cyclone. Cont. Shelf Res. **23**, 93–111 (2003)

Mellor, G.L., Yamada, T.: Development of a turbulence closure model for geophysical fluid problems. Rev. Geophys. Space Phys. **20**, 851–875 (1982)

Pauly, D., Christensen, V.: Primary production required to sustain global fisheries. Nature **374**, 255–257 (1995)

Qiu, Y., Lin, Z., Wang, Y.: Responses of fish production to fishing and climate variability in the northern South China Sea. Prog. Oceanogr. **85**, 197–212 (2010)

Redfield, A.C., Ketchum, B.H., Richards, F.A.: The influence of organisms on the composition of sea-water. In: Hill, M.N. (ed.) The Sea, vol. 2, pp. 26–77. The composition of sea-waterInterscience, New York (1963)

Sassa, C., Tsukamoto, Y., Nishiuchi, K., Konishi, Y.: Spawning ground and larval transport processes of jack mackerel *Trachurus japonicus* in the shelf-break region of the southern East China Sea. Cont. Shelf Res. **28**, 2574–2583 (2008)

Shiah, F.-K., Chung, S.-W., Kao, S.-J., Gong, G.-C., Liu, K.-K.: Biological and hydrographical responses to tropical cyclones (typhoons) in the continental shelf of the Taiwan Strait. Cont. Shelf Res. **20**, 2029–2044 (2000)

Siswanto, E., Morimoto, A., Kojima, S.: Enhancement of phytoplankton primary productivity in the southern East China Sea following episodic typhoon passage. Geophys. Res. Lett. **36**, L11603 (2009)

Tsai, Y., Chern, C.-S., Jan, S., Wang, J.: Numerical study of cold dome variability induced by Typhoon Morakot (2009) off northeastern Taiwan. J. Mar. Res. **71**, 151–164 (2013)

Wang, K.-Y., Liao, C.-H., Lee, K.-T.: Population and maturation dynamics of the swordtip squid (*Photololigo edulis*) in the southern East China Sea. Fish. Res. **90**, 178–186 (2008)

Wu, C.-R., Lu, H.-F., Chao, S.-Y.: A numerical study on the formation of upwelling off northeast Taiwan. J. Geophys. Res. **113**, C08025 (2008)

Zhao, H., Tang, D., Wang, Y.: Comparison of phytoplankton blooms triggered by two typhoons with different intensities and translation speeds in the South China Sea. Mar. Ecol. Prog. Ser. **365**, 57–65 (2008)

Zheng, G.M., Tang, D.: Offshore and nearshore chlorophyll increases induced by typhoon winds and subsequent terrestrial rainwater runoff. Mar. Ecol. Prog. Ser. **333**, 61–74 (2007)

Chapter 14
Typhoon Impacts on Fishery in the South China Sea

DanLing Tang, Jie Yu, SuFen Wang and Gang Pan

Abstract Marine phytoplankton and primary production can be greatly impacted by typhoons, as discussed in the previous chapters; the present chapter introduces new observations about typhoon impacts on fishery, focusing on fish abundance. A fish monitoring program was conducted in the northern region of the South China Sea from March 2009 to December 2010. During this period, two typhoons, GONI and Koppu, hit this region consecutively in August and September 2009. The fish and satellite data were analyzed to understand the influence of the typhoons on fish activities. The results showed that the fish species number (FSN) increased by approximately 14.29 and 14.81 % after the two typhoons, GONI and Koppu, respectively. The five increased fish species included three estuarine species and two shallow sea species. However, one shallow sea species was also absent. In the nearshore (near the Pearl River Estuary) and offshore (along the typhoon's track) regions after GONI, the FSN increased by approximately 24 % (nearshore) and 52.63 % (offshore), with estuarine species accounting for 42.86 % (nearshore) and 33.33 % (offshore) of the fish species; after Koppu, the FSN increased by approximately 15.38 % (nearshore) and 163.64 % (offshore), with estuarine species accounting for 60 % (nearshore) and 26.32 % (offshore) of the

D. L. Tang (✉) · J. Yu · S. F. Wang · G. Pan
Research Center for Remote Sensing of Marine Ecology and Environment, State Key Laboratory of Tropical Oceanography, South China Sea Institute of Oceanography, Chinese Academy of Sciences, Xingang Xi Road 164, Guangzhou 510300, People's Republic of China
e-mail: lingzistdl@126.com
URL: http://lingzis.51.net/

D. L. Tang · J. Yu
Graduate University of Chinese Academy of Sciences, Beijing 100049, People's Republic of China

J. Yu
Scientific Observing and Experimental Station of South China Sea Fishery Resources and Environments, Ministry of Agriculture, South China Sea Fisheries Research Institute, Chinese Academy of Fishery Sciences, Guangzhou 510300, People's Republic of China

D. L. Tang and G. J. Sui (eds.), *Typhoon Impact and Crisis Management*, Advances in Natural and Technological Hazards Research 40,
DOI: 10.1007/978-3-642-40695-9_14,

fish species. In the increased records, small and medium-sized fish species were dominant nearshore, and small fish species were dominant offshore. The FSN increased to a maximum value between the 5th and the 10th days after the typhoon nearshore and between the 3rd and 8th days after the typhoon offshore. The results indicated that river discharge, triggered by the typhoon's nearshore rainfall, as well as offshore upwelling nutrients, also triggered by the typhoons, and may have played important roles in the variability of fish species. This research found that the increase in the FSN was associated with the typhoons in the northern South China Sea. At the end, the chapter summaries the ecological mechanisms of typhoon impacts on fishery.

Keywords Typhoon · Upwelling · River discharge · Fish species number · Shannon Weiner index · South China Sea

14.1 Introduction

The marine environment can be disturbed by large-scale weather events, such as typhoons (Zheng and Tang 2007; Yang and Tang 2010), cyclones/hurricanes (Kundu et al. 2001; Gautam et al. 2005; Hagy et al. 2006; Lohrenz et al. 2008), earthquakes and tsunamis (Singh et al. 2001, 2007; Tang et al. 2004a, b, 2009; Yan and Tang 2008), as were introduced in the previous chapters. Changes in the habitat of aquatic animals are attributed to direct or indirect damage to the stability of the community (Andrews 1973; Boesch et al. 1976; Jury et al. 1995; Locascio and Mann 2005).

In estuaries, the strong wind and rainfall that are caused by typhoons result in increased freshwater inflow, low dissolved oxygen (Guo et al. 2000) and changes in the water quality and water temperature, all of which may affect the fish community. The immediate changes in the aquatic habitat, cause marine species to escape or perish and estuarine species to increase in number, can alter the fish community and composition (Greenwood et al. 2006; Switzer et al. 2006). Similar results were also found near coral reefs (Kaufman 1983; Adams 2001). The broken coral reefs can cover some sedentary species, leaving some less visible fishes exposed, changing the fish community and composition accordingly (Letourneur et al. 1993; Adams and Ebersole 2004; Hernández et al. 2008).

Weak typhoons can also induce upwelling and bring nutrients to the surface waters from the bottom (Walker and Leben 2005; Zhao et al. 2009; Sun et al. 2010). Our previous researches also indicated that phytoplankton bloom after typhoons and enhance primary production (Tang et al. 2004a, b; Zheng and Tang 2007; Zhao et al. 2009; Chen and Tang 2012). It is still unclear whether typhoons can affect the Pearl River Estuary (PRE) and offshore, in the northern South China Sea (SCS). Specifically, it is still not known what effect typhoons may have on the FSN.

In this research, by taking advantage of a long-term monitoring program in the northern SCS, we not only analyzed the in situ monitoring data to investigate the impact of the typhoons on the fisheries but also analyzed the satellite data to understand the oceanic parameters for interpreting the mechanisms. Two typhoons (GONI and Koppu) triggered rainfall and upwelling nearshore (just on the mouth of the PRE) and offshore and were compatible with the extent of the fish monitoring program. These factors offer us a valuable opportunity to complete the research (Yu et al. 2013).

Typhoons GONI and Koppu, with different intensities, hit the SCS on August 3rd and September 13th, 2009 (Fig. 14.1b). This chapter focuses on the changes in the FSN during these two typhoons.

14.2 Study Area, Data and Methods

14.2.1 Study Area

The SCS is located in the western Pacific Ocean, covering an area of 3.5 million km^2. This area is not only a typhoon-dominated tropical sea with high primary productivity but also a crucial fishing region in China (Zheng and Tang 2007; Tang et al. 2008; Zhao et al. 2008). According to its geological features and traditional fish distribution, the region is divided into the northern SCS continental-shelf fishing area, the northern SCS continental slope fishing area, the southern SCS continental shelf fishing area and the Xisha and Nansha islands fishing area (Ma et al. 2007). Our study area includes the coastal waters in the northwestern SCS and extends to both the northern SCS continental shelf and slope fishing areas, which are important traditional fishing grounds with water depths from 20 to 200 m (Fig. 14.1a). Our research mainly focuses on the western Guangdong Fishing Ground (Box FG, Fig. 14.1b), where we conducted a fisheries monitoring program from March 2009 to December 2010 (black triangle in Fig. 14.1b).

14.2.2 Data and Methods

14.2.2.1 Analysis of Fish Data

We conducted a long-term tracking program (Oceanic Fishing Information Dynamical Collection, the work in the SCS was conducted by the South China Sea Fisheries Research Institute, SCSFRI) to evaluate the changes in fishery resources from March 2009 to December 2010. To standardize the records for each net's production, we designed a special fishing log with the 43 most common commercial fish species in the SCS. Considering the differences in the scientific

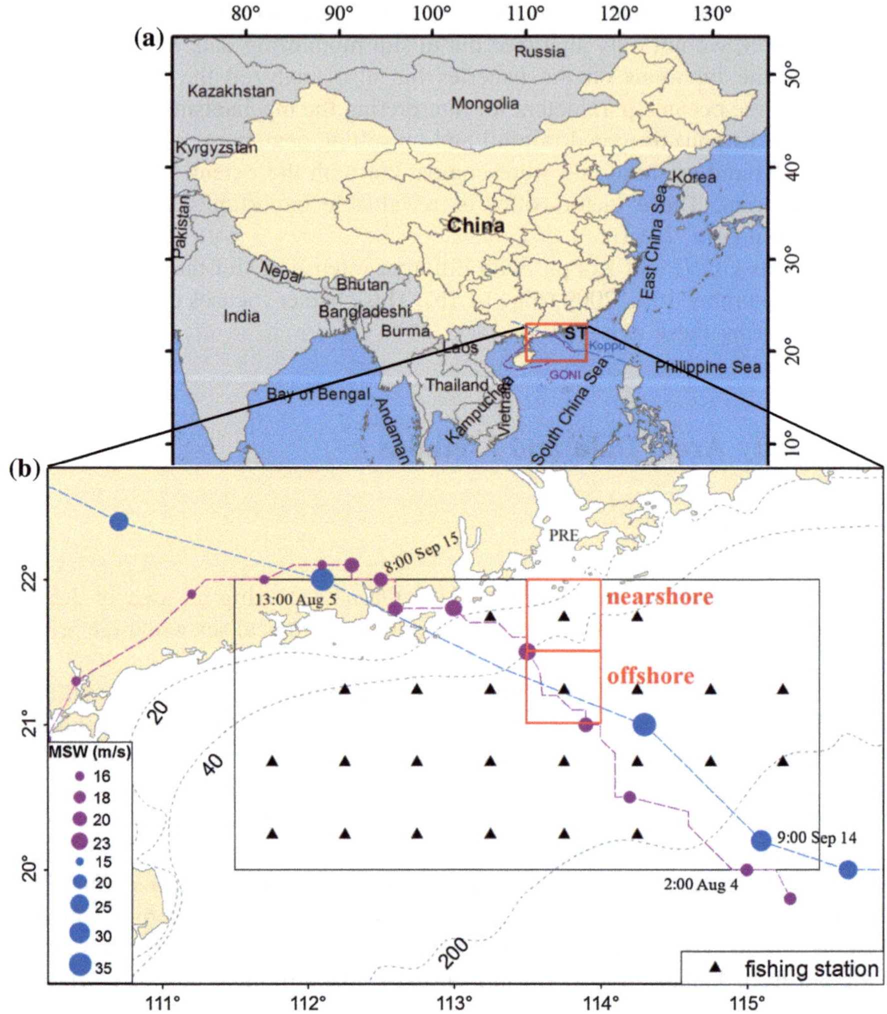

Fig. 14.1 **a** Location of study area (*box ST*) with two typhoons tracks. **b** Fishing stations and tracks of typhoons GONI and Koppu. *FG* Fishing ground; *black triangles* fishing stations; *dashed lines* isobathic lines; *MSW*: maximum sustained wind (m/s); *purple disks* and *blue circles* with the *dashed line* typhoon paths; *GT1* and *KT1* the time typhoons GONI and Koppu arrived at *FG*, *GT2* and *KT2* the time typhoons GONI and Koppu arrived offshore, *GT3* and *KT3* the time typhoons GONI and Koppu left the offshore region, *GT4* and *KT4* the time typhoons GONI and Koppu left the *FG*

and popular names of the same fish, we used popular names in the log for the convenience of the fishermen and then converted these popular names into scientific names when they were entered into the database. The fishing logs were distributed to the fishermen at the beginning of 2009. An associated training was also conducted to help the fishermen learn how to fill in the fishing logs.

The fishing data recorded in the logs and some auxiliary information such as date, trawling time and production value were entered into a network database at the beginning of each month. The original logs were then sent to SCSFRI by mail. To obtain the correct production data, fishing place and trawling time, all fishermen followed this fishing practice.

Fishing data from dual trawl boats (DTBs) in the study area were collected from March 2009 to December 2010, excluding the periods from the middle of May to the end of July in both 2009 and 2010 due to the Chinese fishing moratorium in the SCS. We used the central point of each small fishing region in an area of 0.25° × 0.25° to represent the fishing stations.

Initially, the FSN and Shannon Weiner index in box FG (Fig. 14.1b) were distributed by an average of three days to investigate the long-term variability. To understand how the FSN was affected by the typhoons, we analyzed the spatial distribution of the FSN before and after the typhoons. To understand the short-term effects of the typhoon, we compared the daily variation in the FSN nearshore and offshore before and after the typhoons.

Fish species were divided into two groups: estuarine species (Li et al. 2000; Wang and Lin 2006; Yang et al. 2005) and shallow sea species (Lei et al. 1981; Ji et al. 2007; Ou 2009). Furthermore, we divided the catches into three size groups to evaluate the effect of typhoons on different length groups: fish with average body length (AVL) lower than 20 cm (S type), fish with 20–30 cm AVL (MS type) and fish with AVL larger than 30 cm (Table 14.3). The FSN was estimated for each size class.

14.2.3 Information of Typhoon Tracks and Related Environment

The typhoon data obtained from the Wenzhou Typhoon Delivery System (http://www.wztf121.com/) were used to represent the central position of the typhoons and the maximum sustained wind speeds.

The sea surface wind (SSW) data came from QuickScat and WindSat. QuickScat data from March 2nd to November 21st, 2009, and WindSat data from November 22nd to December 23rd, 2010, were obtained from JPL (http://podaac.jpl.nasa.gov/ and http://www.remss.com). To understand the temporal variation in wind, the time series of the SSW in the key region was evaluated every three days on average. The spatial and daily distribution of the SSW before and after the typhoons was used to indicate the typhoons' position and wind strength.

Rainfall data with a resolution of 0.25 × 0.25° were obtained from NASA (http://mirador.gsfc.nasa.gov/). The spatial and daily variation in rainfall before and after the typhoons was used to study the effect of the typhoons' precipitation.

The sea surface temperature (SST) data, with a resolution of 25 km, was merged from two microwave radiometers: Aqua AMSR-E, TRMM microwave imager (TMI), on board the Tropical Rainfall Measuring Mission (TRMM), and MODIS infrared (IR) data with a 9 × 9 km grid (http://www.remss.com). The three-day average SST over box FG (Fig. 14.1b) was calculated to illustrate the temporal change in SST from March 2009 to December 2010. The spatial and daily distribution of the SST was mapped to indicate the changes in water temperature affected by the two typhoons.

We obtained chlorophyll a (Chl-a) data from EESA (http://www.globcolour.info/). The resolution of this data is 4 km, merged from Meris, Modis, and SeaWiFs. The spatial distribution of Chl-a was measured to study daily variations during the two typhoons. The sea surface current (SSC) data were obtained from NOAA (http://www.oscar.noaa.gov/). Satellite altimeter and scatterometer *data* provided by the OSCAR Project Office have a resolution of 100 km. The six-day average data was mapped to compare current conditions before and after the two typhoons.

14.3 Results

14.3.1 Boats, Fishing Station, and Two Typhoons

DTBs maintained stable fishing activities in our study area, FG. There were 13 DTBs participating in the research, with powers ranging from 426 to 1,716 h (Table 14.1). All these boats with mesh size are about 40 mm and belong to the same province, they have similar gear selectivity (Yang et al. 2003). The length of fishing periods and the number of working days for each boat were unequal. Four DTBs (Y00040/00145, Y00088/00812, Y00572/00573 and Y00813/00608) provided data before the first typhoon, GONI, which helped us to accurately assess the fish species. The other nine DTBs provided data during the typhoon periods for 71–237 working days. The average drawl time (ADT) changed smoothly, except for boat Y47148/47149, which had an ADT of 11.06 h.

We had 24 fishing stations west of the Pearl River Estuary (PRE), uniformly distributed along isobathic lines from 20 to 200 m (Fig. 14.1b).

GONI was a weak, Class-II typhoon and originated in the northern SCS (19.8 N, 115.3 E) at 20:00 on August 3rd. It moved to the northwest with an initial maximum speed of 18 m/s. After following a half circle over China and the Beibu gulf, GONI ended over the southeastern HNI at 14:00 on August 9th (Fig. 14.1a).

Koppu began north of the Philippines and south of the Bashi Channel at 2:00 on September 13th. It was a Class-IV typhoon with an initial wind speed of 18 m/s. Koppu traveled northwest and was upgraded to a tropical storm between 10:00 and 16:00 on September 14th. It then became a typhoon, lasting until 9:00 on September 15th, and ending over the Chinese mainland at 5:00 on September 16th (Fig. 14.1a).

Table 14.1 Information about DTBs

No	Boat name	Power (hr)	Fishing period	Working date (d)	Average drawling time (h)
1	Y00040/00145	899	April 5 2009 to May 6 2009	12	5.21
2	Y00088/00812	721	March 2 2009 to May 13 2009	24	5.02
3	Y00572/00573	1,368	March 2 2009 to May 11 2009	28	4.94
4	Y00813/00608	1,244	March 16 2009 to May 9 2009	28	4.96
5	Y47148/47149	884	April 7 2009 to November 12 2010	71	11.06
6	Y04201/04202	1,189	March 17 2009 to August 19 2010	75	4.89
7	Y04198/04199	1,716	March 9 2009 to April 29 2010	80	4.58
8	Y00701/00702	779	August 1 2010 to December 23 2010	80	4.63
9	Y00595/00596	1,354	March 4 2009 to December 22 2010	134	4.38
10	Y00008/00137	1,058	April 1 2009 to December 20 2010	163	5.11
11	Y00413/00414	645	March 2 2009 to December 21 2010	199	4.19
12	Y42108/42109	1,050	March 8 2009 to December 29 2010	219	6.28
13	Y47145/47146	426	March 5 2009 to November 25 2010	237	8.62

The two typhoons, GONI and Koppu, passed box FG in Fig. 14.1b with speeds of 20–23 m/s from 13:00 to 20:00 on August 4th, 2009 and 33–38 m/s from 21:00 on September 14th to 3:00 on September 15th, 2009, respectively.

14.3.2 Temporal Variations in the SSW, SST, Shannon Weiner Index and FSN

The three-day average SSW was variable from March 2nd, 2009, to December 22nd, 2010. From August 2nd to September 15th, 2009, there were two peaks with high wind speeds due to typhoons GONI and Koppu on August 3rd and September 14th (Fig. 14.2a).

The three-day average SST increased in the beginning of the summer and then decreased to a low value by the end of the year in both 2009 and 2010. Many small serrations can be seen on the curve, including two valleys on August 6th and September 15th, 2009, just a few days after the two typhoons (Fig. 14.2b).

The three-day average Shannon Weiner index exhibit two peaks with value of 2.28 and 2.41 on August 14th and September 19th, 2009 (Fig. 14.2c). The three-day average FSN from March 2nd, 2009 to December 22nd, 2010, displayed an irregular distribution. Two peaks, with values of 30 and 29, appeared on August 14th and September 19th, 2009, respectively (Fig. 14.2d).

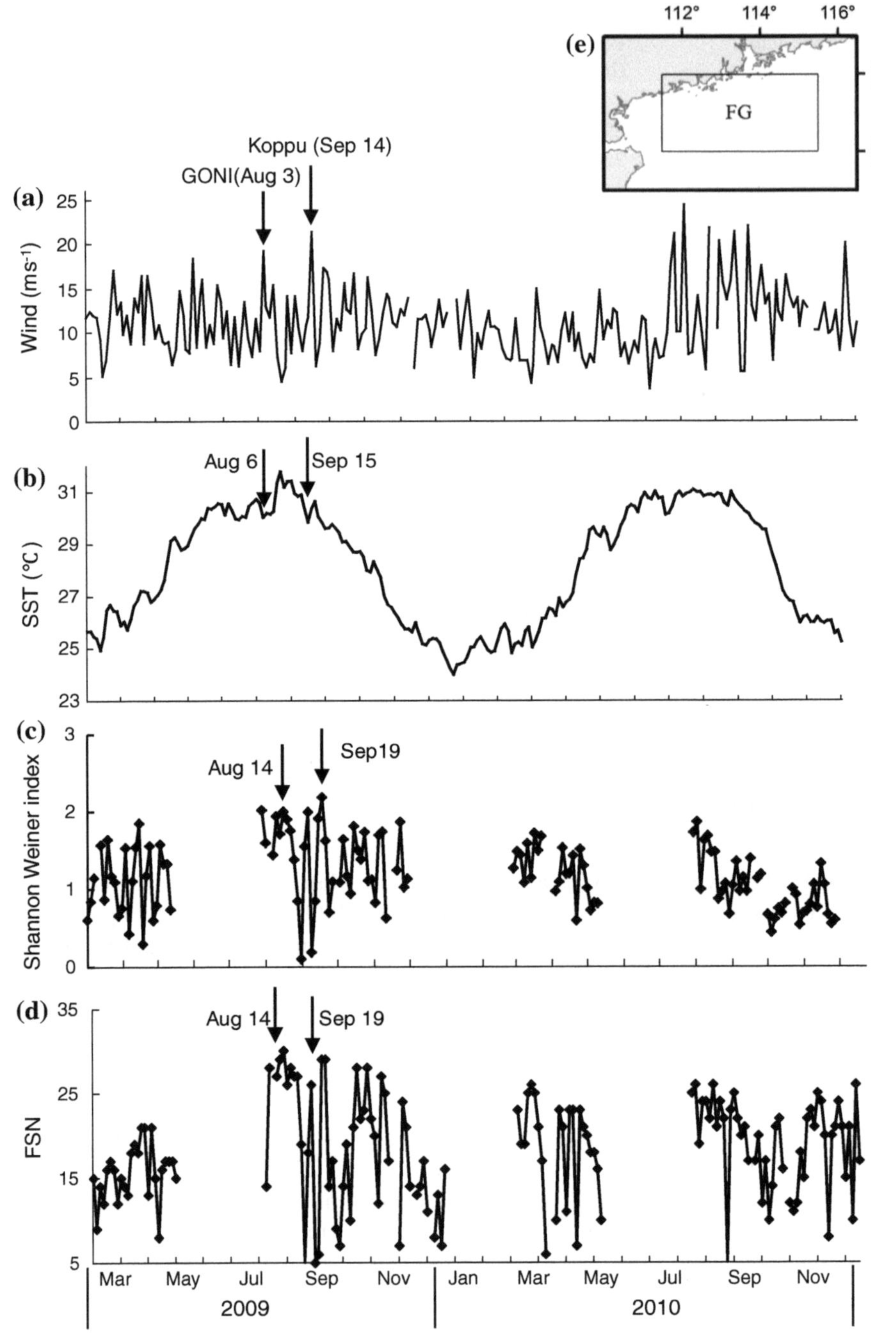

Fig. 14.2 Time series of three-day averaged FSN. **a** Wind speed (ms^{-1}). **b** SST (°C). **c** Shannon Weiner index. **d** FSN. **e** Data averaged for the FG. *Black arrows* the date of the two typhoons' occurrences (**a**), lowest SST (**b**) and maximum Shannon Weiner index (**c**) and FSN (**d**)

14.3.3 Spatial Variations in the Marine Ecological Environment Before and After the Two Typhoons

The SSW data for August 1st indicated that there was a weak wind in the FG region (Fig. 14.3*A1*). On August 4th, an SSW higher than 20 m/s gathered near the water, with 40–100 m isobaths outside the PRE (Fig. 14.3*B1*). Daily data from September 13th–14th, 2009, indicated changes in the SSW before and on the day of the typhoon. An extremely high wind occurred on the day of the typhoon that was not observed on September 13th, 2009 (Fig. 14.4*A1*–*B1*). Before GONI, there was little rainfall in the FG (Fig. 14.3*A2*). On the day of the typhoon, August 4th, rainfall was greater than 130 mm in the FG (Fig. 14.3*B2*). A significant rainfall event occurred in the FG on September 14th compared to the rainfall on September 13th (Fig. 14.4*A2*–*B2*).

The SSC from August 5th–10th (after GONI) indicated a river discharge out of the PRE, while a river discharge was not observed in the SSC from July 25th–30th (Fig. 14.3*A3*–*B3*). The SSC had a stronger discharge from September 15th–20th (after Koppu) than from September 5th–10th (Fig. 14.4*A3*–*B3*). Cold water developed near the typhoon's track in the FG after the typhoon (Figs. 14.3*B4*), 14.4*B4*) that was not present before the typhoon (Figs. 14.3*A4*, 14.4*A4*), indicating that the upwelling emerged on the right side of the typhoon's track.

After GONI, the average Chl-a from August 4th–20th was higher at the mouth of the PRE and where the 40 m isobath and the typhoon's track intersected (Fig. 14.3*A5*–*B5*). After Koppu, a tongue-style increase in Chl-a was observed moving across the typhoon's path from the inner PRE to the coastal area (Fig. 14.4*A5*–*B5*). After GONI, the FSN increased from 25 to 31 nearshore and from 19 to 29 offshore (Fig. 14.3*A6*–*B6*). After Koppu, the FSN increased from 26 to 30 nearshore and from 11 to 29 offshore (Fig. 14.4*A6*–*B6*).

14.3.4 Daily Variations in the FSN and Related Environmental Elements Nearshore and Offshore

The SSW obviously strengthened during the typhoons, with maximum values of 14.12 and 15.9 m/s on August 4th, when the typhoon GONI passed nearshore and offshore, respectively (Fig. 14.6*A1*). During Koppu, the SSW increased to 19.16 and 20.47 m/s nearshore and offshore, respectively (Fig. 14.6*B1*).

Before GONI, there was less than 10 mm of rainfall within 2 days. Nearshore, rainfall increased to the maximum value (60 mm) on August 5th, one day after the typhoon occurred, while offshore, rainfall increased to 90 mm on August 4th, the day the typhoon arrived (Fig. 14.6*A2*). Before Koppu, there was less than 25 mm of rainfall in 4 days. The rainfall over the two regions reached its maximum value (40 and 60 mm) on September 14th, the day the typhoon occurred (Fig. 14.6*B2*).

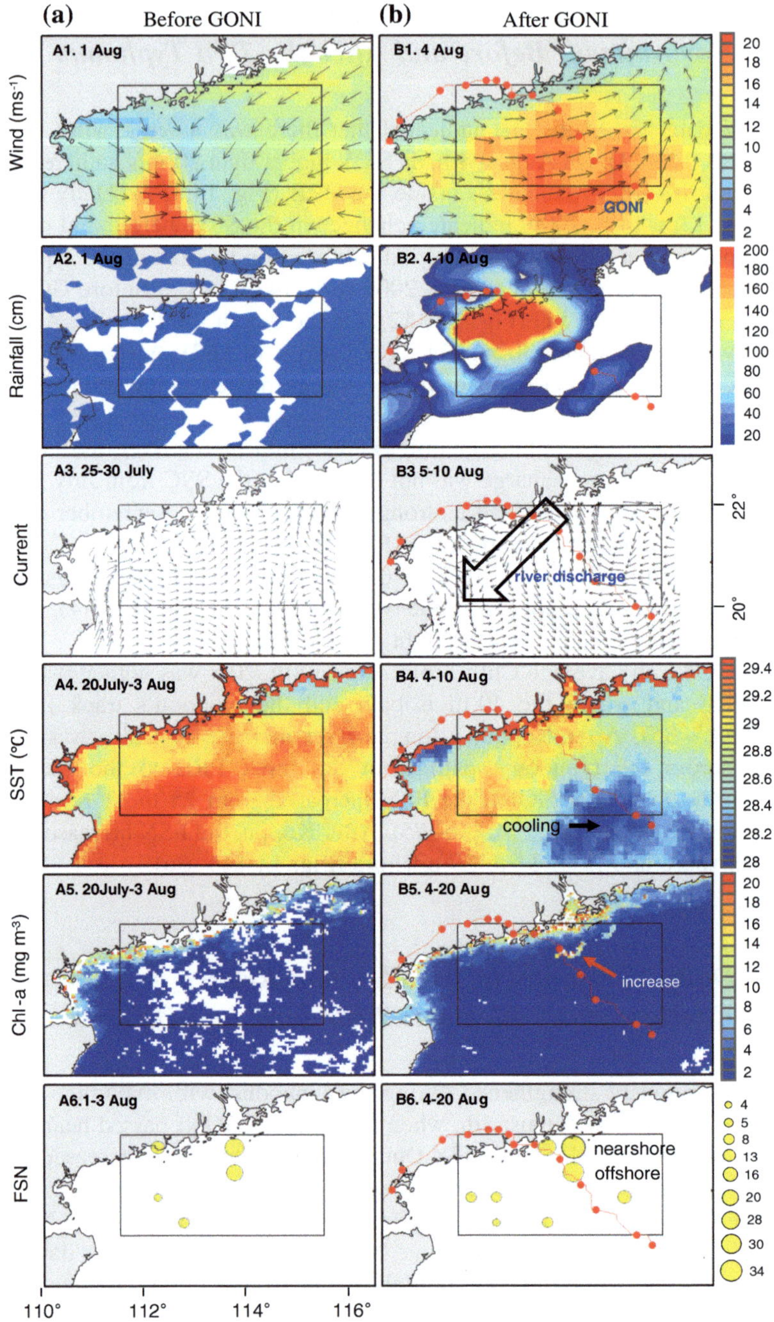

Fig. 14.3 Spatial distribution of wind (ms^{-1}), rainfall (cm), SSC, SST (°C), Chl-a (mg m^{-3}) and FSN. **a** Are before typhoon GONI (*A1*, *A2*, *A3*, *A4* and *A5*). **b** Are after GONI (*B1*, *B2*, *B3*, *B4* and *B5*)

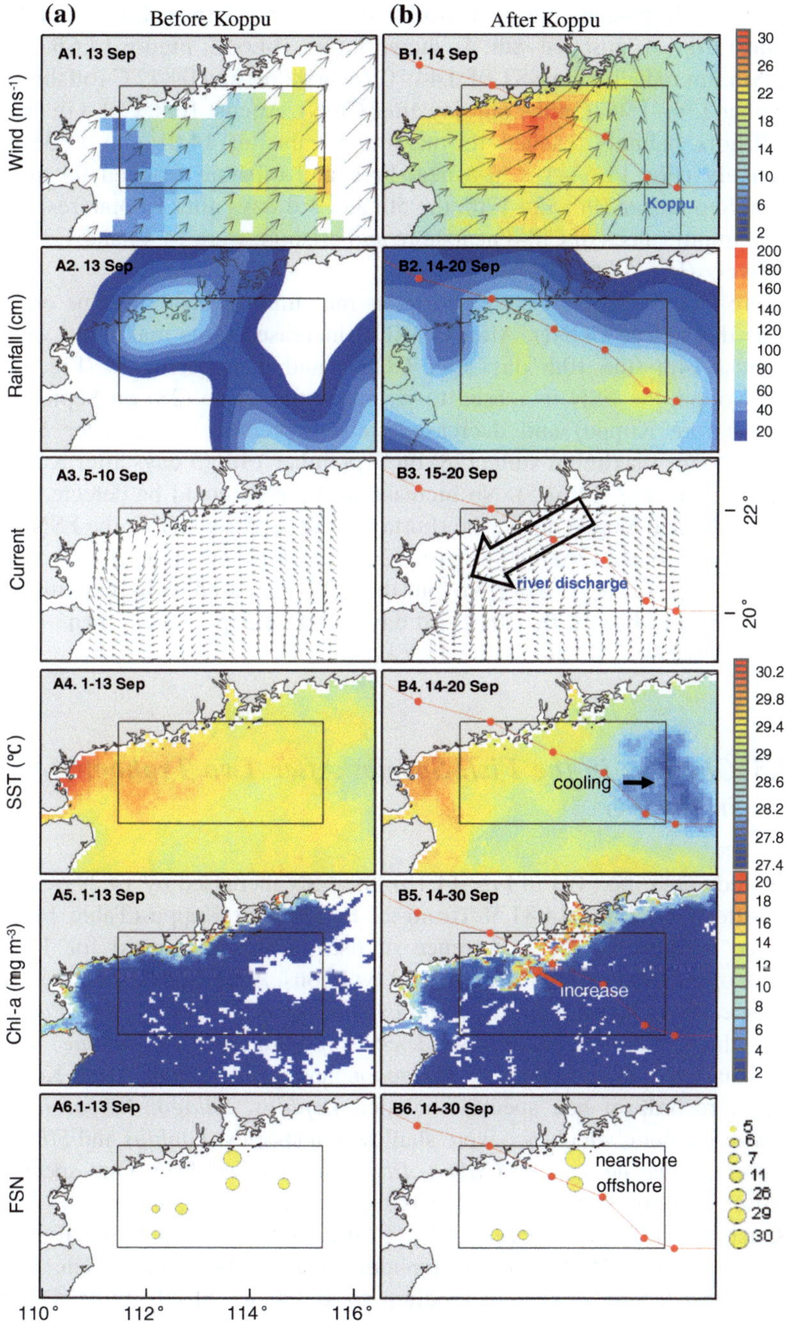

Fig. 14.4 Spatial distribution of wind (ms^{-1}), rainfall (cm), SSC, SST (°C), Chl-a ($mg\ m^{-3}$) and FSN before typhoon Koppu (*A1*, *A2*, *A3*, *A4* and *A5*) and after typhoon Koppu (*B1*, *B2*, *B3*, *B4* and *B5*)

After GONI passed, the SST declined to 0.65 °C (nearshore) and 0.43 °C (offshore) from August 3rd–8th, as shown in Fig. 14.6*A3*. Figure 14.6*B3* displays the maximum reduction in SST of 0.44 °C (nearshore) and 0.84 °C (offshore) from September 13th–15th. On the 6th day after GONI, August 10th, Chl-a in nearshore and offshore habitats reached its maximum value of 14.01 and 17.35 mg/m^3, respectively (Fig. 14.6*A4*). Chl-a nearshore and offshore reached its maximum value on September 19th and 17th, the 5th and 3rd days after Koppu, respectively. The maximum values of Chl-a nearshore and offshore were 14.44 and 1.31 mg/m^3, respectively (Fig. 14.6*B4*).

Before GONI, the FSN increased to its maximum nearshore value on August 2nd (the day before GONI). After GONI, it increased to its maximum value (27) on August 14th (the 10th day after GONI) and then decreased (Fig. 14.6*A5*). Before Koppu, the FSN increased to its maximum value (28) on September 8th (4 days before Koppu) and decreased until September 13th. After Koppu, it increased to the maximum value (28) on September 19th (5 days after Koppu) and then decreased (Fig. 14.6*B5*). No increase in the FSN could be detected offshore before GONI and Koppu due to the shortage of data. After GONI, the FSN reached its maximum value (26) on August 12th (the 8th day after GONI) and then decreased (Fig. 14.6*A5*). After Koppu, the FSN increased to its maximum value (28) on September 17th (the 3rd day after Koppu) and then decreased (Fig. 14.6*B5*).

14.3.5 Changes in the Fish Species After Two Typhoons in the FG

In the study area (box FG in Fig. 14.1b), the FSN increased by 14.29 % (from 28 to 32) after GONI and 14.81 % (from 27 to 31) after Koppu (Table 14.2). Five increased records for the occurrence of fish species accounted for 17.86 and 18.52 % of the total records after the two typhoons. After GONI, increased records of fish species included *Lophius*, *Johnius*, *Harpodon*, *Collichthys* and *Portunidae*, among which two were shallow sea species (*Lophius* and *Portunidae*) and three were estuarine species (*Johnius, Harpodon* and *Collichthys*). After Koppu, the increased records of fish species included *Lophius*, *Sillago*, *Nibea*, *Ilisha* and *Collichthys*, among which two were shallow sea species (*Lophius* and *Sillago*) and three were estuarine species (*Nibea*, *Ilisha* and *Collichthys*). One species (*Katsuwonus*) was not found after either typhoon (Tables 14.2, 14.3).

Nearshore, the FSN increased by 24 % (after GONI) and 15.38 % (after Koppu) compared to the FSN before the typhoons (Tables 14.2, 14.3). The increased records consisted of small and medium-sized species (Table 14.4). During the period with the increased FSN, estuarine species accounted for 42.86 % of the species after GONI (*Harpodon*, *Nibea*, *Johnius*, *Muraenesox*) and 60 % after Koppu (*Ilisha*, *Nibea*, *Collichthys*), while the other species were shallow sea

Table 14.2 Changes in FSN

Typhoon	FR	Before	After								
		Total	Total		Increased records					Lost records	
		Number	Number	P (%)	Total		Shallow sea	PRE main		Total	
					Number	P (%)	Number	Number	P (%)	Number	P (%)
GONI	WGFG	28	32	14.29	5	17.86	2	3	60	1	3.57
	Nearshore	25	31	24.00	7	28.00	4	3	42.86	1	4.00
	Offshore	19	29	52.63	9	47.37	6	3	33.33	1	5.26
Koppu	WGFG	27	31	14.81	5	18.52	2	3	60	1	3.70
	Nearshore	26	30	15.38	5	15.38	2	3	60.00	0	0.00
	Offshore	11	29	163.64	19	163.64	14	5	26.32	0	0.00

FR fishing region, *Total* the number of fish species, *Increased record* increased records of fish species, *Lost records* fish species that are not caught, *Shallow sea* fish species mainly living in shallow sea areas, *PRE main* fishes that are mainly found in the Pear River Estuary, *P (%)* percent increase compared to the original FSN

Table 14.3 Fish information

No.	Fish	Distribution	Size type
e1	*Harpodon*	Estuary	MS
e2	*Ilisha*	Estuary	MS
e3	*Nibea*	Estuary	MS
e4	*Collichthys*	Estuary	MS
e5	*Johnius*	Estuary	S
e6	*Argyrosomus*	Estuary	S
e7	*Siganus*	Estuary	S
s1	*Katsuwonus*	Shallow sea	M
s2	*Lophius*	Shallow sea	M
s3	*Muraenesox*	Shallow sea	M
s4	*Scomberomorus*	Shallow sea	M
s5	*Decapterus*	Shallow sea	MS
s6	*Epinephelussp*	Shallow sea	MS
s7	*Scomber*	Shallow sea	MS
s8	*Saurida*	Shallow sea	MS
s9	*Chondrichthyes*	Shallow sea	MS
s10	*Sphyraena*	Shallow sea	MS
s11	*Tetraodontidae*	Shallow sea	MS
s12	*Leiognathus*	Shallow sea	S
s13	*Thamnaconus*	Shallow sea	S
s14	*Nemipterus*	Shallow sea	S
s15	*Parargyrops*	Shallow sea	S
s16	*Priacanthus*	Shallow sea	S
s17	*Psenopsis*	Shallow sea	S
s18	*Sardina*	Shallow sea	S
s19	*Sillago*	Shallow sea	S
s20	*Pampus*	Shallow sea	S
s21	*Trichiurus*	Shallow sea	S
s22	*Upeneus*	Shallow sea	S
o1	*Octopodidae*		
o2	*Portunidae*		
o3	*Sepiidae*		
o4	*Penaeidae*		
o5	*Loliginidae*		

In 'No.' column, *e* estuarine fish species, *s* shallow sea fish species, *o* other species. In 'Distribution' column, *estuary* main species in the PRE, *shallow sea* main species in the northern SCS continental shelf. In 'size' column, *S* fishes with an average body length less than 20 cm, *MS* fishes with an average body length of 20–30 cm, *M* fishes with an average body length greater than 30 cm. The name of shark and ray are defined in order term, the name of cephalopod and crab are defined to family term, the name of other fishes are defined in genus term

species (*Scomber*, *Chondrichthyes* and *Portunidae* after GONI; *Priacanthus* and *Sillago* after Koppu) (Tables 14.2, 14.5, 14.6).

Offshore, the FSN increased by 52.63 % after GONI and 163.64 % after Koppu, compared to the FSN before the typhoons (Tables 14.2, 14.3). The increased records

Table 14.4 Distribution of increased records of fish species according to their body sizes

Typhoon	FR	S		MS		M	
		Number	P (%)	Number	P (%)	Number	P (%)
GONI	Nearshore	2	28.57	3	42.86	1	14.29
	Offshore	5	55.56	3	33.33	0	0
Koppu	Nearshore	2	40	3	50	0	0
	Offshore	9	47.37	5	26.32	3	15.79

FR fishing region, *S* (small), *MS* (small and medium) and *M* (medium), see Table 14.3

were generally small species (Table 14.4). During the period with the increased FSN, estuarine species accounted for 33.33 % after GONI (*Harpodon*, *Ilisha* and *Argyrosomus*) and 26.32 % after Koppu (*Ilisha*, *Nibea*, *Johniu*, *Argyrosomus* and *Siganus*), while the other species were shallow sea species (*Psenopsis*, *Sardina*, *Sillago* and *Sepiidae* after GONI; *Lophius*, *Muraenesox*, *Scomberomorus*, *Epinephelussp*, *Scomber*, *Sphyraena*, *Leiognathus*, *Parargyrops*, *Sardina*, *Sillago*, *Psenopsis*, *Sepiidae*, and *Penaeidae* after Koppu) (Tables 14.2, 14.7, 14.8).

14.3.6 The Effect of Typhoons on the FSN and Related Environmental Factors

Nearshore, the two typhoons passed with maximum speeds of 14.12 m/s (after GONI) and 19.17 m/s (after Koppu) and rainfalls of 57.51 cm (after GONI) and 40.19 cm (after Koppu). The SST decreased by 0.92 °C (from August 2nd–8th) after GONI and by 0.44 °C (from September 13th–15th) after Koppu, and the FSN increased by 3 (from August 2nd–14th) after GONI and 9 (September 9th–19th) after Koppu (Table 14.9).

Offshore, the two typhoons passed at maximum speeds of 15.89 m/s (GONI) and 20.47 m/s (Koppu), with rainfalls of 86.19 cm (GONI) and 59.12 cm (Koppu). The SST decreased by 0.93 °C (from August 2nd–8th) after GONI and 0.84 °C (from September 13th–15th) after Koppu. The FSN increased by 7 (from August 2nd–14th) after GONI and 20 (from September 9th–17th) after Koppu (Table 14.9).

14.3.7 Spatial Variations in the Chl-a and FSN in August 2009–2010, and September 2009–2010

In August 2009, GONI passed through the northern SCS, and induce an increase in Chl-a in nearshore and offshore compared that in August 2010 (Fig. 14.5*A1*, *B1*).

Table 14.5 Fishes caught before and after typhoon GONI in August, nearshore

No	1	2	3	4 to 8	9	10	11	12	13	14	15	16	17	18	19	20
e2		√	√		√	√		√	√	√				√		
e6		√	√		√	√	√	√	√	√	√	√	√	√	√	√
e7	√	√				√		√	√	√						
s4		√	√		√	√	√	√	√	√	√	√	√	√	√	√
s5	√	√	√			√	√	√	√	√	√	√	√	√	√	√
s6		√							√					√		
s8	√	√	√		√	√	√	√	√	√	√	√	√	√	√	√
s10		√	√		√	√	√	√	√	√	√	√	√	√	√	√
s11	√	√			√	√	√	√	√	√	√	√	√	√	√	√
s12	√						√	√	√	√			√			
s13	√	√	√				√	√	√	√			√	√	√	
s14	√	√	√			√	√	√	√	√	√	√	√	√	√	√
s15		√	√	G		√		√	√	√	√		√	√	√	√
s16		√	√	O		√	√	√	√	√	√			√		
s17		√	√	N		√	√		√	√			√	√		
s18		√	√	I		√			√					√		
s19		√	√			√		√	√	√				√		
s20		√	√		√	√	√	√	√	√	√	√	√	√	√	√
s21	√	√	√			√	√	√	√	√	√		√	√	√	√
s22	√	√	√			√	√	√	√	√	√		√	√	√	
o1		√				√	√		√	√			√	√	√	
o3		√	√			√		√	√	√				√		
o4		√	√		√	√	√	√	√	√	√			√	√	
o5	√	√	√			√	√	√	√	√	√	√	√	√	√	√
s1		√														
e1					√	√	√	√		√	√					
e4										√						
e5								√		√	√					
s3				Increased		√		√	√	√	√					
s7				records		√	√	√	√	√	√		√	√	√	
s9						√					√		√	√		
o2						√	√									

The *gray column* indicates the period when the typhoon hit. (*e* estuarine fish species, *s* shallow sea fish species, *o* other species)

Table 14.6 Fishes caught before and after typhoon GONI in August, offshore

No	1	2	3	4 to 8	9	10	11	12	13	14	15	16	17	18	19	20
e7		√									√					
s3		√						√								
s4		√						√							√	√
s5		√						√	√	√	√	√	√	√	√	√
s7		√						√	√	√	√	√	√	√	√	√
s8		√						√	√	√	√	√	√	√	√	√
s10		√						√							√	
s11		√		G				√	√						√	√
s12		√		O				√			√					
s13		√		N				√	√	√	√	√	√	√	√	
s14		√		I				√	√	√	√	√	√		√	√
s15		√						√				√	√		√	√
s16		√						√			√		√		√	√
s21		√						√				√			√	√
s22		√						√		√		√	√	√	√	
o1		√						√	√	√	√	√	√	√	√	
o4		√						√							√	
o5		√						√	√	√	√	√	√	√	√	√
s9		√														
e1								√								
e2								√								
e6								√							√	√
s6				Increased				√	√				√	√		√
s17				records				√		√					√	√
s18								√							√	√
s19								√								√
s20								√			√				√	√
o3								√				√	√		√	√

The *gray column* indicates the period when the typhoon hit. (*e* estuarine fish species, *s* shallow sea fish species, *o* other species)

Accordingly, an increase in FSN in August 2009 showed compared that in August 2010 (Fig. 14.5*A2*, *B2*). In September 2009, an obviously and highly increase in Chl-a and FSN also can be seen after Koppu compared to that in September 2010 (Fig. 14.5*A2*, *B2*).

Table 14.7 Fishes caught before and after typhoon Koppu in September, nearshore

No	4	5	6	7	8	9	12	13	14 to 15	16	17	18	19	20	21	22	23	24	25	26
e1					√								√	√						√
e5					√					√	√	√	√			√	√	√		
e6	√	√	√	√	√	√	√			√	√	√	√	√	√				√	√
e7						√				√	√		√	√	√					
s1					√															
s3					√					√	√		√	√						
s4	√	√	√	√	√	√	√			√	√		√	√	√		√		√	√
s5		√	√	√	√	√		√		√	√	√	√	√	√		√	√		
s7				√	√			√				√	√	√	√		√	√		
s8	√	√	√	√	√	√		√		√	√	√	√	√	√	√	√	√	√	√
s10	√	√	√	√	√	√				√		√	√	√	√					√
s11	√	√	√	√	√	√			K	√	√	√	√	√	√	√			√	√
s12				√					o	√		√	√	√	√	√	√			
s13				√	√	√		√	p			√	√	√	√					
s14	√	√	√	√	√	√			p	√	√	√	√	√	√	√	√		√	√
s15				√	√	√			u	√	√	√	√	√	√	√				
s16					√	√							√	√	√					√
s17	√	√	√	√	√	√				√			√	√	√					
s18			√		√	√								√	√					
s20	√	√	√	√	√	√	√			√	√	√	√	√	√	√	√		√	√
s21	√	√	√	√	√	√				√	√	√	√	√	√	√	√		√	√
s22				√	√	√		√		√	√	√	√	√	√					
o1				√	√	√							√	√						
o3	√				√	√				√	√	√	√	√	√	√	√			
o4		√	√	√			√			√	√	√	√	√	√	√	√		√	√
o5	√	√	√	√	√	√	√	√		√	√	√	√	√	√	√	√		√	√
e2													√	√					√	√
e3					Increased								√							
e4					records															√
s6										√	√	√	√	√						
s19													√	√	√					

The *gray column* indicates the period when the typhoon hit. (*e* estuarine fish species, *s* shallow sea fish species, *o* other species)

14.4 Discussion

14.4.1 Increases in the FSN After the Two Typhoons in Both Nearshore and Offshore

In estuaries, typhoons can affect fish habitats by inducing hypoxia and decreasing salinity through dilution with freshwater discharge. Typhoons may also cause resident fish species to disappear (Collins et al. 1981). However, the increased

Table 14.8 Fishes caught before and after typhoon Koppu in September, offshore

No	1	2	3	4	5	6	7	8	9	14 to 15	16	17	18	19	20	25	26
s5	√		√						√	Koppu	√	√	√	√	√	√	√
s8	√								√		√	√	√	√	√		√
s11	√		√						√		√	√	√	√	√		√
s13	√		√						√			√	√	√			
s14	√		√						√		√	√	√	√	√		
s16	√		√						√		√	√	√				√
s17	√		√									√	√				
s21	√										√	√	√	√	√		√
o1			√						√		√	√	√				
o5	√		√						√		√	√	√	√	√		√
e2												√					
e3													√				
e5												√	√				√
e6												√	√	√	√		
e7											√	√	√	√	√		√
s2											√						
s3											√	√	√	√			
s4						Increased records					√	√	√	√		√	
s6												√	√				
s7											√	√	√	√			
s10											√	√	√	√			
s12												√	√	√	√		
s15												√	√	√			
s18											√	√	√				
s19												√	√				
s20												√	√	√			
s22											√	√	√	√	√		
o3											√	√	√	√	√		
o4												√	√	√	√		

The *gray column* indicates the period when the typhoon hit. (*e* estuarine fish species, *s* shallow sea fish species, *o* other species)

river discharge also brings nutrients to nourish the mouth of the estuary, where the increased freshwater fishes are found following the increased flow (Greenwood et al. 2006) and marine species accumulate for feeding. The changes in the FSN depend on the difference between the number of additional species and the number of species lost (Paperno et al. 2006; Switzer et al. 2006).

Table 14.9 Comparison of the maximum wind, sustained time and maximum rainfall while the two typhoons passed through the FG and the decrease in the SST and increase in the FSN in the FG region

Cofference	Nearshore		Offshore	
	GONI	Koppu	GONI	Koppu
Maximum wind (ms^{-1})	14.12	19.17	15.89	20.47
Maximum rainfall (cm)	57.51	40.19	86.19	59.12
Sustain time (h)	Unpass	Unpass	7	5
Decrease in SST (°C)	0.92	0.44	0.93	0.84
FSN before typhoon	24 (2 Aug)	19 (9 Sep)	19 (2 Aug)	9 (1 Sep)
Maximum FSN after typhoon	27 (14 Aug)	28 (19 Sep)	26 (12 Aug)	28 (17 Sep)
Increase in FSN	3	9	7	20

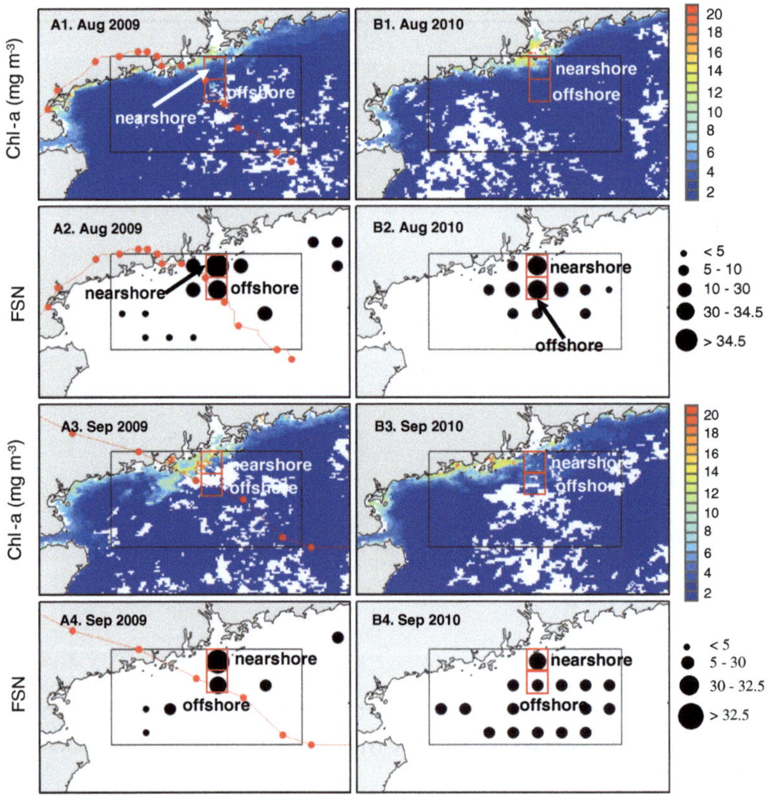

Fig. 14.5 Spatial distribution of Chl-a (mg m^{-3}) in August 2009 (*A1*), August 2010 (*B1*), FSN in August 2009 (*A2*), August 2010 (*B2*), Chl-a (mg m^{-3}) in September 2009 (*A3*) and September 2010 (*B3*), FSN in August 2009 (*A4*), August 2010 (*B4*)

In this study, the Shannon Weiner index and FSN increased in the northwestern SCS after the two typhoons (Fig. 14.2d). In the increased species records, the large proportion of resident species in the PRE indicated that the PRE's discharge played an important role in the accumulation of the fishes. There were two reasons for the increase in Chl-a after the typhoons. One was that a substantial quantity of nutrients was transported with the river's discharge, and the other was the increase in Chl-a from upwelling [also observed in this study (Figs. 14.3*B5*), 14.4*B5*)] due to the strong wind (Lin et al. 2003; Zheng and Tang 2007; Sun et al. 2010), which brings nutrients up from the deeper waters (Zhao et al. 2009). The increase in Chl-a may attract increased shallow sea and estuarine fish species after typhoons (Figs. 14.3*B6*, 14.4*B6*).

The FSN obviously changed in the two regions, nearshore (in the southern PRE, near the PRE) and offshore (in the southern PRE on the typhoon track) (Figs. 14.3*B6*), 14.4*B6*). After the two typhoons, Chl-a increased in the two regions (Figs. 14.3*B5*), 14.4*B5*). This increase in Chl-a indicated that the increase in the FSN in the two regions was associated with the typhoons.

After the two typhoons, the FSN offshore increased to its maximum value two days earlier than nearshore (Fig. 14.6*A5*, *B5*). This result indicates that the two typhoons affected the FSN in regions that were close to the typhoon track initially and then extended to the regions near the PRE.

Nearshore, estuarine species were dominant in the increased records of fish species (Table 14.2), indicating that river discharge may play a more important role than upwelling in changing the FSN. Offshore, shallow sea species were dominant in the increased records of fish species (Table 14.2), indicating that upwelling may play a more important role than river discharge in the increase in the FSN.

14.4.2 Increased Records of Fish Species in the Two Regions After Two Typhoons

Although typhoons can harm fish habit, increased records of fish species can still be found. Previous research found that increased records of freshwater species in estuary areas occurred in high-nutrient areas with increased freshwater inflow due to typhoons (Montane and Austin 2005; Houde et al. 2005; Pizza and Peyre 2009). In this research, seven and nine increased records of fish species in the nearshore and offshore regions, respectively, were found after GONI. Five and nineteen increased records of fish species were found in nearshore and offshore regions, respectively, after Koppu. Obviously, these increased records were associated with the typhoons. The high nutrient concentrations and upwelling, caused by river discharge and typhoons, respectively, may be the main reasons for the increase in the FSN.

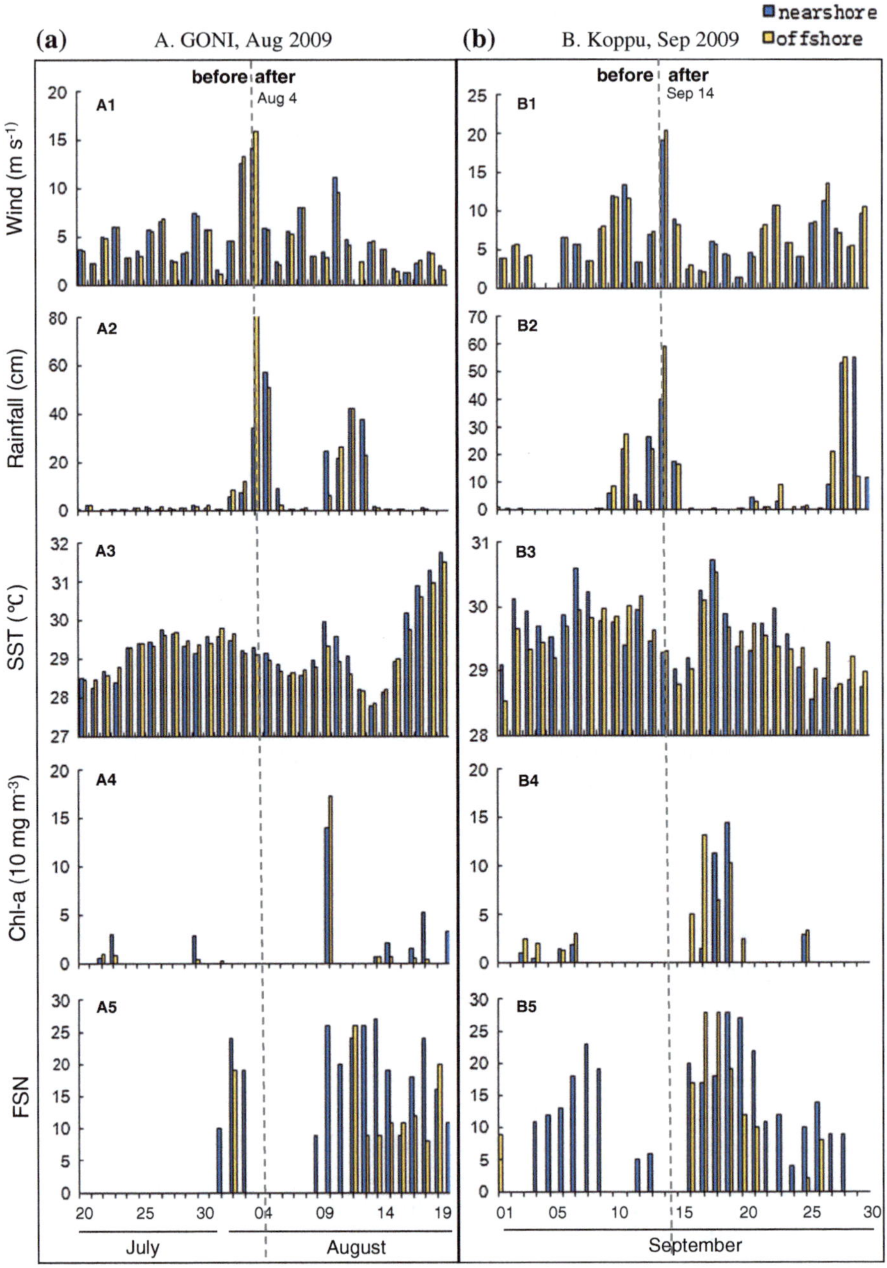

Fig. 14.6 Time series of wind (ms^{-1}), rainfall (cm), SST (°C), Chl-a (mg m^{-3}) and FSN. **a** During typhoon GONI from July 20th–August 20th (*A1*, *A2*, *A3*, *A4* and *A5*). **b** During typhoon Koppu from September 1st–30th (*B1*, *B2*, *B3*, *B4* and *B5*). *Blank bar* represents the nearshore changes, *black bar* denotes the offshore distribution, *gray dashed lines* indicate the occurrence of typhoons

Nearshore, small and medium-sized fish species were dominant in the increased records. Offshore, however, small fish species were dominant in the increased records (Table 14.4).

14.4.3 Difference in the FSN After Two Typhoons

GONI and Koppu caused different changes in the aquatic environment and FSN. Nearshore and offshore, Koppu brought stronger wind, less rainfall, a shorter duration in the FG and a smaller decrease in the SST than GONI. In contrast, the increase in the FSN after Koppu was greater than that after GONI. Above all, a more powerful typhoon, Koppu, caused less rainfall and a decrease in the SST. These factors indicated that fewer nutrients were imported through river discharge and upwelling during Koppu than during GONI. A greater increase in the FSN after Koppu indicated that other factors may also affect the movement of fishes.

According to the Chinese fishing moratorium for the SCS, fishing activities were forbidden from May 15th to July 30th in 2009 and 2010. An associated increase in fish resources after the fishing moratorium has been documented (Shi et al. 2008). Because of the fishing moratorium, the FSN before GONI (on August 2nd) was 24 nearshore and 19 offshore, while before Koppu, it was 19 (on September 9th) nearshore and 9 (on September 1st) offshore. After GONI, the FSN was 27 nearshore (on August 14th) and 26 offshore (on August 12th). After Koppu, the FSN was 28 nearshore (on September 19th) and 28 (on September 17th) offshore. Therefore, before GONI, the FSN was larger than before Koppu, but after the two typhoons, the difference in the FSN was small. The increase in the FSN after GONI was smaller than that after Koppu.

14.4.4 Possible Ecological Mechanisms

This research described the variation in the FSN near the PRE and offshore, far from the PRE, during the typhoons GONI and Koppu. The influence of the typhoons on fish in the nearshore and offshore areas is depicted in Fig. 14.7.

1. In the atmosphere: When the typhoons hit, rainfall and wind are the two main factors causing environmental changes in the water.
2. In the ocean: Rainfall can cause nutrient enrichment via two mechanisms: atmospheric deposition and river discharge, which bring nutrients into the water. Wind is the other key factor that increases the nutrient concentrations by bringing the nutrient-rich water to the surface. Chl-a increases during the process of these two nutrient enrichment mechanisms.
3. Fish: Schools of fish accumulate in two ways. Active feeding behavior encourages fishes to congregate in the nutrient-enriched area that is formed by

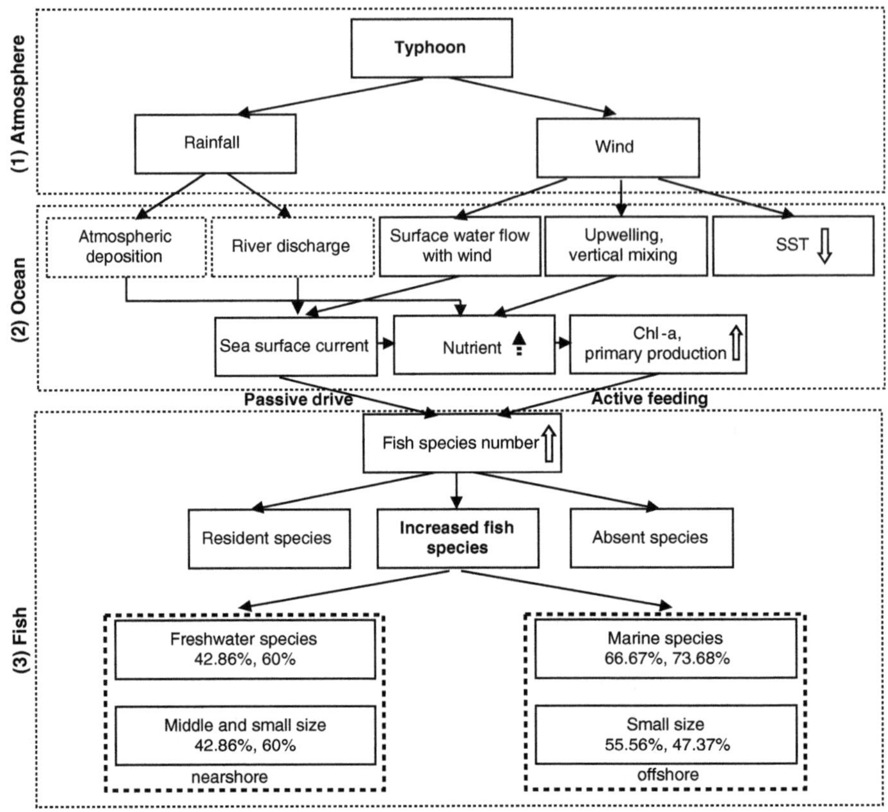

Fig. 14.7 Mechanism of the increase in the FSN after typhoon in the northern SCS. *Blank up* and *down arrows* indicate increases and decreases, respectively, according to the results of this research; *Black arrow* with tail in *dashed line* indicates increases according to references; *Black arrow* indicates correlation between every two boxes; Boxes with *dashed borders* in (*2*) Ocean indicate that it is not discussed in this research; Boxes with *dashed borders* in (*3*) Fish represent results from nearshore and offshore

the typhoons. Due to the typhoons, the strong river discharge and sea surface currents drive some small, sick and old fish with weak swimming abilities into the fishing region. Nearshore, the increase in the FSN is mainly caused by an increase in estuarine species (largely the small and medium-sized species), which follow the increased river discharge from the typhoon's rainfall. Offshore, the increase in the FSN is mainly caused by shallow sea species (predominantly the small species), which gather to feed.

14.5 Conclusion

1. *The FSN increases in both nearshore and offshore environments*

The FSN increases both nearshore (near the river's estuary) and offshore (near the typhoon's track), exhibiting an enhanced Chl-a that is associated with typhoons. Nearshore, estuarine species dominate the increased records of fish species, while offshore, shallow sea species dominate the increased records of fish species.

2. *The effect on the FSN was stronger offshore than nearshore*

The FSN increases more and faster in offshore (near the typhoon's track) than in nearshore (near the river's estuary). This finding indicates that upwelling plays a more critical role close to the typhoon's track, while river discharge is the major factor in the estuary.

3. *Increased records of fish species* are predominantly small and medium-sized in nearshore environments and small in offshore environments.

Nearshore (near the river's estuary), increased records of fish species are mainly small and medium-sized estuarine species that can overcome the strong river discharge. Offshore (near the typhoon's track), increased records of fish species are mainly small, shallow sea fish species.

Acknowledgments This research was supported by: (1) National Natural Sciences Foundation of China (31061160190, 40976091, NSFC–RFBR project awarded to DL Tang and Dmitry Petrenko), (2) Guangdong Natural Science Foundation, China (8351030101000002, 2010B031900041), (3) Innovation Group Program of State Key Laboratory of Tropical Oceanography (LTOZZ1201) (4) National finance special project program (Dynamic acquisition about information of oceanic fish catch in South China Sea), (5) Science and Technology Planning Project of Guangdong Province, China (2010B030800008, 2011B031100001), (6) The Special Project for the Social Common wealth Research of the National Science Research Institute (2010ZD01, 2011YD03). The OSCAR Project Office provided sea surface current data. Thanks to Yongsong Qiu, Zuozhi Chen, Yuezhong Wang and Xuehui Wang for assisting in interpreting the fish data.

References

Adams, A.: Effects of a typhoon on two assemblages of coral reef fishes: multiple-year analysis reverses a false 'snapshot' interpretation. Bull. Mar. Sci. **69**, 341–356 (2001)

Adams, A.J., Ebersole, J.P.: Resistance of coral reef fishes in back reef and lagoon habitats to a Hurricane. Bull. Mar. Sci. **75**, 101–113 (2004)

Andrews, J.D.: Effects of tropical storm Agnes on epifaunal invertebrates in Virginia estuaries. Chesapeake Sci. **14**(4), 223–234 (1973)

Boesch, D.F., Diaz, R.J., Virnstein, R.W.: Effects of tropical storm Agnes on soft-bottom macro benthic communities of the James and York estuaries and the lower Chesapeake bay. Chesapeake Sci. **17**(4), 246–259 (1976)

Chen, Y.Q., Tang, D.L.: Eddy-feature phytoplankton bloom induced by tropical cyclone in the South China Sea. Int. J. Remote Sens. **33**(23), 7444–7457 (2012)

Collins, J.P., Young, C., Howell, J., Minckley, W.L.: Impact of flooding in a Sonoran Desert Stream, including elimination of an endangered fish population (*Poeciliopsis o. occidentalis*, Poeciliidae). Southwest. Nat. **26**(4), 415–423 (1981)

Gautam, R., Singh, R.P., Kafatos, M.: Changes in ocean properties associated with Hurricane Isabel. Int. J. Remote Sens. **26**(3), 643–649 (2005)

Greenwood, M.F.D., Stevens, P.W., Matheson Jr. R.E.: Effects of the 2004 Hurricanes on the fish assemblages in two proximate Southwest Florida estuaries: change in the context of interannual variability, Estuaries Coasts **29**(6A), 985–996 (2006)

Guo, W.D., Yang, Y.P., Gong, Z.W., Hu, M.H.: Variation of salinity, pH, dissolved oxygen and COD in Xiamen Bay during typhoon and rainstorm process. Mar. Sci. **25**(7), 1–5 (2000)

Hagy, J.D., Lehrter, J.C., Murrell, M.C.: Effects of hurricane Ivan on water quality in Pensacola Bay Florida. Estuaries Coasts **29**(6), 919–925 (2006)

Hernández, L., Balart, E.F., Reyes-Bonilla, H.: Effect of Hurricane John (2006) on the invertebrates associated with corals in Bahía de La Paz, Gulf of California. In: Proceedings of the 11th International Coral Reef Symposium, Ft. Lauderdale, Florida, 7–11 July (2008)

Houde, E.D., Bichy, J., Jung, S.: Effect of Hurricane ISABEL on fish production and communities in Chesapeake Bay. In: Sellner, K.G. (eds.) Hurricane Isabel in Perspective, pp. 193–199. Chesapeake Research Consortium, CRC Publications 05–160, Edgewater, Maryland (2005)

Ji, G.L., Gao, T.X., Liu, B.Z.: Morphological differences and isozyme variations between Lophius litulon of Yellow Sea and Japan Sea. Mar. Fish. Res. **28**(3), 73–79 (2007)

Jury, S.H., Howell, W.H., Watson, W.H.: Lobster movements in response to a Hurricane. Mar. Ecol. Prog. Ser. **119**, 305–310 (1995)

Kaufman, L.S.: Effects of Hurricane Allen on reef fish assemblages near Discovery Bay Jamaica. Coral Reefs **2**(1), 43–47 (1983)

Kundu, S.N., Sahoo, A.K., Mohapatra, S., Singh, R.P.: Change analysis using IRS-P4 OCM data after the Orissa super cyclone. Int. J. Remote Sens. **22**(7), 1383–1389 (2001)

Lei, Q.C., Fan, N.C., Zheng, C.W.: A primary observation on the morphological characteristics of embryos, larvae and juveniles on *Nibea albiflora* (*Richardson*). Mar. Fish. Res. **2**, 77–84 (1981)

Letourneur, Y., Harmelin-Vivien, M., Galzin, R.: Impact of hurricane "Firinga" on fish community structure on fringing reefs of Reunion Island, S.W Indian Ocean. Environ. Biol. Fish. **37**, 109–120 (1993)

Li, Y.Z., Chen, G.B., Sun, D.R.: Analysis of the composition of fishes in the Pearl River estuarine waters. J. Fish. Chin. **4**, 312–317 (2000)

Lin, I., Liu, W.T., Wu, C.: New evidence for enhanced ocean primary production triggered by tropical cyclone. Geophys. Res. Lett. **30**, 1718 (2003)

Locascio, J.V., Mann, D.A.: Effects of Hurricane Charley on fish chorusing. Biol. Lett. 22, **1**(3), 362–365 (2005)

Lohrenz, S.E., Cai, W.J., Chen, X., Tuel, M.: Satellite assessment of bio-optical properties of northern Gulf of Mexico coastal waters following Hurricanes Katrina and Rita. Sensors **8**, 4135–4150 (2008)

Ma, C.H., You, K., Chen, D.G., Li, F.Q.: The study on the fishery resource and fishery divisions in the South China Sea. Trans. Oceanol. Limnol. **2**, 127–134 (2007)

Montane, M.M., Austin, H.M.: Effects of Hurricanes on Atlantic croaker (Micropogonias undulatus) recruitment to Chesapeake Bay. In: Sellner, K.G. (ed.) Hurricane Isabel in Perspective, pp. 185–192. Chesapeake Research Consortium, CRC publication 05–160, Edgewater, Maryland (2005)

Ou, Y.J.: Artificial breeding technique rockfishes. Ocean Fish. **3**, 14–15 (2009)

Paperno, R., Tremain, D.M., Adams, D.H., Sebastian, A.P., Sauer, J.T., Dutka-Gianelli, J.: The disruption and recovery of fish communities in the Indian River lagoon, Florida, following two Hurricanes in 2004. Estuaries Coasts **29**(6A), 1004–1010 (2006)

Pizza, B.P., Peyre, M.K.L.: The effect of Hurricane Katrina on nekton communities in the tidal freshwater marshes of Breton sound Louisiana, USA, Eutuarine. Coast. Shelf Sci. **83**(1), 97–104 (2009)

Shi, Y.R., Li, Y.Z., Sun, D.F., Lu, W.H.: Effects of a ten-year summer closed fishing on fishery resource, ecological protection, commercial value and social influence in the South China Sea. J. Fish. Sci. Chin. **9**, 14–16 (2008)

Singh, R.P., Bhoi, S., Sahoo, A.K.: Significant changes in chlorophyll concentration after Gujarat earthquake. Curr. Sci. **80**(11), 1376–1377 (2001)

Singh, R.P., Cervone, G., Kafatos, M., Prasad, A.K., Sahoo, A.K., Sun, D., Tang, D.L., Yang, R.: Multi-sensor studies of the Sumatra earthquake and tsunami of 26 December 2004. Int. J. Remote Sens. **28**(13–14), 2885–2896 (2007)

Sun, L., Yang, Y.J., Xian, T., Lu, Z., Fu, Y.F.: Strong enhancement of chlorophyll a concentration by a weak typhoon. Mar. Ecol. Prog. Ser. **404**, 39–50 (2010)

Switzer, T.S., Winner, B.L., Dunham, N.M., Whittington, J.A., Thomas, M.: Influence of sequential hurricanes on nekton communities in a southeast Florida estuary: short-term effects in the context of historical variations in freshwater inflow. Estuaries Coasts, **29**(6A), 1011–1018 (2006)

Tang, D.L., Kawamura, H., Doan-Nhu, H., Takahashi, W.: Remote sensing oceanography of a harmful algal bloom (HAB) off the coast of Southeastern Vietnam. J. Geophys. Res. (Ocean) **109**, 1–7 (2004a)

Tang, D.L., Kawamura, H., Dien, T.V., Lee, M.A.: Offshore phytoplankton biomass increase and its oceanographic causes in the South China Sea. Mar. Ecol. Prog. Ser. **268**, 31–41 (2004b)

Tang, D.L., Yan, Z.Z., Zhao, H.: Remote sensing of ocean hazard influences on marine ecosystem. J. Biotechnol. **136**(Supplement), 1 (2008). doi:10.1016/j.jbiotec.2008.07.1227:519-526

Tang, D.L., Zhao, H., Satyanarayana, B., Zheng, G.M., Singh, R.P., LV, J.H.: Enhancement of chlorophyll-a in the Northeastern Indian Ocean after the 2004 South Asian Tsunami. Int. J. Remote Sens. **30**(17), 4553–4565 (2009)

Walker, N.D., Leben, R.R.: Hurricane-forced upwelling and chlorophyll a enhancement within cold-core cyclones in the Gulf of Mexico. Geophys. Res. Lett. **32**(18), 1–5 (2005)

Wang, D., Lin, Z.J.: Spatial and temporal variations of fish community structure in the Pearl River Estuary waters. S. Chin. Fish. Sci. **2**(4), 37–45 (2006)

Yan, Z.Z., Tang, D.L.: Changes in Suspended sediments associated with 2004 Indian Ocean Tsunami. J. Adv. Space Res. **43**, 89–95 (2008)

Yang, X.X., Tang, D.L.: Location of sea surface temperature cooling induced by Typhoons in the South China Sea. J. Tropical Oceanogr. Chin. **29**(4), 26–31 (2010)

Yang, L., Zhang, X.F., Tan, Y.G., Zhang, P.: Selectivity study on minimum mesh size of bottom trawl codend in South China Sea. J. Fish. Sci. Chin. **10**(4), 325–332 (2003)

Yang, L., Zhang, X.F., Zhang, P., Tan, Y.G.: Composition of by-catch of shrimping beam trawl in the Pearl River estuary China. S. Chin. Fish. Sci. **1**(1), 27–34 (2005)

Yu, J., Tang, D.L., Li, Y.Z., Huang, Z.R., Chen, G.B.: Increase in fish abundance during two typhoons in the South China Sea. Adv. Space Res. **51**, 1734–1749 (2013)

Zhao, H., Tang, D.L., Wang, Y.: Comparison of phytoplankton blooms triggered by two typhoons with different intensities and translation speeds in the South China Sea. Mar. Ecol. Prog. Ser. **365**, 57–65 (2008)

Zhao, H., Tang, D.L., Wang, D.X.: Phytoplankton blooms near the Pearl River estuary induced by Typhoon Nuri. J. Geophys. Res. **114**, 1–9 (2009)

Zheng, G.M., Tang, D.L.: Offshore and nearshore chlorophyll increases induced by typhoon winds and subsequent terrestrial rainwater runoff. Mar. Ecol. Prog. Ser. **333**, 61–72 (2007)

Part IV
Typhoon Impacts and Global Changes

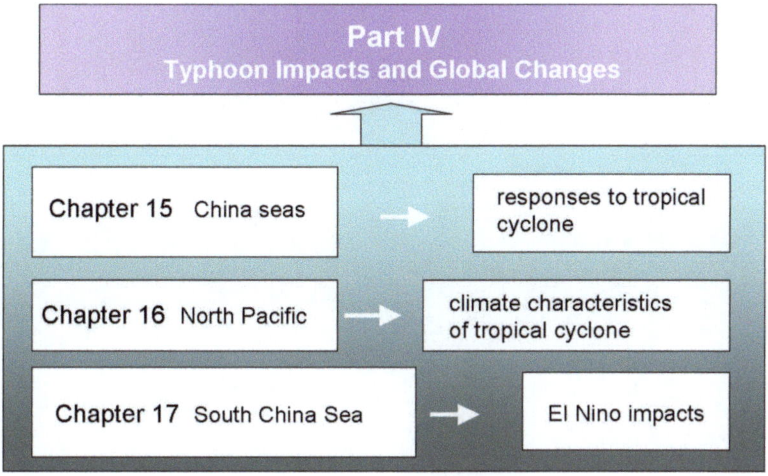

Chapter 15
Responses of the China Seas to Tropical Cyclone

Chau-Ron Wu

Abstract In this chapter the ocean responses to tropical cyclone (TC) in the China seas are discussed. The powerful Kuroshio, mesoscale eddy activity, and complicated bathymetry in the area modulate the responses to the passage of TCs. Upwelling, entrainment and near-inertial oscillation trailed behind TCs are well simulated by ocean models. Sea surface temperature (SST) cooling caused by upwelling and vertical mixing is observed using satellite observations and three-dimensional models. Observational and high-resolution model analyses are carried out to study the favorable conditions and relevant physical processes that cause the SST cooling. The result demonstrates that upwelling dominates vertical entrainment in producing the surface cooling for a subcritical storm. Further, surface chlorophyll-a concentration often increases significantly after TC's departure. Upwelling and mixing bring nutrient-rich subsurface water to the sea surface, causing enhancement of phytoplankton bloom.

Keywords Air-sea interaction · Tropical cyclone · Kuroshio · Mesoscale eddies · Ocean model · SST cooling · Upwelling · Vertical entrainment · Ocean thermal structure · Propagation speed

15.1 Introduction

Western North Pacific Ocean is the entity that receives the most tropical cyclone (TC) events on earth. The Kuroshio, a western boundary current, is the main current in the area. It carries large amounts of heat, salt, and mass polewards from the tropical ocean and plays an important role in the global climate. TC intensification

C.-R. Wu (✉)
Department of Earth Sciences, National Taiwan Normal University, Taipei, Taiwan
e-mail: cwu@ntnu.edu.tw

D. L. Tang and G. J. Sui (eds.), *Typhoon Impact and Crisis Management*,
Advances in Natural and Technological Hazards Research 40,
DOI: 10.1007/978-3-642-40695-9_15,

over the strong current, such as the Loop Current and Gulf Stream has been recognized and studied. E.g. Shay et al. (2000) noted that sudden intensification may occur when TCs pass over the warm Loop Current or warm eddy in the Gulf of Mexico. On the other hand, less is known about the effects of a typhoon on the powerful Kuroshio in Pacific, the counterpart of the Gulf Stream in Atlantic. Here an unusual TC is selected to demonstrate the interactions between typhoon and the Kuroshio.

The track of TC Nari (Sep/6-16/2001) is unique to the region. Nari was a moderate category-3 typhoon on the Saffir-Simpson scale. The storm's minimum central pressure was 944 hPa and its maximum wind speed (MWS) was 50 m/s. The storm formed off the northeastern Taiwan at ~25.0°N, 124.7°E on Sep/06/00Z (Fig. 15.1). It first moved eastward following a mid-latitude atmospheric through, turning northwestward and westward after passing Okinawa Island as another typhoon (Danas) approached southern Japan. On Sep/09/00Z, it turned southward and then southeastward to complete a loop. Nari lingered for about 2 days around 26.4°N, 127°E before proceeding northwestward. After Danas moved eastward away from Japan and a weak high pressure system developed over central China,

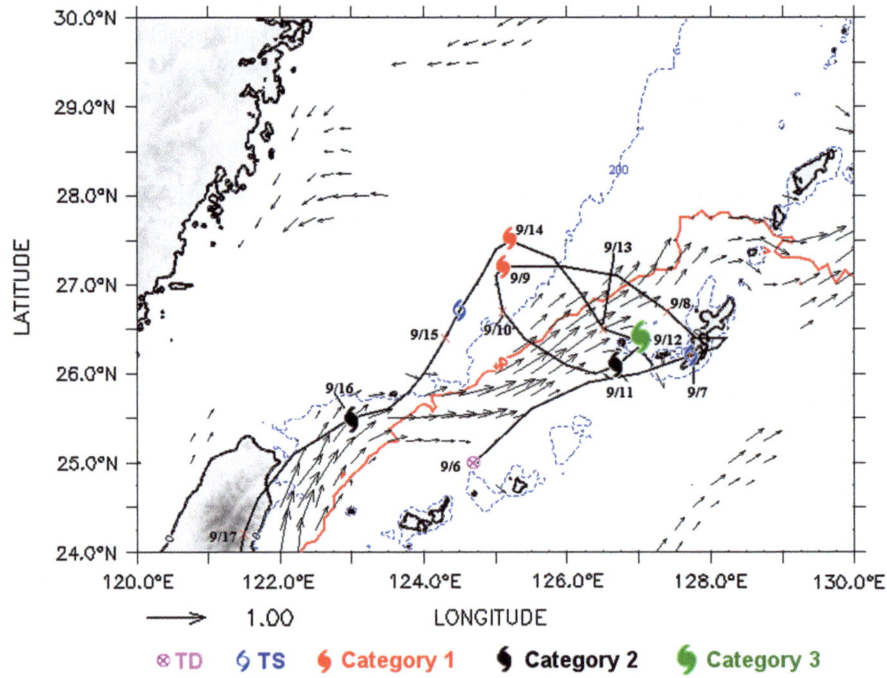

Fig. 15.1 Track and intensity of Nari (*thick-black line*) together with the modeled Kuroshio velocity field (vectors) where the magnitude larger than 40 cm/s is presented. *Thick-red contour* indicates OHC = 50 kJ/cm^2. The 200 m isobath is also overlaid (*blue contours*)

Nari began to move slowly southwestward and finally made landfall at the northern Taiwan coast on Sep/16/18Z (Wu et al. 2008).

Nari remained within a restricted area of the southeastern East China Sea (ECS) for a long 11-day period. Often a near-stationary typhoon weakens and dissipates because of the lack of the energy supplied by the ocean as the sea surface temperature (SST) cools (Schade and Emanuel 1999). Nari vacillates about the Kuroshio for about one week. As Fig. 15.1 shows, there appears to be an interesting interaction between the storm and the Kuroshio front. Before Nari's arrival, the southern ECS is characterized by warm SST above 29 °C: this was one of the warmest summers on record based on satellite observations. In addition, Nari intensified as it crossed over the Kuroshio onto the shelf on Sep/7–9, and continued to strengthen as the storm crossed back to the Kuroshio (Sep/10–12). However, on its second crossing onto the shelf, Nari weakened (Sep/13–15). The wobbling of Nari in the vicinity of Kuroshio appears to cause variations in the typhoon intensity.

The interplay between the Kuroshio and TC aside, several model experiments are carried out to study the favorable conditions and relevant physical processes that cause the SST cooling. Cold SSTs in the wake of TC are well known. The combined action of upwelling and entrainment (vertical mixing) induced by TC brings deep cold waters to the sea surface, producing cold patches with SST's several degrees cooler than the ambient waters. The maximum SST drop caused by TCs could be as much as 5–6 °C, but rarely exceeds 6 °C (Wentz et al. 2000). In July 2000, however, an unusually intense surface cooling of about 10.8 °C was observed in the northern south china sea (SCS). The extreme cooling was induced by a relatively weak typhoon, Kai-Tak (July 3–11, 2000)—a category-1 typhoon on the Saffir-Simpson hurricane scale with a maximum wind speed of 38.6 m s^{-1} and a minimum central pressure of 965 hPa.

Kai-Tak formed off the northwestern Luzon Island at ~118.1°E, 15.7°N on July 3, 2000. It moved northeastward and turned northwestward near 120.9°E, 18.8°N on July 5. During July 6–8, it was impeded by typhoon Kirogi (July 2–July 9, 2000), a fast, northward-moving category-4 typhoon formed approximately 1,000 km east of Philippines. From July 6–8, Kai-Tak lingered and traced an anticlockwise loop around 119°E, 20°N. Kai-Tak began to move rapidly northward after July 8 and finally made landfall at the eastern coast of Taiwan on July 9. The minimum SST after typhoon was 20.3 °C at 119.1°E, 20.1°N, while the SST at the same location before typhoon was 31.1 °C, leading to a maximum SST drop of −10.8 °C. The area of intense surface cooling (SST drop >9 °C) is about 1 × 1° (118.8–119.8°E, 19.5–20.5°N).

Price (1981) suggested that entrainment is the primary mechanism that decreases the SST. In addition, for a slow-moving typhoon, strong upwelling may also reduce the SST. In the case of Kai-Tak, the maximum Ekman pumping estimated from the QuikSCAT/NCEP blended wind on July 7, is large, about 10^{-3} m s^{-1}. However, wind intensity and propagation speed alone cannot account for the drastic SST drop induced by Kai-Tak. Thus, a fine-resolution SCS model driven by satellite wind product is used to understand the various mechanisms that could cause the intense SST cooling by Kai-Tak (Chiang et al. 2011).

15.2 Satellite Observations and Model Description

The intensity and track of Typhoons Nari and Kai-Tak are obtained from the best-track data from the joint typhoon warning center (JTWC). The cloud-penetrating SST from TRMM/TMI is used to study the upper ocean response to TCs. Satellite data also include NASA QSCAT ocean surface wind vectors. Chlorophyll-a concentration (Chl-a) is obtained from the SeaWiFS sensor (Sea-viewing Wide Field-of-view Scanner). The gridded geostrophic velocity (GSV) is from the archiving, validation and interpretation of satellite oceanographic data (AVISO).

The East Asian marginal seas (EAMS, domain is 99–140°E and 0–42°N) model used here is based on the Princeton Ocean Model (POM) (Mellor 2004) with realistic topography and forcing at a horizontal resolution of 1/8° embedded in a north Pacific model at 1/4° resolution, also using POM. The EAMS model was continuously run for 26 years from 1980 through 2005, and the period covering typhoon Nari is used in the present study. Additional information and validation on the EAMS model is given by Wu and Hsin (2005) and Hsin et al. (2008). The vectors in Fig. 15.1 are the modeled surface velocity fields averaged over Nari's duration from September 6 to 16, 2001. For clarity, only velocities larger than 40 cm/s are presented.

The SCS model used in this study is also based on the sigma-coordinate POM. The model domain extends from 99 to 124°E in longitude and from 2 to 27°N in latitude. The horizontal grid size is 1/16° and there are 26 sigma levels in the vertical. A larger-scale EAMS model is used to specify the open boundary condition of the SCS model at daily time interval. The SCS model is driven by the 6-hourly 0.5° $\times$ 0.5° QuikSCAT/NCEP blended wind product (Milliff et al. 1999) and is nudged at the sigma grid nearest to the sea surface by the weekly one-degree advanced very high resolution radiometer (AVHRR) SST data. The SCS model has been validated by the observed temperature from a long-term mooring in the northern SCS as well as current velocity data from several mooring stations in the SCS. These validations and more detailed descriptions of the SCS model are given in Wu and Chiang (2007) and Chiang et al. (2008). The SCS model simulation period is from 1992 to 2005. For this study, we use the results from July 1, 2000 and July 1, 2001 as initial conditions in the later experiments. More details of these experiments will be given below. The control run (CTL, see Table 15.1) is initialized from July 1, 2000 of the SCS model, and continued for one month with 6-hourly wind forcing but without SST-nudging to simulate the ocean response due to typhoon Kai-Tak.

Table 15.1 List of experiments

Case	Wind forcing	Thermal structure	Note	SST drop (°C)
CTL	2000 (Kai-Tak)	2000-July	Control run	10.3
EX1	2000 (Kai-Tak)	2000-July	$K_M = 0.003$ and $K_H = 0.0037\ \mathrm{m^2\ s^{-1}}$	6.4
EX2	2000 (Kai-Tak)	2000-July	1D	3.2
EX3	2000 (Kai-Tak)	2001-July	–	5.9
EX4	2000 (Kai-Tak)	WOA01-June	–	7.2
EX5	2001 (Utor)	2000-July	–	6.7
EX6	2001 (Utor)	2000-July	1-day stop on July 5	9.5

15.3 The Interaction Between the Kuroshio and Nari

15.3.1 Effects of Ocean on Nari

The apparent effect of Kuroshio and adjacent shelf waters on Nari is examined using TMI/SST images (Fig. 15.2) from Sep 7 to 17, 2001. The modeled surface velocity fields averaged over Nari's duration are overlaid on the SST images. In pre-typhoon conditions (Sep/7), SST contours roughly parallel the Kuroshio path and are oriented southwest-northeast with lower temperatures near the Chinese coast and higher temperature offshore. A tropical depression off northeastern Taiwan moved northeastward along the Kuroshio. It turned eastward and became a tropical storm near Okinawa Island (Fig. 15.2a). It strengthened and became a category-1 typhoon on Sep/8 as it traversed northwestward over the warm Kuroshio (Fig. 15.2b). Nari strengthened further while it moved southeastward and crossed the Kuroshio again; it slowed down and became a category-3 typhoon on Sep/11/12Z (Fig. 15.2c). Nari weakened and became a category-1 typhoon on Sep/13 when it approached a cold dome, centered at ~27.2°N, 125°E, which was apparently due to upwelling produced by the storm during its first encounter with the shelf water (Fig. 15.2d). On Sep/14/18Z, Nari weakened further and became a tropical storm at ~27°N, 124.6°E (Fig. 15.2e). Subsurface cold waters were entrained up to the sea surface through upwelling and mixing, and the process leads to a decrease in SST that can weaken Nari. The maximum SST cooling occurred on Sep/17 (Fig. 15.2f). The Nari-induced cold SST patch (123.5–126°E, 26.6–29°N) has the dimension of around 250–300 km. The minimum SST of 24.5 °C is found in the cold pool center (125°E, 28°N), increasing outwards to ~27.5 °C around the periphery. Nari strengthened into a category-2 typhoon as it moved southwestward and passed over the warm Kuroshio again on its way to Taiwan (Fig. 15.2f).

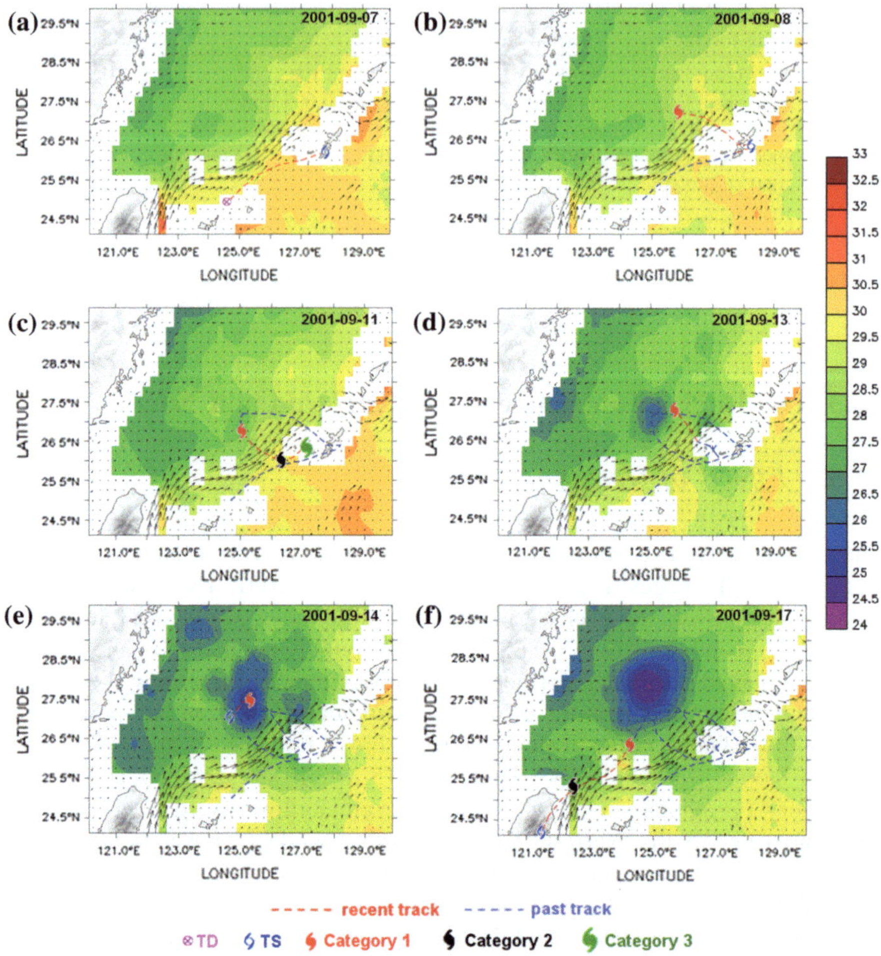

Fig. 15.2 The sequential representative TMI/SST images during the passage of Nari. Typhoon track and modeled Kuroshio averaged over Nari's duration are overlaid on **a** Sep/07, **b** Sep/08, **c** Sep/11, **d** Sep/13, **e** Sep/14, and **f** Sep/17

15.3.2 Oceanic Response Induced by Nari

The oceanic response induced by Nari is investigated using the model simulation. Although the counter-clockwise rotating winds favor divergence of the surface water and upwelling of cooler water from the subsurface, the SST cooling does not occur throughout the trail of Nari. A cyclonic eddy is well developed only when Nari is located north of Kuroshio (e.g. on Sep/08/18Z, Sep/09/00Z, Sep/13/12Z, and Sep/14/12Z). In contrast, the cyclonic circulation is diffused, and does not

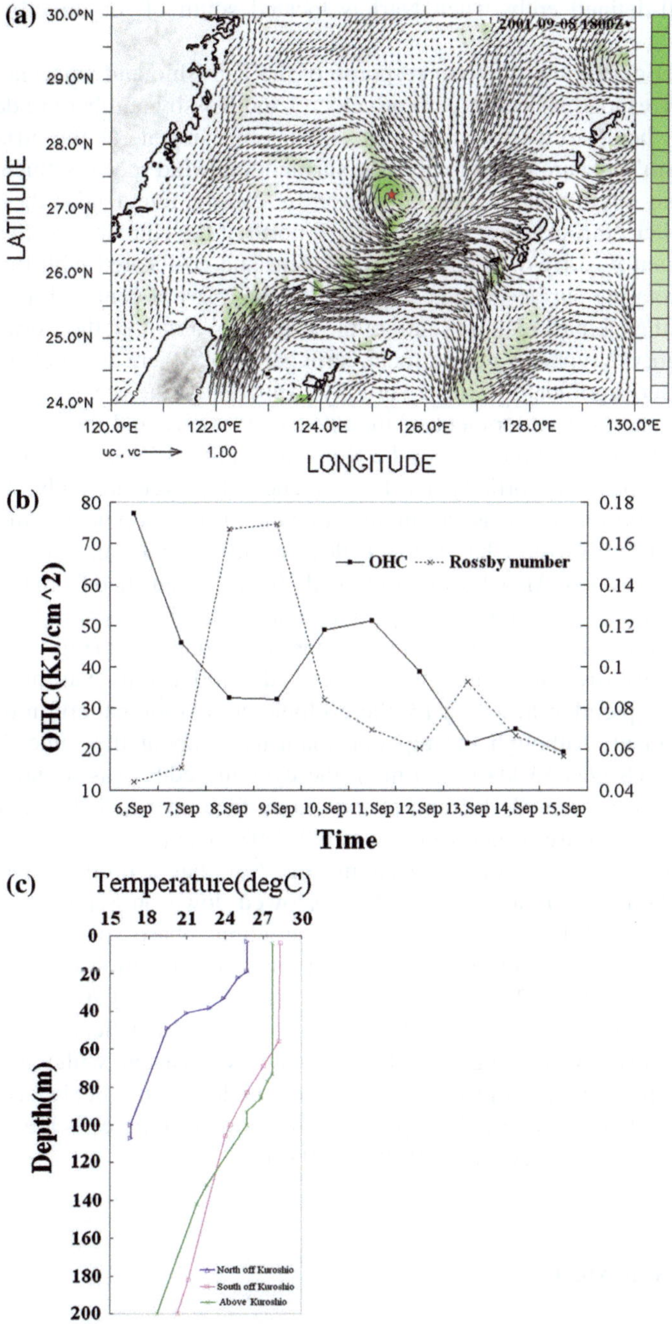

Fig. 15.3 **a** Current velocity vectors and Rossby number (ζ/f, positive values only are shown, *green shading*) on Sep/08/18Z, **b** the daily OHC (*solid squares*) together with Ross by number (*cross*) during Nari's passage. **c** GTSPP temperature profiles from shelf waters north of Kuroshio (29°N, 125.99°E), south of Kuroshio (25.9°N, 128.15°E), and above Kuroshio (28.16°N, 127.23°E)

form a well-defined eddy when Nari is located south of, or directly over the Kuroshio.

To further clarify the different response of the Kuroshio and adjacent waters to Nari, the corresponding ocean current velocity and Rossby number are depicted in Fig. 15.3a, when Nari was situated north of Kuroshio. In general, the surface winds near the storm's eye are weaker while the surrounding wind speed is much stronger. The cyclonic circulation of current vectors is clearly found near the center of the storm when Nari is at ~27.2°N, 125.5°E north of the Kuroshio on Sep/08/18Z. The significant positive vorticity is also evident and co-located with the counter-clockwise rotating surface winds. On Sep/11/18Z, Nari is centered at ~26.1°N, 127°E near the Okinawa Island south of the Kuroshio; although the storm intensify into a category-2 typhoon with a much stronger MWS there is no clear cyclonic circulation and significant positive vorticity. Apparently, the presence of Okinawa Island prevents the development of the positive vorticity and the formation of a cyclonic eddy. Nari is directly over the Kuroshio on Sep/15/18Z, but there is again no significant positive vorticity and the evidence of a cyclonic eddy. The strong currents and vorticity of the Kuroshio may prevent the development of the cyclonic circulation. The Kuroshio has a much deeper thermocline, and cooling due to upwelling is lessened. Also, heat is continually being replenished from the south by strong advection. These processes together explain why there is no significant SST cooling observed near the Kuroshio axis after Nari's departure (Fig. 15.2f).

The vorticity structure around the cyclonic eddy induced by Nari has also been studied. On September 8, 9, and 13, the cyclonic circulation pattern in the surface ocean is evident. Although the typhoon intensity is about the same these days when the MWS was 65 kts (~35 m/s), the cyclonic eddy was found above the depth of 20 m on Sep/08/18Z and at the sea surface on Sep/13/12Z, whereas it penetrates to the entire ocean depth on Sep/09/00Z. The propagation speed, U, of the storm is responsible for the phenomenon. Nari has a propagation speed of 2.2 m/s on both Sep/08 and Sep/13, but it slowed down on Sep/09 with a propagation speed of 1.4 m/s. The first-mode baroclinic phase speed, C, computed from the observed temperature and salinity profiles from National Oceanographic Data Center (NODC) is 2.2 m/s, so that Nari was subcritical ($U/C < 1$) on Sep/09 and near-critical or supercritical ($U/C > 1$) on Sep/08 and Sep/13. The deeper influence depth on Sep/09 suggested by our model is consistent with the theoretical expectation that localized upwelling is prevalent under a subcritical storm whereas a wave-like pattern of alternating upwelling and downwelling cells are left behind a supercritical storm (Geisler 1970; Price 1981).

15.3.3 Discussion

During the passage of Nari, the significant SST cooling took place only in shelf and slope regions north of the Kuroshio where the water depths are around 100–170 m. The cooling in the upper ocean is in part caused by mixing as well as upwelling of the subsurface cooler water. Figure 15.3b shows daily ocean heat content (OHC) (Leipper and Volgenau 1972) and Rossby number (Ro = ζ/f) near 27.2°N, 125.5°E (indicated by an asterisk in Fig. 15.3a) from the model simulation. The OHC tends to decrease whenever Ro peaks, indicating that significant cooling of near-surface water during those dates (on September 8, 9, and 13). The strongest sea surface cooling is observed to the north of the Kuroshio where Nari had already weakened, which suggests that the significant vertical mixing is not wholly associated with Nari's MWS or initial SST. Rather, the significant cooling is also caused by the shallower thermocline that existed on the shelf prior to Nari. No observed temperature profiles exist during the period of the storm, but from October 6 to 8, 2001, three temperature profiles in the region from GTSPP (Global Temperature-Salinity Profile Program) are available and are shown in Fig. 15.3c. The thermocline is shallow (20–30 m) north of the Kuroshio over the shelf; it is deep (80–100 m) over and south of the Kuroshio. Thus cool waters beneath the thermocline are closest to the sea surface on the shelf, and typhoon-induced vertical mixing would be expected to reduce the SST more effectively than in regions where the thermocline is deeper.

Internal waves were also generated after Nari's departure, as seen from the large vertical displacement of the thermocline simulated by the model. Temporal variation of water temperature at a fixed location (indicated by an asterisk in Fig. 15.3a) is characterized by near-inertial oscillations. The period estimated from the model is ~26 h in agreement with the theoretical value at a latitude of 27.2°N.

The SeaWiFS ocean color sensor has been suggested as a feasible tool in quantifying the contribution from TCs for enhanced ocean primary production. e.g., Lin et al. (2003) indicated that the contribution of TCs to the annual new production in the SCS may reach as much as 20–30 %. The biological response to the passage of Nari is also observed in the present study. The pre-typhoon condition is illustrated in the SeaWiFS image composite (August 5–28) of surface Chl-a with a concentration of about 0.2 mg/m^3 in the study region (Fig. 15.4a). After Nari's passage, a significant increase of surface Chl-a concentration, up to 1.0 mg/m^3 is visible in the SeaWiFS composite (September 6–29) (Fig. 15.4b). The maximum surface Chl-a concentration of 1.0 mg/m^3 occurred on Sep/14 at ~28°N, 125°E. It is centered near the maximum SST cooling, revealed by TMI (Fig. 15.2f). The following sequence of events may be inferred: the large and positive vorticity was associated with the strong counter-clockwise rotating TC surface winds on September 8, 9 and 13 during Nari's arrival. The maximum SST cooling occurred a few days later on Sep/16. The combined action of upwelling and mixing brings cold and nutrient-rich water from deeper layers to the sea surface, and significantly increases the surface Chl-a concentration (Fig. 15.4c).

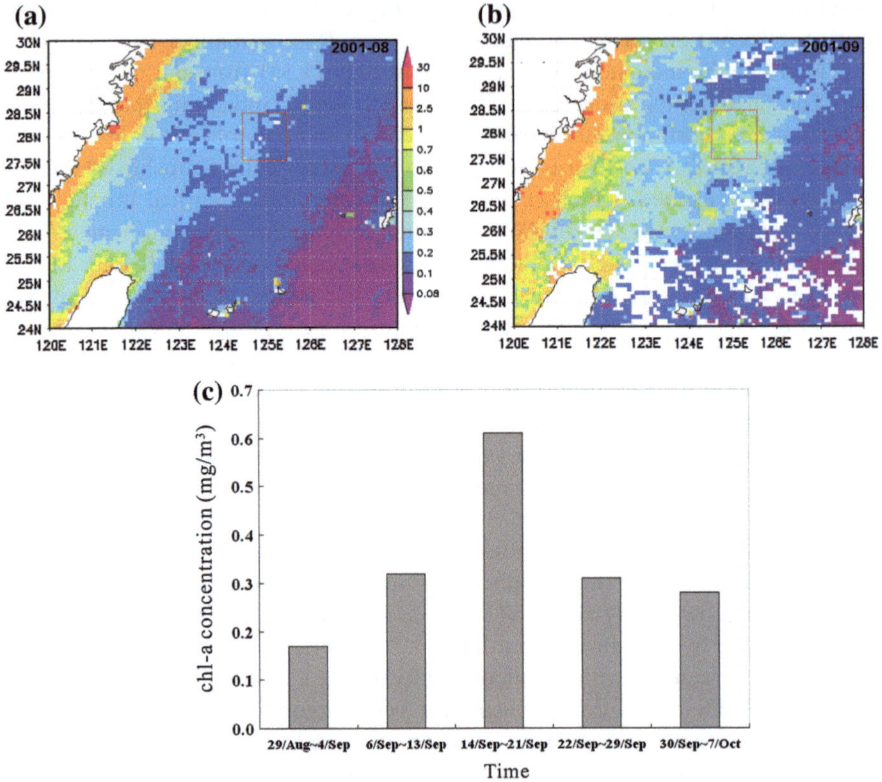

Fig. 15.4 **a** The SeaWiFS composite of Chl-a in mg/m^3 from August 5–28, 2001 before Nari arrival, **b** the SeaWiFS image composite from September 6–29, 2001 after Nari departure, **c** the temporal variations of chlorophyll-a concentration averaged over the *red box* in Fig. 15.4b

15.4 Intense SST Cooling by Kai-Tak

15.4.1 Relative Importance of Upwelling and Entrainment

As Price (1981) suggested, for a slow-moving cyclone, strong upwelling together with entrainment mixing may significantly decrease the SST. The relative importance of upwelling and entrainment in realistic simulations of ocean response to tropical cyclones is rarely estimated due to the inherent difficulty in separating the two processes. Here, two experiments are conducted to evaluate the relative importance of these two processes. Experiment 1 (EX1, see Table 15.1) is an upwelling experiment. We set the values of vertical kinematic viscosity (K_M) and vertical diffusivity coefficient (K_H) to be constants = 0.003 and 0.0037 m^2 s^{-1}, respectively. These are the smallest values chosen to keep the model numerically stable under the strong winds. The chosen K_H-value is about two orders of

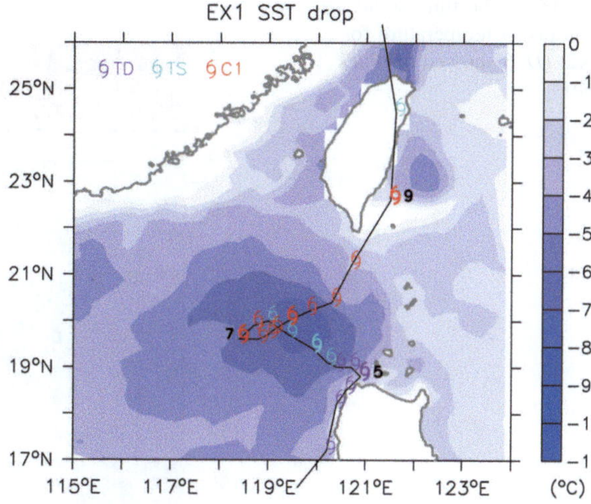

Fig. 15.5 SST drop for EX1 between July 2 and July 9 with Kai-Tak track with selected dates indicated and intensity from the JTWC best-track data

magnitude smaller than typical values found in CTL during Kai-Tak, and ensures that entrainment is small in this experiment. Other model's parameters and forcing remain unchanged. Since Ekman pumping is independent of K_M, this experiment therefore simulates the upwelling component of the cooling process. Other temperature fluctuations caused by horizontal variations, e.g. meso-scale eddies, near-inertial internal waves etc., are also included. However, eddies generally evolve at longer time scales than typhoon-induced ocean response, and inertial waves are reduced by the daily averaging that we apply to the model results. Figure 15.5 shows that the ΔSST pattern in EX1 is similar to that in CTL, but with a weaker magnitude. The maximum ΔSST in EX1 is at the same location as in CTL (~119.4°E, 19.9°N), and is about 6.4 °C. We have repeated this experiment (EX1) by doubling the values of K_M and/or K_H, the results are virtually identical to Fig. 15.5, suggesting that the chosen K_H is sufficiently small, and that the upwelling is largely independent of K_M.

Experiment 2 (EX2, see Table 15.1) is designed to simulate entrainment only. To eliminate upwelling caused by Ekman pumping, a one-dimensional (1-D) numerical model is used. The 1-D model is derived from the three-dimensional (3-D) POM, and the array structure, the vertical positions of variables, and other model parameters, including K_M and K_H, are kept unchanged. The Mellor-Yamada turbulence closure sub-model and the treatment of vertical advection and diffusion are identical with the 3-D POM. Note that the 1-D model has the advantage of not only eliminating effects of upwelling by Ekman pumping, it eliminates also other temperature fluctuations related to horizontal variations, e.g. eddies, near-inertial internal waves etc. In other words, EX2 simulates purely the vertical entrainment and mixing processes. Figure 15.6 shows the time-depth contours of temperature for EX2. The ΔSST in EX2 is about 3.2 °C.

Fig. 15.6 The time-depth contours of temperature for EX2. (*D*26 showed by *thick line*)

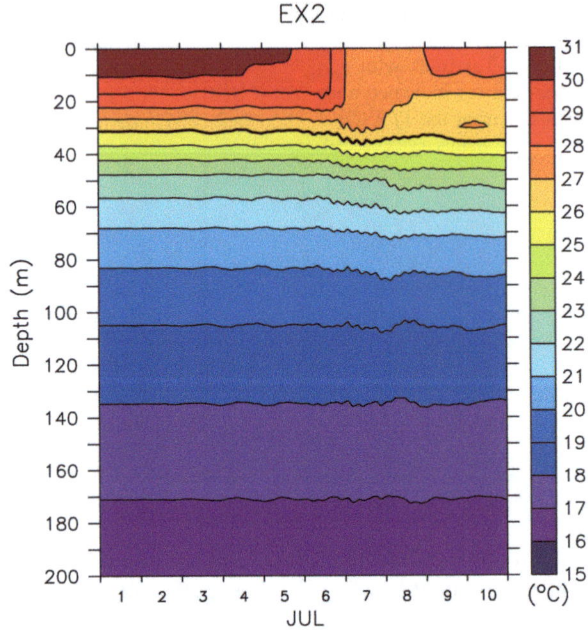

Therefore, 62 % (6.4 °C) and 31 % (3.2 °C) of the SST drop by Kai-Tak are caused by upwelling and entrainment, respectively. In the case of typhoon Kai-Tak, upwelling dominates vertical entrainment in producing the simulated intense surface cooling. The "upwelling experiment" (EX1) and "entrainment experiment" (EX2) together account for 93 % of the total ΔSST. As mentioned above, the two processes interact and there also exist other processes. Nonetheless, the close agreement of the summed (i.e. upwelling + entrainment) contribution to ΔSST to the control experiment's ΔSST indicate that (1) the two contributions are dominant, and that (2) our designs of EX1 and EX2 to separate them are appropriate.

When Kai-Tak was at 119°E, 20°N on July/06/18Z–July/08/00Z, its propagation speed was 0.65–1.96 m s^{-1}. The first-mode baroclinic phase speed computed using the July temperature and salinity profiles from WOA01 is ~2 m s^{-1}, so that Kai-Tak's propagation was subcritical during this period. Localized upwelling is prevalent under a subcritical storm whereas lee waves with alternating upwelling and downwelling cells are left behind a supercritical storm (Geisler 1970). Strong upwelling is often observed under slowly-moving (subcritical) storms (e.g. Price 1981; Oey et al. 2006, 2007). However, a ΔSST of more than 10 °C, such as that occurred after Kai-Tak, is uncommon. There may therefore be other factor(s) involved in the excessive cooling.

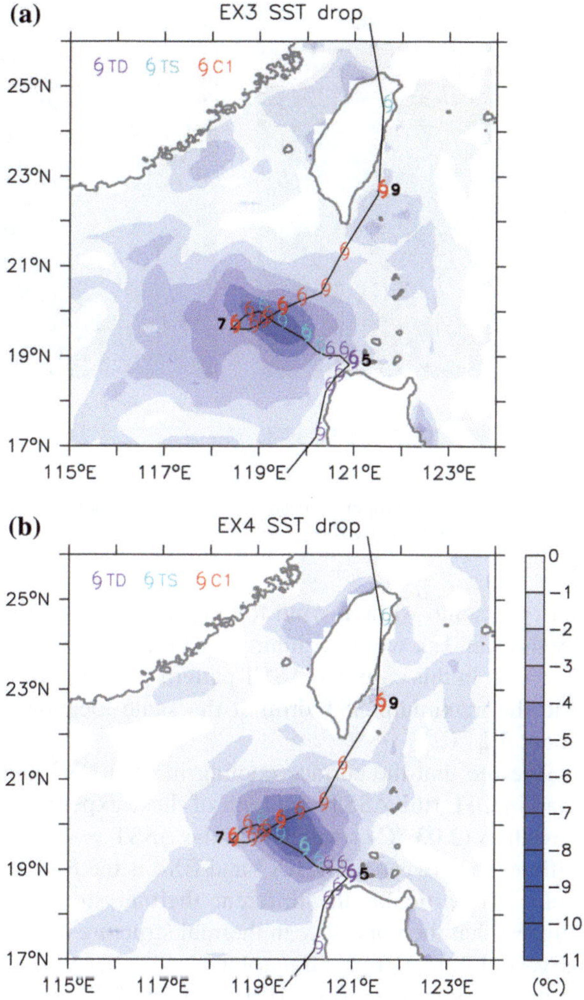

Fig. 15.7 SST drops for **a** EX3 and **b** EX4 between July 2 and July 9 with Kai-Tak track with selected dates indicated and intensity from the JTWC best-track data

15.4.2 Effects of Background Thermal Structure

The excessive cooling is most likely due to the unusual upper-ocean state of summer 2000. Experiments 3 and 4 (EX3 and EX4, see Table 15.1) were therefore conducted to investigate the sensitivity of cooling due to different background fields. Experiment 3 is initialized with the SCS model's July 1, 2001 field (instead of July 1, 2000). Kai-Tak forcing and other model parameters are the same. Figure 15.7a shows that the ΔSST pattern in EX3 is similar to that in CTL, but the maximum ΔSST is weaker, about 5.9 °C.

Experiment 4 is initialized with the June climatological temperature and salinity (World Ocean Atlas 2001, WOA01; Conkright et al. 2002), and spun-up from rest

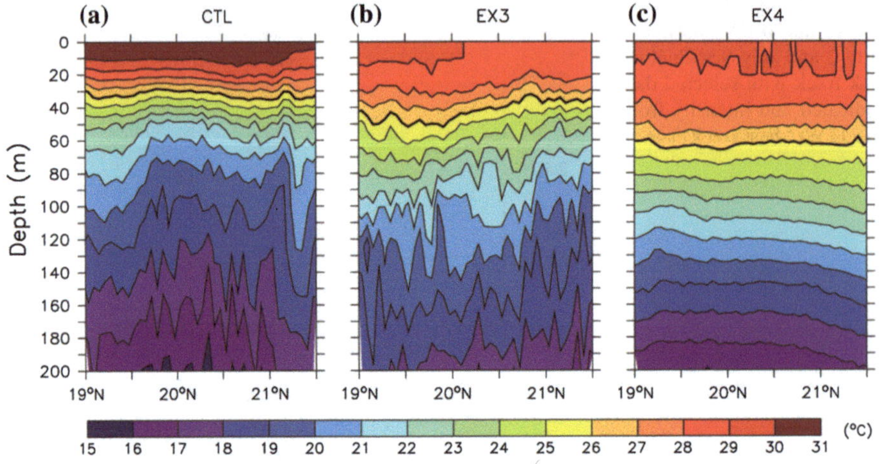

Fig. 15.8 Initial temperature structures in the upper 200 m at 119°E for **a** CTL, **b** EX3, and **c** EX4. (*D*26 showed by *thick line*)

for one month without wind forcing. It is then continued for another month forced by the Kai-Tak wind beginning on June 23, 2000 approximately one week before Kai Tak. In this case, the ΔSST pattern (Fig. 15.7b) is again similar to that in CTL, and the maximum SST drop at the same location as for the CTL experiment is about 7.2 °C.

We see that the surface cooling in both EX3 and EX4 is significantly weaker than in CTL; the ΔSST's in both of these experiments are more than one standard deviation (2.03 °C) smaller than the ΔSST = 10.3 °C of the CTL. Since the only difference between CTL, EX3 and EX4 is the background temperature and salinity fields, in particular the different thermal structures, the above sensitivity tests suggest that the upper-ocean thermal structure during July 2000 (prior to Kai-Tak) is very different from climatology and other years. Figure 15.8 shows the initial temperature structures at 119°E for CTL, EX3, and EX4. In 2000 (Fig. 15.8a), the near-surface temperature is 1–2 °C warmer than in 2001 (Fig. 15.8b) and also than climatology (Fig. 15.8c). The thermocline in 2000 is 20–80 m shallower, so that the corresponding depth of the 26 °C isotherm (*D*26) is the thinnest among the three experiments CTL, EX3 and EX4. Near 119°E, 20°N (maximum SST drop due to Kai-Tak) for example, the 19–20 °C isotherms are at $z \approx 85$ m in CTL, but are depressed to a depth of $z \approx 155$ and 140 m in EX3 and EX4, respectively. For the CTL experiment, the pre-typhoon depth-averaged temperature (0–200 m) over the intense cooling area on July 1, 2000 is 21.0 °C, which is ~2 °C cooler than in 2001 (23.3 °C) and climatology (22.8 °C). It is clear that the shallow thermocline in 2000 allows cooler water to be more easily brought up to the surface, by upwelling or mixing, leading to the excessive surface cooling by Kai-Tak. Because of the correspondingly stronger stratification in summer 2000, mixing may be less effective than Ekman pumping. This is consistent with the results of EX1 and EX2.

15.4.3 Effects of Propagation Speed

Bender et al. (1993) distinguished three groups of SST cooling observed after the passage of 16 tropical cyclones in various regions. The three groups are categorized according to their propagation speeds: slow (<4 m s^{-1}), medium (4–8 m s^{-1}), and fast (>8 m s^{-1}), and the corresponding maximum drops in SST (ΔSST's) are 5.3, 3.5, and 1.8 °C, respectively. The maximum ΔSST's due to various typhoons in the northern SCS fall roughly into Bender et al categories and criteria except Kai-Tak (ΔSST = 10.8 °C).

To examine the effects of propagation speeds, we conduct two experiments both with the same background thermal (and salinity) fields as in summer 2000 (thinner

Fig. 15.9 SST drops for **a** EX5 between July 3 and July 7 with Utor track and intensity from JTWC best-track data and **b** EX6 with 1-day stop track of Utor

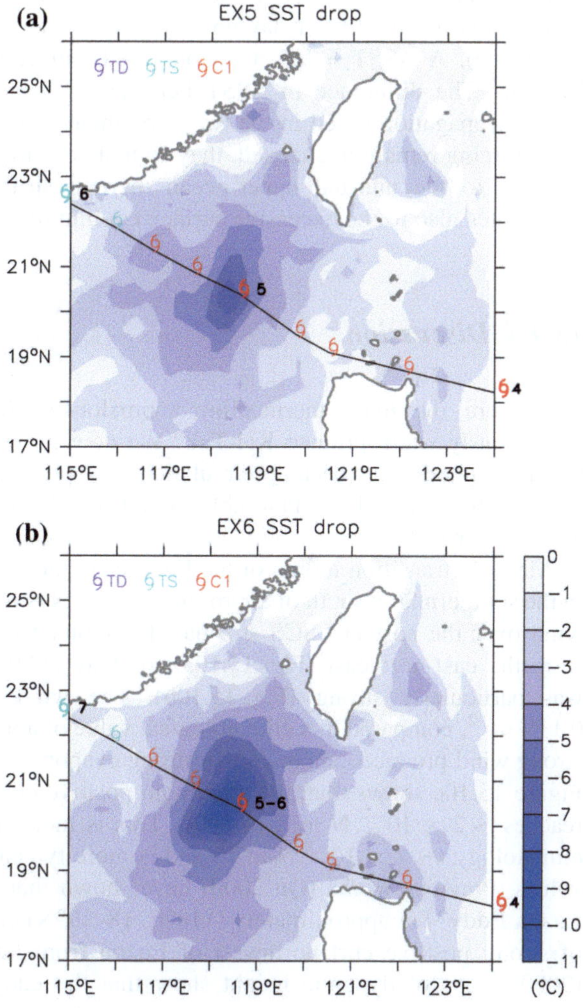

*D*26 and colder seawater beneath) for typhoon Kai-Tak, but we replace the wind field with that of typhoon Utor (June 30–July 7, 2001). Similarly to Kai-Tak, Utor is a category-1 typhoon with a maximum wind speed of 41.2 m s^{-1}. The storm also originated north of Luzon and during the first half of its life followed a northwestward path. It was a faster typhoon however ($U < 8$ m s^{-1}) during July 5–6, 2001, and the maximum SST drop was about 3.8 °C; it therefore belongs to Bender et al. medium-speed category storm. In experiment 5 (EX5), Utor's wind is used, while in experiment 6 (EX6), Utor's propagation speed is artificially slowed (see below). The maximum SST cooling in EX5 is about 6.7 °C (Fig. 15.9a) which is approximately double the Bender's criteria for the medium moving storms (3.5 °C). Therefore, the thinner *D*26 and colder upper-ocean can enhance typhoon-induced SST cooling. However, the SST drop in EX5 is still significantly lower than that induced by Kai-Tak, indicating that there are additional factors involved.

Experiment 6 (EX6, see Table 15.1) is the same as EX5 except that Utor's propagation is halted for 1 day on July 5 when the storm was near (119°E, 20°N; Fig. 15.9b). A ΔSST = 9.5 °C is observed and comparable to that induced by Kai-Tak. The difference in ΔSST between EX5 and EX6 is purely caused by Utor's propagation speed in the two experiments. Since the thermal structure and other forcing remain unchanged, they cannot contribute to the difference. Thus, in addition to the unusual upper-ocean thermal structure, a slow-moving storm contributes also to the excessive surface cooling in summer 2000 during Kai-Tak.

15.4.4 Discussion

Why were the main thermocline anomalously shallow and the sea surface anomalously warm prior to Kai-Tak? We now show how various components of the atmospheric and oceanic circulations in spring to early summer of 2000 fit together to create these favorable conditions for the extreme SST drop when Kai-Tak arrived.

May is a transition month of the East Asian monsoon. The climatological wind in the southern SCS south of approximately 16°N is from the west-southwest while wind over the northern SCS still has the remnant of the winter monsoon and is from the east-northeast (Fig. 15.10a). In May 2000, the east-northeasterly wind was particularly strong (Fig. 15.10b). The wind stress magnitude exceeded 0.1 N m^{-2}, compared to a climatological value of approximately 0.04 N m^{-2}. The strong wind produced anomalously strong evaporative heat loss at the sea surface. Figure 15.10c shows that, the maximum positive curl off the northwestern Luzon reaches $\sim 2 \times 10^{-7}$ N m^{-3} in 2000. This is more than ~5 times stronger than climatology ($\sim 4 \times 10^{-8}$ N m^{-3}) and is more typical of winter (January–March) values. Previous works (Qu 2000) have shown that a cyclonic eddy (the "West Luzon Eddy") at approximately (118°E, 18–19°N) coincides well with the region of strong positive curl northwest of Luzon from late fall to early spring. Qu's (2000) maps of dynamic height show that the eddy appears in December and

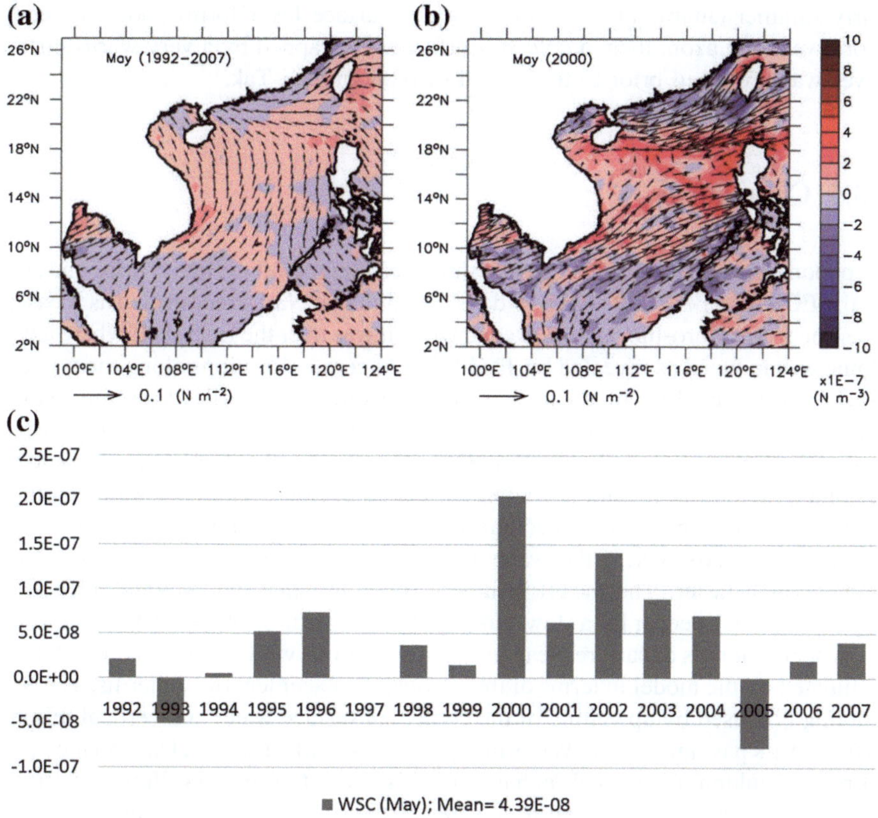

Fig. 15.10 **a** May climatological wind stress vectors superimposed on wind stress curl (WSC) (*color*); **b** same for May, 2000; **c** WSC (N m^{-3}) averaged over 116°–120°E and 16°–20°N for May from 1992–2007

strengthens through the season till April when its center shifts northward to about the 19°N. The shift follows also a northward shift of the wind stress curl, and brings the eddy's center closer to the location of Kai-Tak's maximum ΔSST near (119°E, 20°N). However, because the eddy is large ~300 km, the slight shift is not central to our argument. The important point here is that the positive wind stress curl correlates well with the West Luzon Eddy (Qu 2000). The eddy weakens in May–June. However, the anomalously strong wind stress curl in May of 2000 prolongs the eddy's strength well into June prior to the arrival of Kai-Tak.

In summary, the anomalously strong positive wind stress curl in late spring of 2000 spins up a strong West Luzon Eddy northwest of Luzon. The eddy lifts up isotherms and brings cool water nearer to the surface (Fig. 15.8). At the same time, the strong eddy also transports warm coastal waters west of Luzon northward and westward. Together with increased northward Ekman drift and solar radiation in early summer (June), an anomalously warm surface layer forms above the eddy. Northwest of Luzon, then, a cold subsurface eddy capped by a very warm surface layer was produced prior to the arrival of typhoon Kai-Tak.

15.5 Conclusion

Typhoon Nari circled around the Kuroshio over a region off the outer-slope and shelf of the ECS for a period of about 11 days. We find that Nari strengthened as it passed over the warm Kuroshio, and weakened when it was over the ECS shelf where a cold dome was induced as revealed by TMI/SST. At certain locations along the typhoon track, surface cooling of up to 5 °C is observed after Nari's departure. Model simulation indicates that positive vorticities and cyclonic current vectors do not occur throughout the trail of Nari. Only regions north of Kuroshio provide a favorable condition for the development of the cyclone. The cool water north of Kuroshio is maintained for longer than a week. The conditions are very different near the Kuroshio axis, where the water temperatures are restored to their pre-storm values much faster. The model results also confirm that the cyclonic circulation penetrates much deeper for a slow-moving storm regardless of the typhoon intensity.

After typhoon's departure, near-inertial oscillations with period of about 26 h are simulated by the model in terms of the vertical displacement of isotherms. The SST cooling is caused by upwelling of the cooler subsurface water and vertical mixing. After Nari's passage, the SeaWiFS image composite illustrates a clear enhancement of phytoplankton bloom. This enhancement is a result of the upwelling and mixing which bring subsurface nutrient-rich water to the sea surface (Wu et al. 2008).

By a careful design of model experiments, we successfully determine that 62 and 31 % of the SST drop by Kai-Tak are caused by upwelling and entrainment, respectively. Kai-Tak belongs to a subcritical storm and upwelling dominates entrainment in producing the intense surface cooling. Moreover, because of the shallow thermocline and correspondingly stronger stratification in summer of 2000, mixing contributes less SST drop than Ekman pumping. A unique combination of Kai-Tak's slow propagation speed and favorable climatic and oceanic conditions prior to the storm's arrival contributes to the large ΔSST. The unusually strong northeasterly monsoon wind over the northern SCS in spring of 2000, in combination with southwesterly wind in the southern SCS, spins an anomalously strong cyclonic wind stress curl hence also a strong west Luzon Eddy northwest of Luzon. The thermocline is up-lifted near the eddy's center some 300 km northwest of Luzon, near 119°E, 20°N where Kai-Tak passed. The eddy and Ekman drift by southerly wind in June also transport surface water northward along the Luzon west coast where strong solar radiation and weak winds produces anomalously warm SST's. These climatic and oceanic processes created a cold eddy that was capped by a thin surface layer of anomalously warm water. Kai-Tak was the ocean's perfect storm in passing over the eddy at the "right time", producing the record SST drop and high chlorophyll-a concentration (Chiang et al. 2011).

References

Bender, M.A., Ginis, I., Kurihara, Y.: Numerical simulations of tropical cyclone-ocean interaction with a high-resolution coupled model. J. Geophys. Res. **98** (D12), 23245–23263 (1993). doi: 10.1029/93JD02370

Chiang, T.-L., Wu, C.-R., Chao, S.-Y.: Physical and geographical origins of the South China sea warm current. J. Geophys. Res. **113**, C08028 (2008). doi:10.1029/2008JC004794

Chiang, T.-L., Wu, C.-R., Oey, L.-Y.: Typhoon Kai-Tak: An ocean's perfect storm. J. Phys. Oceanogr. **41**, 221–233 (2011). doi:10.1175/2010JPO4518.1

Conkright, M.E., Locarnini, R.A., Garcia, H.E., O'Brien, T.D., Boyer, T.P., Stephens, C., Antonov, J.I.: World Ocean Atlas 2001: Objective Analyses, Data Statistics, and Figures, CD-ROM Documentation. National Oceanographic Data Center, Silver Spring, MD, p. 17 (2002)

Geisler, J.E.: Linear theory of the response of a two-layer ocean to a moving hurricane. Geophys. Fluid Dyn. **1**, 249–272 (1970). doi:10.1080/03091927009365774

Hsin, Y.-C., Wu, C.-R., Shaw, P.-T.: Spatial and temporal variations of the Kuroshio east of Taiwan, 1982–2005: A numerical study. J. Geophys. Res. **113**, C04002 (2008). doi:10.1029/2007JC004485

Leipper, D.E., Volgenau, D.: Hurricane heat potential of the Gulf of Mexico. J. Phys. Oceanogr. **2**, 218–224 (1972)

Lin, I.-I., Liu, W.T., Wu, C.-C., Wong, G.T.F., Hu, C., Chen, Z., Liang, W.-D., Yang, Y., Liu, K.-K.: New evidence for enhanced ocean primary production triggered by tropical cyclone. Geophys. Res. Lett. **30**(13), 1718 (2003). doi:10.1029/2003GL017141

Mellor, G.L., Users guide for a three-dimensional, primitive equation, numerical ocean model, manual. Program in Atmosphere and Oceanic Sciences, Princeton University, Princeton, p. 56 (2004)

Milliff, R.F., Large, W.G., Morzel, J., Danabasoglu, G., Chin, T.M.: Ocean general circulation model sensitivity to forcing from scatterometer winds. J. Geophys. Res. **104** (C5), 11337–11358 (1999) doi: 10.1029/1998JC900045

Oey, L.-Y., Ezer, T., Wang, D.-P., Fan, S.-J., Yin, X.-Q.: Loop current warming by hurricane wilma. Geophys. Res. Lett. **33**, L08613 (2006). doi:10.1029/2006GL025873

Oey, L.-Y., Ezer, T., Wang, D.-P., Yin, X.-Q., Fan, S.-J.: Hurricane-induced motions and interaction with ocean currents. Cont. Shelf Res. **27**, 1249–1263 (2007). doi:10.1016/j.csr.2007.01.008

Price, J.F.: Upper ocean response to a Hurricane. J. Phys. Oceanogr. **11** (2), 153–175 (1981). doi: 10.1175/1520-0485(1981)011<0153:UORTAH>2.0.CO;2

Qu, T.: Upper-layer circulation in the South China sea. J. Phys. Oceanogr. **30**, 1450–1460 (2000). doi: 10.1175/1520-0485(2000)030<1450:ULCITS>2.0.CO;2

Schade, L.R., Emanuel, K.A.: The ocean's effect on the intensity of tropical cyclones: results from a simple coupled atmosphere-ocean model. J. Atmos. Sci. **56**, 642–651 (1999)

Shay, L.K., Goni, G.J., Black, P.G.: Effects of a warm oceanic feature on hurricane Opal. Mon. Wea. Rev. **128**, 1366–1383 (2000)

Wentz, F.J., Gentemann, C., Smith, D., Chelton, D.: Satellite measurements of sea surface temperature through clouds. Science **288**, 847–850 (2000). doi:10.1126/science.288.5467.847

Wu, C.-R., Hsin, Y.-C.: Volume transport through the Taiwan strait : a numerical study. Terr. Atmos. Oceanic Sci. **16**(2), 377–391 (2005)

Wu, C.-R., Chiang, T.-L.: Mesoscale eddies in the Northern South China sea. Deep Sea Res. II **54**, 1575–1588 (2007). doi:10.1016/j.dsr2.2007.05.008

Wu, C.-R., Chang, Y.-L., Oey, L.-Y., Chang, C.-W.J., Hsin, Y.-C.: Air-sea interaction between tropical cyclone Nari and Kuroshio. Geophys. Res. Lett. **35**, L12605 (2008). doi:10.1029/2008GL033942

Chapter 16
An Inferential Statistical Study on the Climate Characteristics of Tropical Cyclones over the Northwestern Pacific

ZhangJin Wei, DanLing Tang and GuangJun Sui

Abstract This chapter presents an inferential statistical study on the climate features of cyclone activities in the Northwestern Pacific (WNP) using data from 2,029 tropical cyclones (TCs) obtained within 60 years (1949–2008). The analysis shows the following results. (1) The annual genesis frequency of TC from 1949–2008 exhibits a statistically significant downward trend. The frequencies of tropical storms, typhoons, and strong typhoons have also declined, although less significantly than that of TCs. The interannual occurrence frequency of TCs has a statistically significant high value from 1960–1975, whereas its value is low from 1990–2008. (2) The highest frequency occurs at 14°–19°N and 115°–120°E within the 5° × 5° area, which is a narrow ocean surface to the west of Luzon Island in the South China Sea. (3) In general, the more southward and eastward the location of the TC genesis is, the stronger the TC is. Comparatively speaking, the effect of longitude on TC intensity is greater than that of latitude. Moreover, the interaction impact on TC intensity exists between latitude and longitude. This study provides new information that can improve the understanding of TC climate features over the WNP.

Keywords Tropical cyclone · Climate feature · Analysis of variance · Non-parametric test

---

Z. J. Wei
Information College in Guangdong University of Foreign Studies, Guangzhou, China

D. L. Tang (✉)
Research Center for Remote Sensing of Marine Ecology and Environments, State Key Laboratory of Tropical Oceanography, South China Sea Institute of Oceanology, Chinese Academy of Sciences, Guangzhou, China
e-mail: Lingzistdl@126.com
URL: http://lingzis.51.net/

G. J. Sui
Guangdong Research Institute for International Strategies, Guangdong University of Foreign Studies, Guangzhou, China

D. L. Tang and G. J. Sui (eds.), *Typhoon Impact and Crisis Management*, Advances in Natural and Technological Hazards Research 40,
DOI: 10.1007/978-3-642-40695-9_16,

16.1 Introduction

The Northwestern Pacific (WNP) is the area with the most frequent typhoon activities worldwide, wherein tropical cyclone (TC) frequency accounts for 36 % of the global total (Wang 1998). Moreover, TCs over the WNP are among the strongest in the world, with typhoons possessing a central pressure of 870 hpa (Wang 1998).

A number of studies on TC climate characteristics over the WNP have been reported, including those on interannual and interdecadal frequency variation (Chan and Shi 1996; Chan 2005; Yan et al. 2007; Chen and Zheng 1999), genesis location (Chen 1990; Chen et al. 1999; Wang et al. 2007), intensity (Webster et al. 2005; Yuan et al. 2008; Wei et al. 2012), and the effect of global warming on TC activities (Chan and Liu 2004; Walsh 2004). Previous researches mainly employed the descriptive statistics method which can not indicate whether the results are derived by randomness nor provide further explanation on the statistical significance. Existing data derived from the inferential statistical method are regarded as random samples and the characteristics of the population are inferred by excluding randomness. At present, however, studies adopting the inferential method in investigating TC climate features are limited. Therefore, research on TC activities based on inferential statistics should be further explored. Moreover, the results of previous studies would be more reasonable and less controversial if the data from the early period were complete and accurate. Expanding the sample size to investigate climate features by supplementing TC data from the year 2000 and beyond is necessary. Therefore, the present study provides new data and adopts several inferential statistical methods to examine the characteristics of interannual variation, seasonal variation of genesis locations, and the relationship between genesis location and intensity. Moreover, this study provides new information on TC climate features and improves the understanding of TC activities over the WNP.

This chapter is organized into five sections. Section 16.2 describes the data set and methods used in the study. Section 16.3 examines interannual and seasonal variations in intensity and genesis location, as well as the relationship between genesis location and intensity. Section 16.4 presents the discussion of the results. The conclusion is provided in Sect. 16.5.

16.2 Data and Methods

This study used the TC best track route data set (1949–2008) from the China Meteorological Administration–Shanghai Typhoon Research Institute (CMA–STI). This data set was compiled according to operation and service rules. We extracted 12 parameters from this data set to form a new TC data set for statistical analysis. Extracted data include TC series number, symbol, maximum velocity, and intensity category; minimum and average pressure in the center of the cyclone;

genesis location longitude and latitude; etc. The TCs selected in this study are divided into six categories according to CMA standards (Zhang 2007). The classification of TCs is provided in Table 16.1.

One objectives of this study is to find the trends in TC frequency using the sign test presented by Cox and Stuart (1955). A series of observations is determined to exhibit an upward trend if the magnitudes of later observations tend to be greater than those of earlier observations. By contrast, the results exhibit a downward trend if earlier observations tend to be greater than later observations. Therefore, the Cox–Stuart test has been adopted in this study to determine whether observed trends are significant.

We have also employed the Kruskal–Wallis test, Jonckheere–Terpstrav distribution test, and median test to examine variations in TC distribution. These tests can determine whether multi-independent samples from a multi-population distribution are significantly different and whether numerous samples are from identical populations. The Kruskal–Wallis test is a non-parametric (distribution-free) test developed by Kruskal and Wallis. In this test, the null hypothesis assumes that samples are from identical populations, whereas the alternative hypothesis assumes that samples come from different populations. The Jonckheere–Terpstra distribution test is a distribution-free test for ordered alternatives under a one-way layout. Jonckheere (1954) and Terpstra (1952) independently suggested this test. In this test, all *k* populations are assumed to have the same continuous distribution. This test has the advantages of not being predicted upon the normal distribution and being insensitive to outliers (Sneyers 1990). The median test is a special case of Pearson's Chi square test. The latter is used to determine whether the observed frequencies in each group differ from the expected frequencies derived from a distribution combining the two groups. This non-parametric test checks the null hypothesis which states that the medians of the populations from which the two samples are drawn are identical. The data in each sample are assigned to two groups. The first group consists of data with values higher than that of the median of the combined two groups. The second group consists of data with values at the median or below it. The formulas of the three test statistics refer to *statistical inference* (Casella and Berger 2008). The three tests are used in mutual corroboration to identify trends in TC frequency.

In most studies regarding TC activity, analysis of variance (ANOVA) models were seldom adopted. The present study explores several categorical factors

Table 16.1 Classification of tropical cyclones

Categories	Acronyms	Maximum mean wind speed at the bottom near the center (m/s)
Super strong typhoon	Super TY	≥51.0
Strong typhoon	STY	41.5–50.9
Typhoon	TY	32.7–41.4
Super tropical storm	STS	24.5–32.6
Tropical storm	TS	17.2–24.4
Tropical depression pressure	TD	10.8–17.1

affecting the numerical variables and statistical significance of tests. ANOVA models, including one-way and two-way factor tests, have been conducted to determine differences among two or more independent groups.

16.3 Climate Characteristics of TC

16.3.1 Genesis Frequency and Interannual Variation

From 1949 to 2008, 2029 TCs were formed over the WNP. The annual average is 34. Approximately one out of five TCs is tropical depression; and half of these TCs can develop into typhoons. TC frequency was higher in the 1960s and 1970s. It decreased in the late 1970s and slightly increased in the 1980s. A downward trend in TC frequency was observed in the 1990s. The frequencies of tropical storms (tropical depressions are not included), typhoons, and strong typhoons exhibit similar characteristics (Fig. 16.1).

The Coax–Stuart trend test is a non-parametric statistical method used to investigate trends in a time series. We have adopted this method to test whether the TC time series exhibits a statistically significant trend. According to the variation in TC frequency, we have conducted a significance test in which the alternative hypothesis is a downward trend. The results of the test show that:

1. The *P*-Value for TC frequency is 4.21517E-06;
2. The *P*-Value for TC of above TS is 0.008;
3. The *P*-Value for TC of above TY is 0.0007; and
4. The *P*-Value for TC of above STY is 0.049.

The downward trends in frequency for all TC categories are statistically significant from 1949 to 2008. The discrepancy in *P*-Value reveals that TC frequency declined more significantly than other TC categories. The downward trends in frequency are similar for TY and STY. The trend in frequency for TS is the most insignificant. The last finding is contrary to the result of the study of Yan et al. (2007) who observed that TS frequency increased.

The linear regression of frequency reveals the same results using the Coax–Stuart test.

1. TC frequency decreased at a rate of 1.8 per decade (*P*-Value is 8.21E-05);
2. TS frequency declined at a rate of 0.71 per decade (*P*-Value is 0.0367);
3. TY frequency declined at a rate of 0.83 per decade (*P*-Value is 0.008); and
4. TC above STY frequency declined at a rate of 0.87 per decade (*P*-Value is 0.001).

In the past 60 years, TC frequency generally presents a significant downward trend. A discrepancy in the downward TC rate for different categories is worth noting. The downward trend in TCs stronger than TSs is less significant, which

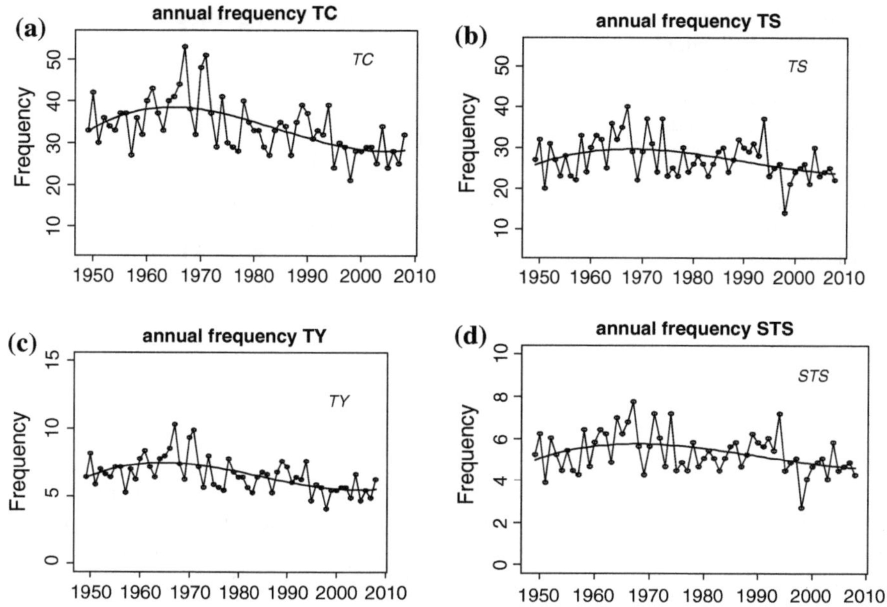

Fig. 16.1 Annual frequency of different categories of TCs over the WNP from 1949 to 2008. The *horizontal coordinate* "1950" refers to the year 1950, "1960" indicates the year 1960, and so on. The *heavy line* shows the fifth-polynomial fits of the times series

implies that the frequency of stronger TCs exhibits the least reduction compared with other TC categories.

The frequency variation of TCs which made landfall in China from 1949 to 2008 is displayed in Fig. 16.2.

The trend in the frequency for TCs which made landfall in China was also tested using the Cox–Stuart method. The *P*-Value of the significance test is 0.03. The result of the linear regression reveals that the frequency of TCs which made landfall decreases at a rate of 0.4 per decade ($p = 0.042$). This rate is smaller than the rate of 1.8 for the linear regression on the frequency of TCs over the WNP. This result implies that the downward trend in the frequency for TCs which made landfall in China is less significant than that of the genesis frequency for all TCs over the WNP.

Interdecadal TC frequency variation is usually analyzed every 10 years (Chan 2005; Chan and Liu 1996; Webster 2005). Thus, we divided the past 60 years into six intervals: 1950–1959, 1960–1969, 1970–1979, 1980–1989, 1990–1999, and 2000–2008. Interdecadal TC frequency variation was calculated, and the interdecadal averages presented several discrepancies. ANOVA methods were used to test whether the discrepancy is statistically significant. When a pairwise comparison was conducted, no significant peak or valley was observed using the least significant discrepancy (LSD) method. We divided 60 years into four periods (1949–1959, 1960–1975, 1976–1994, and 1995–2008) to determine the significant

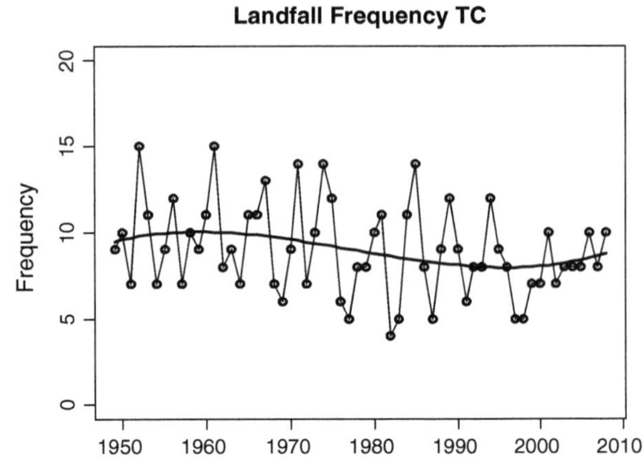

Fig. 16.2 Time series for TCs which made landfall during the last 60 years (1949–2008). The "1950" in the *horizontal coordinate* indicates the year 1950, the "1960" denotes the year 1960, and so on. The *heavy line* shows the fifth-polynomial fits of the times series

statistical interval. The average values of the frequencies for the four periods are 34.27, 39.81, 33.10, and 27.57, respectively. High values were obtained for the period from the 1960s to the 1970s; whereas low values were obtained for the late 1990s. A pairwise comparison of these values reveals a significant discrepancy (Table 16.2). TC frequencies exhibited an upward trend in the 1960s and 1970s, and a downward trend in the 1990s. In general, such trend variation can also be verified in the frequencies of TSs, STs, TYs, and STYs. The ANOVA results of our study are similar to those of Chan (1985) and Chan and Shi (1996), in which they found that annual TC frequency exhibits significant interannual and interdecadal variations. The frequencies reached a relative maximum in the 1960s, which decreased to a minimum in the late 1970s to the mid-1980s. The frequencies reached another peak in the mid-1990s and then decreased thereafter. However, our results are based on statistical inference, and thus, provide new insights.

16.3.2 Seasonal Variation of Genesis Location

The TC genesis region is located in the low latitude ocean surface over the WNP, which is within the area bounded by 0.5°–32°N and 106°–180°E. The genesis location where TCs develop into typhoons has relatively concentrated zones in latitude and longitude directions. From east to west, TCs over the WNP are mainly concentrated in four ocean regions: The central north ocean surface of the South China Sea (SCS), the ocean surface east of the Philippines and nearby Ryukyu, the nearby Mariana surface, and the nearby Marshall Islands surface. The statistical result for TC genesis location distribution from 1949 to 2008 is presented in Fig. 16.3.

For the latitude of genesis location, the frequency of TCs occurring within the 10°–15°N latitude belts comprises 30.56 % of the total frequency. For the longitude, most TCs occur on the longitude belts 130°–135°E, 135°–140°E, 140°–145°E,

Table 16.2 Multiple comparisons analysis

Interval	Interval	Mean difference (I − J)	Sig.
1949–1959	1960–1975	−5.53977*	0.005
	1976–1994	1.16746	0.531
	1995–2008	6.70130*	0.001
1960–1975	1949–1959	5.53977*	0.005
	1976–1994	6.70724*	0.000
	1995–2008	12.24107*	0.000
1976–1994	1949–1959	−1.16746	0.531
	1960–1975	−6.70724*	0.000
	1995–2008	5.53383*	0.002
1995–2008	1949–1959	−6.70130*	0.001
	1960–1975	−12.24107*	0.000
	1976–1994	−5.53383*	0.002

* The mean difference is significant at the 0.05 level

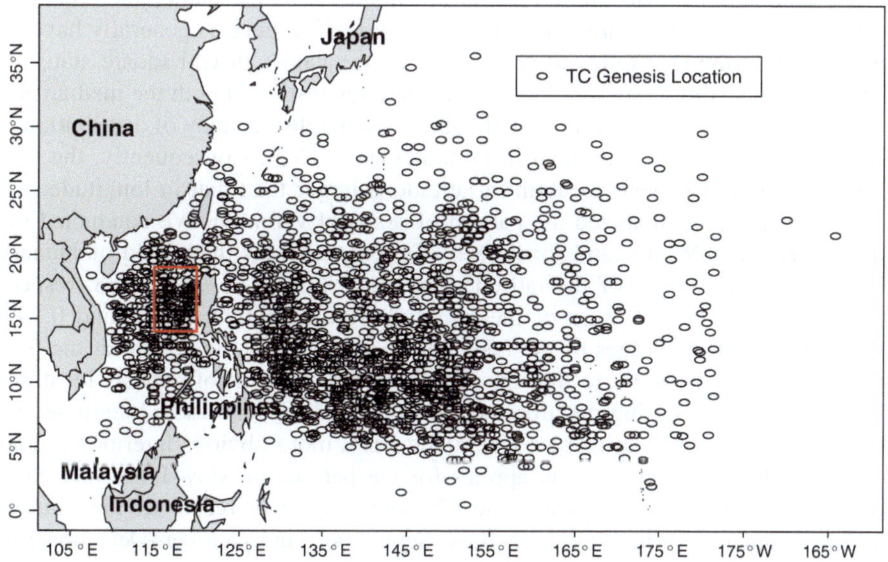

Fig. 16.3 Genesis location distribution of TCs from 1949 to 2008. The *black dots* are the locations of TC occurrences. *One black dot* stands for one TC

and 145°–150°E. The frequencies in these four longitude belts were 232, 222, 239, and 221, respectively. In terms of the rectangular area over the WNP, the longitude belts bounded by latitude 5°–25°N and 110°–160°E is the area where most TCs originate, i.e., 1677 TCs originated from that area, accounting for 82.65 % of the total number. For the 5° × 5° area, most TCs occur in the region bounded by 14°–19°N and 115°–120°E. This area is located in the SCS near the east narrow ocean surface west of Luzon, where 105 TCs occurred from 1949 to 2008.

Table 16.3 Non-parametric test for interseasonal genesis region

	Median test		Kruskal–Wallis distribution test		Jonckheere–Terpstra distribution test	
	Chi square	*P*-value	Chi square	*P*-value	Chi square	*P*-value
Longitude	139.30	0.000	95.54	0.000	14,942.25	0.001
Latitude	494.79	0.000	691.16	0.000	14,941.44	0.004

TCs over the WNP occurred each month during the aforementioned period, and the number of TCs that occurred over the WNP varied seasonally. TCs seldom occur from January to April. The frequency during this period was only 6.8 % of the total number. The months when TCs occur most frequently are July, August, and September, during which TC frequency comprised 65.5 % of the total number.

For the location of TC genesis, a multi-independent sample distribution tests, including Kruskal–Wallis, Jonckheere–Terpstra, and median tests were performed on multi-group independent samples. The null hypothesis assumes that the longitude and latitude distributions for each month were the same; therefore, *P*-Values for all three tests are 0.000. The locations of TC genesis generally have an intermonth variation. Comparatively speaking, the values of Chi square statistics for longitude testing is smaller than that for latitude testing in both the median and Kruskal–Wallis tests, thus implying that the seasonal discrepancy of distribution is less significant in longitude than in latitude (Table 16.3). Consequently, the seasonal variation of genesis location in latitude is larger than that in longitude.

Furthermore, we analyzed the pattern of seasonal variation. We conducted the median, Kruskal–Wallis, and Jonckheere–Terpstra tests for the period of January to April. The *P*-Values of the latitude test are 0.315, 0.148, and 0.988, respectively; whereas the *P*-Values of the longitude test are 0.097, 0.137, and 0.299, respectively. The *P*-Values of the LSD comparison are not significant at the 5 % level. These results indicate that the distribution distinction of genesis locations was not significant during the aforementioned period. In contrast to longitude, the location distribution with respect to latitude has a more obvious intermonth consentience. The same conclusion applies for the periods of May–June, July–September, and October–December. The 12 months were therefore divided into 4 periods (January–April, May–June, July–September, and October–December) to identify the seasonal variation in TC genesis locations. Non-parametric test and ANOVA were conducted to determine the interseasonal variation among the different periods. The *P*-Values of each non-parametric test are all approximately zero, and the central genesis locations of all four periods exhibit significant differences. The LSD results are shown in Table 16.4.

In terms of interseasonal variation in genesis locations within the entire year, TCs originate more frequently from the south and the east during the period of January to April, and the scope of genesis is relatively more concentrated. The central genesis location moves northward for the May to June period, whereas it shifts southward during the October to December period. TC frequency is higher and TCs have a more extensive range during the July to September period (Table 16.5).

Table 16.4 Results of interseasonal variation of genesis regions

Month category		Longitude		Latitude	
		Mean difference	*P*-value	Mean difference	*P*-value
January–April	May–June	15.10613*	0.000	−4.69767*	0.000
	July–September	7.51494*	0.000	−8.99229*	0.000
	October–December	4.96917*	0.001	−3.70060*	0.000
May–June	January–April	−15.10613*	0.000	4.69767*	0.000
	July-September	−7.59119*	0.000	−4.29462*	0.000
	October-December	−10.13696*	0.000	0.99706*	0.008
July–September	January-April	−7.51494*	0.000	8.99229*	0.000
	May–June	7.59119*	0.000	4.29462*	0.000
	October–December	−2.54577*	0.003	5.29169*	0.000
October–December	January–April	−4.96917*	0.001	3.70060*	0.000
	May–June	10.13696*	0.000	−0.99706*	0.008
	July–September	2.54577*	0.003	−5.29169*	0.000

* 0.05 significant level difference

16.3.3 Seasonal Variation of Intensity

The variables of maximum velocity, average velocity, and minimum average pressure may generally represent TC intensity. Therefore, we employed single-factor ANOVA to test the interseasonal variation of TC intensity using data of these variables. TC intensity variation within 12 months in a year cannot display significant differences. The genesis locations of TCs have a similar distribution pattern within the periods of January–April, May–June, July–September, and October–December. Thus, we divided the entire year into these four periods to test the seasonal variation of intensity among them. We initially calculated the average intensities of TCs within each period (Table 16.6). Then, the corresponding inferential method was adopted to test the variation of TC intensities with respect to the descriptive statistical results shown in Table 16.6.

TCs are strongest during the October to December period, and the discrepancies among all monthly categories are all statistically significant. Although the genesis region moves southward and eastward during the January to April period, TCs during this period are not the strongest (Table 16.7).

Table 16.5 TC genesis location of each category of months

Month category	Central Location		Standard Deviatiojn	
	Longitude	Latitude	Longitude	Latitude
Jan-April	7.60° N	146.45° E	3.30°	14.30°
May-June	12.30° N	131.35° E	4.60°	14.40°
July-Sep	16.59° N	138.94° E	5.15°	16.13°
Oct-Dec	11.30° N	141.49° E	4.24°	16.78°

Table 16.6 TC intensities of each category of months

Month category	Maximum velocity (m/s)	Mean velocity (m/s)	Minimum pressure (mb)	Mean pressure (mb)
January–April	32.46	21.57	979.15	994.54
May–June	33.09	21.61	975	992.22
July–September	34.68	22.73	972.14	990.08
October–December	38.6	25.41	967.13	988.46

Table 16.7 Results of ANOVA on TC intensity

Month category		Longitude		Latitude	
		Mean difference	*P*-value	Mean difference	*P*-value
January-April	5–6 month	−0.03707	0.971	2.32156	0.086
	7–9 month	−1.15901	0.170	4.45443*	0.000
	10–12 month	−3.83963(*)	0.000	6.07712*	0.000
May–June	1–4 month	0.03707	0.971	−2.32156	0.086
	7–9 month	−1.12194	0.100	2.13288*	0.019
	10–12mnth	−3.80256(*)	0.000	3.75556*	0.000
July–September	1-4 month	1.15901	0.170	−4.45443*	0.000
	5-6 month	1.12194	0.100	−2.13288*	0.019
	10-12 month	−2.68062(*)	0.000	1.62268*	0.014
October–December	1-4 month	3.83963(*)	0.000	−6.07712*	0.000
	5-6 month	3.80256(*)	0.000	−3.75556*	0.000
	7-9 month	2.68062(*)	0.000	−1.62268*	0.014

Moreover, we conducted a contrast analysis between the first half of the year and second half of the year. The *P*-Value of the contrast test is 0.000. The frequency and intensity of TCs which occur over the WNP during the second half of the year are respectively higher and stronger.

16.3.4 Relationship Between Genesis Location and Intensity

The average minimum pressure and maximum velocity of TCs over the WNP from 1949 to 2008 are 971 mb and 18.24 m/s, respectively. The ANOVA method was used to test the effect of categorical variables on numerical variables. Thus, we divided TC genesis location into four independent longitude areas, namely, 110°–120°E, 125°–135°E, 139°–150°E, and 160°–174°E, which respectively correspond to the central north ocean surface of the SCS, the ocean surface east of the Philippines and nearby Ryukyu, the nearby Mariana surface, and the nearby Marshall Islands surface. The frequency of TCs which occur within these four areas is 1,392, which accounts for 68.61 % of the total frequency. ANOVA was

Table 16.8 Intensity and genesis region of a single factor ANOVA

Main genesis		Maximum velocity		Minimum pressure	
		Mean difference	*P*-value	Mean difference	*P*-value
SCS central north	The Philippines islands	−9.08295*	0.000	11.64059*	0.000
	Mariana	−18.33638*	0.000	26.75319*	0.000
	The Marshall islands	−21.51982*	0.000	33.25392*	0.000
The Philippines islands	SCS central north	9.08295*	0.000	−11.64059*	0.000
	Mariana	−9.25343*	0.000	15.11260*	0.000
	The Marshall islands	−12.43687*	0.000	21.61333*	0.000
Mariana	SCS central north	18.33638*	0.000	−26.75319*	0.000
	The Philippines islands	9.25343*	0.000	−15.11260*	0.000
	The Marshall islands	−3.18344*	0.003	6.50073*	0.002
The Marshall islands	SCS central north	21.51982*	0.000	−33.25392*	0.000
	The Philippines islands	12.43687*	0.000	−21.61333*	0.000
	Mariana	3.18344*	0.003	−6.50073*	0.002

* The mean difference is significant at the 0.05 level

performed to examine the relationship between genesis and intensity of TCs over the WNP with the process average pressure and maximum average velocity data (Table 16.8). The *P*-Values are all almost 0.000, thus indicating that the average maximum velocity and minimum pressure in the four main genesis regions exhibit significant differences. The strongest TC in the four main genesis regions originated from the ocean surface near the Marshall Islands. The general trend is TCs are stronger from west to east. The Cox–Stuart test reveals the same significant trend, which is verified by several previous studies (Chan 2005). Namely, TCs originating from the west travels longer distances and absorbs a high amount of latent energy, which makes them stronger.

We calculated the mean maximum wind velocity and mean minimum pressure for each 2° latitude distance within the 0°–30°N latitude belt (Fig. 16.4). The mean maximum wind velocity generally declines with the increase in latitude of the genesis region. By contrast, the mean minimum pressure increases with the increase in latitude of the genesis region. The Cox–Stuart trend test on latitude variation shows that the *P*-Value is 0.000. This result implies that intensity generally increases as the genesis region moves southward. The mean maximum velocity and mean minimum pressure by each 5° longitude within the 0°E–180°E longitude belt are also shown in Fig. 16.4. The *P*-Values of the Cox–Stuart trend test on longitude variation are also 0.000, which reveals that the farther west the genesis region lies, the stronger the TC is (Fig. 16.4).

The longitude belt area of TCs occurring over the WNP was divided into eight categories by a 10° longitude span; whereas the latitude belt area was divided into six categories by a 5° latitude span to determine whether an interactive effect exists

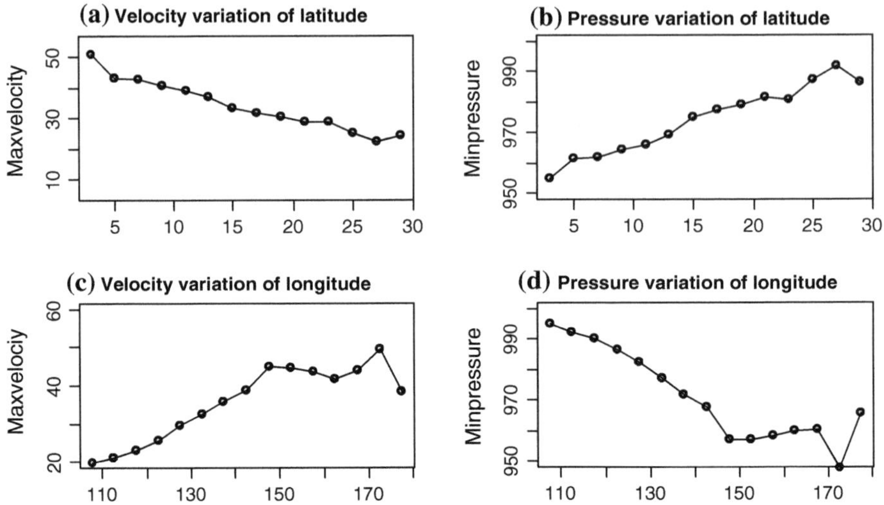

Fig. 16.4 Variation of maximum velocity and minimum pressure in different longitude and latitude belts. **a** Maximum velocity varies with the latitude location. **b** Minimum pressure varies with the latitude location. **c** Maximum velocity varies with the longitude location where TCs occurred. **d** Minimum pressure varies with the longitude location

between the latitude and longitude of genesis regions and TC intensity. We selected the variables of maximum and mean wind velocity; and then, we conducted double-factor ANOVA (Table 16.9).

where:

1. The F statistics of longitude effect is 34.76; P-Value is 0.000;
2. The F statistics of latitude effect is 14.571; P-Value is 0.000.

The result implies that the effects of latitude and longitude are statistically significant. However, TC intensity varies more significantly with longitude variation compared with that of latitude. In addition, the F statistics of the interactive effect is significant although it remains smaller (Fig. 16.5).

In Fig. 16.5, the lower the latitude of the TC genesis location is, the stronger the TC is in terms of mean intensity; and the more eastward the genesis region is, the stronger the TC is in terms of average intensity. However, as the TC genesis location moves eastward in the same latitude belt, the mean TC intensity does not get smaller in any region. In the same longitude belt, the TC intensity does not get smaller in any region as the latitude of TC genesis location increases. For example, considering the small area within several 5° × 10° regions in Fig. 16.5, then TC intensity within the area bounded by 5°–10°N and 150°–160°E is stronger than that in the area bounded by 5°–10°N and 160°–170°E. The TC intensity in the area bounded by 20°–25°N and 160°–170°E is weaker than that in the areas bounded by 20°–25°N and 150°–160°E, 20°–25°N and 140°–150°E, and 20°–25°N and 130°–140°E. This finding implies that an interaction exists between latitude and

Table 16.9 TC genesis and intensity two-way factors of ANOVA

Genesis	Type III sum of squares	DF	Mean square	F	Sig.
Corrected model	189,608.851*	45	4,213.530	17.313	0.000
Intercept	152,121.530	1	152,121.530	625.059	0.000
Latitude category	21,276.591	6	3,546.098	14.571	0.000
Longitude category	58,732.928	7	8,390.418	34.476	0.000
Latitude* longitude category	19,097.851	32	596.808	2.452	0.000
Error	480,171.795	1973	243.371		
Total	3,213,196.000	2019			
Corrected total	669,780.646	2018			

* R squared = 0.283 (Adjusted R squared = 0.267)

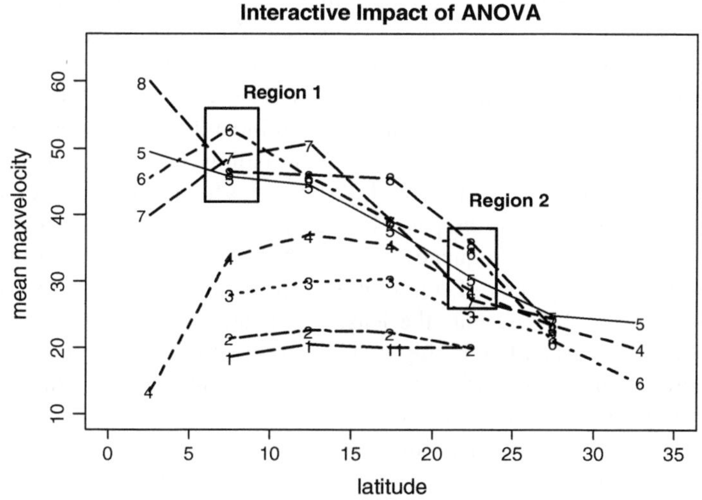

Fig. 16.5 Interactive effect of ANOVA. Each node in the line refers to an area bounded by 5° latitude and 10° longitude. The location surrounded by circles refers to several abnormal areas

longitude to a certain extent. The same conclusion can be drawn from the observed mean TC wind velocity.

16.4 Discussions

16.4.1 *Inferential Method and Randomness*

Previous studies were mainly conducted using the descriptive statistical approach. Descriptive statistics is efficient and direct, thus arriving at the variations of TC activities. However, several variations may result from the randomness of the observed samples. This randomness can be excluded by using the statistical test approach. Therefore, inferential statistics is required in investigating TC activities. Studies based on inferential statistics are still rarely reported in emerging literature. One of characteristics of this study is the use of inferential statistical methods.

16.4.2 Variation in Frequency

In terms of TC frequency and intensity, numerous researchers suggested that the buildup of greenhouse gases (Watson et al. 2001) would likely result in a rise in sea surface temperature (SST), thus subsequently increasing both the number and maximum intensity of TCs. However, a number of previous statistical studies were unable to present consistent evidence because of accuracy issues and different sources of data. Our results which were derived using the inferential statistical method show that the annual frequency of TCs over the WNP exhibited a downward trend from 1949 to 2008 and had significantly high values from 1960 to 1975, and low values from 1995 to 2008. Such findings are in accordance with the results of Chan (2005) and Yeung et al. (2006) who found that the annual number of overall TCs decreased in the past. This decrease occurred as a circle variation. Since the mid-1990s, TC frequency in the WNP has been exhibiting low values.

For the annual frequency of different TC categories, our results reveal that the downward trend of TY presents the least significance in the regression and non-parametric tests. In 2005, two highly publicized scientific papers reported evidence for an increase in TC activities based on observation records. Emanuel (2005) provided evidence for a substantial increase in the strength of TCs (denoted by the integral of the cube of the maximum winds over time) for the West Pacific and Atlantic basins during the last 50 years. A number of authors primarily attributed the reported increases to data reliability issues (Landsea et al. 2004, 2006; Landsea 2005). Our results provide a new probable explanation for the controversy surrounding the study of Emanuel (2005) to a certain extent. Compared with the decrease in TC annual frequency, the annual frequencies of strong TCs with intensities higher than those of TSs and TCs which made landfall present no significant change, which likely led Emanuel (2005) to arrive at his conclusion. In addition, our results also imply that typhoon risks are not significantly alleviated even if annual TC frequency decreases.

16.4.3 Variation of Genesis and Intensity

The location with the highest occurrence of TC genesis with a size of $5° \times 5°$ is a narrow ocean surface east of the SCS and west of Luzon. This finding corroborates the result of Zhang's study (2008). We further identified the exact location, which is bounded by 14°–19°N and 115°–120°E.

The seasonal variation in TC genesis location demonstrates a north–south movement and no west–east movement. This information should be given adequate attention to ensure validity. The probable explanation is that SST influences TC formation.

Several studies using descriptive statistics also noted that TCs become stronger as the genesis location moves eastward (Yan et al. 2007). However, further studies regarding this issue are lacking in emerging literature. The relationship between TC location and intensity is complicated because of complex meteorology and geography factors. Our results reveal that the average TC intensity is varied in different genesis locations. An interactive effect exists between longitude and latitude of genesis locations, which implies that several areas have extreme TC intensity. The physical mechanism pertaining to this finding needs to be further investigated.

16.5 Conclusions

This chapter examines the characteristics of TC activities over the WNP using inferential statistics. The analyses show that:

1. TC annual frequency decreased in a circle variation in the past 60 years. The annual frequency from 1960 to 1975 is high and statistically significant on the average; by contrast, TC annual frequency is low from 1995 to 2008 on the average.
2. TCs originating in the area of 5°–25°N and 110°–160°E comprise 82.65 % of the total number of TCs. For the 5° × 5° area, the largest frequency occurs in the area bounded by 10°–15°N and 115°–120°E, where 105 TCs occurred in the past 60 years.
3. TC genesis locations move southward and eastward, and their distribution ranges are relatively concentrated from January to April. The center of genesis location moves northward from May to June, whereas the center of the genesis region moves southward from October to December.
4. TC intensity over the WNP varies seasonally. The strongest TCs occur from September to November in terms of average intensity. The weakest TCs occur from January to February.
5. In general, the more the TC genesis location moves to the south and the east, the stronger the TC is. The effect of longitude on TC intensity is greater than the effect of latitude. However, an interactive effect exists between latitude and longitude.

Numerous studies on the climate characteristics of TCs are limited to the descriptive statistical method. The present study adopted the inferential statistical method to attempt to obtain meaningful information. However, only data from the past 60 years were used in this study. If a larger sample was used, then the corresponding conclusion would be more convincing.

Acknowledgments This study was supported by the following research projects awarded to DL Tang: (1) National Natural Sciences Foundation of China (31061160190, 40976091, NSFC-RFBR Project-41211120181 of DL Tang and D. Pozdnyakov); (2) Guangdong Sciences Foundation (2010B031900041, 8351030101000002); (3) Innovation Group Program of State Key Laboratory of Tropical Oceanography (LTOZZ1201).

References

Casella, G., Berger, R.L.: Statistical inference, pp. 140–170. Brooks/Cole (2008)

Chan, J. C.L.: Tropical Cyclone Activity in the Northwest Pacific in Relation to the El Niño/ Southern Oscillation Phenomenon. Mon. Wea. Rev. **113**, 599–606 (1985)

Chan, J.C.L., Liu, K.S.: Global warming and western North Pacific typhoon activity from an observational perspective. Climate **17**, 4590–4602 (2004)

Chan, J.C.L.: Interannual and interdecadal variations of tropical cyclone activity over the Western North Pacific. Meteorol. Atmos. Phys. **89**, 143–152 (2005)

Chan, J.C.L., Shi, J.E.: Long-term trends and interannual variability in tropical cyclone activity over the Western North Pacific. Geophys. Res. Lett. **23**, 2765–2767 (1996)

Chen, S.R.: Storm source regions over northwest pacific. Meteorology **16**, 23–26 (1990)

Chen, M., Zheng, R. G., Tan, Z. Y.: Tropical cyclone climate feature analysis over northwest pacific in recent 50 years. Trop. Meteorol. J. **16**(2), 23–27 (1999)

Cox, D., Stuart, A.: Some quick sign tests for trend in location and dispersion. Biometrika **42**, 80–95 (1955)

Emanuel, K.A.: Increasing destructiveness of tropical cyclones over the past 30 years. Nature **436**, 686–688 (2005)

Jonckheere, A. R.: A Distribution-Free k-Sample Test Against Ordered Alternatives. Biometrika Trust **41**(6), 133–145 (1954)

Kruskal, W. H., Wallis, W. A.: Use of Ranks in One-Criterion Variance Analysis. J. Amer. Statist. Assoc. **47**(12), 583–621 (1952)

Landsea, C.W.: Meteorology: hurricanes and global warming. Nature **438**, E11–E12 (2005). doi:10.1038/nature04477

Landsea, C.W., Harper, B.A., Hoarau, K., Knaff, J.A.: Climate change. Can we detect trends in extreme tropical cyclones? Science **313**(5786), 452–454 (2006)

Landsea, C.W., Anderson, C., Charles, N., Clark, G., Dunion, J., Partagas, J.F., Hungerford, P., Neumann, C., Zimmer, M.: The Atlantic Hurricane Database Re-analysis Project: Documentation for 1851–1910 Alterations and Additions to the HURDAT Database, pp. 179–217. Columbia University Press (2004)

Sneyers, R.: On the Statistical Analysis of Series of Observations, pp. 182–192. World Meteorological Organization (1990)

Terpstra, T. J.: The asymptotic normality and consistency of Kendall's test against trend, when ties are present in one ranking. Indag. Math. **14**(3), 328–333 (1952)

Walsh, K.: Tropical cyclones and climate change unresolved issues. Clim. Res. **27**(1), 77–83 (2004)

Wang, X.L.: Storm tide hazard forecast and prevention. Ocean Forecast **15**(3), 26–31 (1998)

Wang, G., Su, J., Ding, Y., Chen, D.: Tropical cyclone genesis over the South China Sea. J. Marine Syst. **68**(3–4), 318–326 (2007)

Watson, R.T., Albritton, D.L.: Climate change 2001: Synthesis report: Third assessment report of the Intergovernmental Panel on Climate Change, pp. 125–225. The Press Syndicate of the University of Cambridge (2001)

Webster, P.J., Holland, G.J., Curry, J.A., Chang, H.R.: Changes in tropical cyclone number, duration, and intensity in a warming environment. Science **309**, 1844–1846 (2005)

Wei, Z.J., Tang, D.L., Sui, G.J.: Landfalling probability fitting of tropical cyclone based on Logistic model. J. Appl. Stat. Manage. **21**(3), 389–397 (2012)

Yan, L.J., Huang, X.X., Yu, Y.B., Chen, X.Y., Chen, Z.F.: Tropical cyclone frequency variation feature in recent 58 years. Meteorol. Res. Appl. **28**(S2), 63–67 (2007)

Yeung, K.H., Wu, M.C., Chang, W.L., Leung, Y.K.: Long-term change in tropical cyclone activity in the Western North Pacific. EOS Trans., AGU **87**(48), 537–538 (2006)

Yuan, J.L., Lin, A.L., Liu, C.X.: Change characters of tropical cyclones with different intensities over the western North Pacific during the last 60 years. Acta Meteorologica Sinica **66**(2), 213–223 (2008)

Zhang, R.R.: New classification and relative terminology of tropical cyclone. Chin. Sci. Technol. Terms (2), 56 (2007)

Zhang, J.C., Rao, Z.X., Xie, Q.J., Shi, M.H.: An analysis of tropical cyclone activities in 1884–2006. Guangdong Meteorol. **30**(2), 24–26 (2008)

Chapter 17
Comparisons of Two Types of El Niño Impacts on TC Genesis over the South China Sea

DongXiao Wang, Xin Wang, Wen Zhou and ChongYin Li

Abstract This chapter examines the impacts of warm pool (WP) El Niño on the tropical cyclone (TC) genesis over the South China Sea (SCS) during 1965–2004. We check the relationships between WP El Niño and SCS TC genesis in Sept-Oct-Nov (SON) when the interannual variability of TC genesis is significant. Compared with the cold tongue (CT) El Niño, the relationship of TC genesis with WP El Niño is significant. In recent a couple of decades with more frequent WP El Niño, the significant coherent variations between TC genesis and WP NINO index on the timescales of approximately 4 years are displayed. The distinct different atmospheric teleconnection patterns related to CT and WP El Niño are responsible for these relationships. The WP El Niño could result in a dipolar pattern with anticyclone anomalies over the SCS and cyclone anomalies over the western tropical Pacific warm pool at low- and mid-level, which is different from the CT El Niño-induced anticyclone anomalies over the SCS and western tropical Pacific warm pool. These WP El Niño-related large-scale circulation anomalies could enlarge the low-level northerlies over the SCS, which consequently enhances the vertical wind shear and suppresses the TC genesis over the SCS.

Keywords South China Sea · Tropical cyclone · Warm pool El Niño

D. X. Wang (✉) · X. Wang
State Key Laboratory of Tropical Oceanography, South China Sea Institute of Oceanology, Chinese Academy Science, Guangzhou, China
e-mail: dxwang@scsio.ac.cn

W. Zhou
Guy Carpenter Asia–Pacific Climate Impact Centre, School of Energy and Environment, City University of Hong Kong, Hong Kong, China

C. Y. Li
LASG, Institute of Atmospheric Physics, CAS, 100029 Beijing, China
Institute of Meteorology and Oceanography, PLA University of Science and Technology, 211101 Nanjing, China

D. L. Tang and G. J. Sui (eds.), *Typhoon Impact and Crisis Management*, Advances in Natural and Technological Hazards Research 40,
DOI: 10.1007/978-3-642-40695-9_17,

17.1 Introduction

The South China Sea (SCS) is located between the equator and 22°N and from 105 to 120°E, where high frequency of tropical cyclone (TC) genesis occurs (Camargo et al. 2007; Wang et al. 2007). Many studies have been done on relationships between the El Niño-Southern Oscillation (ENSO) and the interannual variability of the TC genesis over the SCS (Li 1988; Wang and Chan 2002; Lee et al. 2006; Wang et al. 2007; Zuki and Lupo 2008; Huang et al. 2009). Wang and Chan (2002) suggested that ENSO only have weak impacts on TC formation in the SCS. However, other studies revealed that there were more (less) TC activity over the southern SCS in La Niña (El Niño) years in the second half 20th century (i.e., Li 1988; Lee et al. 2006; Wang et al. 2007; Zuki and Lupo 2008).

Great strides have been made in ENSO research in recent years. Evidences revealed a new type of El Niño different from the canonical El Niño in term of the location of the maximum sea surface temperature (SST) anomalies, referred as "El Niño Modoki" (Ashok et al. 2007), "CP El Niño" (Yu and Kao 2007) or "Warm pool El Niño" (Kug et al. 2009). The tropical-mid latitude teleconnections associated with the two types of El Niño are significantly different (Ashok et al. 2007; Kao and Yu 2009; Weng et al. 2009; Yu and Kim 2010; Wang and Wang 2013). Many studies have shown that the impacts of the new type El Niño on TC over the North Atlantic and western North Pacific are distinct from the canonical El Niño (Kim et al. 2009; Lee and McPhaden 2010; Chen and Tam 2010; Kim et al. 2010). Motivated by previous results, we attempt to compare the impacts of two types of El Niño on TC genesis over the SCS, and investigate the physical mechanisms in this study.

17.2 Datasets and Methodology

This study uses the TC datasets from the website of International Best Track Archive for Climate Stewardship (IBTrACS) Project (http://www.ncdc.noaa/gov/oa/ibtracs/). We merely selected the TC (maximum sustained wind speed is larger than 34 knots) generated in the SCS, i.e., the TCs first recorded by the IBTrACS dataset falling within the SCS domain (0–22°N, 105–120°E). The dataset covers the whole weather satellite era (1965–2004). The monthly horizontal wind and geopotential height data are from the National Centers for Environmental Prediction/National Center for Atmospheric Research (NCEP/NCAR) reanalysis datasets (Kistler et al. 2001) at the same period.

According to the definitions of warm pool (WP) and cold tongue (CT) El Niño proposed by Ren and Jin (2011), two NINO indices are used, which are able to identify the two types of El Niño events separately, and capture SST characteristics of the two types of El Niño. More details about these indices can be found in Ren and Jin (2011).

17.3 Results and Discussions

The TCs over the SCS are more frequent between June and November (Lee et al. 2006; Wang et al. 2007; Zuki and Lupo 2008). Due to the impacts of the East Asian monsoon, a seasonal location shift of the SCS TC birthplace was found (Lee et al. 2006; Wang et al. 2007). They found that TC usually occurs at north of 12°N in JJA and south of 18°N in Sept-Oct-Nov (SON). In addition to the location shift, the temporal variability in summer is distinct from that in autumn (Fig. 17.1). The significant interannual variability of TC genesis is shown in autumn but not in summer. Moreover, the results of wavelet analysis exhibits that the interannual variability of autumn TC genesis with about 4-year periodicity is robust after 1980 (Fig. 17.1e). Because of the strong interannual variability of ENSO, we merely analyzed its relationship with the autumn TC genesis over the SCS.

The CT El Niño is distinct from the WP El Niño with the warming SST anomalies location during ENSO developing phase (boreal summer and autumn) and peaks (Yeh et al. 2009; Yu and Kim 2010). In this study we investigate the simultaneous relationships between two types of El Niño and TC during SON. Figure 17.2 gives the variations of the two NINO indices. It is found that the TC genesis relates significantly with WP NINO index in SON, but not with CT NINO index, indicating that the interannual variability of TC genesis in autumn may relate to changes in the central tropical Pacific SST anomalies. From Fig. 17.2, it is also found that the WP El Niño occurrence has become more frequent in recent couple decades, which is consistent with previous results (Ashok et al. 2007; Kao and Yu 2009; Yeh et al. 2009). If the significantly negative relationship between WP NINO index and TC is true, it is inferred that (1) the TC genesis over the SCS may be more robust, and (2) this relationship should be much closer in recent decades.

The variations of TC genesis are studied in the two time period. Figure 17.3 shows that there is a significant interannual variability of TC genesis during 1979–2004 (Fig. 17.3a) but not during 1965–1977 (Fig. 17.3b), which is consistent with the wavelet analysis results in Fig. 17.1e. The results of wavelet coherence analysis between WP NINO index and TC genesis confirm our hypothesis (Fig. 17.4). Figure 17.4 demonstrates that these two time series exhibit statistically significant coherent variations on the timescales of around 4 years, particularly from the 1980s and onwards. The left orientation of the arrows in red area in the upper part of Fig. 17.4 suggests their negative relationship on interannual timescales, indicating that the decrease of TC genesis generally corresponds to WP El Niño events.

The above results show evidences of the closer linkage of the SON TC genesis with WP NINO index, rather than CT NINO index. The following attempts to explain why the influences of WP and CT El Niño events on TC formation in the SCS in autumn are different. Figure 17.5 gives large-scales atmospheric circulation anomalies at high- and low-level troposphere associated with WP and CT NINO indices. At high-level, the significant westerly anomalies related to the two

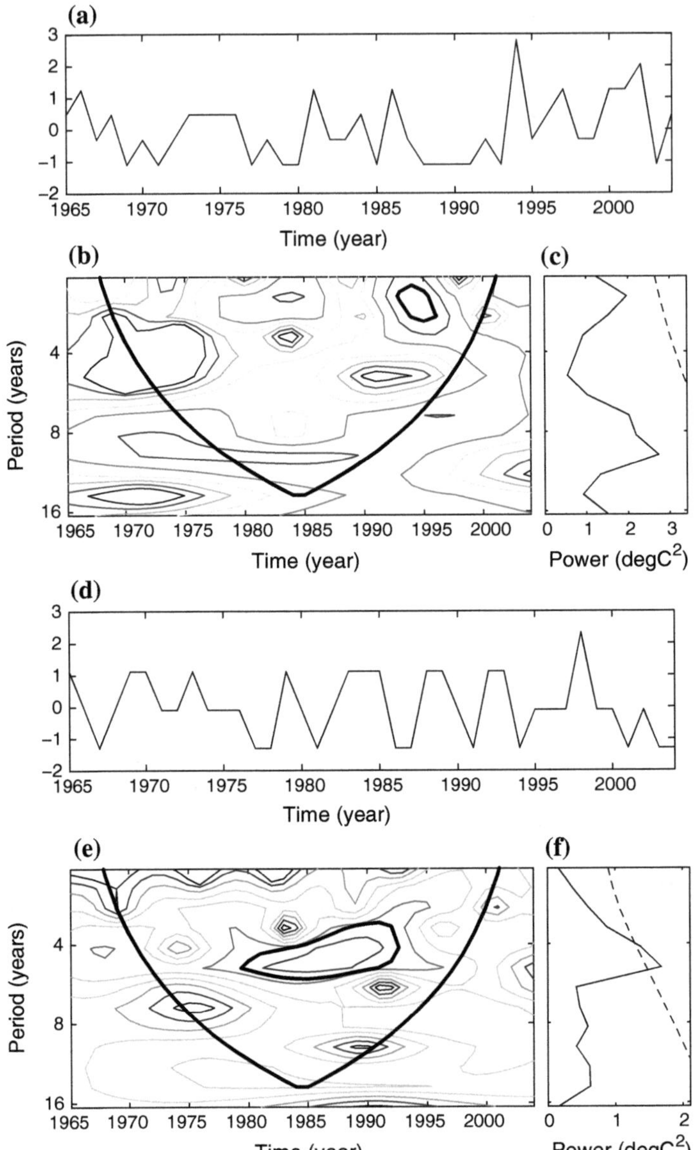

Fig. 17.1 The standardized time series of JJA-mean (**a**) and SON-mean (**e**) anomalies of TC genesis over the SCS during 1965–2004. (**b**) and (**e**) are the continuous wavelet power spectrum of standardized JJA-mean and SON-mean TC genesis. The *thick contours* indicate the 10 % significance level against red noise and the cone of influence where edge effects might distort the picture. (**c**) and (**f**) are the global wavelet power spectrum of standardized JJA-mean and SON-mean TC genesis. The *dashed lines* are the 10 % significance level

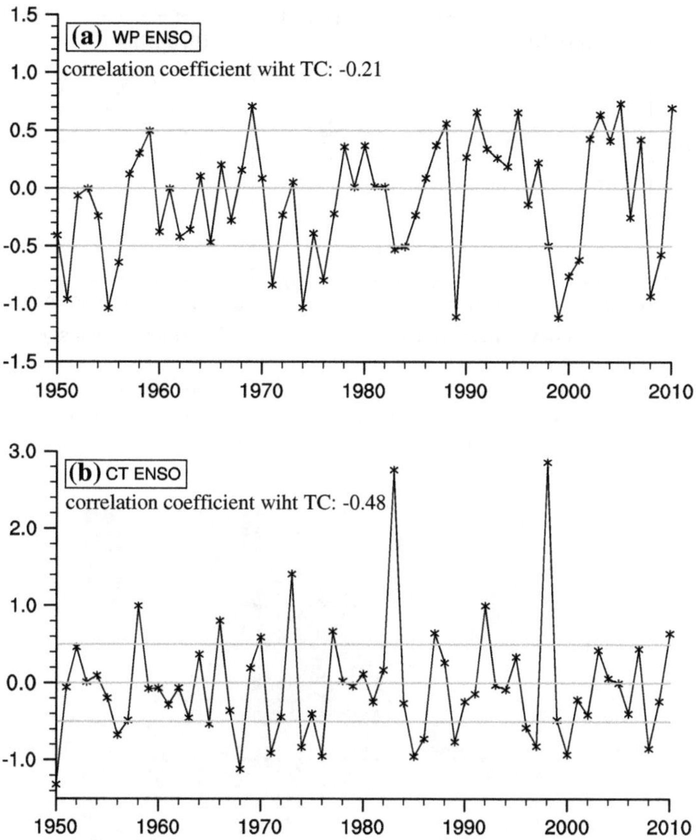

Fig. 17.2 The standardized time series of autumn-mean WP NINO (**a**) and CT NINO (**b**) index according to Ren and Jin (2011)

types of NINO indices are observed over the SCS (Fig. 17.5a, c). Although both WP and CT El Niño could result in the significant anticyclone anomalies over the western North Pacific, it is noted that the anticyclone anomalies associated with WP NINO index are stronger and more westward than those related to CT NINO index. At low-level, the two types of El Niño could lead to the anticyclone anomalies over the SCS. However, there are some differences. The center of significant anomalous anticyclone related to CT NINO index is located in the SCS (Fig. 17.5d), while that associated with WP NINO index shifts westward, which in turn results in the anomalous northerlies enhancement over the SCS (Fig. 17.5b). These large-scale circulations anomalies associated with WP NINO index are consistent with both the observations and model results (Ashok et al. 2007; Chen and Tam 2010). Figure 17.5 manifests that the most distinct features of large-scale atmospheric circulations over the SCS associated with two types of NINO indices are the wind vector at low-level. Compared with CT El Niño, the WP El Niño

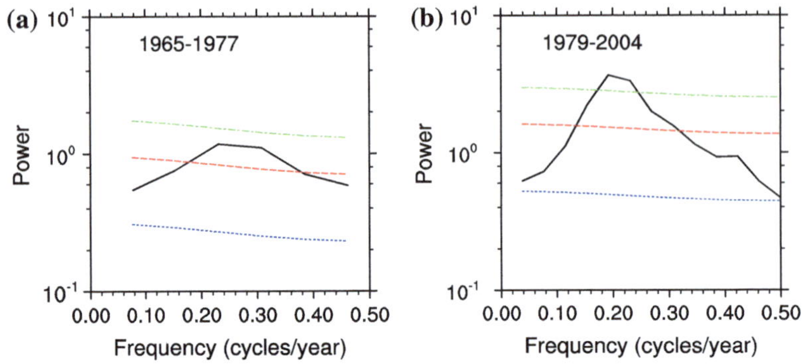

Fig. 17.3 The spectrum analysis for SON-mean TC genesis during 1965–1977 (**a**) and 1979–2004 (**b**). The *dashed red, dotted blue*, and *dash-dotted green lines* represent Markov red noise spectrum, 10 % confidence level, and 90 % confidence level, respectively

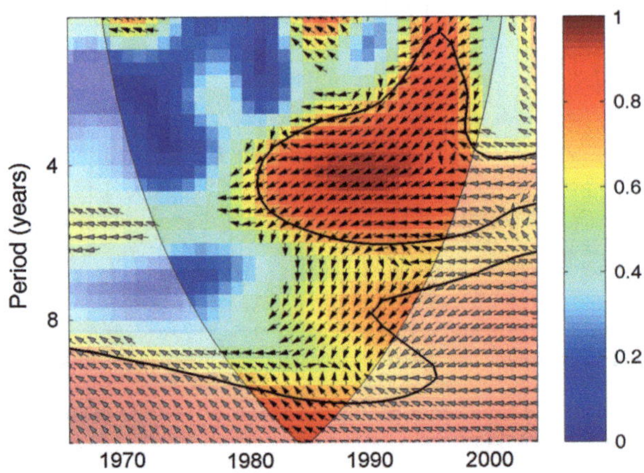

Fig. 17.4 Squared wavelet coherence between the SON TC genesis over the SCS and the WP Niño index. *Thick contours* enclose the areas with correlations statistically significant at 95 % confidence level against red noise. Semitransparent areas indicate the ''cone of influence'' where the edge effects become important. The relative phase relationship is shown as *arrows*, with in-phase pointing *right* and anti-phase pointing *left*

accelerates significantly northerlies at low-level over the SCS, which results in the increase of vertical wind shear and thus depresses TC formation.

The composite analysis is employed to elucidate the comparisons of dynamic controls of vertical wind shear in the SCS in SON related to WP and CT NINO indices. According to the two NINO indices, the five strongest CT and WP El Niño events are composited, respectively. The WP El Niño events include 1968/1969, 1990/1991, 1994/1995, 2002/2003, and 2004/2005 events, and CT El Niño events include 1965/1966, 1972/1973, 1982/1983, 1991/1992, and 1997/1998 event. These WP El Niño events are identical to the so-called El Niño Modoki (Ashok

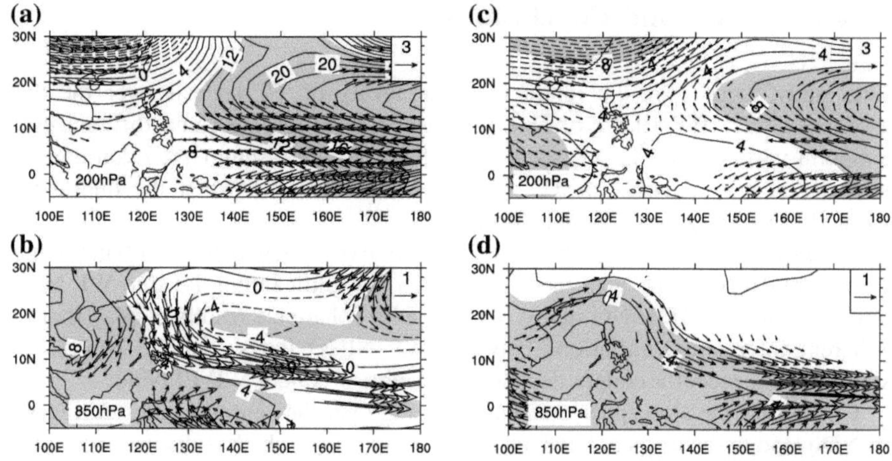

Fig. 17.5 The regression coefficients of the wind (vector) and geopotential height (contour) with respect to WP Niño index (**a** and **b**) and CT Niño index (**c** and **d**). The *shadings* represent the geopotential height anomalies exceeding 95 % significant level. Only the regression coefficients of the wind exceeding 95 % significant level are plotted

et al. 2007) or Central-Pacific El Niño (Yu and Kim 2010). Because the higher correlation between CT Niño index and Niño3 index (Ren and Jin 2011), based on CT Niño index the six strongest La Niña events are selected, including 1970/1971, 1973/1974, 1975/1976, 1988/1989, 1998/1999, and 1999/2000 event. Significant positive vertical wind shear are clearly exhibited in the difference pattern between WP El Niño and La Niña events (Fig. 17.6a), but not in the difference pattern between CT El Niño and La Niña events (Fig. 17.6b). Therefore, the WP El Niño could result in significant decrease in TC formation over the SCS than CT El Niño.

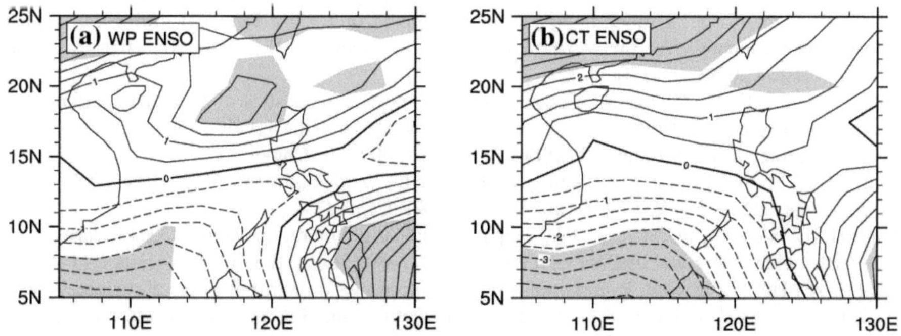

Fig. 17.6 The composite differences of vertical wind shear of La Niña events with WP El Niño (**a**) and CP El Niño (**b**). The shadings are exceeding 90 % significant level. The vertical wind shear is calculated as the magnitude of the vector difference between wind at 200-hPa and 850-hPa

17.4 Summary and Conclusions

This study focuses on the interannual variability of TC genesis over the SCS associated with WP El Niño. Due to seasonal location shift and differences in temporal variation in TC genesis over the SCS in boreal summer and autumn, the impacts of WP El Niño on TC are analyzed in autumn when the interannual variability of TC genesis is significant.

Increasing evidence support a fact that there exist a new type of El Niño named warm pool El Niño. TC genesis over the SCS is significantly related to WP NINO index, but not to CT NINO index. The wavelet coherence results display that the coherent interannual variations between WP NINO index and TC genesis are significant in recent a couple decades with more frequent WP El Niño.

Changes in the atmospheric diabatic forcing over the central tropics during WP El Niño lead to the tropical teleconnections modifying, which is distinct from these during CT El Niño. The CT El Niño-related anticyclone anomalies over the SCS and western tropical Pacific warm pool are replaced by WP El Niño-related dipolar pattern with anticyclone anomalies over the SCS and cyclone anomalies over the western tropical Pacific warm pool at low- and mid-level (Fig. 17.5). These WP El Niño-related atmospheric teleconnection enlarges the low-level northerlies over the SCS, which in turn enhances the vertical wind shear and thus suppresses the TC genesis over the SCS. This work can help improving seasonal prediction of the SCS TC activity.

References

Ashok, K., Behera, S.K., Rao, S.A., Weng, H., Yamagata, T.: El Niño Modoki and its possible teleconnection. J. Geophys. Res. **112**, C11007 (2007). doi:10.1029/2006JC003798

Camargo, S.J., Emanuel, K.A., Sobel, A.H.: Use of a genesis potential index to diagnose ENSO effects on tropical cyclone genesis. J. Climate **20**, 4819–4834 (2007). doi:10.1175/JCLI4282.1

Chen, G., Tam, C.-Y.: Different impacts of two kinds of Pacific Ocean warming on tropical cyclone frequency over the western North Pacific. Geophys. Res. Lett. **37**, L01803 (2010). doi:10.1029/2009GL041708

Huang, Y., Li, C.-Y., Wang, Y.: Further study on the characteristics of frequency variation of tropical cyclones activity over western north pacific and the relationship with sea surface temperature. J. Trop. Meteorol. **25**(3), 273–280 (2009) (in Chinese)

Kao, H.- Y., Yu, J.- Y.: Contrasting eastern-Pacific and central-Pacific types of El Niño. J. Climate **22**, 615–632 (2009). doi:10.1175/2008JCLI2309.1

Kim, H.M., Webster, P.J., Curry, J.A.: Impact of shifting patterns of Pacific Ocean warming on North Atlantic tropical cyclones. Science **325**, 77–80 (2009). doi:10.1126/science.1174062

Kim, J.-H., Ho, C.-H., Chu, P.-S.:Dipolar redistribution of summertime tropical cyclone genesis between the Philippine Sea and the northern South China Sea and its possible mechanisms. J. Geophys. Res. **115**, D06104 (2010). doi: 10.1029/2009JD012196

Kistler, R. et al.: The NCEP–NCAR 50-year reanalysis: monthly means CD-ROM and documentation. Bull. Amer. Meteor. Soc. **82**, 247–267 (2001)

Kug, J.-S., Jin, F.-F., An, S.-I.: Two types of El Niño events: Cold tongue El Niño and warm pool El Niño. J. Climate **22**, 1499–1515 (2009). doi:10.1175/2008JCLI2624.1

Lee, C.-S., Lin, Y.-L., Cheung, K.K.W.: Tropical cyclone formations in the South China Sea associated with the Mei-Yu Front. Mon. Wea. Rev. **134**, 2670–2687 (2006)
Lee, T., McPhaden, M.J.: Increasing intensity of El Nino in the central-equatorial Pacific. Geophys. Res. Lett. **37**, L14603 (2010). doi:10.1029/2010GL044007
Li, C.: Actions of typhoons over the western Pacific (including the South China Sea) and El Nino. Adv. Atmos. Sci. **5**, 107–115 (1988)
Ren, H.-L., Jin, F.-F.: Niño indices for two types of ENSO. Geophys. Res. Lett. **38**, L04704 (2011). doi:10.1029/2010GL046031
Wang, B., Chan, J.C.L.: How strong ENSO events affect tropical storm activity over the western North Pacific. J. Climate **15**, 1643–1658 (2002)
Wang, C., Wang, X.: Classifying El Niño Modoki I and II by different impacts on rainfall in Southern China and typhoon tracks. J. Climate **26**, 1322–1338 (2013)
Wang, G., Su, J., Ding, Y., Chen, D.: Tropical cyclone genesis over the South China Sea. J. Marine. Syst. **68**, 318–326 (2007)
Weng, H., Behera, S.K., Yamagata, T.: Anomalous winter climate conditions in the Pacific Rim during recent El Niño Modoki and El Niño events. Climate Dyn. **32**, 663–674 (2009). doi:10.1007/s00382-008-0394-6
Yeh, S.-W., Kug, J.-S., Dewitte, B., Kwon, M.-H., Kirtman, B.P., Jin, F–.F.: El Nino in a changing climate. Nature **461**, 511–514 (2009). doi:10.1038/nature08316
Yu, J.-Y., Kao, H.-Y.: Decadal changes of ENSO persistence barrier in SST and ocean heat content indices: 1958–2001. J. Geophys. Res. **112**, D13106 (2007). doi:10.1029/2006JD007654
Yu, J.-Y., Kim, S.T.: Three evolution patterns of Central-Pacific El Niño. Geophys. Res. Lett. **37**, L08706 (2010). doi:10.1029/2010GL042810
Zuki, Z.M., Lupo, A.R.: Interannual variability of tropical cyclone activity in the southern South China Sea. J. Geophys. Res. **113**, D06106 (2008). doi:10.1029/2007JD009218

Part V
Typhoon Impacts on Economics

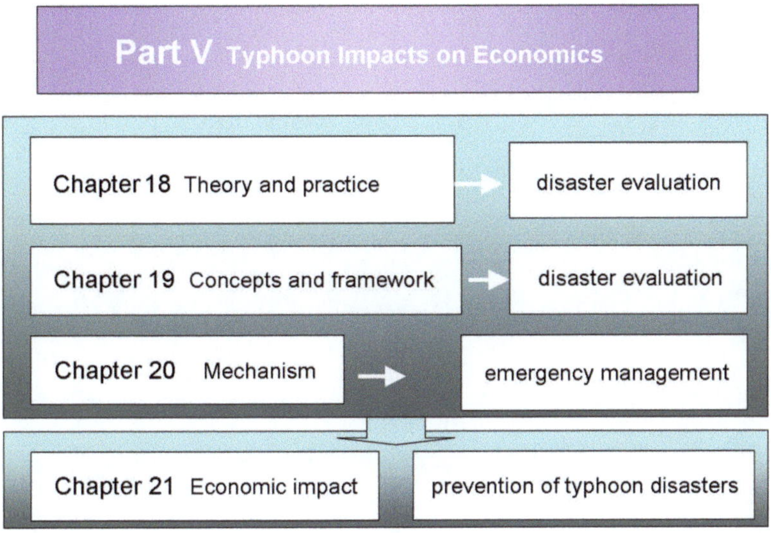

Chapter 18
Theory and Practice for Typhoon Disaster Assessment

GuangJun Sui, ZhangJin Wei and DanLing Tang

Abstract There has been much progress in the study of typhoon disaster assessment in recent years. A comprehensive overview of typhoon disaster assessment from different sources is still missing from the literature of the specialization. Therefore, this chapter reviews and analyzes research on the issues of disaster system prevention theory, risk and damage assessment for typhoon disasters and assessment practice in some countries. On the basis of the analysis of the evolvement of assessment approaches, this study also points out the problems related to typhoon disaster assessment, i.e. the grade classification for assessment, lack of uniform standards and the selection of indicators that cannot take all the factors into full consideration. At the end of this chapter the investigators also give some suggestions for typhoon disaster assessment.

Keywords Typhoon risk assessment · Disaster damage assessment · Economic loss assessment approach · Disaster decision system · Disaster system theory · Outlook

G. J. Sui
Guangdong Research Institute for International Strategies, Guangdong University of Foreign Studies, Guangzhou, China

Z. J. Wei
Information College in Guangdong University of Foreign Studies, Guangzhou, China

D. L. Tang (✉)
Research Center for Remote Sensing of Marine Ecology and Environments, State Key Laboratory of Tropical Oceanography, South China Sea Institute of Oceanology, Chinese Academy of Sciences, Guangzhou, China
e-mail: Lingzistdl@126.com
URL: http://lingzis.51.net/

D. L. Tang and G. J. Sui (eds.), *Typhoon Impact and Crisis Management*,
Advances in Natural and Technological Hazards Research 40,
DOI: 10.1007/978-3-642-40695-9_18,

18.1 Introduction

Tropical cyclones are among the most devastating of natural disasters in terms of the damage to the environment and the loss of human life they cause. The economic losses they induce and the potential dangers of tropical cyclones are becoming more pronounced because of the population growth in tropical coastal regions (Xiao and Xiao 2010; Sui et al. 2012). Some related research has shown that the global losses due to tropical cyclones has risen dramatically in recent decades which demonstrates an increasing awareness of the impact of typhoon disasters on society (Zhang 2009; Schmidt et al. 2010; Chen et al. 2012; Sui and Pu 2012; Sui and Tang 2012).

Disaster assessment is the process of determining the potential impact which a hazard has on society. The assessment can provide the information that focuses on the needs and priorities of that population related to immediate emergency measures and helps to make those measures optimize procedures to assure that the resources are designed and deployed in a manner that will save and sustain the lives of survivors are effectively deployed. The assessment of these procedures and plans are the foundation of decision-making, which contributes directly to effective planning and control of the organized response (Sui et al. 2012). Such a high quality assessment is also beneficial to raising the possibilities for facilitating and expediting long-term recovery and development of affected areas.

Typhoon disaster assessment arose from the period of the 1980s. Much research has been conducted on this issue from diverse perspectives. A review and summary can help identify the development of different perspectives of typhoon assessment all of which can promote a comprehensive study of the field.

This chapter includes seven sections: Sect. 18.2 outlines the systems theoretical study of typhoon disasters; Sects. 18.3, 18.4, and 18.5 present a summary of risks, damage and economic loss assessment procedures as well as existing problems of typhoon disaster assessment, and Sect. 18.6 reviews decision-making systems; Suggestions and recommendations are made as a concluding remarks in Sect. 18.7.

18.2 Typhoon Disaster System Theory

It is generally believed that the research object of Catastrophology is disaster and disaster system prediction and relief. It is a new subject, in which the causes, characteristics, distribution, classification, prediction, disaster prevention, and monitoring of all kinds of disasters is examined. The ultimate goal of this discipline is to find suitable ways to minimize and/or avoid the consequences of disasters and the attendant losses.

Typhoon disasters are one type of natural hazards that is repeated throughout history. In the 1920s geographers focused on specific disaster species and attributed the cause of such disasters to hazard formative factors. Re-appearance probability and exceeding probability calculations were investigated within the framework of risk assessment (Tucker et al. 1994). With the increasing trend of disaster damage assessment and the impact of global and regional environment changes related to disaster occurrences, this specialization became a research topic. The formative environment of disasters was considered to be one of factors influencing the risk assessment of disasters (Park 1995). The vulnerability of society at the point of impact of natural disasters can enlarge their impact, both in human terms and economically. Consequently, some researchers (Turner and Meyer 1991; Carram and Guzzeth 1995) paid particular attention to disaster-affected bodies and conducted that the vulnerability assessments of disaster-affected bodies deserved further attention.

Along with deepening research in this area, researchers gradually viewed disasters as an interactive whole; namely, regarding disasters as a systematic occurance. Blaikie et al. (1994) viewed disasters as an integrated effect of interactions between disaster-affected bodies and hazard-inducing factors. Ma emphasized the need for a comprehensive survey and illustrated disasters from the viewpoint of a systematic occurrence (Zhao and Ma 1993; Ma and Yang 1994).

Shi (1991, 1996) proposed *the disaster system theory*, and proposed that natural disaster damage could be viewed as a synthesized impact among environment factors, hazard inducing factors, and disaster-affected body. The severity of disaster was determined by the disaster-formative environment, hazard formation factors, and disaster-affected bodies. Consequently, he thought that natural disaster risk assessment should include three parts, namely; Assessment of the stability of the disaster environment; Assessment of the fatal potential of hazard factors; and an assessment of the vulnerability of the disaster-affected bodies.

At present, the consensus is seen that as a situation in which the disaster occurrence is an interactional result of different factors. The discrepancy among researchers focuses on points that are not alike. Some researchers emphasize disaster formation factors, while others attributed the disaster partly to the disaster-affected body.

For the disaster formative environment, most researchers view it is an indirect factor, albeit one that is capable of increasing or reducing the severity of the disaster (Table 18.1).

Taking an overall view of typhoon disasters, the system theory seems to well-document the related research (Shi 1996). The disaster formative environment has been described as a natural environment, including climate, geological surroundings, hydrology, soil, vegetation, etc., with all factors interacting mutually via diverse means within the area impacted by the typhoon. Shi (1996) pointed out that the environment played a decisive role in the overall impact of typhoons. Complexity, strength, loss, cybotaxis and mass appearance of typhoon disasters were influenced by the disaster formative environment.

A disaster-affected body is referred to by reference to all objects attacked by the typhoon. This includes personnel, crops, water resource facilities, communications, traffic and buildings, etc. In terms of disaster formative factors, main disaster

Table 18.1 Main theories on causes of disasters

Name	Contents of research	Characteristics	References
The theory of hazard formative factors	Risk and damage assessment of disasters	Attributes that cause disasters or hazards	Tucker et al. (1994)
Disaster-formative environmental theory	Change of disaster environment	Emphasis on the environmental impact of disasters	Park (1995)
The theory of disaster-affected bodies	Vulnerability of disaster-affected bodies	Emphasis on the role of disaster-affected bodies	Turner and Meyer (1991), Carrara and Guzzeth (1995)
The system theory	Risk and damage assessment of disaster	1. Regards disasters as a system 2. Attribute the cause of the disaster to an interaction among actual hazard, environment and disaster-affected bodies	Shi (1996)

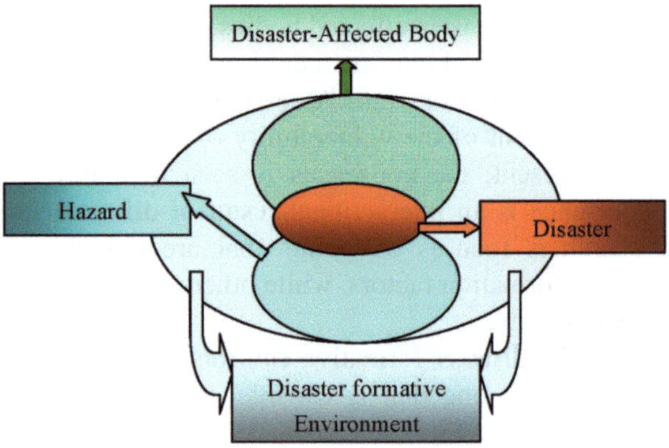

Fig. 18.1 Typhoon disaster system

factors inland consist of gales, rainstorms, and flooding induced by rainstorms. One of the primary features is *multeity*. Another characteristic of typhoon disasters is the effect of chain-reaction (Shi et al. 2006).

The four key factors determining the severity of disaster are the stability of the disaster formative environment, the estimation of the fatalness of the disaster formative factors, the vulnerability of the disaster-affected bodies, and the ability of the disaster-prevention (Chen et al. 2007) (Fig. 18.1).

18.3 Assessment of Typhoon Risks

It is very difficult to accurately predict incidents related to typhoons due to many factors. Risk is defined as the expected loss of life, number of persons injured, property damage and the economic activity disrupted due to a particular hazard in a given area and as related to a specific time period. To be sure, uncertainty always exists, therefore, risk is inevitable.

Disaster risk assessment includes analysis of risk disaster formation factors, assessment of vulnerability of disaster-affected bodies and the assessment of the stability of the disaster formative environment. In a narrow sense, assessment of risk is mainly aimed at assessment of the potential damage that might result from hazard formative factors, namely, from the identification and recognition of the total danger involved (Bai and Huang 2000). The approaches to assessment of typhoon risk in research literature may be summarized as index assessment methods, structural assessment methods and statistical simulation methods.

18.3.1 Index Method of Risk Assessment

The index method is a comprehensive statistical evaluation method in which the synthesis of multiple indexes is used to form an index used to measure the potential of the risk. The main index assessment includes a scoring method, a vague mathematical method, the Grey level analysis, an analytic hierarchy process (AHP). For example, Zeng (1996) conducted preliminary study of typhoon disaster risk involved in South-Eastern China by scoring various elements. Ding and Shi (2002) adopted this vague approach to evaluate typhoon disaster risks in Guangdong province in China. The objective of this analysis was to assess a regional typhoon exposure assessment elaborated via three indicators: Population density; A GDP per capita calculation and an Agriculture proportion/to GDP estimation of the probability of hazardous formative factors and vulnerability in a studied area. All these were incorporated with the exposure to the affected-bodies and were analyzed to measure the typhoon risk. Davidson and Lambert (2001) adopted an analytic hierarchy process (AHP) to develop a composite index to compare the risk of potential hurricane disaster to U.S. coastal counties. Zhou (2004) also adopted the same method (i.e. AHP) to evaluate typhoon disaster risk in which, not only were hazard inducing factors calculated, but also disaster-affected bodies were considered. Additionally, this approach was used to evaluate typhoon disaster risks but, nevertheless more indicators were employed (Chen 2007). It shall be noted in the following that formative disaster environment impact estimates related to hazard severity, in Chen's study, were examined while risk composite indexes were calculated.

In general, factors in risk assessment are typically considered more comprehensive when viewed from a unilateral disaster-causing point-of-view as compared to the vulnerability of hazard bearing bodies or from disaster-breeding environmental stability perspectives. The second characteristic for consideration is that statistical models are adapted from the simple to the complex; such as if from a scoring method toward an analytic hierarchy process, employing a fuzzy comprehensive evaluation and complex grey level method. The basic common-point in this kind of literature is that a composite index is calculated and a risk-zone division is made.

The index is provided for the public to provide a simple risk assessment in an easy-to-understood format. In addition, the resource allocation and high-level planning decisions in typhoon disaster prevention often require a summary of the relative risk to each county (or state or other region), and a general understanding of the causes or reasons for that risk assessment, but does not have to be an extremely fine resolution. Under these conditions, the index method is suitable.

18.3.2 Structural Assessment of Disaster Risk

Here is another distinctive approach in risk measurement that pays attention to the infrastructure, such as buildings, bridges and dams, of the disaster-affected body and this is called "structural risk assessment" (Unanwa et al. 2000). For example, Yoshida (1998) proposed simplified models for analyzing the elastoplastic response of steel dampers and evaluating fatigue-damage induced by winds. Jin et al. (2000) used statistical data obtained in China between 1949 and 1990 to analyze the frequency and strength of typhoons. The method employs the relation between them and the average global temperature. In this study, the probability distribution of wind speed and the load of low buildings, respectively, were obtained on the basis of historical data. The finite element method and experimental comparisons were used to obtain the probability distribution of the lateral force resistance of a three-story brick/concrete building. Finally, the reliability index of this kind of building is calculated. Moreover, Li et al. (2002) aimed at evaluating the effect of a typhoon as fatigue damage on the steel decks of long-span bridges. The strain-time–histories, at critical locations of deck sections of long-span bridges, were investigated while typhoons passed the bridge area, by using on-line strain data acquired from the structural health monitoring system permanently installed on the bridge. The fatigue damage models based on Miner's law and the continuum damage mechanics were applied to calculate increments of fatigue damage attributed to the action of a typhoon.

18.3.3 Statistical Simulation Method of Risk Assessment

The third method is based on statistical simulation, which is then overlaid on the limited historical data. Instead of an index, the monetary value is used to measure risk and loss using a simulation method. With simulation of frequency coupled with intensity, and generally including path simulation, wind fields of typhoon force, and various types of building structure disaster-bearing structure simulations, a vulnerability matrix is established to calculate the expected loss. Another category of wind risk simulation utilizes only simulation of the typhoon wind field, the intensity, and occurrence probability…and without considering building structures…is another in which, some statistical methods and models are employed, such as, the Monte Carlo simulation method (Huang et al. 2001), unsteady Poisson process and the Markov chain random path model (Rumpf et al. 2009), and symmetric equilibrium model (Emanuel 2006).

The characteristics and a comparison of these three methods are shown in Table 18.2.

Index assessment, structural assessment and statistical simulation are three main methods in risk assessment; Their main features are shown in Table 18.3.

Table 18.2 Methods and characteristics of typhoon risk assessment

Three major methods	Method	References	Characteristics
Index assessment	Scoring method	Zeng (1996)	• List related factors of risk • Represent risk with scoring • Simple, but the subjective optional
	Fuzzy evaluation	Li et al. (2009)	• Data of small sample index • Calculates the fuzzy risk • Poor accuracy
	Grey level analysis	Cheng (2004)	•Combine the grey relational analysis and AHP •Some randomness in determining the weight of factors
	Analytic hierarchy process (AHP)	Chen (2007)	• Combine the grey relational analysis and AHP • A comprehensive consideration of various factors: disaster environment, hazard, prevention ability, and disaster-affected body
Structural assessment	Uses historic data	Yoshida (1998)	• Use historical typhoon data • Conducts the risk assessment of the infrastructure • High precision
Statistical simulation	Use simulating data to estimate the damage of infrastructure	Hamid et al. (2010)	• Use stimulating typhoon data • High precision • Not consider rainfall in typhoon risk
	Use simulating data to estimate the risk of typhoon-prone regions	Rumpf (2009)	• Overcome the limitation of historical data • High precision • Not consider rainfall in typhoon risk • Estimate risk in typhoon-prone area by simulating path frequency, intensity, and occurrence probability of typhoons

Table 18.3 Comparison of index and simulation assessment

Method	Flooding caused by typhoon	Object	Result	Advantages	Disadvantages
Index assessment method	Including	Damage grade in macroscopic region; Some specific typhoons	Composite index	Easy to understand; Provide summary information	Poor accuracy
Structural risk assessment	Not including	Damage of concrete building structure	Bearing ability of the structure	Provide reference for engineer designing	No consider rainfall risk
Statistical simulation method	Not including	Infrastructure and houses in typhoon-prone areas	Expected insurance loss	Overcome the limitation of historical data	No consider rainfall risk

18.4 Assessment of Typhoon Damage

In contrast to typhoon risk assessment, assessment of typhoon disaster loss has a different emphasis. Generally, risk assessment is an estimate of latent damage should a typhoon strike in the future. In contrast, disaster loss assessment refers to a specific damage caused by an occurring typhoon or a definite typhoon that is in the wind. Similarly, the assessment or measurement is based on plentiful data of historical and real typhoon disasters.

Obviously, disaster loss can be estimated by a statistical survey. Albeit based on historical datasets, some statistical models may be established to evaluate and forecast disaster situations instead of using surveys. Such models economize on manpower and save physical resources to some extent.

At present, the most frequently used approaches and contents of assessment for typhoon disaster loss rely on the use of combined hazard formative factors with factors of disaster-affected bodies to establish models used to estimate disaster losses. This estimation may be conducted during the disaster or afterwards, or even before the disaster. Similarly, the assessment made during the typhoon disaster was carried out by adopting hazard formative factors and elements of disaster-affected factors. Frequency and intensity of hazard factors, combined with vulnerability estimates and were calculated and used to form statistical models in order to monitor or forecast disaster damages. However, it is worthwhile to note that the assessment is not aimed at a specific typhoon-prone area but rather a specific typhoon. While post-disaster estimations are conducted by actual hazard indicators, in terms of approaches to assessment, regression methods, fuzzy comprehensive assessments and analytic hierarchy processes were mainly used.

18.4.1 Regression Method to Estimate Damage

For regression analysis approach, hazard formative factors including typhoon central nearby maximum wind velocity (TCNMWV), typhoon central ground minimum pressure (TCGMP), rainfall etc. are used to establish assessment models and forecast disaster grades, while hazard factors including causality, collapsed housed and drowned farmland may be used to simulate disaster damages or used directly to form a composite index after disaster.

Some empirical researches on typhoon loss assessment in the past literature were examined to measure the severity of typhoon disasters. Yu and Chen (1993) selected some indicators including the number of collapsed houses, farmland disaster measures and casualties as dependent variables; he used typhoon central pressure, process rainfall, landfall, typhoon wind power grade, bottom central nearby maximum velocity (BCNMV) as endogenous variables to establish simultaneous equations allowing estimates and a forecast of the economic losses. Lin et al. (1995) adopted power function to establish an assessment and forecasting model by means of analyzing and studying the disaster caused by the tropical cyclones landing on Guangdong Province in China. Liu et al. (2003) conducted a multiple linear regression to analyze typhoon disaster damage statistically with datasets from 1950 to 2000 within the area of Guangdong Province, China.

A typical feature in the above regression is directly simulating loss indicators, such as collapsed houses, and casualty by meteorological factors. However, the loss factors can be compressed into a synthesized index, and then the index can be used in the econometric model as a dependent variable. Lu (1995) adopted disaster loss factors corresponding to causalities, flooded farmland, and collapsed houses. Three TC disaster indices have been evaluated by using the mathematical statistics from Shanghai, China from the period 1949 to 1990. Based on these indices, five grades of storms have been objectively established. Other econometric models were used to evaluate and forecast disaster losses using damage composite indices as the dependent variable (Qian et al. 2001; Wei et al. 2012).

18.4.2 Index Assessment Method to Measure Damage

The first index approach examined is the Fuzzy comprehensive assessment method, which applies the Fuzzy mathematic theory to typhoon disaster assessment. It was used to evaluate the damages caused by landfall typhoons in Guangdong, China during the period of 1990–1996 (Fan and Liang 2000), and to assess the damages caused by typhoons in Chinese Taiwan (Chen et al. 2011). Furthermore, the Fuzzy comprehensive method was applied to assessment damage grading using ten factors as disaster impact factors. For evaluation and proof, an assessment test was carried out for landed typhoons during the years from 2000 to 2006 (Ma et al. 2008).

The second index method is the Analytic hierarchy process (AHP) for risk assessment and to evaluate the impact of typhoon disasters (Li et al. 2002). In Li's study, the indices involved in the assessment model are divided into three hierarchies. The uppermost in the hierarchy is the synthetical influence index of typhoons which can be calculated using four factors on the second hierarchy: the intensity parameter of tropical cyclone, synthetical geographic parameter, synthetical wind parameter and the synthetical rain parameter. The basic hierarchy consists of 17 single item indices. After the relative importance of every item, or factor, is decided via Experts Scoring, the influence degree of the TC disaster could be judged and the quantitative direct loss could be estimated.

Other index assessment methods include the Extension method (Liu et al. 2010) and Grey correlation method (Zhou 2004, Wang et al. 2010). Both of these methods select indicators to form index. The former employs an extension set theory and the matter element theory, while the latter adopts the Grey system theory.

18.4.3 Data Fitting Method for Damage Assessment

Due to the complicated nonlinear relationships between the influencing factors and disaster damage, some nonlinear fitting techniques, such as neural networks have also been used for disaster assessment. For example, Lou et al. (2012) employed the BP neural network model to assess damage of typhoons affecting Zhejiang Province in China in 2007 and 2008. In general, the neural network system has a high level of precision.

18.4.4 Other Methods Used for Damage Assessment

In typhoon disaster assessment research, some simple empirical research has also appeared in related literature. For example, Yao and Liu (2001) applied Feng's quantitative calculation method (Feng 1999) for disaster loss to obtain disaster grades. A simplified synthesized grade was obtained by compositing three indices of causality, drowned farmlands and collapsed houses as incurred in Zhengjian Province in China.

It is worth noting that the Index method, Econometric regression method and Data fitting methods may all be applied to forecast before and after disaster damages. Namely, if meteorological factors were selected as an independent variable to be input in a model, the output of the model would play a role in forecasting disaster grades. Instead of meteorological factors, post-disaster indicators, such as the number of collapsed houses and casualties, were selected as input variables in statistical model, therefore, this kind of model is applied after the disaster. The major assessment methods are summarized in Table 18.4.

Table 18.4 Index assessment approach of damage

Assessment method		Pre-assessment			Post-disaster assessment		
		Selection of variable and index	Index of assessment result	References	Selection of variable and index	Index of assessment result	References
Regression analysis	Multiple linear	• Affected-bodies density index • Precipitation index • Wind index	•Economic losses •Collapsed houses •Damage to the houses	Lin et al. (1995)	•The number of injuries •The number of houses collapsed •Flooded farmland area	•Damage index	Qian et al. (2001)
	Power function	• Affected-population • Affected-crops area • Direct economic loss	• Composite index	Zhao et al. (2011)	• Post-disaster index	–	Absent[a]
	Simultaneous equation	• Rainfall of process • Maximum wind speed • Minimum pressure	• The affected area • Number of caustic houses • Number of injuries	Chen et al. (1997)	• Post-disaster index	–	Absent
Fuzzy evaluation		• The maximum process rainfall • Disaster vulnerability • Other ten factors	• Composite index	Ma et al. (2008)	•Agricultural disaster area • The death toll • Other 9 post-disaster indicators	• Composite index	Fan and Liang (2000)

(continued)

Table 18.4 (continued)

Assessment method	Pre-assessment			Post-disaster assessment		
	Selection of variable and index	Index of assessment result	References	Selection of variable and index	Index of assessment result	References
AHP	• Seventeen indicators for: wind, rain, geographical, cyclone intensity	• Composite influencing index	Li et al. (2006)	• Indicators of post-disaster • AHP comprehensive evaluation	–	Absent
Extension method	• Rainfall • Rainfall intensity • Maximum wind velocity	•Grade of disaster loss	Liu et al. (2010)	• Using data of post-indicators	–	Absent
Grey correlation	• Establishing grey correlation coefficients using disaster indicators	–	Absent	• The death toll • Collapsed or damaged houses • Affected crops area • Direct economic loss	•Integrated disaster loss grade	Wang et al. (2010)
Neural network	• Process rainfall of each region •Process maximum wind speed of each region	•Direct economic loss index	Lou et al. (2012)	• Training by using data of post-disaster	–	Absent

[a] "absent" refers to the method is feasible in theory, but there are no corresponding results

18.5 Research on Typhoon Disaster Economic Losses

Different from the risk or damage levels, the economic losses of typhoon disaster can be directly quantified in terms of monetary value. At present, beside estimation and forecast of direct economic loss, other topics include classification of economic loss by category and research on indirect economic loss.

18.5.1 Estimation and Forecast of the Value of Losses

Typhoon disaster assessment tends to provide a disaster grade on typhoons that have occurred or may be established for an area that suffered from a typhoon disaster. It aims to evaluate the grade or grade levels of the disaster as can be determined via the conclusions of empirical studies. A disaster grade is calculated using a composite index and stands for the severity of the typhoon disaster. Albeit the grade is beneficial information about a disaster, it is not or cannot be a totally accurate result. And, as a consistent standard is lacking in the calculation of the grade, there exists a failure to compare the features or results among disasters. Following this logic, an estimation of direct economic losses caused by a typhoon is more efficient.

Some meteorological factors or disaster consequence factors have been adopted as the independent variable in statistical models. Direct economic losses may be estimated and forecasted by the model. However, such studies are still quite scarce. In typhoon disaster assessment, statistical forecast of direct economic losses deserves to be assigned more importance.

18.5.2 Classification of Economic Losses

Defining economic losses and their classifications is also one of important elements of disaster economics. Yang and Xu (2008) made a classification of losses caused by natural disasters according to features at different periods during the process of the disaster. On this basis, economic losses for pre-disaster, ongoing disaster and post-disaster were evaluated and probed. Huang and Yang (1994) investigated the classifications of indirect economic losses. They classified the types of indirect losses related to the disaster, and defined a new type of indirect damage, namely, the indirect cost of shutting down or slowing down production due to the linkage of industrial sectors, the cost of replacement products and the cost of investment premiums.

18.5.3 Indirect Economic Losses Related to Disasters

Over time, researchers have examined the economic impacts related to disasters and have measured the indirect economic losses of the disaster as well (Huang and Yang 1994; Shao et al. 1994). Shao et al. (1994) used the input–output data to point out the linkage between industrial sectors and have forecast the indirect costs due to a recent seismic disaster in Shanghai. Lu et al. (2002) discussed the way to express direct and indirect economic losses using the input–output table; he formulated a model to evaluate the impact of natural disasters on the national economy, in which, not only the losses in terms of gross output were pinpointed, but also the added value in different departments of the economic system could be evaluated. Empirically, the losses of agricultural output due to natural disasters were estimated, and the effects of these agricultural losses on other departments and areas of the national economy were analyzed based on the input–output tables used in 1995 and 1997. Zhang and Sheng (1995) concentrated on the relationship between disaster losses and economic growth. In their research, the influences of disaster on economic growth rate are estimated on the basis of the Harrod-Dorma Model and a new method for evaluating the indirect losses of disasters was developed.

The object, or focus, of disaster economics research tends to be natural disasters. Related research is quite scarce on specific disaster types such as typhoon disasters. Typhoon disasters have their particularities and estimations of their macroeconomic impacts due to typhoon indirect economic losses still needs to be investigated in the future.

As mentioned above, the typhoon disaster evaluation theory focused on risk assessments and disaster damage assessments as well as on the classification and estimation of economic losses. The three assessments are compared in Table 18.5.

18.6 Practice for Assessment Model and Decision Support System

In practice, the assessments and forecasts of typhoon hazards have made some important advance including making rules and establishment of a climate atlas used for estimations of damages by using various statistical models and employing information technology to design the decision support system to forecast the disaster.

Table 18.5 Comparison of risk, damage, and value of loss assessment

	Contents	Objectives	Time-disaster	Variable of representing result	Object	Approach
Risk assessment	Loss probability; severity of loss (i.e. the risk)	• Frequent and planned assessment • Provide reference for risk management	Pre-disaster	Risk index; monetary value.	Typhoon-prone region	Statistical simulation; Statistical comprehensive evaluation; Structural assessment method
Grade of damage assessment	Grade of the damage of specific disaster process	• Assessment during typhoon • Emergency assessment • Provide service in emergency management	Middle-disaster Post-disaster	Grade of disaster; Monetary value of damage	Specific typhoon; Disaster-affected region	Statistical comprehensive evaluation method
Estimation of loss value	Calculation of loss value in risk; Damage assessment	• Daily or emergence assessment	Pre-disaster During-disaster Post-disaster	Monetary value	Typhoon-prone region Specific typhoons	Statistical survey after the disaster Estimate the value of loss by statistical model

18.6.1 Typhoon Atlas and Rules for the Services

In China, The Central Weather Bureau formulated the *Typhoon Loss Statistical Standard* and established rules for typhoon-related professions and services, for data collection of meteorological hazards and set up trials for regulations of assessment. Another agency in China, The Shanghai Typhoon Research Institute, compiled CMA-STI best track datasets of tropical cyclones and set National standards for the tropical cyclone category. They also drew up a Chinese tropical cyclone climatic atlas after each wind season, by year, according to the rules from the typhoon profession and services. Too, they established a retrieval system for TC in the Northwestern Pacific.

18.6.2 Actual Models for Disaster Assessment

Meteorological and disaster prevention agencies in different countries have developed assessment models and adopted them in route to disaster prevention efforts. Once assessment models are established, whenever strong tropical cyclones occur, meteorological departments may, in due course, assess and release hazard forecasts for disaster prevention and mitigation. For example, the Chinese National Meteorological Center and local meteorological sectors established some pre-assessment models to assess typhoon hazards and provided the bases for disaster prevention and mitigation. To these the Zhejiang Meteorological Center applied the Fuzzy mathematical approach, the Meteorological Bureau in Fujiang Province and Jiangxi Province adopted the regression model, Guangdong Province applied the analytic hierarchy process to a quantitative pre-estimation of damages, and the Shanghai Meteorological Bureau adopted the empirical orthogonal function.

The above assessment models are practiced in routine forecasts of typhoons move forward with some problem. Indicators for selection of the assessment model show discrepancies and the categories of assessment are not consistent. For example, Jiangxi and Guangdong Provinces in China obtain actual reference values of damages, while the National Meteorological Center and Shanghai Meteorological Bureau calculated hazards beforehand using pre-assessment grades. Secondly, vulnerability to typhoons has discrepancies related to diverse industries. For example, agriculture is more sensitive to typhoon disaster. Suitable diverse industry typhoon damage assessment models haven't really got going yet. Finally, these pre-assessment models were aimed at landfall typhoons in one district before a typhoon disaster happened, more large-scale national research models still need to collect economic data within the National scope.

In America, homeowner insurance rate-making is an important public policy issue. To encourage the use of sophisticated actuarial methods to assure consumers are charged lawful insurance rates, the Florida Office of Insurance Regulation (OIR), sponsored the development of a Florida Public Hurricane Loss Model (FPHLM) to assess the risk and project insured residential losses (Powell et al. 2005; Chen et al. 2009). Hamid et al. (2010) describe the basic structure of FPHLM (Fig. 18.2).

Theoretical models are only able to estimate and forecast disaster damage to an extent. When a typhoon disaster occurs, effective response actions can significantly reduce the damage caused. Emergency managers require plenty of information before setting up appropriate strategies. The data gathering and analyzing processes is complex and the time is constrained during an emergency. Thus, assessment using theoretical models can only partly treat the problem. Generally, in the process of a typhoon disaster, the challenges confronting local authorities, who must decide if and when to initiate evacuations from tropical cyclones and what measures to take for disaster mitigation. To reduce the damages and losses

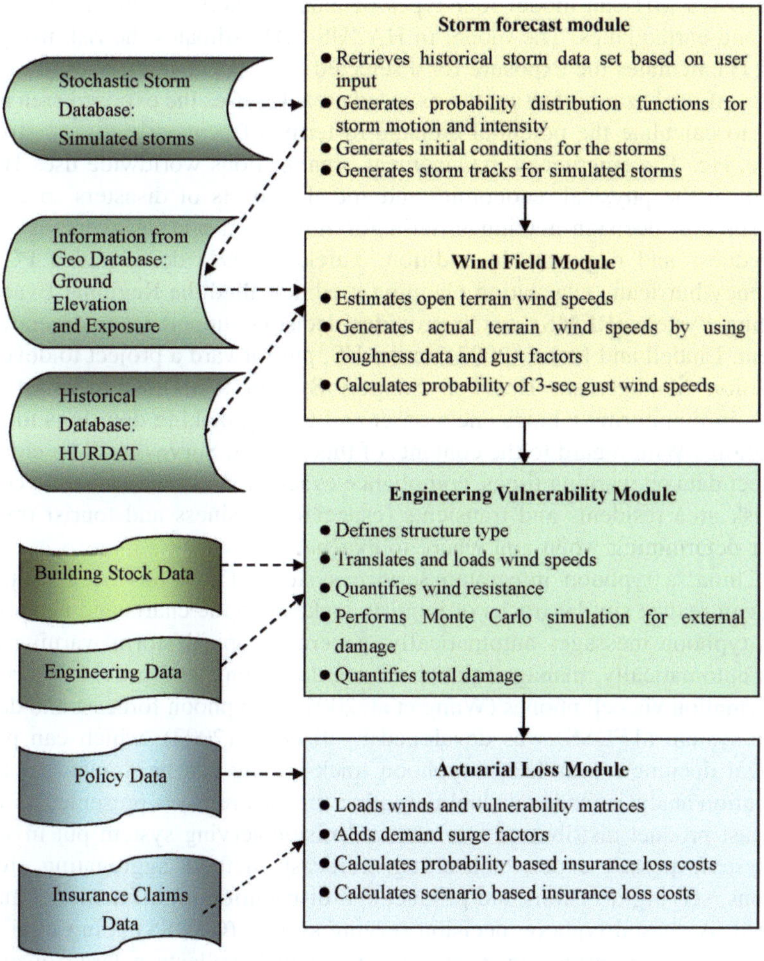

Fig. 18.2 The basic structure of FPHLM

caused by typhoon hazards, an integrated and complete decision support system for decision-makers is necessary. Consequently, a typhoon decision support system was developed during past decades.

18.6.3 Decision Support System for Typhoons

Decision support system (DSS) was developed in recognition of managers, whose job is to apply intervention techniques, and to assist them to make decisions. One of the famous hazard DSS is *HAZUS* which is a GIS-based natural hazard loss estimation software package developed and freely distributed by the Federal Emergency Management Agency (FEMA) in the USA since 1997. The new version HAZUS-MH can model four types of hazards: flooding, hurricanes, coastal surge, and earthquakes. The model in HAZUS-MH estimates the risk using three steps. (1) calculates the exposure for a selected area; (2) characterizes the level or intensity of the hazard affecting the exposed area; (3) uses the exposed area and the hazard to calculate the potential damage in terms of economic losses, structural damage, etc. The emergency management organizations worldwide uses HAZUS to estimate the physical, economic, and social impacts of disasters so they can make suitable decision-making criteria for mitigation and recovery as well as preparedness and response. In addition, Tufekci (1995) developed a PC-based emergency hurricane evacuation planning module called the Regional Evacuation Modeling System (REMS) which provides decision support to emergency management. Lindell and Prater (2007), in the US, put forward a project to develop an Evacuation Management Decision Support System (EMDSS) to assist public officials in monitoring a hurricane's onset and to help making decisions to initiate evacuations. With regard to the contents of this project, Surveys will be conducted to collect data on warning times, compliance evacuation rates, evacuation costs for both risk area residents and transients (especially business and tourist travelers) and for determining when and where to evacuate.

In China, a typhoon integrated service system (TISS) has been established which can realize similar tracks of inquiry, make synoptic charts and nephograms, search typhoon messages automatically, generate tropical storm warning manuscripts automatically, manage related service documents and provide short bursts of information via cell phones (Wang et al. 2001). A typhoon forecast and decision service system (TFDSS) was developed by Ji et al. (2003), which can perform historical document searches, a typhoon track-tracing can be drawn, a real-time information analysis made available, weather forecast releases presented as well as a forecast product distribution interactive decision serving system put in service. The system posses expert consulting, forecast method aggregating, forecast decisions, serving decisions and product and distribution multifunctional analyses prepared. A coastal typhoon decision serving system (CTDSS) (Ling et al. 2004) has been put into service in Zhoushan city, which collects a large quantity of meteorological information and also on hydrology, the important flood prevention facilities, and water resources etc., It also retrieves the most closely-related historical typhoon reports and obtains decision- service reporting documents.

In Chinese Taiwan, the National Science and Technology Center for Disaster Reduction (NCDR) developed a GIS-based decision support system (DSS) to help the emergency managers in setting up efficient response strategies. The DSS consists of the basic geographical database, the real-time monitor and scans the forecasted rainfall database, the inundation potential database and the debris and landslide potential database. There are four main modules integrated into the subsystem including the rainfall monitoring and forecasting, the estimation of potential inundation areas, the estimation of potential landslide and debris flows, and the management of disaster information. The results of hazard risk analysis which include potential rainfall distribution, inundation and landslide risk areas, early warning messages, and total suggestions covering the next 24 h are finally demonstrated by a decision support subsystem and are designed to help the commander make the right decisions in disaster preparedness and to set the sequence of the response phases (Hsu et al. 2011 ; Su et al. 2010) (Fig. 18.3).

In summary, there are three levels incorporated into the prevention service for typhoon disaster: (1) Provides simple weather information on typhoon. (2) Establish models to forecast disaster damage and then release the information to selected points. (3) The decision-making support system that can provide results of inquiries of the historical information bases, the forecast of disaster damage and the comprehensive service for decision-making support. Some decision-making systems cannot only realize wind disaster assessment, but also realize potential flood damage

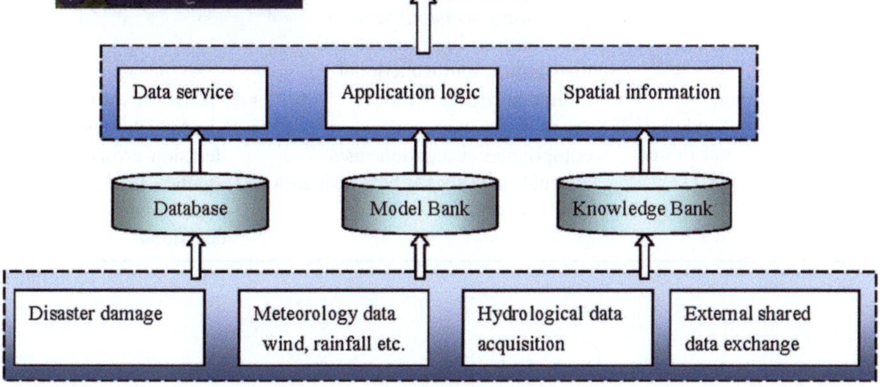

Fig. 18.3 A typical logic structure of a typhoon DSS

Table 18.6 Typical systems for typhoon disaster prevention

Name	Country/ region	System function	Characteristics
NCDR System	Chinese Taiwan (NCDR 2010)	• Rainfall monitoring and forecasting • Estimation of potential inundation areas • Estimation of potential landslide and debris flows • Early warning • Suggestion to preparedness and response	• Forecast over 24 h early • Mainly secondary disaster, such as flooding, landslide and debris flows
TISS	China, Taizhou (Wang et al. 2001)	• Similar track inquiry • Make synoptic chart and nephogram • Search typhoon message automatically • Generate tropical storm warning manuscript automatically • Manage related service document • Provide short information of cell phone	• Real time • Forecast typhoon track and intensity • Not making disaster pre-assessment
CTDSS	China, Zhoushan (Lin et al. 2004)	• Collect information of meteorology, hydrology, flood prevention facilities, and water resource etc • Retrieve the most related historical typhoon • Obtain decision service reporting document	• Provide decision service • Provide hydrologic data and information for facilities
TFDSS	China, Wenzhou (Ji et al. 2003)	• Historical document search • Track tracing • Real-time information analysis • Weather forecast release, forecast product, posse expert consulting • Provide forecasting method, forecasting decision, serving decision, serving product and distribution multifunctional	• Advanced means of product distribution • Mainly track forecast • Not provide decision suggestion • Real time
EMDSS	America (Lindell and Prater 2007)	• Monitor hurricane's onset • Initiate evacuations warning times, compliance evacuation rates, evacuation costs for both risk area residents and transients	• Evacuation warning • Calculate the cost of decision errors • Calculate the economic benefits of correct decisions

(continued)

Table 18.6 (continued)

Name	Country/ region	System function	Characteristics
HAZUS	America (Vickery et al. 2006)	• Provide map templates to support rapid impact assessment and disaster response	• Multi-hazard assessment • Comprehensive assessment for affected-body and hazard
		• Calculate the exposure for a selected area • Characterize the level or intensity of the hazard affecting the exposed area • Calculate the potential economic losses, structural damage, etc	• Calculate loss
REMS	America, Florida (Tufekci 1995)	• Hurricane emergency management • Forecasting of evacuation time • Provide information of the traffic flow on a given transportation road network	• Real time • Make evacuation decision

assessments. However, the factors of geology, geography, society and economics are not effectively integrated into any of the systems. The decision service system for typhoon disaster is still not mature (Table 18.6).

18.7 Summary

A number of attributions have been obtained related to typhoon disaster assessments, all of which are reviewed and addressed in this chapter. Typhoon disaster system theory has been elucidated via research of various bodies of literature, many approaches, including statistical simulating and other comprehensive evaluation theories have been applied in risk and damage assessments. Analyses of pre-disaster, during-disaster and post-disaster damages caused by typhoons have been performed for the purpose of disaster prevention and mitigation. These are especially important when confronted with the decisions of emerging crises. The decision-support systems that have been adopted play an important role in the practice of disaster prevention. However, there are still some problems:

1. Identifying key selection of indicators: Some assessments consider only hazard-formative factors. Other adopted indicators reflect only the vulnerability of disaster bodies and hazard formative indicators. Some studies use three aspects of indicators and add the capacity of disaster prevention and reduction. Albeit failure to effectively integrate these factors with typhoon tracking, strength and other factors could hamper the comprehensive assessment via incorporating natural and societal attributes effectively and still are not taken into the formula.
2. Quantification: Assessment is aimed at quantifying damage and at allowing results to be compared among diverse disasters, however; the quantification of

typhoon risk or damage does not share consistent criteria across the board. Diverse composite indices were selected in the assessment models thus it is hard to make comparisons among different periods and different regions.

3. The assessment model: In early methodologies simple scoring models were undertaken in damage assessment. Then more complex approaches were applied such as the analytic hierarchy process and the Fuzzy mathematical assessment. In all these models, the resulting assessments are somewhat subjective due to weighting coefficients being selected subjectively. In addition, the econometric model, designed to measure indirect economic loss or macroeconomic impact is not fully established and no empirical studies have been conducted. Overall, a comprehensive statistical model for assessment needs to be improved.

Focusing on the above problems, we suggest:

1. Combing macro-assessment with micro-assessment. Such combinations of micro-assessment and macro-assessment are worth investigation which can provide more information for disaster prevention.
2. Assessment aiming at specific industries should be undertaken. The adaptive capacity among diverse industries varies widely. It is likely meaningful to establish the assessment model for typhoon damage in order to satisfy the requirements of diverse industries.
3. Reinforcing research on decision support systems is needed. It is imperative to improve the function between man-and-machine to optimize the interactive qualities of DSS for typhoons. The aim would be to make it to play a more pertinent role and make it perform to the greatest advantage.
4. Cross-disciplinary assessments are essential. Typhoon disasters bring about economic losses and have a significant direct or indirect impact on macroeconomic element of society such as population, housing, human settlement environment, education and health and on the health and life infrastructures in society. It is worthwhile to further examine other fields such as that of disaster sociology, of calamities and disaster psychology to address the cross-disciplinary assessment of these important issues.
5. Typhoon disaster assessment shall likely include the important hazard assessments of risk, loss, ecological damage and benefit assessment for disaster reduction and prevention. Research in the future in these aspects needs to be given more importance.

Acknowledgments This study was supported by the following research projects awarded to DL Tang: (1) National Natural Sciences Foundation of China (31061160190, 40976091, NSFC-RFBR Project-41211120181 of DL Tang and D. Pozdnyakov); (2) Guangdong Sciences Foundation (2010B031900041, 8351030101000002); (3) Innovation Group Program of State Key Laboratory of Tropical Oceanography (LTOZZ1201).

References

Bai, H.L., Huang, C.F.: Natural disaster fuzzy risk. J. Nat. Disasters **9**(1), 47–53 (2000)

Blaikie, C., Davis, I.P.T., Wisner, B.: At Risk: Natural Hazards, People's Vulnerability and Disasters, pp. 140–148. Routledge, London (1994)

Carram, A., Guzzeth, F.: Geography Information System in Assessing Natural Hazards, pp. 15–25. Kluwer Academic Publisher, Dordrecht (1995)

Chen, X., Sheng, J.R., Cheng, J.: Application of DELI in disaster economic loss assessment-taking analysis of trend of economic loss in typhoon disaster in Fujian province as an example. Sci. Disaster **22**, 31–35 (1997)

Chen, W.K., Sui, G.J., Tang, D.L.: Predicting the Economic Loss of Typhoon by case Base Reasoning and Fuzzy Theory. In: Proceeding of the 2011 International Conference on Machine Learning and Cybernetics, pp. 254–257, Guilin (2011)

Chen, S., Chen, M., Zhao, N., Hamid, S., Chatterjee, K., Armella, M.: Florida public hurricane loss model: Research in multi-disciplinary system integration assisting government policy making. Gov. Inf. Q. **26**, 285–294 (2009)

Chen, X.: Risk assessment and zonation of typhoon disasters in Fujian Province. Chin. J. Ecol. **26**(6), 961–966 (2007)

Chen, Y.Q., Tang, D.L., Wang, S.F.: Eddy-feature phytoplankton bloom induced by tropical cyclone in the South China Sea. Int. J. Remote Sens. **33**(23), 7444–7756 (2012)

Davidson, R.A., Lambert, K.B.: Comparing the hurricane disaster risk of US coastal counties. Nat. Hazards Rev. **8**, 132–142 (2001)

Ding, Y., Shi, P. J.: Fuzzy risk assessment model of typhoon hazard. J. Nat. Disasters. **11**(1), 34–43 (2002)

Emanuel, K., Ravela, S., Vivant, E., Risi, C.: A statistical deterministic approach to hurricane risk assessment. Bull. Am. Meteorol. Soc. **87**(3), 299–314 (2006)

Fan, Q., Liang, B.J.: The evaluation of disastrous losses caused by tropical cyclones. Acta Geographica Sinica **55**(11), 52–56 (2000)

Feng, L.H.: Tropical cyclone risk assessment. Ocean Bull. **18**(2), 40–43 (1999)

Hsu, M.H., Chen, A.S., Lee, C.S., Chen, L.C., Lin, F.T., Huang, C.J.: A GIS-based decision support system for typhoon emergency response in Taiwan. Geotech. Geol. Eng. **29**, 7–12 (2011). doi:10.1007/s10706-010-9362-0

Hamid, S., Golam Kibria, B.M., Gulati, S.: Predicting losses of residential structures in the state of Florida by the public hurricane loss evaluation model. Stat. Methodol. **7**(5), 552–573 (2010)

Huang, Y.X., Yang, Z.Y.: Measurement of indirect economic loss. J. Disaster Sci. **9**, 7–11 (1994)

Huang, Z., Rosowsky, D.V., Sparks, P.R.: Long-term hurricane risk assessment and expected damage to residential structures. Reliab. Eng. Syst. Saf. **74**, 239–249 (2001)

Ji, C.X., Ye, Z.Q., Du, Y.Q., Zou, J.: Typhoon forecast and decision-making system. Zhejiang Meteorol. **24**(2), 7–10 (2003)

Jin, W.L., LI, H.B., Hu, Y.: Reliability assessment of building structures under typhoon calamity. J. Zhejiang Univ.—Sci. A **1**(1), 48–55 (2000)

LI, C.M., Luo, X.L., Liu, J.H., He, J.: Application of analytical hierarchy process in the assessment model on tropical cyclone disaster's influence. J. Trop. Meteorol. **22**(3), 223–228 (2006)

Lin, J.S., Luo, J.L.: The evaluation and forecasting models of tropical cyclone disaster landing on Guangdong Province. J. Nat. Disasters **4**(1), 92–97 (1994)

Lindell, M. K., Prater, C. S.: A hurricane evacuation management decision support system (EMDSS). Nat Hazards **40**, 627–634 (2007)

Ling, W., Li, K.Q., Wang, Z.W., Li, X.L.: Typhoon decision-making support system for coastal region. Meteorol. Sci. Technol. **32**(2), 129–131 (2004)

Liu, Y.H., Tang, X.C., Song, L.L.: Typhoon disaster damage research in Guangdong Province. Trop. Geogr. **23**(2), 119–122 (2003)

Liu, S. J., Zhang, J. H., He, Z. W.: Danger assessment of typhoon hazard based on extension method. Yunnan Geogr. Environ. Res. **22**(4), 100–104 (2010)
Li, Z.X., Chan, T.H.T., KO, J.M.: Evaluation of typhoon induced fatigue damage for Tsing Ma Bridge. Eng. Struct. **24**, 1035–1047 (2002)
Lou, W.P., Chen, H.Y., Qiu, X.F., Tang, Q.Y.: Assessment of economic losses from tropical cyclone disasters based on PCA-BP. Nat. Hazards **60**, 819–829 (2012)
Lu, W.F.: Assessment and prediction of disastrous losses due to tropical cyclones in Shanghai China. J. Nat. Disaster **4**(3), 40–45 (1995)
Lu, C., Wei, Y. M., Fan, Y., Xu, W. X.: Quantitatively analytic model for the impact of natural disaster on national economy. J. Nat. Disasters **11**(3), 15–20 (2002)
Ma, Z.J., Yang, H.T.: Natural disaster economic characteristics and societal development. Sci. Technol. Rev. **7**, 61–64 (1994)
Ma, Q. Y., Li, J. Y., Wang, X. Y., Wang, W. G., Gao, X. Y.: A Fuzzy Synthetic Evaluation Model for Typhoon Disaster. Meteorogocal Monthly **34**(5), 20–25 (2008)
Park, C.: Environment Issues. Prog. Phys. Geogr. **19**(3), 379–390 (1995)
Powell, M., Soukup, G., Cocke, S., Gulati, S.: State of Florida hurricane loss projection model: atmospheric science component. J. Wind Eng. Ind. Aerodyn. **93**(8), 651–674 (2005)
Qian, Y.Z., He, C.F., Yan, Y.Q.: Wang Jizhi an assessment of damage index for tropical cyclones. Meteorology **27**(1), 14–18 (2001)
Rumpf, J.H.W., Hoppe, P., Rauch, E., Schmidt, V.: Tropical cyclone hazard assessment using model-based track simulation. Nat. Hazards **48**, 383–398 (2009)
Schmidt, S., Kemfert, C., Höppe, P.: The impact of socio-economics and climate change on tropical cyclone losses in the USA. Reg. Environ. Change **10**, 13–26 (2010)
Shao, Y.H., Huang, Y.U., Yang, Z.Y.: Disaster indirect economic science of disaster. J. Nat. Disasters **3**, 8–11 (1994)
Shi, P.J.: Theory and practice of disaster research. Nanjin Univ. Acta, Naural Disaster Album **3**, 26–31 (1991)
Shi, P.J.: Theory and practice of disaster study. Nat. Disaster Acta **5**, 6–17 (1996)
Shi, P.J., YE, T., Wang, J.A., Zou, M., He, F.: Integrated governance of natural disaster risk. Beijing Normal Univ. Acta **5**, 30–136 (2006)
Su, W.R., Wu, S.Y., Hsu, P.H., Lin, F.T., Chou, H.C.: Development of safe Taiwan Information System (SATIS) for typhoon early warning in Taiwan. J. Syst. **5**, 30–34 (2010)
Sui, G.J., Tang, D.L.: Typhoon disaster and regional economic divergence: evidence form Guangdong. Reform **6**, 18–25 (2012)
Sui, G.J., Pu, H.Y.: Index system of vulnerability by typhoon in coastal areas and emergency management strategies. Reform **3**(145), 154 (2012)
Tufekci, S.: An integrated emergency management decision support system for hurricane emergencies. Saf. Sci. **20**, 39–48 (1995)
Tucker, B.E., Erdik, M., Hwang, C.N.: Issues in Urban Earthquake Risk, pp. 23–28. Kluwer Academic Publishers, Dordrecht (1994)
Turner II, B.L., Meyer, W.B.: Land use and Landcover in global environmental change consideration for study. Int. Soc. Sci. J. **43**(4), 668–679 (1991)
Unanwa, C.O., McDonald, J.R., Mehta, K.C.: The development of wind damage bands for buildings. J Wind Eng. Ind. Aerod. **84**(1), 119–149 (2000)
Vickery, P.J., Skerlj, P.F., Lin, J.X., Twisdale, L.A., Young, M.A., Lavelle, F.M.: HAZUS-MH hurricane preview model methodology. II: damage and loss estimation. Nat. Hazards Rev. **7**(2), 94–103 (2006)
Wang, Y.Y., Guo, X.Q., Liu, Z.: System of typhoon synthesize service. Proc. Acad. Appl. Database Technol. **5**, 33–36 (2001)
Wang, X. R., Wang, W. G., Ma, Q. Y.: Model for general grade division of typhoon disaster and application. Meteorol. Mon **36**(1), 66–71 (2010)
Wei, Z.J., Sui, G.J., Tang, D.L.: Typhoon disaster statistical assessment based on cluster and econometric model. J. Appl. Stat. Manag. http://www.cnki.net/kcms/detail/11.2242.O1.20121026.1125.002.html (2012)

Xiao, F.J., Xiao, Z.N.: Characteristics of tropical cyclones in China and their impacts analysis. Nat. Hazards **54**, 827–837 (2010)

Yang, J., Xu, G.: Disaster economic assessment approach research. Econ. Res. Guide **23**, 81–82 (2008)

Yao, D.R., LIU, X.L.: The assessment of disastrous losses caused by tropical cyclones in Zhejiang Province. J. Zhejiang Univ. (Sci. Edn) **28**(3), 344–348 (2001)

Yoshida, M.: Evaluation of fatigue damage to a damper induced by a typhoon. J. Wind Eng. Ind. Aerod. **74**, 955–965 (1998)

Yu, C.L., Chen, X.H.: Econometric model for estimation and forecasting of economic losses in meteorological disasters. J. Nanjing Inst. Meteorol. **11**(1), 61–72 (1993)

Zeng, L.F.: Risk evaluation of typhoon disasters in the coastal areas in Gangxi. J. Disaster Sci. **11**, 43–47 (1996)

Zhang, Q., Wu, L.G., Liu, Q.F.: Tropical cyclone damages in China 1983–2006. Bull. Am. Meteorol. Soc. **4**, 489–495 (2009)

Zhang, X.D., Sheng, R.F.: The quantitative study on the relationship between disasters and economic growth. J. Nat. Disasters **4**(4), 23–26 (1995)

Zhao, A.X., Ma, Z.J.: Appraising study for the loss evaluation system of natural disasters. J. Nat. Disasters **2**(3), 1–7 (1993)

Zhou, J.H.: Typhoon disaster risk comprehensive study, pp. 10–30. Beijing Normal University, Beijing (2004)

Zhao, F., Liao, R. F., Zhang, L. N., Zhang, Y. X., Chenyang, S. Q.: A Pre-evaluation Model for Typhoon Disasters in China. Journal of Catastrophology **26**(2), 81–85 (2011)

Chapter 19
Concepts and a Framework for Typhoon Disaster Assessment

GuangJun Sui, ZhangJin Wei and DanLing Tang

Abstract Good progress has been made in typhoon disaster assessment in recent years. However, related terminologies are still vague and the framework for typhoon disaster assessment is lacking in the literature. This chapter focuses on the concepts of typhoon disaster evaluation, clarifying the definition of typhoon disaster assessment, and outlining a framework for typhoon disaster assessment. Based on the current evaluation, this chapter classifies composite indices into two types (i.e., absolute and relative), summarizes the definition of these indices in typhoon disaster damage measurement, and points out problems in the typhoon damage grade classification and indicator system, that is, previously, various factors were not fully considered in index selection and the uniform standards are still lacking in disaster grade classifications. This chapter provides a reference for disaster prevention and mitigation by forming the basic concepts and framework for typhoon disaster assessment.

Keywords Typhoon disaster assessment · Concept · Framework · Indicator

G. J. Sui
Guangdong Research Institute for International Strategies, Guangdong University of Foreign Studies, Guangzhou, China

Z. J. Wei
Information College in Guangdong University of Foreign Studies, Guangzhou, China

D. L. Tang (✉)
Research Center for Remote Sensing of Marine Ecology and Environments, State Key Laboratory of Tropical Oceanography, South China Sea Institute of Oceanology, Chinese Academy of Sciences, Guangzhou, China
e-mail: Lingzistdl@126.com
URL: http://lingzis.51.net/

D. L. Tang and G. J. Sui (eds.), *Typhoon Impact and Crisis Management*, Advances in Natural and Technological Hazards Research 40,
DOI: 10.1007/978-3-642-40695-9_19,

19.1 Introduction

Disaster management is important to realize sustainable societal development all over the world. Disaster assessment is an important step and the basis for all decisions made regarding disaster management. Disaster assessment is basic to the foundation for evaluating disasters in all respects, confirming goals for disaster reduction, optimizing measures for disaster prevention, resistance, and rescue, evaluating the benefits of disaster reduction and making decisions regarding disaster reduction (Sui and Pu 2012). Disaster assessment is also an important reference for land exploitation and utilization planning as well as social and economic development planning (Sui and Tang 2012).

The destructive impact of typhoon disasters on many regions of the world has caused increased attention on the study of typhoon disasters (Walsh 2004; Wei and Zhang 1996). Proper typhoon risk and damage assessment is one part of disaster studies. Regrettably, although typhoon disaster assessment has been conducted in many studies and in actual practice, several basic theoretical problems remain unresolved. For example, the concepts and terminology for typhoon assessment remain vague, and a workable framework for typhoon assessment has not been established.

Disaster assessment is a fundamental concept in the science of disasters. Thus, the meaning of the term should be determined and analyzed because the connotations of the term and the extensions of the concept provide the foundation for assessment practice.

This chapter mainly focuses on the basic problems of typhoon disaster assessment. After the introduction to the definitions of typhoon and typhoon disasters in Sects. 19.2, and 19.3 addresses the concepts and framework for assessment in which the distinction of related terminologies is analyzed, and a framework is outlined. Section 19.4 discusses the nature and definition of several composite indices of disaster loss. In Sect. 19.5, the grade classification and the indicator system of typhoon disaster assessment are summarized. The final section is devoted to summary and conclusions.

19.2 Definition of Typhoons and Typhoon Disasters

'Tropical cyclone' is the generic term for a non-frontal, synoptic-scale, low-pressure system originating from tropical or subtropical oceans, systems with organized convection and definite cyclonic surface wind circulation (Schmidt et al. 2009). Technically, a tropical cyclone is principally driven by heat transfer from the ocean (Sellers et al. 1998). Tropical cyclones are categorized according to their maximum wind speed defined as the maximum speed of the wind at an altitude of 10 m, averaged over 10 min (over 1 min in the United States) (Table 19.1).

Table 19.1 Tropical cyclone intensity classification

Classification	The largest wind velocity near the center of the bottom (ms^{-1})	Wind power index
Tropical depression	10.8–17.1	6–7
Tropical storms	17.2–24.4	8–9
Strong tropical storms	24.5–32.6	10–11
Typhoon (or hurricane)	32.7–41.4	12–13
Strong typhoon	41.5–50.9	14–15
Super-strong typhoon	≥51.0	16 or more

Table 19.2 Appellations for tropical cyclone

	Regions	Appellations
Ocean	Northeastern Asia, Northwestern Pacific	Typhoon
	North America, Caribbean	Hurricane
	Indian ocean	Cyclones
	Philippines	Bagyo
Country or region	Australia	Willies
	Haiti	Tylo
	Mexico	Cordonazo

Moreover, tropical cyclones have different appellations in different regions (Table 19.2).

According to the natural disaster system theory, a typhoon disaster system is complicated because of typhoon hazard formative factors, hazard formative environmental factors, and diverse hazard-affected body reciprocity in the typhoon influencing area. Hazard formative factors for typhoon disasters primarily include strong winds, torrential rainfall and storm surges, etc., strength and influencing radius of which are preconditions and motivity to induce typhoon hazards. Typhoon hazard formative environment factors mainly include typhoon movement track and climatological, geographical, hydrological, soil, and vegetation conditions in the typhoon landfall area. All these environmental factors reciprocate the typhoon and can strengthen or lessen typhoon hazards and secondary disasters related to a particular extent thereby influencing the magnitude of the disaster. The hazard formative environment also varies with time. Its stability may be examined through the frequency and intensity of hazard formative factors. The hazard-affected body of the typhoon disaster is primarily comprised of the affected regional population, residential buildings, crops, industrial and mining establishments, water resource facilities, cultivated beaches and transportation. Vulnerability of possible landfall areas to typhoon hazards is a key factor for forecasting and preventing typhoon disasters.

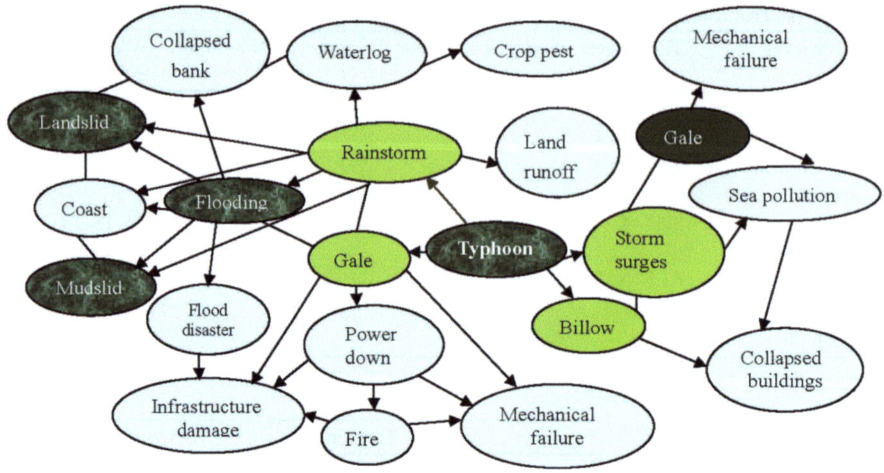

Fig. 19.1 Typhoon-rainstorm chains

Typhoons are characterized by massive energy output, which is mainly released as gales, torrential rains, and storm surges. Gales directly threaten the security of buildings. Torrential rains cause floods, landslides, and debris flows which threaten buildings, roads, and crops and affect the production of crops. Overland run-off caused by typhoon flooding destroys the ecological system near the drainage area.

Hazard chains tend to form other simultaneous chains and cause more severe disasters. The latter is an important feature of hazard chains. The typhoon hazard chain is as follows (Fig. 19.1) (Shi 2002):

19.3 Concept and Framework for Assessment

Typhoon Disaster assessment is an important part of the Science of Disasters and is the basis for disaster forecasts, disaster prevention and rescue. Its concept and approach are derived from generic disaster assessment. It includes the assessment of disaster risk, casualties and economic losses from the perspective of the object and purpose of assessment and is also divided into pre-, ongoing, and post-disaster assessments from the occurrence time of the assessment. However, in emerging literature, these terms easily confuse readers. In much of the emerging literature, all these terms used to be called disaster assessment. Studies have not been able to differentiate these concepts, thereby causing the confusion.

Unequivocally, the basic concept is the foundation for a science. The identification of key definition and connotation is helpful in improving the understanding of the science of typhoon disaster assessment.

19.3.1 Concept Comparison

19.3.1.1 Risk Assessment

Risk is referred to as potential losses associated with a hazard or an extreme event in a given place within a given period of time, and is defined in terms of the adverse consequences (i.e., damages or losses) and the probability of occurrence.

Typhoon risk assessment is the assessment of the likelihood of being struck by a typhoon of great intensity and the quantitative analysis and pre-evaluation of consequences in typhoon-prone regions.

One methodology for identifying the nature and extent of typhoon risk is by analyzing potential hazards and evaluating the existing conditions of vulnerability, both of which could potentially harm exposed people, property, services, sources of livelihoods, and environments. Typhoon risk assessment (and associated risk mapping) includes a review of the technical characteristics of hazards, such as their location, intensity, frequency, and probability; the analysis of exposure and vulnerability in terms of the physical, social, health, economic, and environmental dimensions; and the evaluation of the effectiveness of prevailing and alternative coping capacities with respect to likely risk scenarios. This series of activities is sometimes called the risk analysis process. As described above, a risk may be measured from four aspects, namely; hazard, exposure, vulnerability and prevention capacity.

$$Risk = Disaster \times Exposure \times Vulnerability \times Prevention\,Capacity \quad (19.1)$$

Disaster risk includes different types of potential losses that are often difficult to quantify. Nevertheless, with the information on the prevailing hazards and the patterns of population and socio–economic development, disaster risks can be assessed and mapped, in the broadest sense at least. In general, risk assessment includes following four steps.

- Disaster natural quality: Exceedance probability is the likelihood of a disaster occurrence above an established critical point. Through the identification of occurring frequency or exceedance probability in the relevant region within a period, a relationship can be established between the intensity and the frequency to form a disaster statistical model.
- Vulnerability of the analysis of the hazard-affected body: The capacity for prevention and rescue during the disaster, the resilience capacity after the disaster, and insurance measures are formed into a vulnerability matrix of the hazard-affected body to determine the vulnerability of a hazard-affected body (in a typhoon-prone region) struck by a typhoon of different intensities and probabilities. As long as hazard formative factors occur and amount to a particular strength, disaster probability can be calculated by estimating the vulnerability.
- Risk value modelling and risk loss estimation: The value model is a statistical model of the magnitude value of the affected body in a defined risk area.

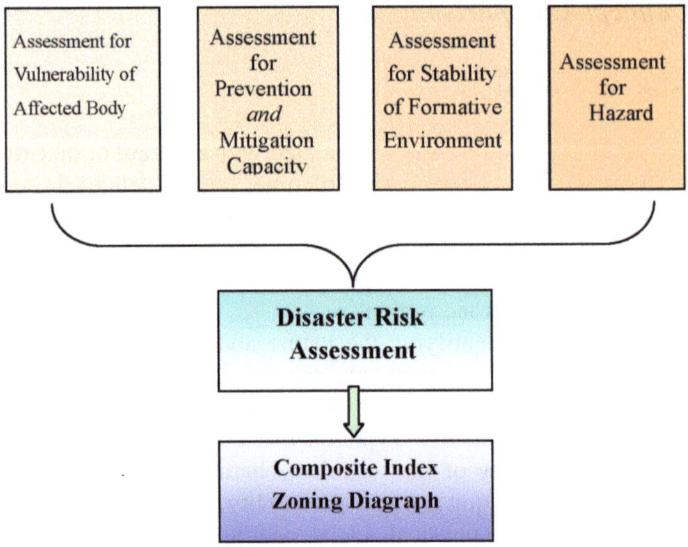

Fig. 19.2 Framework for risk assessment

Possible direct loss, indirect loss, and casualties can be measured in a risk region by establishing the value model for a disaster risk area, that is, by integrating the disaster model and vulnerability of the hazard-affected body. Once a disaster occurs, damage is unavoidable. Therefore, the next procedure of risk assessment is loss estimation. The results of loss assessment also possess probability attributes because of randomicity.

- Risk grade index: Calculating for the hazard occurrence is performed in step one, the vulnerability in step two, and the value of the model display exposure in step three. The magnitude of potential damage may also be calculated using previous results. Once the magnitude of potential damage is obtained, the risk index and its spatial distribution can be solved. Thus, a risk map can be established.

The risk assessment process is outlined in Fig. 19.2.

19.3.1.2 Loss Assessment

Disaster loss (damage) assessment generally evaluates the magnitude of the disaster loss, which is a different concept from risk assessment.

1. Risk assessment is a regular assessment process based on historical documents. It mainly involves estimates of potential damage and possible loss of the disaster-affected body. However, disaster damage assessment evaluates the magnitude of the disaster that is certainly about to occur or of an ongoing event, whose object is a typhoon or disaster-affected body.

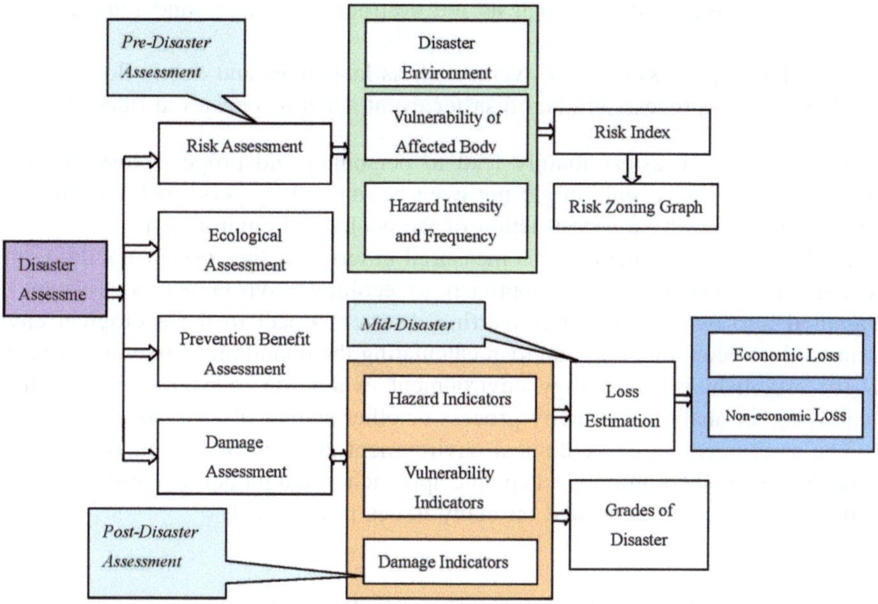

Fig. 19.3 Framework for disaster assessment

2. Disaster risk assessment is latent and anticipative disaster loss assessment. In terms of the period, it is pre-assessment. However, disaster loss assessment may be conducted anytime in the duration of the disaster. Therefore, loss assessment may be classified into pre-, monitoring (during the ongoing disaster), and post-, assessment.
3. Risk assessment is the basis for a long-term disaster prevention project, whereas damage assessment aims at immediate and emergency prevention and rescue.

In terms of concrete items, typhoon disaster loss assessment can be classified into economic loss, personnel casualty and rescue cost assessments. Typhoon disaster economic loss assessment may be classified into direct and indirect economic loss assessments (Fig. 19.3). In terms of approaches, loss assessment is carried out by the following methods: one way is to form a statistical model based on hazard factor indicators and hazard affected-body indicators, which is an assessment done during the disaster. The second way is to use damage indicators after the disaster to form a composite index of disaster severity.

19.3.1.3 Ecology Environmental Assessment

In terms of mechanisms and manifestations, typhoon disasters may be classified into three types:

1. Emergent disaster event, such as the destruction of trees and inundation of farmland
2. Cumulative process disaster event, such as landslides and debris flow
3. Slow-motion process, gradual disaster event, such as ocean red tides (HAB).

The first two disasters mainly lead to personnel and property loss, whereas gradually occurring events might not even be marked by personnel and property loss but is marked by the destruction of the ecological environment.

Ecological environment loss is measured in terms of the cost of rebuilding the ecological environment. The approach to ecology environment assessment is classified into two types: (a) converting disaster impact to the ecological environment into economic loss and then calculating the quantitative assessment result or (b) establishing an ecology environment evaluation index system and then employing an analytic hierarchy process or other statistical analysis approaches.

The assessment of an ecological environment affected by a typhoon is a new issue. Mature and commonly acceptable approaches and modeling are still lacking. Further investigation is needed on many issues in this field.

19.3.1.4 Prevention and Mitigation Benefits Assessment

Disaster prevention engineering involves efforts toward the reduction of probable damage or the likelihood of being struck by a disaster. Typhoon disaster prevention engineering mainly involves flood prevention engineering, mountain reforestation, water conservation in hilly land, dyke strengthening in the lower reaches, dredging waterways, and flood bypass engineering.

Disaster prevention engineering has a definite cost and output. Disaster engineering assessment complies with cost-benefit principles. In terms of approach, reduction beneficial assessment is the analysis of cost-benefit calculated after adopting selected disaster prevention measures. The procedure mainly includes value assessments, beneficial assessment, and opportunity cost calculations.

19.3.1.5 Disaster Tolerance Assessment

Disaster tolerance assessment is the assessment of the risk magnitude that the hazard-affected body can possibly resist. No prevention measure can reduce the disaster occurrence probability to zero, therefore, via perception analysis, the level of risk that the masses can tolerate can be estimated and evaluated, a process called tolerance assessment. Tolerance assessment is used in many disciplines, such as psychology and sociology. In the academy, few studies have explored typhoon disaster tolerance assessment.

19.3.2 Pre-, Mid-, and Post-disaster Assessments

In addition to the classification derived from the content of the evaluation, we can also make the classification according to the period of assessment (i.e., pre-, mid-, or post-disaster assessment).

Disaster pre-assessment involves three factors: (a) hazard strength and frequency; (b) population density, economic development status, and the disaster prevention and mitigation capacity of the disaster area; and (c) disaster degree and occurrence rate in the specified area.

Mid-disaster assessment involves tracking and remote-sense monitoring, the evaluation of the likelihood of losses as the disaster occurs, and further assessments of secondary disasters (done in pre-assessment). Monitoring assessment is the rapid estimation of the severity of the pending disasters as they occur. The monitoring assessment is the provision of a direct basis for rescue decisions and of parameters for loss calculations as the disaster proceeds in order to determine the status of the event and to try to calculate total loss measurements to provide a basis for establishing a prompt comprehensive strategy to manage the aftermath of the event.

Post-disaster assessment refers to measurement estimates and assessments after the disaster and involves the use of a statistical census data to collect information about direct or indirect damage in the disaster region. The result is derived from a district-by-district and item-by-item register. Post-disaster assessment also involves the evaluation of the likelihood of secondary occurrences or disasters, and of the actual results of the event in question, both of which are the basis for the final assessment report. Post-disaster assessment is an important foundation for rescue, general planning, and comprehensive decision-making for follow-up activities.

Given the relationships between the different terms and concepts related to disaster assessment, risk assessment is the assessment made before the disaster (pre-assessment). Disaster damage assessment is the measurement of losses associated with a specific typhoon or event, all of which may be carried out pre-, mid-, or post-disaster. Disaster damage assessments can be a pre-assessment (i.e., before a typhoon disaster), a monitoring assessment (mid-disaster), or the measurement of the final actual damage. The ecological assessment of a typhoon disaster includes both risk and damage assessments, which are undertaken pre-, mid-, or post-event/disaster. Typhoon prevention and mitigation engineering assessments are mainly post-disaster assessments.

19.3.3 Frameworks for Typhoon Disaster Assessments

Given the conceptual differentiation and analysis of typhoon disaster assessments, a proposed framework is outlined in Fig. 19.3.

19.4 A Composite Index for Assessment

Development of a disaster damage quantitative index and grade classification system are the focus of theoretical research related to disaster assessment. One key problem in the quantitative measurement of disaster damage is the delineation of a composite index. Three quantitative measurement approaches are usually adopted during a typhoon disaster assessment: (a) arrangement of an index of damage factors and the subsequent conversion of all indices into a composite index of severity, the value of which is the foundation for grade classification of the severity of the event; (b) the indication of disaster damage in terms of monetization of direct economic losses; (c) the conversion of all damages into economic losses and the subsequent quantitative assessment of the monetary value of such losses.

In these three approaches, the single indicator of direct economic loss that quantitatively expresses the extent of the disaster damage. However, current approaches frequently ignore other damage indicators, such as collapsed houses and casualties, although the third method involves conversion, which changes damages such as casualties and collapsed houses into economic loss figures. Conversion involves subjective judgments, which may lead to a biased result. In most research programs, the first approach, which involves the use of a composite index for severity of disaster, is widely employed in typhoon disaster assessments.

For the purposes of anticipated rescue efforts, all types of damage-related information are comprehensively collected in the effort to guide disaster relief fund allocations. This information is compiled in a detailed report in an effort to estimate disaster damages in a multi-index manner. However, if the multi-index data is converted into a composite index, the magnitude of the disaster in different districts could be more easily compared and can reflect the temporal-spatial distribution of the disaster. Furthermore, a composited index is more suitable in guiding the development of economic planning systems and budgets. Following this assertion, a suitable and comparable composite index is required. However, at present, no uniform composite index structure exists. The various composite indices have appeared in previous studies and it has been determined that these composite indices can be classified into two types: namely, absolute and relative.

19.4.1 The Absolute Composite Index

19.4.1.1 Disaster Degree

Disaster Degree terminology put forward by Ma et al. in (2008), cited a composite index derived from earthquake research and used to reflect earthquake intensity. Subsequently, this index was extensively adopted to indicate the grade of all kinds of disasters (Yu 1993).

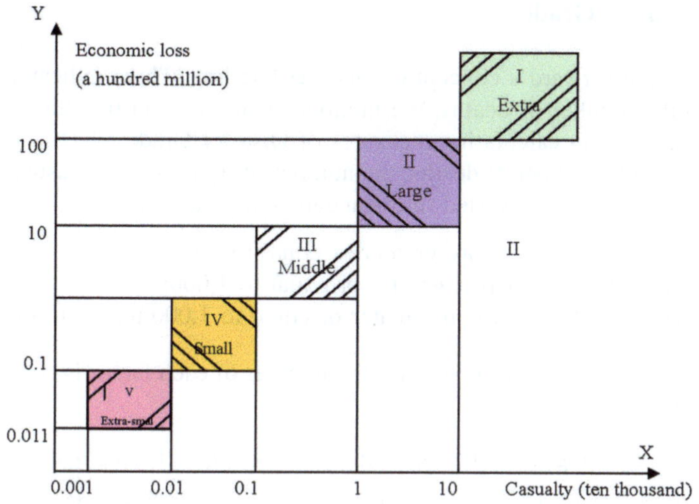

Fig. 19.4 Determination of disaster degree I, II, III, IV, and V represents extra-, large-, middle, small, and extra-small disasters

The *Disaster Encyclopedia* defines disaster degree as the loss quantities of resources resulting from a particular event in a unit, or defined, area. Ma et al. (2008) classified disasters into five grades in order to compare magnitude of a disaster or event according to this definition of disaster degree. These grades are accepted as follows: extra-, large-, middle, small, and extra-small disasters (i.e. Grade I, II, III, IV, and V as seen in Fig. 19.4, respectively). In this definition, *extra*-disaster refers to a casualty rate measured by the hundreds of thousands and direct economic losses beyond RMB10 billion. If the magnitude is reduced by one unit, the grade is correspondingly reduced. In this approach, casualty and loss of property are used as discriminating factors for measures of disaster intensity.

In terms of disaster significance, a single event or disaster is classified according to the magnitude of the disaster damage, a measure which further extends the concept of disaster grades. Disaster degree not only considers the naturality of the disaster, but also the sociality of disaster. This considers natural disaster losses as a bridge and makes interconnections between hazard intensity and society-bearing capacity a unified measure. However, the approach to creating a Disaster Degree is, at present, too simplistic because of the adoption of only two indices relating to disaster severity…namely where casualty and monetary loss values are considered (Fig. 19.4) and, in addition, via this index, one thousand casualties are regarded as one hundred million, which lacks reasonable explanation.

Considering the diversity of natural disasters, disaster situation factors are not alike thus use of only the disaster degree index is not enough to reveal the different types of impacts of natural disasters.

Given the logic of the existent Disaster Degree, most emerging studies adopt numerous indicators and method to obtain a composite index to reveal the magnitude of the disaster.

19.4.1.2 Disaster Grade

Feng (1993) put forward a concept of Disaster Grading. Three indicators, namely casualties, the number of deaths, the number of injured and the direct economic loss were selected in calculating measures of Disaster Grade.

First of all, the number of deaths, the number of injuries and measures of direct economic losses were converted into standardized means. Thus, if:

- The number of deaths (d) are greater or equal to 100;
- The number of injured (h) is greater or equal to 1,000;
- Direct economic losses (e) are greater or equal to 1,000 ten thousand Yuan.

Then, by using algorithm functions, the number of each factor is converted into standardized index:

$$I_d = \log(d) - 1; \quad I_h = \log(h) - 1; \quad I_e = \log(e) - 1 \tag{19.2}$$

In the formula, I_d, I_h, and I_e death, injury, direct economic loss of the standardized index.

As d < 100, h < 1,000, e < 10,000 Yuan.

The number of each factor shall be converted to a standardized index by the linear relationship:

$$I_d = d/100; \quad I_h = h/1{,}000; \text{ and } I_e = e/10{,}000 \tag{19.3}$$

Thus, *Disaster Grade* is

$$G = I_d + I_h + I_e \tag{19.4}$$

Subsequently, the concept of Disaster Gradi has been widely used in the typhoon disaster assessment (Lu 1995; Qian et al. 2001; Yao and Liu 2001; He and Qian 2002; Meng et al. 2007). Standard figures have been put forward for the typhoon disaster index (TDI) in which Feng's disaster loss composite method was applied to determine the magnitude of typhoon disaster. However in typhoon disaster assessments, three indicators were adjusted to include flooded fields, collapsed houses and related casualties. Too, criterion between linear and logarithmic functions were also modified in order to adapt the typhoon disaster estimates to the real parameters. Lei et al. (2009) put forward a typhoon disaster index (ATDI) designed to quantify typhoon disasters. Direct economic losses were added into his study and converted via four single indicators into a composite index, i.e. ATDI.

19.4.1.3 Disaster Index by Weighted Accumulation

Various conversion methods can form different composite indices. Additionally, the Disaster Degree and Disaster Grades of Fan and Liang (2000) elaborated on the original Disaster Index to quantify the results of typhoon disasters. Researchers assigned different weights to each single indicator then added a weighted average which is employed to obtain a composite index.

19.4.1.4 TDCG and HPI

Grey's correlation is also employed to form an absolute composite index which can assess disaster damage. The correlation coefficient is calculated, averaged to obtain a correlation degree, and ranked to determine the magnitude of the disaster. Wang et al. (2010) adopted this method to form a Comprehensive Typhoon Disaster Grade (CTDG). Cheng and Wang (2004) also put forward concepts for Hazard Potential Indices (HPI) to quantify the amplitude of typhoon disasters. It is worth noting that both Grey's correlations and the AHP are adopted in forming the HPI.

19.4.1.5 TCEI and TCDI

The Fuzzy assessment methods is a common method used for comprehensive statistical evaluations in which each single indicator of disaster factors are converted into fuzzy values, and the *Membership degree* is formed a membership matrix. After the determination of the weighing value, a composite index is calculated to quantify disaster damages. For example, Liang et al. (1999) and Ma et al. (2008) adopted the fuzzy evaluation system to establish a comprehensive typhoon disaster index (TCDI) in their work. Zhao et al. (2011) adopted the fuzzy method to establish the comprehensive typhoon evaluation index (TCEI) as well. The weighting values of each single indicator were determined by using AHP and the weighted average programming methods respectively.

19.4.1.6 The Typhoon Comprehensive Influencing Index

Typhoon disaster influencing factors can be formed using a hierarchical index system. AHP can be employed to convert all single indicators into a composite index measuring disaster damages. Li et al. (1991) put forward a TCII based on AHP.

19.4.1.7 Typhoon Economic Loss Index

In addition to the above absolute disaster indices, Chen et al. (2007) put forward a fresh DELI for typhoon disaster assessments. DELI consists of disaster loss degree (DLD) and environmental instability measures (EI).

Typhoon loss degree (DLD) measures involve representative indicators such as death tolls, estimates of the injured population, crop disaster areas, crop damage areas, direct economic losses and indirect economic losses. These indicators are the main elements used to establish measures of typhoon disaster degree which elaborate the model for calculation. This model is as follows:

$$DLD = f(pd, ph, Ec, Dd, Id) \tag{19.5}$$

Disaster formative environment factors can increase or decrease overall disaster damage to a particular extent because they have a significant impact on disaster loss. Thus, EI is elaborated. The concept used to design this system is based on the idea that the gap that may exist between the maximum and minimum economic loss caused by the typhoon disaster in a specified area can reveal the vulnerability of a disaster-affected area and that this can depict the degree of the environment instability of that area. Thus, environment instability may be said to be the ratio of the sum of the minimum disaster economic losses measured during the study period of the disaster compared to the economic loss in the current year added to the sum of the economic loss related to the disaster in the current year *and* the maximum disaster economic loss in the study period. The calculation formula is as follows:

$$EI = (Ee_i + Mine) / (Ee_i + MaxEe) \tag{19.6}$$

Where EI refers to environment instability, Ee_i is the economic total loss caused by the typhoon in the i-th year, $MinEe$ is the minimum economic loss in the study period, and $MaxEe$ is the maximum value of economic loss caused by the typhoon disaster.

DELI is the average value of DLD and EI. The total composite index, DELI = (DLD + EI)/2.

Cheng's *DELI* is a fresh idea because it considers environmental stability. It might be argued that typhoon DELI virtually adopts the degree of hazard formative environment stability so as to diminish the randomicity of economic loss induced by the typhoon disaster. Thus, the DELI can match the actual economic losses of a typhoon disaster to better estimate the damages.

Any scientific standard should reflect an inherent nature and possess relative stability and universal applicability. The results should be comparable using the same criteria. At present, the absolute composite indices have several drawbacks. (a) Economic losses are measured only via monetary values. Since the inflation causes a variation in such values and this factor induces factors that, in the deferent period show that absolute disaster indices cannot be directly compared then, (b) Absolute disaster indices should reflect the absolute magnitude of the damage. However, the bearing capacity of different regions does differ thus absolute disaster indices cannot correctly reflect the relative impacts of the disaster on society.

19.4.2 Relative Composite Index

The ACI can reveal the absolute quantity of the damage to societal property caused by a natural disaster. However, the ACI is not able to reflect the comprehensive ratio of loss caused by a disaster to all societal property and to the gross product of the area. The latter is, perhaps, a more suitable index of the overall societal impact and the relative degree of the total destructive impact caused by the disaster.

From the perspective of disaster rescue and reduction engineering, measuring the impact of natural disasters in terms of gross product impact/loss, gross national wealth and/or remaining production capacity may be more important. As a disaster occurs, there are external parts of the disaster area and international relief and financial aid efforts that should be assessed and used to investigate the resilience and to calculate the direct/indirect disaster losses and also to determine the period of time needed to restore order to the disaster area and to social production and order. A relative index is more likely to be suitable in fulfilling this purpose.

Following the abovementioned reasons, several relative disaster composite indices are elaborated in the emerging literature. The type of composite indices being shown are similar in basic principle, to others but slight and interesting differences exist.

19.4.2.1 Disaster Degree Index

Disaster systems are an interactional and complex system of the hazard formative environment and they must consider hazard formative factors, disaster-affected bodies in which the vulnerability of the latter as related to the hazardous event must be fully considered. The DDI is the societal average quantity related to labor inputs as calculated in the current production efficiency estimates and used to eliminate elements to the overall disaster impact. In the end, the DDI reveals the relative magnitude of disaster damage.

The mathematical expression is as follows (Sun 1993).

$$\mathrm{D} = \frac{(\mathrm{X}_t + \mathrm{X}_n)P}{GDP} \tag{19.7}$$

$$\mathrm{D} = \frac{(\mathrm{X}_t + \mathrm{X}_n)}{GDP/P} \tag{19.8}$$

where D is the DDI, X_t is the societal total loss of property, X_n is the loss converted from the casualties of the population in the disaster area, GDP is the gross domestic product loss estimates in the current year, and GDP/P is the per domestic gross product loss. Formulas (19.7) and (19.8) are established based on the disaster rate index and may be used in post-disaster evaluation. However, statistics cannot be performed on the GDP because of the time lag. The formula was slightly modified to evaluate the scope of the disaster. The substitution of the GDP in the preceding year for the current GDP is employed in pre- and mid-assessment process.

$$\mathrm{D} = \frac{(\mathrm{X}_t + \mathrm{X}_n)\mathrm{P}_{t-1}}{\mathrm{K}_1 \cdot \mathrm{K}_2(GDP_{t-1})} \tag{19.9}$$

$$\mathrm{D} = \frac{(\mathrm{X}_t + \mathrm{X}_n)\mathrm{P}_{t-1}}{\mathrm{K}_1 \cdot \mathrm{K}_2(GDP_{t-1}/\mathrm{P}_{t-1})} \tag{19.10}$$

In Formulas (19.7) and (19.8), *P* refers to the population in the disaster area, GDP_{t-1} refers to the gross national product before the disaster situation occurred; K_1 is the price index calculate in the year before the disaster, and K_2 is the product index in the year before the disaster.

The calculated disaster rate index may be converted into an estimate of the disaster grade. The concrete method is expressed in Formula (19.9). *M* refers to the disaster grade.

$$\mathrm{M} = \mathrm{LOG}(D) - 3 \tag{19.11}$$

Based on Formula (19.11), *M* is determined by the DDI. The merit is used to factor in the impact of the disaster into the hazard-affected body and, as well into the capacity of the hazard-affected body to resist the disaster.

19.4.2.2 Disaster Loss Rate

Another similar relative composite index of disaster loss is noted as the disaster loss rate, a measure which is noted as the proportion of natural disaster losses in comparison to the gross national product. The formula is expressed as follows: DLI = disaster economic loss/(gross product in the preceding year × price index) (Wei and Zhang 1996).

Disaster economic losses not only include direct and indirect economic losses but also the disaster rescue costs. This index reflects the gross national product loss rate as caused by the typhoon disaster.

Considering the complexity of the indirect economic and rescue costs, direct economic losses may be regarded as more important versus as only one point-of-reference value which may be used as the numerator in the formula. This index may be called the direct disaster loss rate: DLI = disaster direct economic loss (gross product in the preceding year × price index).

The gross national product in the preceding year is adapted in the formulas used as the disaster rate and disaster loss rate which imply that the gross national product in the disaster year is affected by the subject natural disaster. Thus, the gross national product in the year preceding the disaster is used in the formula because it can represent the actual production capacity or economic development level during the year in the disaster area. The price index is multiplied by the GDP to make it more accurate.

19.4.2.3 The Hazard Destruction Degree

The relative composite index may also be referred to as the hazard destruction degree in emerging literature (Qing and Gou 1991).

If *G* is the GDP in the country a year ago, and *A* is the disaster total loss value,

$$HDI = A/G \tag{19.12}$$

19.4.3 Comparisons of absolute and relative cost indices

Compared with the index of the ACI, the RCI has two merits in terms of comparability and relative stability. The RCI considers the economic and societal development level of the hazard-affected body by measuring the magnitude of loss via the medium of the gross national product. The index is comparable among different disasters in different periods and regions because it has no dimensional coefficient. Second, instabilities are eliminated based on the fluctuation of price, and because the price index is merged into the design of the relative composite index.

Furthermore, common disasters are representative and tend to cause large losses in a number of less-developed areas. Yet, the aid given by a disaster agencies may not be sufficient because the absolute value of the loss is perhaps estimated as less as compared to the absolute composite index. In this case, the index may indirectly misguide the calculations of the disaster management leadership. Conversely, the use of the RCI can effectively avoid the abovementioned drawbacks and reflect the disaster's actual, or more realistic, impact on people's lives (Table 19.3).

19.5 Grade Classifications and the Indictor System

Disaster grading has an important role in organizing and clearing work duty and successfully implementing disaster relief efforts. Disaster prevention decisions are often made according to estimates of the disaster grade levels decided by disaster prevention management. Disaster grade classifications should meet three rules; those of comparability, maneuverability and transitivity so that the gradient index can supply scientific and exact information for realization of the actual disaster damages.

19.5.1 Grade Classification Methods

Once the disaster index has been calculated, it needs to communicate a clear meaning to recipients so as to guide disaster relief efforts to achieve effective decision-making. Therefore, it must be divided into different levels relative to the index, which reflects the corresponding grade of the disaster.

In some previous studies, the classification of grades are often arbitrary, however, there are a number of researchers who are trying to use some more objective methods and to carry on the classification of the scale of disaster events through use of the actual disaster data.

1. Grey clustering: According to the Grey clustering principle, clustered factors are converted into albinism functions by order. The standard and actual weights

Table 19.3 Typical disaster composite indices

	Classification	Index name	Acronyms	References	Composite method
Composite index	Absolute composite index (ACI)	Disaster degree	DD	Ma (1998); Yu (1993)	The critical interval of grade
		Damage grade	DG	Feng (1993)	Logarithm function
			TDI	Lu(1995); Qian et al. (2001)	
			ATDI	Lei et al. (2009)	
		Disaster index	DI	Fan and Liang (2000)	Weighted accumulation
		Typhoon disaster comprehensive grade	TDCG	Wang et al. (2010)	Grey correlation
		Hazard potential indices	HPI	Cheng and Wang (2004)	Grey correlation and AHP
		Typhoon comprehensive influencing index	TCII	Liang et al. (1999)	Fuzzy assessment
		Typhoon comprehensive disaster index	TCDI	Ma et al. (2008)	Fuzzy assessment
		Typhoon comprehensive evaluation index	TCEI	Li et al. (1991)	AHP
		Typhoon economic loss index	DELI	Chen et al. (2007)	Linear combination of the logarithmic function
	Relative composite index (RCI)	Disaster rate index	DDI	Sun (1993)	Personnel property and casualty loss divided by GDP
		Disaster loss rate	DLI	Wei and Zhang (1996)	Economic losses divided by GDP
		Hazard destruction degree	HDD	Qing and Gou (1991)	Total loss divided by GDP

are fixed, clustered coefficients are calculated, and the disaster grade is determined according to the maximum subjective rules and guidelines (Feng 1997).
2. Fuzzy discriminance: Following the fuzzy model theory, the subjection function is established, the subjection degree is calculated, and the disaster grade is determined by the subjection rule (Ren 1996).
3. Ordered sample cluster: Cluster analysis or clustering is the task of grouping a set of objects in such a way that objects in the same group (called a 'cluster') are more similar (in some sense or another) to each other than to those in other groups (or other clusters). The typhoon disaster index is expressed in terms of a numerical order of size, a number which can be divided and distributed on the interval to reveal the grade of disaster The statistical method; the ordinal samples and cluster technology can be employed. Qian et al. (2001) classified the grade of the typhoon in Zhejiang Province in China, presenting a good example of the use of this method.

19.5.2 Typical Grade Classification

The aspect of the storm is highly important in creating a quantitative criterion for natural disaster grade classification, and in providing a scientific basis for grade classifications

In investigating the criteria for grade classification in disaster assessment, researchers tend to determine the criteria by themselves thus, not applying uniform standards and, as a result, uniformity is lacking and the criteria applied differ significantly among researchers.

Several typical grade classifications in typhoon disaster assessment are summarized as:

1. Three grades: large, medium, and small (Li et al. 1991)
2. Four grades: extra-large, large, medium, and small (Ma et al. 2008); or large, medium, small, and extra-small (Zhao and Ma 1993).
3. Five grades: extra-large, large, medium, small, and extra-small (Liang et al. 1999).

Other grades, such as 6 (Ren 1996), 8 (Gao 1992), 10 (Xu 1997), and 11 (Feng et al. 1994), have also been adopted in typhoon disaster assessment (Feng 1997; Yu and Shen 1995).

19.5.3 Indicator System for Assessment

The magnitude of disaster damage is assessed by the values of the indicators of the disaster, thus a reasonable selection of factors is needed to measure natural disaster damage. In terms of a basic approach, the tendency is to select hazard indicators,

Table 19.4 Typical indicator systems for typhoon assessment

	Number of indicators	Indicators	References
Indicator system	1	Direct economic losses	Zhao and Ma (1993); Sun et al. (1994)
	2	Death toll, direct economic loss	Gao (1992); Ma et al. (1990)
		Casualty, economic loss	Xu (1990)
	3	Death toll, injured toll, direct economic loss	Feng et al. (1994), Feng (1997), Xu (1997)
		Main societal economic indicator, main economic indicator, assisting economic	Ma et al. (1990)
		Casualty, economic loss, environment indicator	Qing and Gou (1991)
	4	Collapsed houses, disaster area, casualty, direct economic loss	Li and Dun (1994)
		Population suffering from the disaster, death toll, disaster area, and damage area;	Li et al. (1991)
		Death toll, collapsed houses, disaster area, grain reduction quantity	Wei and Zhang (1996)
	5	Social function and ecology environment destroyed, death toll, injured toll, synthesized economic loss, disaster loss duration	Liu et al. (1995), Ren (1996)
	More	7 indicators	Yan (1992), Yang (1997)
		12 indicators	Zhang (1996), Yu (1993)
		102 indicators	Wang et al. (1994)

indicators of the hazard-affected body, or post-disaster damage indicators that are dependent on the purpose of assessment. For example, before the disaster, we may select typhoon frequency and intensity to evaluate the risk in typhoon-prone areas. For mid-disaster, we may select economic and societal indicators and wind and rain indicators of the typhoon to input the established model for forecasting and assessing disaster damage. After the disaster, we may examine damage indicators, such as casualties, collapsed houses and inundated crop areas to assess the magnitude of disaster loss.

For post-disaster assessments a composite index is commonly adopted to compare disaster situations or disaster management. Such an index naturally involves a typhoon disaster assessment indicator system. Indicators in any one assessment indicator system may vary in significance among studies without an acceptable uniform standard analogous to disaster grade classification and this leads to confusion. Typical indicators systems are listed in Table 19.4.

19.6 Summary

In recent studies, the various concepts used have puzzled researchers regarding disaster assessment. In addition, the selection standard is characterized by a lack of a universally acceptable rule in terms of disaster grade terminology, indicator systems, and assessment approach. Given the critical value of each grade and assessment index system, researchers have employed different approaches and criteria thus the result will rely heavily on the researcher's subjective judgment.

This chapter addressed these issues. The concepts of disaster events are examined from a fundamental perspective with emphasis on general discrepancies. A framework including typhoon risk, typhoon disaster situation, and disaster assessment is outlined for typhoon disaster assessment. All indices are summarized and classified into two types by their nature, namely, absolute and relative.

The chapter also provides a detailed analysis of a typical composite index in disaster assessment and a comparison of all types of composite indices. Moreover, the study also improves the understanding of the basic theory and present status of disaster assessment.

Acknowledgments This study was supported by the following research projects awarded to DL Tang: (1) National Natural Sciences Foundation of China (31061160190, 40976091, NSFC-RFBR Project-41211120181 of DL Tang and D. Pozdnyakov); (2) Guangdong Sciences Foundation (2010B031900041, 8351030101000002); (3) Innovation Group Program of State Key Laboratory of Tropical Oceanography (LTOZZ1201).

References

Chen, X., Shen, J.R., Chen, J.: Application of DELI in disaster economic loss assessment-taking analysis of trend of economic loss in typhoon disaster in Fujian province as an example. J. Disaster Sci. **22**, 31–35 (2007)

Cheng, S.P., Wang, R.Y.: Analyzing hazard potential of typhoon damage by applying grey analytic hierarchy process. Nat. Hazards **33**, 77–103 (2004)

Fan, Q., Liang, B.J.: The evaluation of disastrous losses caused by tropical cyclones. Acta Geographica Sinica **55**(11), 52–56 (2000)

Feng, L.H.: Grey clustering analysis of disaster grade. J. Nat. Disasters **6**(1), 14–18 (1997)

Feng, L. H.: Quantitative calculation of disaster loss. J. Catastrophology **8**(2), 17-19 (1993)

Feng, Z. C., Hu, Z., He, J.: Loss evaluation of earthquake disaster and classification of the disaster grade. J Catastrophology **9** (1), 13–16 (1994)

Gao, J.G.: The establishment of disaster evaluation standard is China's "international disaster reduction ten years" target is an important content. J. Nat. Disasters **1**(1), 48–55 (1992)

He, C.F., Qian, Y.Z.: An assessment for tropical cyclone disaster in Zhejiang Province in 2000. Zhejiang Meteorol. **23**(2), 4–7 (2002)

Li, Z.R., Dun, X.M.: Study on the evaluation of natural disaster based on matter element analysis. J. Nat. Disasters **3**(2), 28–33 (1994)

Lei, X. T., Chen, P. Y., Yang, Y. H., Qian, Y. Z.: Characters and objective assessment of disasters caused by typhoon in China. Acta Meteorol Sinca **67**(5), 875-883 (2009)

Li, Y.S., Wei, Y.L., Li, H.Z.: A research on the economic and social effects. J. Catastrophology **6**(4), 1–11 (1991)
Liang, B.Q., Fanq, Y.J., Wang, T.M.: A fuzzy mathematic evaluation of the disaster by tropical cyclones. J. Trop. Meteorol. **5**(4), 305–311 (1999)
Liu, Y.H., Li, J.Z., Zhao, Y.L.: Regional characters of natural disaster in the last years in China. Geogr. Res. **14**(3), 14–25 (1995)
Lu, W.F.: Assessment and prediction of disastrous losses due to tropical cyclones in Shanghai. J. Nat. Disasters **4**(3), 40–45 (1995)
Ma, Z.J., Gao, Q.H., Wei, M.H.: Natural disaster and 600 questions and answers, pp. 6–7. Seismological Press, Beijing (1990)
Ma, Q.Y., Li, J.Y., Wang, X.R., Wang, W.G., Gao, L.Y.: A fuzzy synthetic evaluation model for typhoon disaster. Meteorol. Monthly **34**(5), 20–25 (2008)
Meng, F., Kang, J.C., Li, W.J., Wu, T., Wang, T.T., An, Y.: Analysis and evaluation of typhoon disasters In Shanghai in past 50 Years. J. Disaster Sci. **22**(4), 71–76 (2007)
Qian, Y.Z., He, C.F., Yan, Y.Q., Wang, J.Z.: An assessment of damage index for tropical cyclones. Meteorology **24**, 14–18 (2001)
Qing, C.B., Gou, X.H.: Disaster evaluation, risk evaluation and damage evaluation. Probing Nat. **10**(2), 68–69 (1991)
Ren, L.C.: The measurement method of fuzzy disaster degree for classification of disaster loss. J. Nat. Disasters **5**(3), 13–17 (1996)
Schmidt, S., Kemfert, C., Höppe, P.: Tropical cyclone losses in the USA and the impact of climate change—A trend analysis based on data from a new approach to adjusting storm losses. Environ. Impact Assess. Rev. **29**, 359–369 (2009)
Sellers, H.A., Zhang, H., Berz, G., Emanuel, K., Gray, W., Landsea, C., Holland, G., Lighthill, J., Shieh, S.-L., Webster, P., McGuffie, K.: tropical cyclones and global climate change: A post-IPCC assessment. Bull. Am. Meteorol. Soc. **79**(1), 19–38 (1998)
Shi, P.J.: Theory on disaster science and disaster dynamic. J. Nat. Disasters **11**, 1–9 (2002)
Sui, G.J., Pu, H.Y.: Index system of vulnerability by typhoon in coastal areas and emergency management strategies. Reform **3**, 145–154 (2012)
Sui, G.J., Tang, D.L.: Typhoon disaster and regional economic divergence: evidence form Guangdong. Reform **6**, 18–25 (2012)
Sun, W.D.: Relative disaster degree: Meaning and uses. J. Disaster Sci. **8**(3), 88–89 (1993)
Sun, Z.G., Mao, M., Zou, Q.J.: Study on index of disaster situation division of natural disaster. J. Catastrophology **9**(2), 84–87 (1994)
Walsh, K.: Tropical cyclones and climate change unresolved issues. Clim. Res. **27**, 77–83 (2004)
Wang, J.A., Shi, P.J., Zhu, L.: A research on regional distribution of major natural hazards in china. Acta geographica **49**(1), 18–26 (1994)
Wang, X.R., Wang, W.G., Ma, Q.Y.: Model for general grade division of typhoon disasters and application. Meteorol. Monthly **36**(1), 66–71 (2010)
Wei, Q.C., Zhang, Q.H.: Determination of disaster losses and disaster grading. J. Disaster Sci. **11**, 1–5 (1996)
Xu, H.M.: Research on class of disaster, pp. 137–141. Disaster and Society, Seismological Press, Beijing (1990)
Xu, F.Q.: Disaster grade and the definition. J. Catastrophology **12**(1), 16–18 (1997)
Yan, J. P.: Disaster geography. pp. 78–81, Shanxi Normal University press, China (1992)
Yang, S.S.: Research of comparative method for different disastrous conditions of natural disaster. J. Nat. Disasters **6**(1), 8–13 (1997)
Yao, D.R., Liu, X.L.: The assessment of disastrous losses caused by tropical cyclones in Zhejiang Province. J. Zhejiang Univ. **28**(3), 344–348 (2001). (Science Edition)
Yu, Q.D.: The limitation and improvement of the method of disaster situation grade measurement. J. Nat. Disasters **2**, 8–11 (1993)

Yu, Q.D., Shen, R.F.: All evaluation of the model of the absolute status of natural disaster and its application. Syst. Eng. Theor. Methodol.·Appl. **4**, 47–52 (1995)
Zhang, J.G.: comprehensive evaluation and regionalization of natural disaster taking southern Sichuan. J. Catastrophology **6**(2), 1–4 (1996)
Zhao, A.X., Ma, Z.J.: Appraising study for the loss evaluation system of natural disasters. J. Nat. Disasters **2**(3), 1–6 (1993)
Zhao, F., Liao, R. F., Zhang, L. N., Zhang, Y. X., Chenyang, S. Q.: A Pre- evaluation model for typhoon disasters in china. J Catastrophology **26**(2), 81–85 (2011)

Chapter 20
Mechanisms of Emergency Management Against Typhoon Disasters: A Case Study of Typhoon "Bilis"

GuangJun Sui, HuiYing Pu and DanLing Tang

Abstract This chapter analyzes the mechanism of emergency management employed by governments, and typhoon No. 0604 "Bilis" in 2006 has been selected as a typical case for comparative analysis of emergency response plans and precautions among different provinces in China. The analysis shows that Zhejiang Province in China has established a complete system, which can be used as a significant model for the improvement of our emergency management preparation and implementation. Finally, some suggestions are made to improve the system of emergency management against typhoon disasters in Guangdong, China.

Keywords Emergency management · Case study · Typhoon "Bilis"

20.1 Introduction

The southeast coast of China is one of the most vulnerable areas to typhoon disasters which cause billions in economic losses and heavy casualties each year. In China, a total of 28 provinces (10 coastal and 6 inland) are affected by typhoon induced disasters each year. In other words, more than 250 million lives are affected by typhoons, with an associated cost of approximately 60 trillion RMB (Liu et al. 2009). According to the "Natural Disaster Hotspots—A Global Risk

G. J. Sui
College of Economics and Trade, Guangdong University of Foreign Studies, Guangzhou 510420, China

H. Y. Pu
Institute of Industrial Economy, Jinan University, Guangzhou 510632, China

D. L. Tang (✉)
State Key Laboratory of Tropical Oceanography, South China Sea Institute of Oceanology, Chinese Academy of Sciences, Guangzhou 510301, China
e-mail: Lingzistdl@126.com

D. L. Tang and G. J. Sui (eds.), *Typhoon Impact and Crisis Management*, Advances in Natural and Technological Hazards Research 40,
DOI: 10.1007/978-3-642-40695-9_20,

Analysis" (published by the World Bank, 2005), at least 73 % of the land area and population in China are exposed to the threat of more than three types of natural disaster. Therefore, studies of emergency management measures to guard against typhoon disasters have drawn great attention from the government and the academic community.

According to the 2006 tropical weather summary of China, super typhoon Bilis landed on the coasts of China, which had a 150/h duration after landfall. The life and property losses caused by typhoon Bilis have been the most serious in recent years. Taking into account the historical events related to typhoon disasters, this chapter chooses typhoon No. 0604 "Bilis", in 2006, as a typical case for comparative analyses of typhoon anti-disaster preparations in China.

20.2 Case Study of Typhoon "Bilis"

Firstly, a basic introduction to typhoon No. 0604 "Bilis" is in order: After "Bilis" first landed on the Yilan district of Taiwan at 22:00 h on 13 July 2006, it crossed the straits and landed on Fujian Province, China, again at 24:00 h on the 14th, then moved toward the southwest, and disappeared in Jiangxi Province at last (Fig. 20.1).

Because of the combined effects of air circulation and the monsoon, "Bilis" brought heavy rainfalls to vast areas of South China, including Hunan, Fujian, Guangdong, Guangxi, Jiangxi, Zhejiang, and others, all of which led to floods and mountain-area hazards. From Fig. 20.2, showing the remote sensing image of the distribution of accumulated rainfall between July 13th and 17th, 2006, it can be seen that two areas have the heaviest rainfall. The areas bounded by Guangdong and Fujian Provinces and that by Guangdong and Hunan, were hardest hit. According to the records of weather stations, the precipitation in the east and north parts of Guangdong, as well as the Pearl River Delta regions, were the heaviest in over a century. Storms were warned by over 114 stations, of which 51 were heavy rainstorms. The accumulated rainfall was more than 120 mm, especially in the eastern part of Guangdong where the 300–500 mm rainfall ranked as the second highest in history. Consequently, the water levels of the North River and the East River rose rapidly. For example, the flood peak of the Pingshi hydrological station was 165.39 m, which was 5.01 m more than the recorded highest; the biggest flow capacity of the Feilaixia and and Shijiao reservoirs were 17,500 and 17,600 m^3/s respectively, which typically appear once in 50 years.

The heavy rainfall of "Bilis" resulted in disastrous floods. One typical example was Shaoguan city. Many villages were submerged by floods, thousands of people were homeless, and all lifeline infrastructures were cut off. According to statistics, there were 75 counties and 727 villages affected by floods, there

Fig. 20.1 The landfall points and path of typhoon "Bilis" (from http://www.zjwater.com/typhoneweb/)

were 7.79 million homeless people and 123 deaths. Additionally, 3.09 billion m^2 of farmland was affected, 121.2 thousand houses collapsed and 6 middle-scale reservoirs and 125 small-scale ones were damaged causing a total direct economic loss of 15.2 billion Yuan (Li 2006). Therefore, the disaster caused by "Bilis" is the most serious one during the past decade.

From the statistics of Bureau of Civil Affairs (Table 20.1), Hunan and Guangdong Provinces were those most severely affected by "Bilis"; however, though quite near the storm, the losses of Zhejiang Province were much less than those of other provinces. Since Guangdong and Zhejiang Provinces are located at the opposite sides of the landing path (Fig. 20.1), both provinces were not directly attacked by "Bilis" so one asks, "Why do their losses seem so different?"

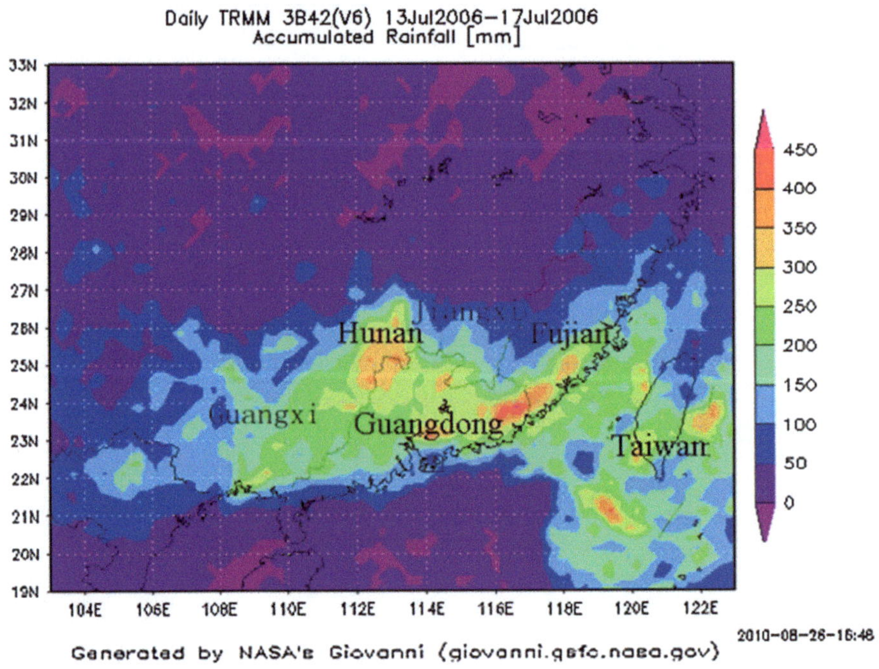

Fig. 20.2 The remote sensing image showing the distribution of accumulated rainfall of "Bilis"

Table 20.1 The statistics on losses and casualties related to "Bilis"

Province	Homeless people (thousand)	Number of deaths	Direct economic loss (100 million)
Hunan	9,300.4	416	129.24
Guangdong	7,790	123	151.77
Fujian	3,720.9	88	46.85
Guangxi	7,174	35	27
Jiangxi	1,870	2	3.87
Zhejiang	1,748	0	6.93

20.3 Comparative Analyses of the Emergency Management Mechanisms Arrayed Against Typhoon Bilis in Two Province

There have been numerous works focused on risk of natural disasters published over the last couple of decades. Barton (1994) suggested that everyone, from senior administration down to grass-root units, should have a predetermined role to play in an anticipated crisis. A well thought-out crisis plan can help management response and facilitate damage control. Drabek (1995) argued that the best way to handle disaster planning and response is to plan appropriate actions ahead of time, resist the temptation to deny responsibility, arrange to have a single person in charge, improve employee-customer communication, anticipate the needs of special populations, recognize family priorities, as well as facilitate good media relations. Faulkner (1999) took a step further providing a framework for analyzing natural disasters, and later also developed a generic model and a set of principles for analyzing and developing disaster management strategies. Based on these earlier papers, Faulkner and Vikulov (2001) developed a more refined model for disaster management. Here we make a case study according to disaster management theories, including the comparative analyses of emergency management mechanisms between different province and make conclusions, as described below.

The damages and losses of typhoon No. 0604 "Bilis" are so huge that it draws people's attention to the emergency management programs of the respective governments. Though human beings cannot resist the occurrence of natural disasters, certain human factors should take some responsibility for such huge losses, and there might be such things as local governments' underestimating the damage, insufficient precautions taken beforehand and so on (Ning 2006; Shi and Peng 2006). Fortunately, under the unified plan of the Central Government, local governments begin to establish emergency management systems to protect citizens and property against natural disasters. But, in practice, what part do these systems actually play? And how efficient are they?

It is of utmost importance for local governments to study and master the regular patterns of typhoons in order to enhance the emergency response capabilities. Being so located, in the path of many tropical storms and typhoons, the southeastern coast of China is attacked every year. With years of experience, Zhejiang Province has established a complete emergency management system (Lu and Yao 2006) (Fig. 20.3).

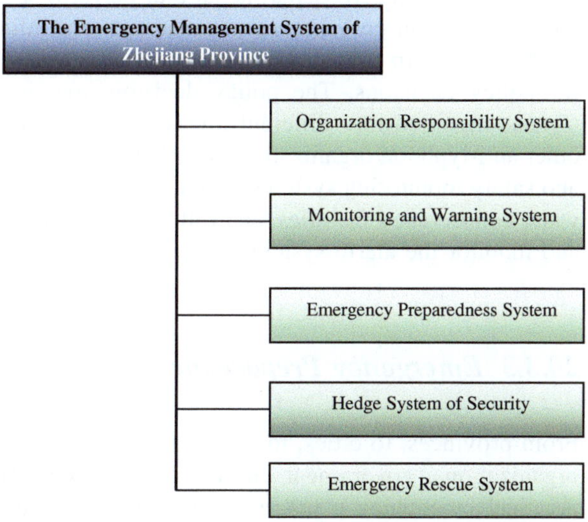

Fig. 20.3 The emergency management system of Zhejiang province, China

20.3.1 Organizational Responsibility System

From province to cities, then to districts and counties, governments at all levels implement the office of Chief Executive Responsibility and set up command centers which are responsible for the arrangements of disaster prevention and relief. During the period of a typhoon, command centers at all levels work simultaneously, and the chief executives take all the responsibility for execution of relevant measures. Too, specialized tasks are implemented by relative departments, such as the Bureau of Meteorology, the Department of Water Resources, the Land and Resources Bureau, the Marine Fisheries Service, the Bureau of Civil Affairs and so on.

The "Grass-Roots" anti-typhoon organization is one of the most important features of the emergency management system of Zhejiang Province. It includes two critical aspects of emergency preparedness:

1. Each town organizes specialized offices in charge of different emergency assignments, such as monitoring and early warning, transferring the masses of residents to safe places, organizing rescue teams, providing logistics services and so on.
2. Each village or community can be divided into several responsibility grids according to terrain, neighborhoods, designation as a farm belt, industrial zone etc. Each responsibility grid establishes special working groups, which are assigned different tasks to resist typhoon disasters.

20.3.2 Monitoring and Warning Systems

In addition to relative departments that perform specialized tasks, there are also those that submit proven data/information to the information platform. This entity has been set up to collect information, collate it and send it to the command centers for policy decisions. The policy decision, together with a system of warning signals classified into different levels, will be distributed to the grass-roots and other anti-typhoon organizations simultaneously. Additionally, regular inspections and safety monitoring systems should set up to monitor reservoirs, dams and other water conservancy facilities, and full-time staff need to be arranged to take charge and monitor the alarm systems.

20.3.3 Emergency Preparedness System

From provinces, to cities, to counties/districts and on to grass-roots anti-typhoon organizations, government entities at all levels also must have particular and practical emergency plans, and organize regular exercises. While other relative departments have their working regulations, during the period of a typhoon, the provincial command center, in consultation with other relative departments, makes policy decisions and assigns defensive tasks to subordinate departments. Thus the emergency plans will be initiated step by step, and if necessary, experts will be sent to the site to direct the preventative and/or rescue work.

20.3.4 The Hedge System of Security

Local command centers are responsible for hedge management. Every year before the flood seasons, local command centers require Administration of Work Safety tasks and assignments to carry out as a pre-inspection hedge test of security measures. Grass-roots anti-typhoon organizations divide their own responsibility grids into safety zones, warning areas and danger zones, and they draw security lines and refuge spots by principles of proximity. Too, the general populous are to be informed of emergency services locations by clear billboards and other means.

20.3.5 Emergency Rescue Systems

Emergency rescue systems and detailed schedules are set up to manage the reserves of rescue materials and equipment. On the one hand, the command centers at all levels are responsible for the operation of the systems and for the organization of professional rescue teams. On the other hand, the grass-roots anti-typhoon organizations set up specialized offices in towns that integrate the forces of public security, fire brigade and militia in the most efficient manner in order to strengthen the rescue teams. In addition, accounting systems and databases are established to manage the use of rescue materials and equipment. As a result, the smooth operation of the emergency rescue system can be guaranteed by access to the comprehensive inspection of rescue materials and equipment carried out every year before the flood season begins.

20.3.6 Comparative Analysis

The most important goal of the emergency management system is to provide relief for direct economic losses, casualties and damage to the greatest extent. Consequently, it's required that the local emergency management systems should be complete and practical. The successful experience of Zhejiang Province shows that its organizational responsibility system is progressively refined from province to city to town and village, all of which lays a solid foundation for the emergency preparedness system and the hedge system of the security and emergency rescue

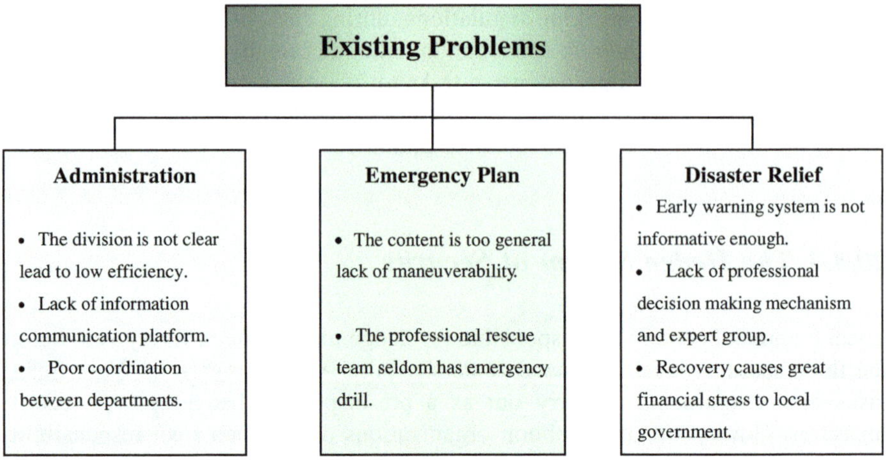

Fig. 20.4 Existing problems of Guangdong's emergency system

systems. However, no matter how perfect the system may be felt to be, it cannot be separated from the implementation capabilities of the governments and relevant departments that implement the plans. Zhejiang Province has paid high attention to carrying out the emergency management system both in planning and in practice, enabling it to improve its emergency response capabilities gradually from experience.

In contrast, though Guangdong Province has already done a lot of work in emergency management against natural disaster, and has established the "Three anti-Emergency Management Platforms", yet generally speaking, the operation of this platform seems immature, and many local governments lack specialized emergency plans and rescue systems (Chen et al. 2010). Existing problems identified are shown in Fig. 20.4.

20.4 Suggestions

Through a comparative analysis of the two provinces, we find that the emergency management mechanisms of Guangdong Province still need to be improved. Here are our suggestions to solve the above-mentioned problems.

1. There are efforts needed to complete the pre-disaster warning system and make emergency management plans more practical. The main function of the pre-disaster warning system is to establish an information and communication platform, the purpose of which is to collect accurate information from relative departments and to send it to the command centers for policy decisions. Meanwhile, the local emergency plans should be made more practical and set up to be more closely coordinated with the various warning signals. Thus, once

disasters approach or break out, the command center of each province can take charge of the distribution of the warning signals and the initiation of related emergency plans, and then the command center of each province can efficiently direct the local governments and relative departments to perform specific tasks.

2. It is significant that the building of emergency rescue systems be strengthened and priority should be given to community participation in the organization of professional local rescue teams. These rescue teams should be based on personnel of the local forces of public security, fire brigades and militias and, as well, counting on the participation of the masses. In addition, the emergency rescue system should also include an efficient material scheduling system, one which can quickly dispatch necessary materials and equipment for emergency purposes.
3. It is important, too, that more post-disaster financial subsidies and tax incentives should be allocated to affected areas. To do this, we highly recommend that financial policy instruments should be introduced into the recovery and compensation system. According to foreign experiences and examples, the forms of financial policy instruments might include, not only the general policy loans, but also other financing tools such as social security deposits, home insurance programs, agriculture/crop insurance, catastrophe bonds and so on. In a word, taking into account a variety of financial policy instruments can make full use of social capital and provide long-term financial aid for reconstruction and compensation schemes with low additional cost.

Acknowledgments This research was supported by the following projects awarded to DL Tang: (1) National Natural Science Foundation of China (31061160190, 40976091); (2) Guangdong Natural Science Foundation Team Research Project, China, "Hazard Evaluation on Typhoons and Examination on Their Biological Effect in the South China Coastal Area Based on Satellite Remote Sensing" (8351030101000002); (3) Guangdong Science and Technology Project, China, "The Development of Integrated Emergency Response System of Serious Typhoon and Fire Disasters" (2010B031900041). Additionallly, other projects jointly supported this study: (4) Youth Project of Philosophy and Social Science Foundation of Guangdong Province, China, "Study on Typhoon Disaster Prediction of Guangdong Province" (GD11YGL03); (5) Guangdong University Young Talents Cultivation Project, "Model of typhoon disaster prediction and disaster degree evaluation: An Empirical Analysis of typhoon disaster in Guangdong Province" (LYM11062). Dr. Dounald Thomas of GDUFS contributed the English editorial work for this chapter.

References

Barton, L.: Crisis management: preparing for and managing disasters. Cornell Hotel Restaurant Adm. Q. **35**, 59–65 (1994)

Chen, H., Pu, H.Y., Shen, M.H.: A research on the investment of disaster prevention and mitigation of typhoon disasters in Guangdong province (in Chinese). Mod. Manage. Sci. **7**, 81–83 (2010)

Dilley, M.: Natural disaster hotspots: a global risk analysis, the world bank publications (2005)

Drabek, T.E.: Disaster planning and response by tourist business executives. Risk Manage. Solutions **36**, 86–96 (1995)

Faulkner, B.: Natural disasters: towards a generic model. In: CRC Work-in-Progress Report Series, Report 6 (1999)

Faulkner, B., Vikulov, S.: Katherine, washed out one day, back on track the next: a post-mortem of a natural disaster. Nat. Dis. Manage. **22**, 331–344 (2001)

Li, G.H.: An Encounter action against typhoon "Bilis" (in Chinese). Disaster Reduction China **9**, 10–14 (2006)

Liu, D.F., Pang, L., Xie, B.T.: Typhoon disaster in China: prediction, prevention, and mitigation. Nat. Hazards **49**, 421–436 (2009)

Lu, Z.P., Yao, Y.Y.: Typhoon disasters in Zhejiang Province and the establishment of emergency response mechanism (in Chinese). J. Catastrophil. **3**, 69–71 (2006)

Ning, M.: Discussion on typhoon "Bilis" (in Chinese). Water Conserv. Hydropower Hunan Province **6**, 70–71 (2006)

Shi, Q., Peng, Q.Y.: Practice and thinking about the defense work of Yongzhou city against typhoon "Bilis" (in Chinese). Water Conserv. Hydropower of Hunan Province **6**, 35–37 (2006)

Chapter 21
Comprehensive Impact and Defensive Measures for Typhoon Disasters: A Case Study of Typhoon Morakot

GuangJun Sui, He Chen and DanLing Tang

Abstract Typhoon disasters are a global issue. With the continuously improving economic development of Mainland China, a comprehensive impact, resulting from development, has been exerted on coastal areas, particularly that of Guangdong Province. Through a comprehensive analysis of the impact of typhoon Morakot on the Taiwan region, and through a similar analysis of the emergency measures taken by the regional government of Taiwan, plus that of Mainland China and, as well, taking into consideration the pros and cons of these measures, this paper tries to draw upon the defensive experience of the typhoon disaster relief programs that might be suitable for Guangdong Province. Results of the follow-up studies show that, as a typhoon-prone province, Guangdong Province can gain some insights and, perhaps, even inspiration from the emergency measures related to Typhoon Morakot. Drawing on experiences from Typhoon Morakot, Guangdong Province should establish a comprehensive, three-dimensional typhoon disaster defensive system, including pre-disaster, during-disaster and post-disaster measures, namely pre-disaster preventative and damage-reduction systems, in-disaster crisis management systems and post-disaster loss evaluation and compensation systems.

Keywords Typhoon disaster · Comprehensive impact · Defensive measures · Crisis management system · Compensation system · Experience

G. J. Sui · H. Chen
Guangdong University of Foreign Studies, 510420 Guangzhou, China

D. L. Tang (✉)
Research Center for Remote Sensing of Marine Ecology and Environment,
State Key Laboratory of Tropical Oceanography, South China Sea Institute of Oceanology,
510301 Guangzhou, China
e-mail: Lingzistdl@126.com

D. L. Tang and G. J. Sui (eds.), *Typhoon Impact and Crisis Management*,
Advances in Natural and Technological Hazards Research 40,
DOI: 10.1007/978-3-642-40695-9_21, © Springer-Verlag Berlin Heidelberg 2014

21.1 Introduction

Currently, natural disasters are global issues which have attracted great attention around the world (Sui et al. 2010; Zheng and Tang 2007). Taking precautions and emergency measures is of great worldwide importance. As a region with frequent typhoons, Taiwan has set up a series of methods and emergency measures to protect against the disasters that might result from typhoons. However, the Taiwan region still suffered from great losses, and even social unrest, despite all the measures taken (Liu et al. 2009a, b) when typhoon Morakot struck there in 2009. This paper investigates the influence of typhoon Morakot on the Taiwan region and the results of the emergency measures adopted by the regional government of Taiwan and aims to extract important typhoon prevention experience that may be suitable for Guangdong Province in Mainland China.

21.2 Impacts of Typhoon Disasters on the Guangdong Economy and Society

The costal areas of Guangdong are often hit by typhoons. Recent years have witnessed an increasing trend in the frequency of the landfall of violent typhoons, many of which have caused enormous economic losses to Guangdong Province (Sui et al. 2010). For example, in 2008, the landfall of typhoon Heigebi on the western region of Guangdong, brought about a direct economic loss of about RMB 6 billion (Song and Ou 2010). In particular, typhoons have destroyed aprons of four major dams and sub-dams in Maoming, have involved one large-scale reservoir in Luokeng and, at other times, have involved two medium-sized reservoirs (typhoon Huangsha and typhoon Reshui). The direct economic loss related to these water conservancy facilities amounted to RMB 1.07 billion, accounting for about 15 % of the total economic loss (Song and Ou 2010).

A typhoon features not only a powerful wind-force, but also strong rainfall which can cause, or be combined with, such disasters as windstorms, floods, mountain torrents, mud-rock flows and landslides (Chen and Chen 2011). Too, they deeply threaten people's lives and properties. Meanwhile, a typhoon is often viewed as a factor involved with social instability and one that can undermine the government's prestige. Therefore, immediate efforts are called for to master the natural laws of typhoons, improve the emergency capability for typhoon damage prevention and reduce subsequent losses and potential social disturbances. The present study is designed to analyze the damage caused by typhoon Morakot in the Taiwan region in 2009 and the actions taken to deal with the results of the disaster, aiming to draw some enlightened recommendations for the establishment of typhoon prevention systems in Guangdong Province.

21.3 Comprehensive Influences of Typhoon Morakot in the Taiwan Region

21.3.1 Economic Impacts of the Disaster

In 2009, the strength of typhoon Morakot caused tremendous economic losses. Statistics from the Taiwan Regional Disaster Emergency Department show that Typhoon Morakot caused 461 deaths, leaving 192 missing and 46 injured (Cui et al. 2010; Yin 2009). Table 21.1 presents detailed information related to this typhoon.

21.4 Disaster Prediction and Prevention

There were deviations and errors related to Typhoon Morakot's prediction, in-disaster tracking and post-disaster observations. First, since Morakot moved slowly at the very beginning, the weather forecast wrongly considered it to be of low destructive power, thus Taiwan regional authorities failed to predict the tremendous amount of rainfall. In some areas, the 2,000 mm rainfall brought by Morakot caused many villages to be engulfed by flash floods and debris flow, both of which resulted in enormous casualties and property loss (Lin et al. 2009). During the disaster, though Taiwan regional authorities strived to provide relief, many implementation problems have come to light, such as a generally slow response to the disaster, inefficiency, chaotic rescue systems, poor infrastructure, a lack of comprehensive planning of alternatives and so on. These problems, together with the mistakes related to the meteorological forecast, unsatisfactory coordination between the super and subordinate governments, and the untimely

Table 21.1 The loss, rescue operations and repayments related to typhoon Morakot in the Taiwan region

Typhoon Morakot		Amount
Economic loss	Crops	NTD 4.4 billion
	Facilities	NTD 4.2 billion
	Fish	NTD 4.1 billion
Rescue operation	Residents evacuation	24,950
	Asylums setting up	56
	Victims accommodation	5,822
Repayments	Water supply	2,393
	Power supply	6,959
	Intra-city telephone	2,035
	Mobile phone stations	81
	Provincial highways	66
	Country roads	34

promulgation of the “Emergency Order”, brought about an immense loss of life on Taiwan Island. The question is: “What we can learn from the experience of typhoon Morakot?” One obvious conclusion is that a sound pre-disaster prediction and prevention system is an important measure against the overall impact of such a disaster.

21.5 Political Crisis Within the Disaster

Typhoon Morakot engendered severe floods in the Taiwan region on the 8th of August but, when the Taiwan regional authorities were confronted with the sudden and unexpected disaster, many problems within the crisis management system came to the surface. Lapses such as insufficient pre-disaster prediction and alarm systems, ineffective emergency treatment facilities and so on, became evident, all of which resulted in strong criticism of authorities' poor work and, finally, the stepping-down of the President of Disaster Relief Administration Department.

After the disaster, because of problems related to communications between the authorities and the media, the media reported this disaster in a dramatic way and the public could not get official or accurate information in a timely manner. As a result, all the victims were unhappy and all the efforts made by the government were completely denigrated by the opponent party and even the print media, that typically support the authorities, frowned on the rescue efforts made. The strongest criticism came from the international media, such as those of America and Japan, both of which questioned the ineffective rescue operations of the Taiwan Regional Authorities. Therefore, it is evident that the government image was severely impaired as the typhoon disaster relief was not properly managed.

21.6 Compensation Measures Following the Disaster

After Typhoon Morakot, having learnt the lesson from previous ineffective rescue efforts, Taiwan Regional Authorities proactively assessed the loss evaluations and set in motion all kinds of compensation activities to ensure a smooth post-disaster reconstruction. The Taiwan Regional Legislative Council immediately issued “Special regulations for post-Morakot reconstruction”, in which it was stated that: “The insurance for the typhoon-stricken insured, participants of the Farmer Health Insurance and Annuity Insurance were to be paid by the government within a fixed period of time following the disaster and, the qualifications, conditions, time length and other relevant regulations should be specified by the internal department (Chen et al. 2010a, b). Similarly, the insurance for the typhoon-stricken parties of the Labor Insurance and Employment Insurance programs were to be paid by the government within a fixed period after the disaster. Further, it was decreed that the insured parties of the Labor Insurance Scheme who were taken ill or were injured during the typhoon, could apply for government-paid wound pensions and the insurance, partial medical care charges and dietary fees of the hospital... for the typhoon-stricken insured of national health insurance plan...– were to be paid by the government within a fixed period after the disaster. Additionally, the qualifications, conditions, time length and other relevant regulations should be specified by the Administration and Health Department".

Due to the appropriate and effective measures on the post-disaster compensation, the discontent emotions of the public in Taiwan region were gradually eased and social order was restored.

21.7 Prevention of Future Typhoon Disasters in the Taiwan Region

In a typhoon-prone area such as the Taiwan region, one which is often hit by such disasters, the typhoon season is generally posted as starting from June and continuing to October, with the highest frequency occurring from July to September.

Typhoon disasters typically cause, not only, huge risks to the safety of people's lives and property, but also significant damages resulting in poverty, economic and political instability, all of which will impede positive social development. In view of these elements, the Taiwan region has already prepared some additional emergency plans.

Taiwan region typhoon emergency plans can be divided into three major steps: Data collection, analysis and forecast, and pre-impact warning programs which are presented below (Chen et al. 2011a, b).

21.7.1 Data Collection

When a typhoon is approaching, the Taiwan Regional Bureau of Meteorology gathers weather observation reports which are reported every 3–6 h from around the world through the meteorological global telecommunication network. Besides, the 24 weather stations and 4 Doppler radar Weather stations distributed throughout the Taiwan region all start intensive observation immediately and send the real-time observation data to the headquarters site in Taipei. Meanwhile, the meteorological center also receives the nephogram data sent by the geostationary meteorological satellite every hour and provides analyzed data related to the position of the typhoon center and the variations in strength of the typhoon to the working group for further analysis and application.

21.7.2 Analysis and Forecast

After a detailed analysis of the available information by the working group, a typhoon forecasting seminar will be held regularly four times a day in which various issues will be discussed in detail, including detailed environmental conditions, the position and strength of the typhoon center, the storm area, the forecasting methods and the situation of combined wind and rainfall around the island (Liu et al. 2009a, b). All these data are then to be provided and used to calculate the basis for the prediction of the typhoon path, its impact on each area and the situation of the wind and rainfall around the island. Finally, the steps and content of warnings, including the involved areas and all relevant matters, are decided and released.

21.7.3 Pre-impact Issuing of the Warning

The warning is issued each 3 h from the ocean typhoon alarming stage; while at the landing typhoon alarming stage, warnings are issued once every hour. Through the automated telephone fax, point-to-point service system and e-mail installations, the real-time alert messages will be sent to various government agencies, disaster prevention and relief units, mass communication organizations and local weather stations. Alternatively, the alarm information can also be reported directly through the weather service phone lines. Meanwhile, daily press briefings about typhoon warnings are regularly held and 30 real-time reports will also be delivered through the radio alerts at fixed times.

The Taiwan Regional Weather Bureau checks typhoons occurring in the Northwest Pacific and South China Sea every year. According to data collected over the past 10 years related to the mean range-error of typhoon tracking predictions, the forecasting skill has been improved, but a significant potential of forecasting error still exists and is likely to increase over time. (Chen et al. 2011a, b) The typhoon forecasting difficulty was particularly obvious in predicting the strength of Morakot in 2009 (Fig. 21.1).

Typhoon Morakot moved slowly at the beginning and thus was considered of low destructive power by the Taiwan Regional Weather Bureau which thereby failed to predict the large amounts of precipitation it brought. There was more than 2,000 mm precipitation in some areas in the Taiwan region, which later triggered flash floods and mudslides that engulfed many villages and caused heavy casualties and property losses.

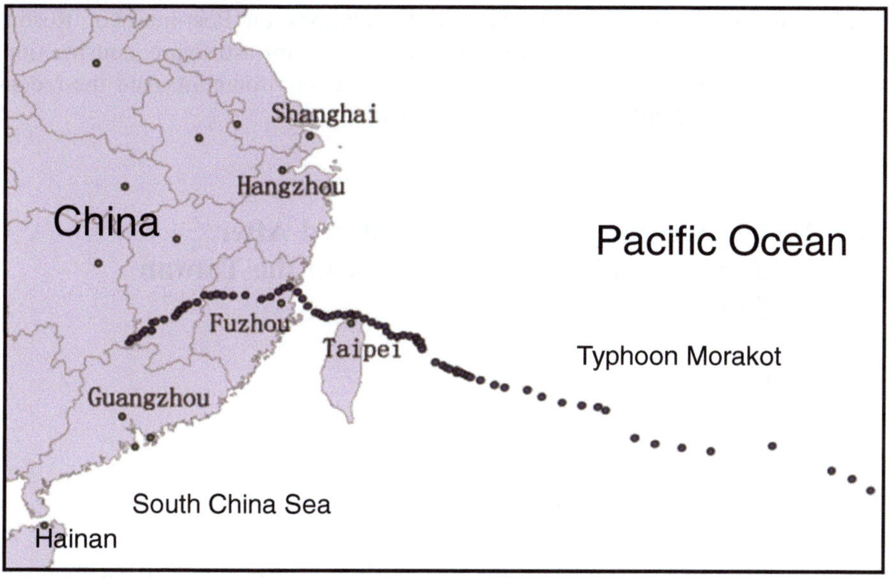

Fig. 21.1 The actual line of Morakot

21.8 The New Emergency Regulations in the Taiwan Region Following Typhoon Morakot

Following the disaster of Typhoon Morakot, Taiwan Regional Authorities took urgent emergency measures (see Table 21.2). The departments concerned, such as the Taiwan Regional Administrative Office, promulgated a series of emergency

Table 21.2 New emergency regulations after typhoon Morakot in the Taiwan region

Issuing department	Promulgated regulations
Administrative office	Special regulations on post-morakot reconstruction
Administrative office	Organization structure of the promotion committee for the post-morakot reconstruction
Treasury department	Various financial measures
Construction bureau of home office	Accommodation measures for homeless victims
Committee on agriculture	Rescuing measures for agricultural natural disaster
Committee on agriculture	Measures on industrial special case counseling
Ministry of economy	Service group for assisting enterprises' post-morakot reconstruction
Labor committee	"8–8 temporary worker" project
Aborigine committee of the administrative office	Relocation plan for the original residents
Ministry of transportation	Emergency repair and reconstruction plan of highway system
Health department of administrative office	Policy reports on health

regulations exemplified by the "Special Regulations on Post-morakot Reconstruction", all of which focused on economic aid, infrastructure construction, scientific research, medical care, education and the environment. And the reconstruction work gradually stepped into right path.

21.9 New Emergency Measures Instituted After the Disaster of Typhoon Morakot in the Taiwan Region

See Fig. 21.2.

21.9.1 Economic Assistance

Facing such a serious disaster as that caused by Morakot, the Taiwan Regional Authorities "Ministry of Finance" issued several orders to assist in post-disaster reconstruction. In one case, the State Tax Bureau, the Local Tax Authorities and the Bureau of State-Owned Property were directed that they should not only help the victims apply for a reduction in taxes and land rents, but should also coordinate with State-owned banks in providing loans for reconstruction. To assist the reconstruction work of the Administrative Office and local governments, the Ministry of Finance took various measures to ensure the needed funds, such as giving priority to using relative funds, diverting funds from less urgent programs and making good use of the integration funds and so on.

In the homeland rebuilding program, settlement plans made by the Construction Bureau of the Home Office provided victims with cash subsidies, of which 90.2 % was distributed, and temporary shelters. "Rescuing Measures for Agricultural Natural Disasters" made by the Committee on Agriculture provided subsidies and

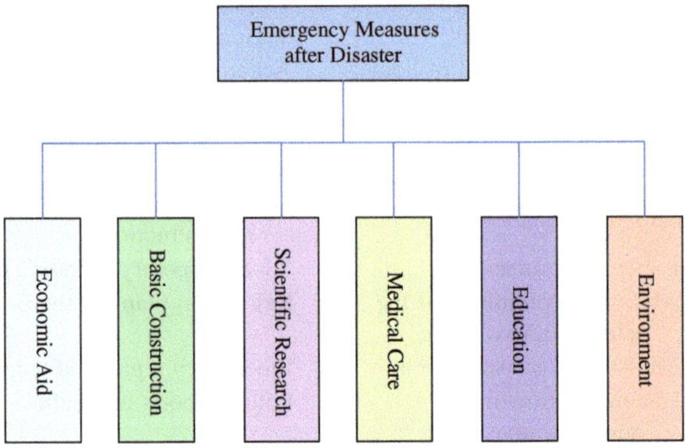

Fig. 21.2 New emergency measures after typhoon Morakot in the Taiwan region

low-interest loans, and the Measures on Industrial Special Case Counseling was sketched out to assist in re-cultivation of damaged lands.

As for the industrial and commercial sector, the Service Group for Assisting Enterprises Post-morakot Reconstruction, a body established by the Ministry of Finance, helped victims with the reconstruction of commercial venues and applications for tax reduction and special loans that were supported by the Small-and-Medium-Sized Enterprise Development Fund. Meanwhile the “8–8 Temporary Worker” Project, promoted by the Labor Committee, made great efforts to revitalize the labor market and help typhoon victims find jobs by implementing programs that offered 14,000 jobs, pushing forward the training programs for the young and providing for an occupational exploration study. There were 63 people participating in the training program and 88 in the exploration study.

21.10 Infrastructure Construction

To deal with reconstruction matters, the Taiwan Regional Authority Administrative Office passed a special regulation for post-morakot reconstruction which stated that New Taiwan Dollar (NTD) 100 billion was to be spent in the 3 years following the disaster to finish reconstruction work.

Morakot destroyed at least 44 provincial and 23 county roads and bridges around the island all of which called for immediate repair. The Taiwan Regional Authority Committee of Economic Construction, immediately approved the emergency repair and reconstruction plans for the road system proposed by the administrative departments for transportation and states that NTD 29.257 billion was to be spent between 2009 and 2112 to repair and rebuild the road system of each province, county and village. As to the issue of permanent housing for victims care, the Construction Bureau of the Home Office surveyed 85 places and selected 42 as village relocation sites so that county and municipal governments could choose sites in accordance with the will of the victims. Immediately following the disaster there were already 3 places undergoing land planning by professional consultant companies. Moreover, some potentially dangerous places not considered suitable for residential areas, were identified from which residents were removed. The Taiwan Regional Authorities purchased these areas and settled the residents on state-or public- owned land with an offer of ex-gratia payment.

21.11 Medical Care and Education

Due to the fact that the possibility of physical disease and psychological distress normally increases after such a disaster, the Health Department of the Administrative Office of Taiwan stressed the importance of continuously monitoring the situation and disease prevention and cure and psychological recovery were put in place within the "Health Policy Address". For example, the annual budget was diverted from less urgent areas to raise money to purchase emergency equipment and materials for the disaster area. Too, support for psychological recovery was planned. Additionally, authorities budgeted 8.66 billion to deal with health and insurance aid needs, psychological consolation, clinic rebuilding and disease prevention in the hope of effectively supporting and keeping the non-stop health and medical care service at full force in the disaster area. Starting from December 2009, the Hemagglutinin1 Neuraminidase1 (H1N1) new influenza vaccinations were especially assigned to the victims, medical staff, pregnant women and babies aged between 1 and 6 years. Meanwhile, the Health Department carried out the health insurance preference policy implemented for the disaster area.

In addition, the Taiwan Regional Authorities added the teaching of disaster prevention to the teaching content of the middle and primary schools. The intention was that the public should learn about disaster prevention through various methods such as television, newspapers, wallboards, outdoor public service ads and others. Meanwhile, in every municipality and county, disaster prevention education museums were built and opened to students and the public in which there were videos of all kinds of disasters, special disaster prevention classrooms for children, command rooms for comprehensive disaster notices, meteorological disaster showrooms, drills for escaping from high-rise buildings, and more.

21.12 Environmental and Academic Research

In order to enhance the function of the information systems of environmental disaster management, the Environmental Protection Department of the Administrative Office conducted training and drills on the use of the system and how to react to emergent ocean pollution. Prior to the flood season, monitoring was started in preparation for each region and for security maintenance. Drills were also integrated related to all resources and environmental protection units, and a system to aid manufacturers in opening contracts based on waste clearing capacity in the region was initiated. In total, a subsidy of 120 million NTD was provided for leasing machinery, hiring the labor force, purchase of cleaning and sterilization equipment as well as the establishment of drinking water testing services.

After the disaster Taiwan regional scientists immediately started research on typhoon disaster programs. The National Scientific Association and the National Cheng Kung University jointly held a seminar on the reconstruction following the "August the 8th Flood Disaster", with more than 200 specialists involved. They initiated a series of profound analyses on issues ranging from explorations into slope lands, water conservancy, seashore and bridge preservation and including suggestions of disaster prevention policies. This was the first comprehensive and in-depth discussion on disaster prevention measures among scholars and specialists following the "August the 8th Flood Disaster". During the congress they made full use of their specialties and proposed various opinions including technical and policy aspects and others, all of which formed the solid basis for the government's future disaster prevention programs.

21.13 The Inspiration of Disaster Emergency Measures and Proposals for the Policy of Defensive System Construction in the Taiwan Region

21.13.1 Inspiration of Disaster Emergency Measures in Taiwan Region

A typhoon-prone province such as the Taiwan region can provide an excellent study example that Guangdong can benefit from. The inspiration related to the emergency measures emerging from the study of Typhoon Morakot can provide an excellent study and starting point.

1. The Government must give high-levels of attention to typhoon disaster programs and establish a comprehensive crisis management system. The establishment of such a system which consists of a number of elements including pre-disaster alarms, in-disaster emergency measures and post-disaster recovery programs can effectively restrain the negative impacts of typhoons. Meanwhile, it is necessary for the government to establish disaster crisis management programs. By improving relevant regulations related to disaster management, management efficiency programs will be enhanced when a disaster might strike.
2. The Government must assist disaster-stricken people and enterprises via relevant policies ensuring the recovery of the economy and a return to normal life as soon as possible. If not, people cannot live a normal life and a great number of enterprises will go bankrupt, which will significantly affect the social harmony and stability within the region.
3. In addition to the potential economic losses brought about by possible typhoons, the reconstruction work in the disaster area also consumes huge amounts of government financial resources. If the government does not take corresponding measures early-on, it will not only face the loss of great volumes of wealth ruined by the disaster, but also have to assume a heavy financial burden. Therefore, it is necessary for all circles to establish corresponding insurance systems to defuse the negative economic impact caused by a typhoon disaster. These programs are powerful guarantees for post-typhoon reconstruction.

21.13.2 Proposals for the Policy of Establishing Typhoon Disaster Defensive Systems

Drawing on experiences from Typhoon Morakot, Guangdong should establish a comprehensive, three-dimensional typhoon disaster defensive system including pre-disaster, in-disaster and post-disaster measures. And these programs must include the three key components, namely pre-disaster prevention and reduction systems, in-disaster crisis management systems and post-disaster loss evaluation and compensation systems.

21.13.2.1 Construction of Pre-disaster Prevention and Reduction Systems

Pre-disaster prevention and reduction systems not only strengthen Guangdong's anti-Typhoon capabilities, but also bring economic growth through the investment of provincial government funds in the prevention of typhoon-related problems (Chen et al. 2010a, b). Though it can't create direct profits, it can bring benefit by ensuring the safety in production and reducing production losses caused by disasters. In a word, the benefit of disaster prevention and recovery programs are closely related to the investment made.

According to some scholars, the proportion of investment to benefit is 1–20 and sometimes even 1–100. Obviously disaster prevention and reduction is an effective measure to increase production and save/preserve resources. In 2008, The State Development and Reform Commission issued the Outline of a Plan for the Reform and Development of the Pearl River Delta, stressing the modernization of fundamental industry sectors, especially that of water conservancy and, later in July, the Guangdong Communist Party of China (CPC) issued The Decision on Accelerating the Construction of Modern Industry, pointing out that the development of fundamental industry should be based on water conservancy. It can be shown from the documents of the State and Guangdong Province that great importance has been attached to fundamental industry development as related to disaster prevention and reduction, which is designed to not only reduce the economic loss caused by disasters, but also coordinate recovery efforts with current positive fiscal policies. Moreover, the investment in relevant fundamental industries can enhance economic growth in many ways.

21.13.2.2 Establishment of In-disaster Crisis Management Systems

The component dealing directly with the example of a typhoon disaster is the in-disaster crisis management system whose effectiveness can minimize the potential damage (Chen and Shen 2010). From the developmental perspective, typhoon disasters can have huge impacts on the development of the overall society of the impacted area which makes the crisis management system an inevitable demand for the social harmony and an indispensable condition for the image of the Provincial Government. Government crisis management is a tool involving pre-crisis alarms, in-crisis emergency handling and post-crisis recovery of the effects of a potential or unexpected crisis. It can restore normal social life by carrying out various effective and foreseeing measures, eliminating emerging crises in time and handling the occurring crisis properly. Therefore, a highly effective in-disaster crisis management system should be the core of a Guangdong typhoon disaster defensive system.

21.13.2.3 Establishment of Post-disaster Loss Evaluation and Compensation Systems

Post-disaster loss evaluation and compensation systems are an important guarantee of economic recovery and home reconstruction, as well as the last component of a comprehensive defensive system. In recent years the frequency and intensity of typhoons that influence Guangdong has continuously increased but a scientific measurement system is still lacking for the computation of predictive impacts and potential economic losses caused by typhoons, all of which makes it impossible to evaluate the losses in a fast and effective manner, so relevant departments cannot predict, adequately prevent or handle the disaster in an adequate manner. Meanwhile, due to the lack of accurate evaluation systems, the corresponding compensation system cannot be set up in time, all of which affects the effective reconstruction work.

Therefore, the establishment of a scientific evaluation system can provide a solid scientific foundation for future defensive work while the establishment of a comprehensive compensation system can minimize the loss of the public resources and can also maintain public confidence and the prestige of the government, and thereby ensure social stability and unity.

Acknowledgments This research was supported by: (1) The Guangdong Natural Science Foundation Team Research Project, China, "Hazard Evaluation on Typhoons and Examination on Their Biological Effect in the South China Coastal Area Based on Satellite Remote Sensing" (8351030101000002); (2) National Natural Science Foundation of China (31061160190, 40976091); (3) Science and Technology Planning Project of Guangdong Province, China, "Development of Emergency Response System of Major Typhoon Disaster and Urban Fire Disaster" (2010B031900041); (4) Philosophy and Social Science Planning Projects in Guangdong Province, China, "Research on Typhoon Disaster Forecasting of Guangdong Province" (GD11YGL03); (5) High-level Research Projects in Guangdong Province Department of Education, China, "Research on Typhoon Disaster Forecasting and Disaster Grade Assessment Model: the Empirical Analysis based on Typhoon Disaster of Guangdong Province" (LYM11062).

References

Chen, S.H., Chen, H.: Research of typhoon rainstorm amount estimation based-on Elman neural network. In: Proceedings of the ITCS 2011, IEEE Xplore, July (2011)

Chen, H., Shen, M.H.: A research on the crisis management of typhoon disaster. Mod. Manage. Sci. **8**, 105–107 (2010). (in Chinese)

Chen W.K., Sui, G.J., Tang, D.L., Cai, D.L., Wang, J.S.: Study of environmental risk management by multivariable analysis of pattern structure. In: Proceedings of The Fourth International Conference on Management Science and Engineering Management. ISBN 978-1-84626-003-2, England, UK. (EI), pp. 3–7 (2010)

Chen, H., Pu, H.Y., Shen, M.H.: A research on the investment of disaster prevention and mitigation of typhoon disasters in Guangdong province. Mod. Manage. Sci. **7**, 81–83 (2010). (in Chinese)

Chen, W.K., Sui, G.J., Tang, D.L.: Predicting the economic loss of typhoon by case base reasoning and fuzzy theory. In: Proceedings of the MLC 2011, IEEE Xplore, vol. 1, pp. 254–257 (2011)

Chen, W.K., Sui, G.J., Tang, D.L (2011) A fuzzy intelligent decision support system for typhoon disaster management. In: Proceedings of the FUZZ 2011, IEEE Xplore, vol. 6, pp. 364–367

Cui, P., Chen, S.Q., Su, H.F., Zhang, J.Q.: A research on the causes and inspiration of mountain disasters induced of Taiwan typhoon Morakot. J. Mt. Sci. **1**, 103–114 (2010). (in Chinese)

Lin, X.R., Gao, Y.T., Chen, R.S.: Why Morakot hit Taiwan heavily? Sci. Cult. **11**, 12–13 (2009). (in Chinese)

Liu, D.F., Pang, L., Xie, B.T.: Typhoon disaster in China: prediction, prevention, and mitigation. Nat. Hazards **49**, 421–436 (2009a)

Liu, S.A., Li, M.J., Sha, X.J.: The situation analysis after the disaster of 2009 Taiwan typhoon Morakot. China Flood Drought Manage. **6**, 11–14 (2009b). (in Chinese)

Song, F.F., Ou, J.P.: Investigation of damage and causes of typhoon Hagupit to city buildings. J. Nat. Disasters **4**, 8–16 (2010). (in Chinese)

Sui, G.J., Tang, D.L., Chen, H.: Economic influence of typhoon catastrophe and its construction of recovery system. Int. Econ. Trade Res **2**, 32–36 (2010). (In Chinese)

Yin, M.X.: Typhoon Morakot hits Taiwan heavily. Relat. Across Taiwan Straits **9**, 12–13 (2009). (In Chinese)

Zheng, G.M., Tang, D.L.: Offshore and nearshore chlorophyll increases induced by typhoon and typhoon rain. Mar. Ecol. Prog. Ser. **333**, 61–74 (2007)

Part VI
Crisis Management for Typhoon Disaster

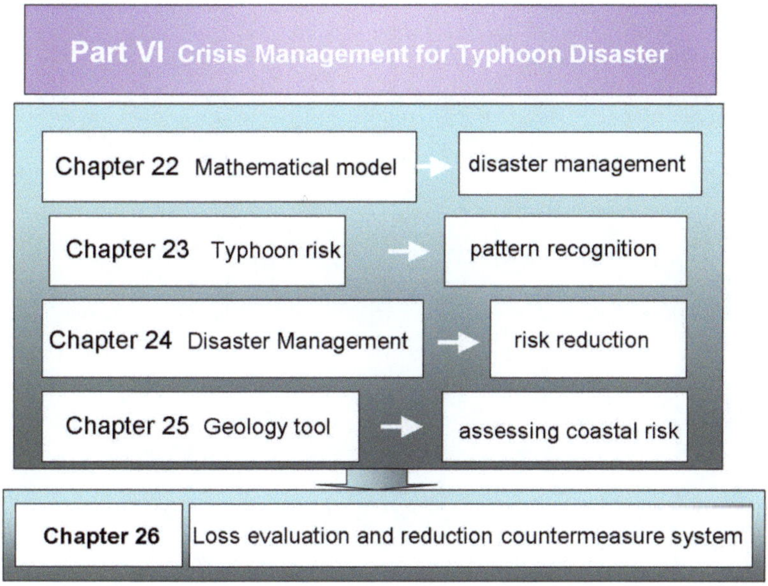

Chapter 22
The Mathematical Model for Typhoon Disaster Management

WangKun Chen, GuangJun Sui and DanLing Tang

Abstract This chapter introduces the management methodology of modeling typhoon disaster with focuses on describing an ideal mathematical way to represent typhoon risk. The mathematical model is based on the pattern structure to estimate the relationship between different characteristics in a typhoon event. The prediction results are calculated by the predictor from the disaster event. The candidate indexes of each pattern are selected from important factors in the literatures. Based on the approach, the relationship between the environmental events, the ecosystem change, the economic loss, and the response measure can be evaluated. The model can be further improved as long as the database of the predictor becomes sufficient and the mathematical scheme is accurate. The development of fuzzy theory, neural-network, and intelligent system can be helpful for the future development of this system.

Keywords Disaster prevention · Pattern recognition · Risk management · Environmental protection · Economic evaluation model

W. K. Chen (✉)
Department of Environmental and Property Management, Jinwen University of Science and Technology, New Taipei 23154 Taiwan, China
e-mail: wangkun@just.edu.tw

G. J. Sui
GuangDong University of Foreign Studies, Guangzhou, China

D. L. Tang
South China Sea Institute of Oceanology, Chinese Academy of Sciences, Guangzhou, China

D. L. Tang and G. J. Sui (eds.), *Typhoon Impact and Crisis Management*,
Advances in Natural and Technological Hazards Research 40,
DOI: 10.1007/978-3-642-40695-9_22,

22.1 Introduction

22.1.1 Environmental Risk Management

Environmental risk is the actual or potential effects on living organisms and environment through discharges, effluents, emissions, and other factors from man-made source to natural disaster. The characteristic of environmental risk needs to be studied further since the environmental problem has become more important recently. Although many studies had been done about the environmental risk management, a systematic approach by the pattern recognition has not been found yet. This chapter suggests a systematic approach by different methods of mathematical analysis in the pattern structure of environmental risk.

Environmental risk management is an important tool because it can reduce much economic loss for society. To prevent the damage and loss from the environmental event, such as the natural disaster or environmental pollution, certain response measure has to be done before or after the event. However, the environmental phenomenon is complex and difficult to predict. A proper measure for risk management is usually very hard on design. Therefore, better assessing an actual phenomenon is an important subject for the environmental risk management.

Among the environmental risk, the disaster of typhoon has been more concerned today. Typhoon is a natural phenomenon comes from certain meteorological conditions. It causes much damage to both the environmental and economical system. In order to prevent the negative effect of typhoon, the emergency management is necessary. Due to the complicated nature of typhoon, a well-designed framework of the emergency management system is the first step toward mitigating typhoon disaster.

The typhoon disaster management system should be able to provide adequate information for the judgments of the response measure. A typical emergency response system has four important categories: the characteristic of typhoon, the damage of environment, the loss of economic, and the measure of control. In this chapter, we try to establish a complete framework of the management system.

The characteristic of typhoon will be discussed first. Then, the method to assess the impact of the marine eco-system will be followed. Thirdly, we will present the evaluation model for assessing a typhoon's economic loess. Finally, we will show the emergency response system for typhoon management.

Recently, researchers have spent a lot of effort in studying the risk encountered in the management process (Chen 2010a; Makuil et al. 2010; Choi et al. 2010; Mojtahedi et al. 2009; Wang and Elhag 2006). Some of them were interested in the project management risk study (Carr and Tah 2001; Chapman and Ward 2004; Patrick et al. 2007; Raftery 1999; Savic and Kayis 2006; Hillson 2002). Some were interested in the product risk management (Cooper 2003; Coppendale 1995; Mojtahedi et al. 2009). While some were focused in the construction risk management (Flanagan and Norman 1993; Chapman 2001; Thevendran and

Mawdesley 2004). For environmental risk, such as typhoon, we need to develop a new tool to clearly describe the phenomenon.

The "pattern recognition" technique is "the act of taking in raw data and taking an action based on the category of the pattern" (Duda et al. 2001; Schuermann 1996). Pattern recognition aims to classify data (pattern) based either on a prior knowledge or on statistical information extracted from the patterns. The patterns to be classified are usually groups of measurements or observations, defining points in an appropriate multi-dimensional space. In this chapter, the concept of pattern recognition technique was introduced to identify the pattern of environmental data set for risk management. Four types of pattern structures have been defined in this chapter to construct the framework of pattern recognition and a multivariable analysis method has been derived to estimate the relationship between each pattern.

22.1.2 Pattern Structure in the Environmental Events

The pattern recognition is the approach to integrate all the important factors within the environmental process. There is a multi dimensional relationship among the functional areas such as environmental events, ecosystem change, economic loss, and response measure. The pattern relationship is defined as how each pattern can influence other patterns. This can be calculated by the multi-variable analysis (MA) or other mathematical functions. Four types of pattern were defined in this paper as the following: (1) characteristic types of environmental event, (2) characteristic types of ecosystem change, (3) characteristic types of economic loss, and (4) characteristic types of response measure.

The level of relationship between each pattern should be well known for the system engineers. The higher degree of relationship between the environmental event and the ecosystem change or economic loss reveals that we need to pay more attention to the possible consequences. Therefore, the pattern of response measure needs to be more carefully designed. The mathematical model determines the total relationships for each pattern and environmental factors. These data are usually obtained from empirical observations and statistical results. Therefore, the data flow for the pattern is important. It is also helpful for the system engineers to find the best responding activities that could reduce or minimize environmental risk in advance.

22.2 The Mathematical Tool for Typhoon Disaster Analysis

22.2.1 Definition of Typhoon Nature by Turbulence Structure

Typhoon is a kind of natural phenomenon which could be described by the atmospheric fluid dynamics. A typhoon is a tropical cyclone that develops in the northwestern part of the Pacific Ocean between 180° and 100°E. For organizational purposes, the northern Pacific Ocean is divided into three regions: the eastern (North America to 140°W), central (140°W to 180°), and western (180° to 100°E). A Pacific typhoon, then, is a tropical cyclone in the northern Pacific Ocean west of 180°. Identical phenomena in the eastern north Pacific are called hurricanes, with tropical cyclones moving into the Western Pacific re-designated as typhoons.

Within the northwestern Pacific there are no official typhoon seasons as tropical cyclones form throughout the year. The majority of storms form between May and December whilst tropical cyclone formation is at a minimum between January and April. The Northwestern Pacific features some of the most intense tropical cyclones on record.

Typhoon can be well defined by its physical properties. The principle of fluid dynamics and the theory of fractal analysis is a good tool for knowing the typhoon properties. The fluid dynamics told us that there are many "eddies" in the atmospheric environment. The scale of eddy is in different "order of magnitude". Typhoon is one of those large eddies exists in the atmosphere. To understand the nature of typhoon, we need to know the characteristic of turbulence with these "eddies".

22.2.2 Understanding of Typhoon Phenomenon by Fractal Analysis

The environmental change of typhoon phenomenon can only be investigated by long term and large area monitoring. The complexity in the time series of typhoon observations has made its interpretation very difficult. Since the typhoon is so complicated, the recent development tool, fractal analysis, has to be employed to describe the nature of typhoon.

Fractal analysis is a contemporary method of applying nontraditional mathematics to pattern that defies understanding with traditional Euclidean concepts. It measures the complexity using the fractal dimension. It has been used in the analysis of the scale dependence of rainfall (Olsson et al. 1992, 1993), air pollutant concentration (Lee 2002; Lee et al. 2003a, b; Lee et al. 2006a, b; Anh et al. 2000), stock price index (Ho et al. 2004; Mantegna and Stanley 2000; Bouchaud and Potters 2000; Evertsz 1995; Skjeltorp 2000; Katsuragi 2000; Turiel and Pérez-Vicente 2003; Mandelbrot 1997), and earthquake (Lee et al. 2006a, b; Godano et al. 1997; Godano et al. 1999; Telesca and Lapenna 2004) distribution in time has been studied intensively in recent years.

The fractal dimension is the key concept of fractal analysis. There are many types of fractal dimension or D_F, but all can be condensed into one category—they are meters of complexity. In fractal analysis, complexity is a change in detail with change in scale. A D_F is, in essence, a scaling rule comparing how a pattern's detail changes with the scale at which it is considered—this is what we mean by *complexity*.

In general, we deduce the scaling rule or fractal dimension, D_F, from knowing how something scales. Formally, this idea is about the relationship between N, the number of pieces and ε, the scale used to get the new pieces. We say that:

$$N \propto \varepsilon^{-D_F} \quad (22.1)$$

To understand how all this fits together in the calculations for scaling rules and fractal dimensions; let's look at things differently than everyday life usually asks us to. First consider something we know, patterns such as the familiar Euclidean shapes of elementary geometry. One shape for which scaling is easy to grasp is a simple line. A line, when scaled by, as an example, 1/3, can be seen to be made up of 3 pieces, each 1/3 the length of the original. Nothing is cosmic to that.

It is interesting to know the N in the scaling relationship, and we can figure out that $D_F = 1.00$ in this situation just by substituting into the equation, because $3 = (1/3)^{-1}$. When scaling a filled square by 1/2, there will always be 4 new pieces, each 1/4 the area of the original, and D would be equal to 2 (e.g., $4 = (1/2)^{-2}$).

For the Koch pattern, we have to choice and calculate the D_F. If we calculate a D_F by solving the general equation for the scaling rule:

$$N = A\varepsilon^{-D_F} \quad (22.2)$$

For its variable, D_F, using logs, which shows that the D_F is the ratio of the log of the number of new parts N, to the log of scale, ε:

$$D_F = \log N / \log \varepsilon. \quad (22.3)$$

22.2.3 Representation of Typhoon Disaster by Fuzzy Theory

Fuzzy logic emerged as a consequence of the 1965 proposal of fuzzy set theory by Lotfi Zadch (1965). Fuzzy theory is an efficient tool for the analysis of typhoon phenomenon. It helps us to abstract the knowledge from the raw data. Fuzzy analysis is a process of clustering. Fuzzy logic is a form of multi-valued logic derived from fuzzy set theory to deal with reasoning that is robust and approximate rather than brittle and exact. In contrast with "crisp logic", where binary sets have two-valued logic, fuzzy logic variables may have a truth value that ranges in degree between 0 and 1 (Novák et al. 1999). Furthermore, when linguistic variables are used, these degrees may be managed by specific functions (Chen 2011, 2013).

22.2.4 Prediction of Typhoon Risk by Neural-Network System

The term neural network was traditionally used to refer to a network or circuit of biological neurons (Hopfield 1982). The modern usage of the term often refers to artificial neural networks, which are composed of artificial neurons or nodes. Thus the term has two distinct usages: biological neural networks and artificial neural networks. This article focuses on the relationship between the two concepts; for detailed coverage of the two different concepts refer to the separate articles: biological neural network and artificial neural network. The neural network uses the input layer and output layer for calculation. Results from neural network are usually much better than that from the traditional statistical regression method (Chen 2010b).

22.3 The Pattern of Typhoon Disaster

22.3.1 Types of Typhoon Disaster Characteristics

In Fig. 22.1, there are four types of pattern in the system: (1) natural disaster itself; (2) change in ecosystem; (3) loss in economic system; (4) response for risk management. The methodology to clarify the patterns becomes more important in the management system. The relationship between the four categories can be determined by different ways. Chen et al. have used the multivariable method, fuzzy theory, and neural network to clarify the relationship (Chen et al. 2010a).

From this framework, the engineers can manage more systematically for a risk management project under various levels of risks. The arrows in the figure represent the direction of influence. For example, the change pattern in ecosystem will influence the loss pattern in ecosystem. Also, the characteristic pattern of environmental event will influence the loss pattern in economic system.

The first category is the characteristics of a typhoon event. An environmental event refers to any observable occurrence or extraordinary occurrence of environmental phenomenon, such as air pollution, river and ocean pollution, typhoon, earthquake, flood, etc. There are a lot of variables that represents the

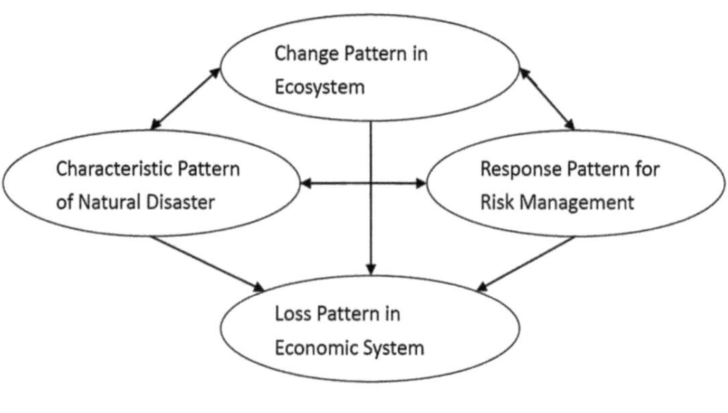

Fig. 22.1 The four types of pattern in the typhoon risk management system

environmental events. For example, the natural disaster of earthquake could be described by strength, location, frequency, time and duration.

The second category is the pattern of change in ecosystem. The environmental event may cause the change of ecosystem. The change will induce the environmental risk. There are a variety of risk types due to the ecosystem change in the environment. By the previous researchers, those risks can be described by the following factors such as: land use and property; social and economic; noise and vibration; visual amenity and urban design; traffic; soil and geology; surface and ground water; flora fauna; air quality; aboriginal heritage; non-indigenous heritage; hazard; resources and waste.

The third category is the pattern of loss in economic system. Both the environmental event itself and the change of ecosystem will cause the loss of economic system. The pattern of loss in economic system includes housing damage; personal injury and death; agricultural loss; roads and bridges damage; indirect economic loss; post-disaster reconstruction costs.

The fourth category is the pattern of response for risk management. Response measure is the activity we take to avoid the damage of ecosystem and the loss of economic system. The entire proper measures combine together to form the response pattern of risk management. The factors of response pattern in risk management system include reinsurance compensation; super fund; major disaster securities market; social public disclosure; education and training; emergency response; human resource.

Each pattern in environmental risk management can be defined as the following equation.

(1) The pattern of environmental event

For an environmental event, there exists a characteristic pattern defined by X,

$$\mathbf{X} = (\mathrm{x}_1, \mathrm{x}_2, \ldots \mathrm{x}_\mathrm{n})^\mathrm{T} \tag{22.4}$$

(2) The pattern of environmental change

For an ecosystem, there exits a change pattern defined by C.

$$\mathrm{C} = (\mathrm{c}_1, \mathrm{c}_2, \ldots, \mathrm{c}_n)^\mathrm{T}. \tag{22.5}$$

(3) The pattern of economic loss

For an economic system, there exits a loss pattern defined by L.

$$\mathrm{L} = (\mathrm{l}_1, \mathrm{l}_2, \ldots, \mathrm{l}_n)^\mathrm{T}. \tag{22.6}$$

(4) The pattern of risk management

For a risk management system, there exits a response pattern defined by R.

$$R = (r_1, r_2, \ldots, r_n)^T \tag{22.7}$$

22.3.2 A Multivariable Model for Estimating of the Environment Risk

The relationship between each pattern leads to a matrix that represents the degree of influence factors. The coefficients of this matrix are the basis for managing the environmental risk. The link between environmental events and change in eco-system is

$$\begin{aligned} c_1 &= a_{11}x_1 + a_{12}x_2 + \cdots + a_{1n}x_n + d_1, \\ c_2 &= a_{21}x_1 + a_{22}x_2 + \cdots + a_{2n}x_n + d_2, \\ &\ldots \\ c_n &= a_{n1}x_1 + a_{n2}x_2 + \cdots + a_{nn}x_n + d_n. \end{aligned} \tag{22.8}$$

The above linear system can be represented by matrix as

$$\begin{pmatrix} c_1 \\ c_2 \\ \vdots \\ c_n \end{pmatrix} = \begin{pmatrix} a_{11} & a_{12} & \ldots & a_{1n} \\ a_{21} & a_{22} & \ldots & a_{2n} \\ \vdots & \ldots & \ldots & \vdots \\ a_{n1} & a_{n2} & \ldots & a_{nn} \end{pmatrix} \begin{pmatrix} x_1 \\ x_2 \\ \vdots \\ x_3 \end{pmatrix} + \begin{pmatrix} d_1 \\ d_2 \\ \vdots \\ d_n \end{pmatrix}. \tag{22.9}$$

Define

$$D = (d_1, d_2, \ldots, d_n)^T. \tag{22.10}$$

The above linear equation can be rewritten as matrix equation as

$$\boldsymbol{C} = \boldsymbol{AX} + \boldsymbol{D}, \tag{22.11}$$

where $\boldsymbol{C}$ represents the total change in the change pattern, $\boldsymbol{AX}$ is the individual contribution of each parameter in the change pattern.

Theorem 1 *The characteristic pattern of environmental events will influence the change pattern in ecosystem, and the change pattern in ecosystem will also influence the characteristic pattern of environmental events.*

From Theorem 1, the equation can be written as

$$
\begin{aligned}
c_1 &= a_{11}\mathrm{x}_1 + a_{12}\mathrm{x}_2 + a_{1n}\mathrm{x}_n + a_{1(n+1)}c_1 + a_{1(n+2)}c_2 + \cdots + a_{1(n+m)}c_m + \delta_1,\\
c_2 &= a_{21}\mathrm{x}_1 + a_{22}\mathrm{x}_2 + a_{2n}\mathrm{x}_n + a_{2(n+1)}c_1 + a_{2(n+2)}c_2 + \cdots + a_{2(n+m)}c_m + \delta_2,\ldots\\
c_n &= a_{n1}\mathrm{x}_1 + a_{n2}\mathrm{x}_2 + a_{nn}\mathrm{x}_n + a_{n(n+1)}c_1 + a_{n(n+2)}c_2 + \cdots + a_{n(n+m)}c_m + \delta_n,\\
\mathrm{x}_1 &= a_{(n+1)1}\mathrm{x}_1 + a_{(n+1)2}\mathrm{x}_2 + \cdots + a_{(n+1)n}\mathrm{x}_n + a_{(n+1)(n+1)}c_1\\
&\quad + a_{(n+1)(n+2)}c_2 + \cdots + a_{(n+1)(n+m)}c_m + \delta_{n+1},\\
\mathrm{x}_2 &= a_{(n+2)1}\mathrm{x}_1 + a_{(n+2)2}\mathrm{x}_2 + \cdots + a_{(n+2)n}\mathrm{x}_n + a_{(n+2)(n+1)}c_1\\
&\quad + a_{(n+2)(n+2)}c_2 + \cdots + a_{(n+2)(n+m)}c_m + \delta_{n+2},\ldots\\
\mathrm{x}_m &= a_{(n+m)1}\mathrm{x}_1 + a_{(n+m)2}\mathrm{x}_2 + \cdots + a_{(n+m)n}\mathrm{x}_n + a_{(n+m)(n+1)}c_1\\
&\quad + a_{(n+m)(n+2)}c_2 + \cdots + a_{(n+m)(n+m)}c_m + \delta_{n+m}
\end{aligned}
\tag{22.12}
$$

The above linear system can be represents by matrix as:

$$
\begin{pmatrix} u_1\\ u_2\\ \vdots\\ u_n\\ u_{n+1}\\ u_{n+2}\\ \vdots\\ u_{n+m} \end{pmatrix}
=
\begin{pmatrix}
a_{11} & a_{12} & \cdots & a_{1n} & a_{1(n+1)} & a_{1(n+2)} & \cdots & a_{1(n+m)}\\
a_{21} & a_{22} & \cdots & a_{2n} & a_{2(n+1)} & a_{2(n+2)} & \cdots & a_{2(n+m)}\\
\vdots & \vdots & \cdots & \vdots & \vdots & \vdots & \cdots & \vdots\\
a_{n1} & a_{n2} & \cdots & a_{nn} & a_{n(n+1)} & a_{n(n+2)} & \cdots & a_{n(n+m)}\\
a_{(n+1)1} & a_{(n+1)2} & \cdots & a_{(n+1)n} & a_{(n+1)(n+1)} & a_{(n+1)(n+2)} & \cdots & a_{(n+1)(n+m)}\\
a_{(n+2)1} & a_{(n+2)2} & \cdots & a_{(n+2)n} & a_{(n+2)(n+1)} & a_{(n+2)(n+2)} & \cdots & a_{(n+2)(n+m)}\\
\vdots & \vdots & \cdots & \vdots & \vdots & \vdots & \cdots & \vdots\\
a_{(n+m)1} & a_{(n+m)2} & \cdots & a_{(n+m)n} & a_{(n+m)(n+1)} & a_{(n+m)(n+2)} & \cdots & a_{(n+m)(n+m)}
\end{pmatrix}
\begin{pmatrix} x_1\\ x_2\\ \vdots\\ x_n\\ c_1\\ c_2\\ \vdots\\ c_n \end{pmatrix}
+
\begin{pmatrix} \delta_1\\ \delta_2\\ \vdots\\ \delta_n\\ \delta_{n+1}\\ \delta_{n+2}\\ \vdots\\ \delta_{n+m} \end{pmatrix}.
\tag{22.13}
$$

Define

$$\mathrm{U} = (\mathrm{c}_1, \mathrm{c}_2, \ldots, \mathrm{c}_\mathrm{n}, \mathrm{x}_1, \mathrm{x}_2, \ldots, \mathrm{x}_\mathrm{n})^\mathrm{T}. \tag{22.14}$$

and

$$\Delta = (\delta_1, \delta_2, \ldots, \delta_\mathrm{n}, \delta_{\mathrm{n}+1}, \delta_{\mathrm{n}+2}, \ldots, \delta_{\mathrm{n}+\mathrm{m}})^\mathrm{T}. \tag{22.15}$$

The above linear equation can be rewritten as matrix equation as:

$$\mathrm{U} = \mathrm{AX} + \Delta, \tag{22.16}$$

where, U represents the total change in the change pattern, AX is the individual contribution of each parameter in the change pattern.

The link between environmental events and loss in economic system is

$$
\begin{aligned}
l_1 &= a_{11}x_1 + a_{12}x_2 + \cdots + a_{1n}x_n + \beta_1,\\
l_2 &= a_{21}x_1 + a_{22}x_2 + \cdots + a_{2n}x_n + \beta_2,\\
l_n &= a_{n1}x_1 + a_{n2}x_2 + \cdots + a_{nn}x_n + \beta_n.
\end{aligned}
\tag{22.17}
$$

The above linear system can be represents by matrix as

$$\begin{pmatrix} l_1 \\ l_2 \\ \vdots \\ l_n \end{pmatrix} = \begin{pmatrix} a_{11} & a_{12} & \dots & a_{1n} \\ a_{21} & a_{22} & \dots & a_{2n} \\ \vdots & \dots & \dots & \vdots \\ a_{n1} & a_{n2} & \dots & a_{nn} \end{pmatrix} \begin{pmatrix} x_1 \\ x_2 \\ \vdots \\ x_n \end{pmatrix} + \begin{pmatrix} \beta_1 \\ \beta_2 \\ \vdots \\ \beta_n \end{pmatrix}. \tag{22.18}$$

Define

$$\mathrm{B} = (\beta_1, \beta_2, \dots, \beta_n)^{\mathrm{T}}. \tag{22.19}$$

The above linear equation can be rewritten to matrix equation as

$$L = AX + B, \tag{22.20}$$

where, L represents the total loss in the loss pattern, AX is the individual contribution of each parameter in the loss pattern.

Theorem 2 *The characteristic pattern of environmental events will influence the loss pattern in economic system, and the loss pattern in economic system will not influence the characteristic pattern of environmental events.*

Theorem 3 *The characteristic pattern of ecosystem change will influence the loss pattern in economic system, and the loss pattern in economic system will not influence the characteristic pattern of ecosystem change.*

From Theorem 2 and Theorem 3, we rewrite the equation to

$$\begin{aligned} l_1 &= a_{11}\mathrm{x}_1 + a_{12}\mathrm{x}_2 + \cdots + a_{1n}\mathrm{x}_n + a_{1(n+1)}c_1 + a_{1(n+2)}c_2 + \cdots + a_{1(n+m)}c_m + \beta_1, \\ l_2 &= a_{21}\mathrm{x}_1 + a_{22}\mathrm{x}_2 + \cdots + a_{2n}\mathrm{x}_n + a_{2(n+1)}c_1 + a_{2(n+2)}c_2 + \cdots + a_{2(n+m)}c_m + \beta_2, \\ &\dots \\ l_n &= a_{n1}\mathrm{x}_1 + a_{n2}\mathrm{x}_2 + \cdots + a_{nn}\mathrm{x}_n + a_{n(n+1)}c_1 + a_{n(n+2)}c_2 + \cdots + a_{n(n+m)}c_m + \beta_{n+m}. \end{aligned} \tag{22.21}$$

The link between environmental events and management system is

$$\begin{aligned} r_1 &= a_{11}x_1 + a_{12}x_2 + \dots + a_{1n}x_{\mathrm{n}} + \gamma_1, \\ r_2 &= a_{21}x_1 + a_{22}x_2 + \cdots + a_{2n}x_n + \gamma_2, \\ r_n &= a_{n1}x_1 + a_{n2}x_2 + \cdots + a_{nn}x_n + \gamma_n. \end{aligned} \tag{22.22}$$

The above linear system can be represents by matrix as

$$\begin{pmatrix} r_1 \\ r_2 \\ \vdots \\ r_n \end{pmatrix} = \begin{pmatrix} a_{11} & a_{12} & \dots & a_{1n} \\ a_{21} & a_{22} & \dots & a_{2n} \\ \vdots & \dots & \dots & \vdots \\ a_{n1} & a_{n2} & \dots & a_{nn} \end{pmatrix} \begin{pmatrix} x_1 \\ x_2 \\ \vdots \\ x_n \end{pmatrix} + \begin{pmatrix} \gamma_1 \\ \gamma_2 \\ \vdots \\ \gamma_n \end{pmatrix}. \tag{22.23}$$

Define

$$\mathrm{G} = (\gamma_1, \gamma_2, \ldots, \gamma_n)^{\mathrm{T}}. \tag{22.24}$$

The above linear equation can be rewritten as matrix equation as

$$\mathrm{R} = \mathrm{AX} + \mathrm{G}, \tag{22.25}$$

where R represents the total response in the risk management pattern, AX is the individual contribution of each parameter in the risk management pattern.

The overall description of the environmental risk management model is

Theorem 4 *The characteristic pattern of environmental events will influence the response pattern for risk management, and the response pattern for risk management will also influence the characteristic pattern of environmental events.*

Theorem 5 *The change pattern in ecosystem will influence the response pattern for risk management, and the response pattern for risk management will also influence the change pattern in ecosystem.*

Theorem 6 *The loss pattern in economic system will influence the response pattern for risk management, and the response pattern for risk management will also influence the loss pattern in economic system.*

Rewrite the equation to a more general form Theorem 4, 5 and 6. And define a new variable to represent the parameters among these four patterns.

$$\Psi = (\varphi_1, \varphi_2, \ldots, \varphi_n)^T. \tag{22.26}$$

$$\begin{aligned} \varphi_1 &= \tau_{11}x_1 + \tau_{12}x_2 + \cdots + \tau_{1n}x_n + \varepsilon_1, \\ \varphi_2 &= \tau_{21}x_1 + \tau_{22}x_2 + \cdots + \tau_{2n}x_n + \varepsilon_2, \\ \varphi_n &= \tau_{n1}x_1 + \tau_{n2}x_2 + \cdots + \tau_{nn}x_n + \varepsilon_n. \end{aligned} \tag{22.27}$$

The above linear system can be represents by matrix as

$$\begin{pmatrix} \varphi_1 \\ \varphi_2 \\ \vdots \\ \varphi_n \end{pmatrix} = \begin{pmatrix} \tau_{11} & \tau_{12} & \cdots & \tau_{1n} \\ \tau_{21} & \tau_{22} & \cdots & \tau_{2n} \\ \vdots & \cdots & \cdots & \vdots \\ \tau_{n1} & \tau_{n2} & \cdots & \tau_{nn} \end{pmatrix} \begin{pmatrix} x_1 \\ x_2 \\ \vdots \\ x_n \end{pmatrix} + \begin{pmatrix} \varepsilon_1 \\ \varepsilon_2 \\ \vdots \\ \varepsilon_n \end{pmatrix}. \tag{22.28}$$

Define

$$E = (\varepsilon_1, \varepsilon_2, \ldots, \varepsilon_n)^T, \tag{22.29}$$

and

$$\Lambda = \begin{pmatrix} \tau_{11} & \tau_{12} & \cdots & \tau_{1n} \\ \tau_{21} & \tau_{22} & \cdots & \tau_{2n} \\ \vdots & \cdots & \cdots & \vdots \\ \tau_{n1} & \tau_{n2} & \cdots & \tau_{nn} \end{pmatrix}. \tag{22.30}$$

The above linear equation can be rewritten as matrix equation as

$$\Psi = \Lambda X + E, \tag{22.31}$$

where, Ψ represents the total response pattern in the risk management system, ΛX is the individual contribution of each parameter in the risk management pattern.

Four types of characteristic pattern were defined in this paper to represent the environmental risk management. The relationship of each pattern was determined by multivariable analysis. By this model, a suitable response measure could be identified. It offers an ideal way for environmental risk management in the early stage or for the post examination.

22.4 Impact Assessment of Marine Ecosystem

22.4.1 Parameter to Represent the Marine Ecosystem

Typhoon influences the ecosystem of marine environment. The ecosystem can be described by several parameters as shown in Table 22.1.

22.4.2 Impact Assessment of Marine Ecosystem by Analytical Hierarchy Process and Delphi Method

The Analytical Hierarchy Process (AHP) and Delphi method are often used to evaluate the extent of an event. The AHP method was proposed by Saaty in 1971. It was originally applied to determine the priority of resource planning, allocation and portfolio etc. The theory was further developed for the systematic application for problem solving. By using the analytical hierarchy, the complex problem can be simplified. Through the quantitative judgments, the contest can be found for the comprehensive assessment. This method can provide the decision maker complete information and chose the appropriate plan to reduce the risk of making wrong decision (Saaty 1980).

The Delphi method is a structured communication technique, originally developed as a systematic, interactive forecasting method which relies on a panel of experts (Linstone, 1975). In the standard version, the experts answer questionnaires in two or more rounds. After each round, a facilitator provides an anonymous summary of the experts' forecasts from the previous round as well as the reasons they provided for their judgments. Thus, experts are encouraged to revise their earlier answers in light of the replies of other members of their panel. It is believed that during this process the range of the answers will decrease and the group will converge towards the "correct" answer. Finally, the process is stopped after a pre-defined stop criterion (e.g., number of rounds, achievement of

Table 22.1 Important parameter to represent the marine ecosystem

Category	Environmental factor	Reference data
Water animal	Type and quantity	Population type, quantity, wavering status, survey method, location, time and scope
	Species diversify	Type, quantity, richness, evenness, sampling volume
	Habitat and habits	Wander characteristics, life cycle, reproduction methods and conditions
	Migration and reproduction	Wander characteristics, life cycle, reproduction methods and conditions
Water plant	Type and quantity	Type, quantity, vegetation and its distribution in the case
	Species diversity	Species richness and evenness, sampling volume
	Distribution of vegetation	Vegetation type, size, distribution, growth conditions
	Community advantage	Dominant species, quantity, distribution
Populations of endangered and protected	Animal	Rare species, endemic species, endangered species and the government announced the conservation of wild animals, protect the control plan
	Plant	Rare species, endemic species, endangered species and rare plants, protection control plan
Ecosystem	Effects of	eutrophication
	Source of nutrients, into the volume and control methods	
	Bioaccumulation	Toxic, harmful or radioactive substances bioaccumulation
Food chain	Ecological resource productivity, food chain relationships	

consensus, and stability of results) and the mean or median scores of the final rounds determine the results (Linstone, 1975).

Delphi method can be applied in the study to evaluate the damage. The environmental factors was collected and used to establish the framework of the evaluation system. The Analytic Hierarchy Process can be used to calculate the weighting factor of each index in this framework. Figure 22.2 shows the analytical hierarchy for assessing the damage of a typhoon.

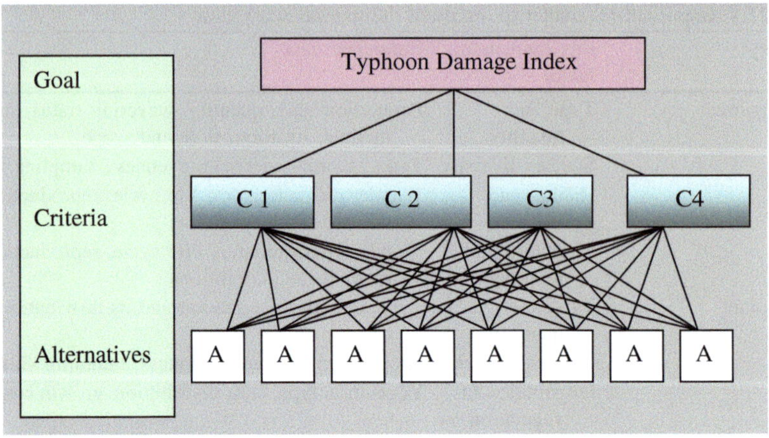

Fig. 22.2 Analytical hierarchies for assessing the typhoon damage

22.5 Economic Loss Evaluation

The economic loss can be evaluated by Input-Output Analysis and CGE model. Input-output model uses a matrix representation of a nation's (or a region's) economy to predict the effect of changes in one industry on others and by consumers, government, and foreign suppliers on the economy. Computable general equilibrium (CGE) models are a class of economic models that use actual economic data to estimate how an economy might react to changes in policy, technology or other external factors. CGE models are also referred to as AGE (applied general equilibrium) models (Chen et al. 2011).

22.5.1 Evaluation by Input-Output Analysis Model

Wassily Leontief is credited with the development of this model. Francois Quesnay developed a cruder version of this technique called Tableau économique. Leontief won the Nobel Memorial Prize in Economic Sciences for his development of this model. Léon Walras's work *Elements of Pure Economics* on general equilibrium theory is both a forerunner and generalization of Leontief's seminal concept. Leontief's contribution was to simplify Walras's piece so that it could be implemented empirically. The International Input-Output Association[1] is dedicated to advancing knowledge in the field of input-output study, which includes "improvements in basic data, theoretical insights and modeling, and applications, both traditional and novel, of input-output techniques".

[1] http://www.iioa.org/

The mathematics of input-output economics is straightforward, but the data requirements are enormous because the expenditures and revenues of each branch of economic activity have to be represented. As a result, not all countries collect the required data and data quality varies, even though a set of standards for the data's collection has been set out by the United Nations through its System of National Accounts[2] (SNA). Because the data collection and preparation process for the input-output accounts is necessarily labor and computer intensive, input-output tables are often published long after the year in which the data were collected. Moreover, the economic "snapshot" that the benchmark version of the tables provides of the economy's cross-section is typically taken only once every few years.

22.5.2 Evaluation by CGE Model

CGE models are descended from the input-output models pioneered by Wassily Leontief, but assign a more important role to prices. CGE models are useful to estimate the effect of changes in one part of the economy upon the rest. They have been used widely to analyze trade policy. More recently, CGE has been a popular way to estimate the economic effects of measures to reduce greenhouse gas emissions of natural disaster. CGE models always contain more variables than equations—so some variables must be set outside the model. These variables are termed exogenous; the remainder, determined by the model, is called endogenous. The choice of which variables are to be exogenous is called the model closure, and may give rise to controversy.

22.6 Emergency Response Measure Management

22.6.1 Response and Control Measure

The decision support system is useful for managing the emergency response system. Many researchers have spent their efforts in the decision support system (Chang and Wang 1996; Chang et al. 1997) Some have applied this idea in the nuclear power plant emergency management (Yeung and Ching 1993; Yoshikawa et al. 1990) Chen et al. has studied the environmental risk management by multi-variable analysis of pattern structure, and submit an idea about the expert system for managing response of disaster (Chen et al. 2010a, b; Chen 2010a, b). However, these researches do not consider the concept of response measure by a systematic thinking. This section tries to combine the concept of management pattern with the

---

[2] http://unstats.un.org/unsd/sna1993/introduction.asp

space-time relationship. A framework of the system including the concept and equation was described detail.

22.6.2 Concept of Disaster Management

The design of environmental disaster system includes the following step as: (1) Identify concept to represent knowledge; (2) Find concept to represent knowledge; (3) Design structure to organize knowledge; (4) Formulate rules to embody knowledge; (5) Validate rules that organize knowledge.

The natural phenomenon can be represented by a mathematical vector. In general, the natural environment is a space-time information phenomenon. The space geographical information can be represented by the following equation:

$$I = \sum_{i=1}^{m} \left[\sum_{j=1}^{n} \begin{bmatrix} S_{ij}(T_{ij}) \\ A_{ij}(T_{ij}) \\ T_{ij} \end{bmatrix} \right] \tag{22.32}$$

$$T_{ij}(t_b, t_e)_{ij}$$

where I is the collection of space geographical information; i is the individual vector for the ith item, j is the state of this item; $S_{ij}(T_{ij})$ and $A_{ij}(T_{ij})$ represent the characteristics of this item in time t_b to t_e. The inferences of the above equation are the following:

【Inference 1】 If i is constant, equation represents the time series data of the same characteristic.

【Inference 2】 If j is constant, equation represents the characteristic distribution in the same time.

22.6.3 Knowledge Bank Analysis and Data Base Design

The knowledge bank of typhoon disaster management is shown as Fig. 22.3.

The knowledge bank contains the model bank, pattern bank, and regulation bank. In designing the expert system for typhoon response management, it is important to choose the appropriate measure. There are two major types of model, meteorological model, and economic model. The meteorological model tells us the behavior of the typhoon. The economic model includes the housing damage model, agriculture loss model, indirect economic loss model, and post disaster reconstruction model.

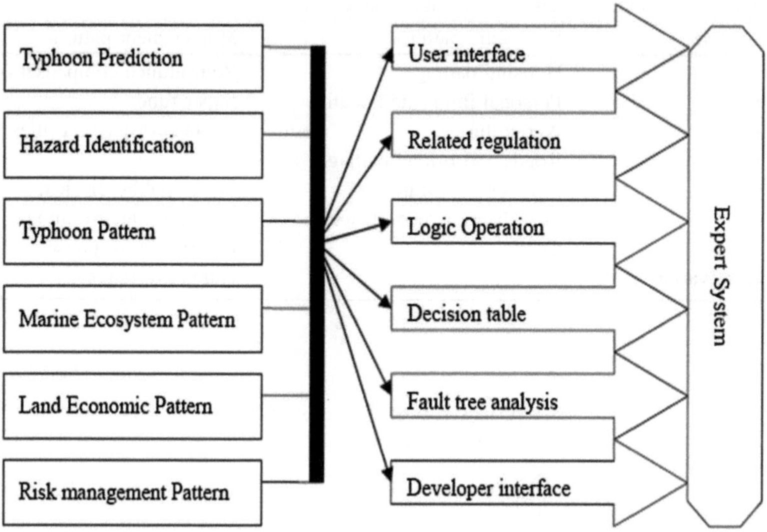

Fig. 22.3 Intelligent knowledge-based expert system

The data base of emergency response system include the meteorological data, spatial data, attribute data (such as population, terrain etc.), source data, etc.

22.6.4 Emergency Response System

The IF/THEN inference mechanism was used in the emergency response system. IF represents the obtained information and THEN represents the action. Table 22.2 is the example of IF/THEN operator.

Different kinds of disaster management are listed in Table 22.3.

The risk management should follow the regulations; therefore, a regulation bank for the response measure is necessary for the system. Following the decision results from the expert system, the decision maker takes action to prevent the economic loss; these actions could be represented by the following models as

Table 22.2 IF/THEN operator for typhoon disaster management

IF	Logical operator	Example	THEN
$>$	Greater than	The wind speed of the typhoon	Action(1)
$<$	Less than	The distance to the center of typhoon	Action(2)
$\geqq$	Greater than or equal to	The flood level	Action(3)
$\leqq$	Less than or equal to	The diameter of the typhoon	Action(4)
$=$	Equal to	The rainfall amount	Action(5)

Table 22.3 Three models in disaster management system

Disaster pattern	Economic pattern	Management pattern
Strength	Housing damage	Reinsurance compensation
Location	Personal injury and death	Super fund
Frequency	Agriculture	Major disaster securities market
Time	Roads and bridge damage	
Duration	Indirect economic loss	Social public disclosure
Terrain	Post disaster reconstruction	Education and training
Wind field	Costs	Emergency response
Temperature variation		Human resource

$$
\begin{aligned}
& M \\
& = (Action_1, Action_2, \ldots, Action_n) \\
& = (m_1, m_2, \ldots, m_n) \\
& = \sum_{i=1}^{n} m_i
\end{aligned}
\tag{22.33}
$$

where M is the collection of emergency response measure; Action i is the individual vector for the ith measure. Each action is represented by a symbol m; and there are m measures in the action domain. If the management system is good enough, there should have enough measure to solve the problem encountered by the typhoon. The inferences are as the following:

【Inference 3】 If i is constant, then for each $S_{ij}(T_{ij})$ and $A_{ij}(T_{ij})$, there exists a measure m_i in the emergency response measures domain.
【Inference 4】 If j is constant, then for each $S_{ij}(T_{ij})$ and $A_{ij}(T_{ij})$, there exists a measure m_i in the emergency response measures domain.

The expert system for typhoon disaster management was established in this section. It utilized the information technology to prevent the possible emergency condition. The system was designed with the knowledge-based decision support system. The system can be used as the emergency preparedness or emergency response guideline.

22.7 Conclusion

A mathematical model for managing the typhoon risk is presented in this chapter. The model considers four natural characteristics of a typhoon: characteristic of typhoon events, change of eco system, change of economic system, and the measure of the emergency response system. Mathematical tools for calculating the relationships between these patterns have been introduced in details, including the multivariable analysis, fuzzy inference, and neural network theory. The Input/output model, CGE (Give the full name) model for economic analysis of typhoon influence can be used as a tool for estimating the variation of economic category due to the influence of typhoon. The systematic method to evaluate the effect of typhoon disaster can provide the decision makers to find the best control strategy for typhoon risk management.

The mathematical pattern relationship of typhoon effects on neighboring area and the corresponding response measures have also been presented in this chapter. With this modeling system, the decision maker can specify the procedure and minimize human errors in the decision process. The improvement of risk response and the quality of management system can be upgraded by this system. In addition, the methodology can also be served as a basis for the future development of risk response.

However, this study is still in the beginning phase. First, more data about typhoon patterns are needed. Second, a new algorithm for estimating the relationships among these patterns is still in development. Third, a quantitative way to calculate the risk is lacking. Finally, the model needs to be validated in more case studies.

Acknowledgments This study was supported by the following research projects awarded to DL Tang: (1) National Natural Sciences Foundation of China (31061160190, 40976091, NSFC-RFBR Project-41211120181 of DL Tang and D. Pozdnyakov); (2) Guangdong Sciences Foundation (2010B031900041, 8351030101000002); (3) Innovation Group Program of State Key Laboratory of Tropical Oceanography (LTOZZ1201).

References

Anh, V.V., Lam, K.C., Leung, Y., Tieng, Q.M.: Multifractal analysis of Hong Kong air quality data. Environmetrics **11**(2), 139–149 (2000)

Bouchaud, J.P., Potters, M.: Theory of Financial Risk. Cambridge University Press, Cambridge (2000)

Carr, V., Tah, J.: A fuzzy approach to construction project risk assessment and analysis. Adv. Eng. Softw. **32**, 847–857 (2001)

Chang, N.B., Wang, S.F.: The development of an environmental decision support system for municipal solid waste management. Comput. Environ. Urban Syst. **20**(3), 201–212 (1996)

Chang, N.B., Wei, Y.L., Tseng, C.C., Kao, C.Y.: The design of a GIS-based decision support system for chemical emergency preparedness and response in an urban environment. Comput. Environ. Urban Syst. **21**(1), 67–94 (1997)

Chapman, R.: The controlling influences on effective risk identification and assessment for construction design. Int. J. Project Manage. **19**(3), 147–160 (2001)

Chapman, C., Ward, S.: Project Risk Management: Processes, Techniques and Insights, 2nd edn. Wiley, Chichester (2004)

Chen, W.K.: A framework for nuclear power plant emergency response system. Nucl. Sci. Technol. **21**, 375–378 (2010a)

Chen, W.K., Sui, G.J., Tang, D. L., Cai, D. L., Wang, J.S.: Study of environmental risk management by multivariable analysis of pattern structure. Paper presented at proceeding of the fourth international conference on management science and engineering management, Chungli, Taiwan, (2010a)

Chen, W.K., Sui, G.J., Tang, D.L.: A fuzzy intelligent decision support system for typhoon disaster management. Paper presented at proceeding of 2011 IEEE international conference on fuzzy systems, institute of electrical and electronics engineers (IEEE), Taipei, Taiwan, (2010b)

Chen, W.K., Sui, G.J., Tang, D.L.: Predicting the economic loss of typhoon by case base reasoning and fuzzy theory. Paper presented at proceeding of 9th international conference on machine learning and cybernetics (ICMLC), institute of electrical and electronics engineers (IEEE), systems, man, and cybernetics society (IEEE SMC society), Guilin, China, (2011)

Chen, W.K.: An approach to pattern recognition by fuzzy category and neural network simulation. Paper presented at proceeding of the 8th international conference on machine learning and cybernetics (ICMLC), institute of electrical and electronics engineers (IEEE), systems, man, and cybernetics society (IEEE SMC society), Qingdao, China, (2010b)

Chen, W.K.: Environmental applications of granular computing and intelligent systems. In: Pedrycz, W., Chen, S. (eds.) MGranular Computing and Intelligent Systems: Design with Information Granules of Higher Order and Higher Type, Reference Library, vol. 13, pp. 275–301. Springer, Berlin, Germany (2011)

Chen, W.K.: Fuzzy forecasting with fractal analysis for the time series of environmental pollution. In: Pedrycz, W., Chen, S. (eds.) MTime Series Analysis, Modeling and Applications Intelligent Systems, Reference Library, vol. 47, pp. 199–213, Springer, Berlin, Germany (2013)

Choi, H., Ahn, J., et al.: A framework for managing risks on concurrent engineering basis. Int. J. Manage. Sci. Eng. Manage. **5**(1), 44–52 (2010)

Cooper, L.: A research agenda to reduce risk in new product development through knowledge management: a practitioner perspective. J. Eng. Tech. Manage. **20**(1–2), 117–140 (2003)

Coppendale, J.: Manage risk in product and process development and avoid unpleasant surprises. Eng. Manage. J. **5**(1), 35–38 (1995)

Duda, R.O., Hart, P.E., Stork, D.G.: Pattern Classification, 2nd edn. Wiley, New York (2001)

Evertsz, C.J.G.: Fractal geometry of financial time series. Fractals **3**, 609–616 (1995)

Flanagan, R., Norman, G.: Risk Management and Construction. Blackwell Science Pty Ltd, Victoria, Australia (1993)

Godano, C., Alonzo, M.L., Vilardo, G.: Multifractal approach to time clustering of earthquakes; application to Mt Vesuvio seismicity. Pure Appl. Geophys. **149**(2), 375–390 (1997)

Godano, C., Tosi, P., Derubeis, V., Augliera, P.: Scaling properties of the spatio–temporal distributionof earthquakes: a multifractal approach applied to a Californian catalogue. Geophys. J. Int. **136**, 99–108 (1999)

Hillson, D.: Extending the risk process to manage opportunity. Int. J. Project Manage. **20**, 235–240 (2002)

Ho, D.S., Lee, C.K., Wang, C.C., Chuang, M.: Scaling characteristics in the Taiwan stock market. Phys. A **332**, 448–460 (2004)

Hopfield, J.: Neural networks and physical systems with emergent collective computational abilities. Proc. Nat. Acad. Sci. USA **79**, 2554–2558 (1982). Biophysics

Katsuragi, H.: Multiaffinity and entropy spectrum of self-affine fractal profiles. Phys. A **278**, 275–281 (2000)

Lee, C.K.: Multifractal characteristics in. Air pollutant concentration time Series. Water Air Soil Pollut. **135**(1–4), 389–409 (2002)

Lee, C.K., Ho, D.S., Yu, C.C., Wang, C.C.: Fractal analysis of temporal variation of air pollutant concentration by box counting. Environ. Model. Softw. **18**(3), 243–251 (2003a)

Lee, C.K., Ho, D.S., Yu, C.C., Wang, C.C., Hsiao, Y.H.: Simple multifractal cascade model for air pollutant concentration (APC) time series. Environmetrics **14**(2), 255–269 (2003b)

Lee, C.K., Juang, L.C., Wang, C.C., Liao, Y.Y., Liu, Y.C., Ho, D.S.: Scaling characteristics in ozone concentration time series (OCTS). Chemosphere **62**(6), 934–946 (2006a)

Lee, C.K., Yu, C.C., Wang, C.C., Hwang, R.D., Yu, G.K.: Scaling characteristics in aftershock sequence of earthquake. Phys. A **371**(2), 692–702 (2006b)

Linstone, H.A., Murray, T. (eds.): The Delphi Method: Techniques and Applications, Addison-Wesley (1975)

Makuil, A., Mojtahedi, S., Mousavi, S.: Project risk identification and analysis based on group decision making methodology in a fuzzy environment. Int. J. Manage. Sci. Eng. Manage. **5**(2), 108–118 (2010)

Mandelbrot, B.B.: Fractal and Scaling in Finance: Discontinuity, Concentration, Risk. Springer, NY (1997)

Mantegna, R., Stanley, N.H.E.: An Introduction to Econophysics: Correlations and Complexity in Finance. Cambridge University Press, Cambridge (2000)

Mojtahedi, S., Mousavi, S., et al.: A nonparametric statistical approach for analyzing risk factor data in risk management process. J. Appl. Sci. **9**, 113–120 (2009)

Novák, V., Perfilieva, I., Močkoř, J.: Mathematical Principles of Fuzzy Logic Dodrecht. Kluwer Academic, Boston (1999)

Olsson, J., Niemczynowicz, J., Berndtsson, R., Larson, M.: An analysis of the rainfall time structure by box counting—some practical implication. J. Hydrol. **137**, 261–277 (1992)

Olsson, J., Niemczynowicz, J., Berndtsson, R.: Fractal analysis of high-resolution rainfall time series. J. Geophys. Res. **98**, 23265–23274 (1993)

Patrick, X., Zhang, G., et al.: Understanding the key risks in construction projects in China. Int. J. Project Manage. **25**, 601–614 (2007)

Raftery, J.: Risk Analysis in Project Management. E&FN Spon, London (1999)

Saaty, T.L., The analytic hierarchy process, McGraw-Hill, New York, (1980)

Savic, S., Kayis, B.: Knowledge elicitation for risk mapping in concurrent engineering projects. Int. J. Prod. Res. **44**(9), 1739–1755 (2006)

Schuermann, J.: Pattern Recognition. Wiley, New York (1996)

Skjeltorp, J.A.: Scaling in the Norwegian stock market. Phys. A **283**, 486–528 (2000)

Telesca, L., Lapenna, V., Macchiato, M.: Mono and multi-fractal investigation of scaling properties in temporal pattern of seismic sequences. Chaos Soliton Fractals **19**, 1–15 (2004)

Thevendran, V., Mawdesley, M.: Perception of human risk factors in construction projects: an exploratory study. Int. J. Project Manage. **22**, 131–137 (2004)

Turiel, A., Pérez-Vicente, C.J.: Multifractal geometry. Phys. A **322**, 629–649 (2003)

Wang, Y., Elhag, T.: Fuzzy TOPSIS method based on alpha level sets with an application to bridge risk assessment. Expert Syst. Appl. **31**, 309–319 (2006)

Yeung, M.R., Ching, E.M.: RADIS—a regional nuclear accident consequence analysis model for Hong Kong. Nucl. Technol. **101**, 123–139 (1993)

Yoshikawa, T., Kimura, F., Koide, T., Kurita, S.: An emergency computation model for the wind field and diffusion during accident nuclear pollutants release. Atmos. Environ. **24A**, 2739–2748 (1990)

Zadeh, L.A.: Fuzzy sets. Inf. Control **8**(3), 338–353 (1965)

Chapter 23
A Study on Typhoon Risk Prediction by Different Methods of Pattern Recognition

Wang-Kun Chen, GuangJun Sui and DanLing Tang

Abstract This chapter presents the economic loss prediction of typhoon by different pattern recognition methods, such as multivariable statistics (MS), case base reasoning (CBR), fuzzy theory (FT), and neural network model (NN). The typhoon records in Taiwan before 2000 were used as the database for reference, and the records after the year 2000 were predicted using the pattern derived from the database. Six scenarios were calculated using these methods. The first scen-eries include the parameters: maximum wind speed, minimum atmospheric pressure, maximum wind speed in typhoon center and lowest atmospheric pressure near typhoon center. The second scenario includes the previous four parameters with rainfall and calculated by CBR. The third scenario uses the fuzzy calculation with five parameters. The successful rate of prediction for the three methods was 12.5, 37.5, and 57 %. The results reveal that the fuzzy calculation can significantly increase the prediction rate than the traditional CBR method. On the other hand, five neural network methods were compared, which were back propagation network (BPN), extend neuron networks (ENN), fuzzy neural network (FNN), analysis adjustment synthesis network (AASN), and genetic algorithm neural network (GANN). The result reveals that the BPN is the best choice, because the error is the lowest among the five schemes in this study.

Keywords Case base reasoning · Damage evaluation · Nature disaster · Fuzzy theory · Neural network · Typhoon risk

W.-K. Chen (✉)
Department of Environment and Property Management, Jinwen University of Science and Technology, New Taipei, Taiwan
e-mail: wangkun@just.edu.tw

G. J. Sui
GuangDong University of Foreign Studies, Guangzhou, People's Republic of China

D. L. Tang
South China Sea Institute of Oceanology, Chinese Academy of Sciences, Guangzhou, People's Republic of China

D. L. Tang and G. J. Sui (eds.), *Typhoon Impact and Crisis Management*, Advances in Natural and Technological Hazards Research 40,
DOI: 10.1007/978-3-642-40695-9_23,

23.1 Introduction

Prediction of the economic loss is the major concern in managing the typhoon disaster. Since typhoon is a very complicated system, it is not stable, therefore the prediction of its behavior become extremely difficult. On the other hand, the economic loss is influenced by many factors such as the land usage, population density etc. To predict the risk of typhoon is not only a nature science, but also have to consider the social interaction.

Estimating the typhoon losses immediately is important to allocate and deliver necessary resources for helping people during the period of typhoon and post-disaster. Therefore, a new approach for calculating typhoon losses based on different mathematical method is presented in this chapter by means of typhoon characters data, geographical spatial information and historic typhoon loss statistics.

Taiwan is a small island located within the subtropical area, and is the main typhoon track of the northwestern Pacific Ocean. By the historical records, typhoons hit the island more than three times per year. The yearly damage of typhoon include the agricultural losses, building losses, and construction losses. For example, the agriculture losses caused by typhoons are about US$ 250 million. Due to the economic losses of typhoon will seriously influence the economic development of Taiwan, it is important to properly estimate the possible economic losses and submit an useful management plan. Therefore, the goal of this study is to investigate the influence of typhoons on the total economic losses with the appropriate method.

This section focus on the prediction of economic loss of typhoon by different mathematical methods. However, the characteristic of typhoon includes uncertainty, complexity and nonlinearity which result in the difficulty to establish of models. In order to overcome above-mentioned difficulties, the researches start from constructing the economic losses model based on different mathematical methods. Then, determine input and output variables. The pattern recognition technique was used in this section to calculate the loss value. The historical statistics of typhoon in Taiwan were used as the reference data. Finally, the numerical experiment using historic typhoon data in Taiwan is made to test the feasibility of methods approached in this chapter. The result indicates that these models are effective and prediction results are also receptive.

23.2 Mathematical Tool for Pattern Recognition of Typhoon Disaster

23.2.1 Multi Variable Statistical Analyses

Statistical pattern recognition is the most extensively studied and used in practice among the various frameworks of pattern recognition. The procedure of statistical pattern recognition includes the following issues: definition of pattern class, pattern recognition, feature extraction, cluster analysis, classifier design, machine learning, and performance evaluation (Jain et al. 2000).

The multi variable statistical analysis used different parameters to estimate the economic losses of typhoon. Physical properties of typhoon can be selected as the predictors. The wind speed, rainfall, temperature, and atmosphere pressure are typical examples (Chen 2010a; Chen et al. 2010b).

23.2.2 Case Base Reasoning

Typhoon can be estimated from the previous data with different estimation methods. If we choose the most similar case in previous experience, then the economic loss can be estimated by the similar case. This is the basic concept of case base reasoning, CBR. The successful rate of prediction for this method depends on the amount of previous data. If the data is enough, then the prediction results will be good, however, if the database can not provide enough case for reference, then the successful rate of prediction will not be fine (Aamodt and Plaza 1994; Schalkoff 1992; Schuermann 1996; Chen et al. 2011).

Case base reasoning (CBR) is the process to solve new problems based on the solutions of similar past problems. Scientist fixes the prediction rule by recalling previous example that exhibited similar characteristics. So, the CBR process just resembles the elements of natural phenomenon, and treats then as a solution database to predict the appropriate results (Cheetham and Goebel 2007; Lebowitz 1983; Watson and Gardingen 1999). The process to conduct the CBR process should follow the following four-step procedure: (1) retrieve; (2) reuse; (3) revise; (4) retain.

23.2.3 Fuzzy Prediction Method

Fuzzy theory is a new field of research constructed by the fuzzy logic, which is a form to deal with uncertainty and ambiguous situation. Fuzzy set theory was proposed by Zadeh in 1965. Now it has been widely used in many field including science and technology (Zadeh 1965).

Fuzzy theory is useful in predicting the uncertainty of natural phenomenon (Chen 2011, 2013; Chen et al. 2010a). Since typhoon is not easy to describe, so the fuzzy theory can play an important role in describing this phenomenon. We also use the fuzzy theory to predict the economic loss of typhoon in this chapter.

23.2.4 Neural Network Simulation

Neural network is a new area of computation; it changes the experience in the human brain to a quantitative factor or value. It has been used by many researches to systematically solve these kinds of problems. The expert system and artificial intelligence are the typical example.

Five neural network methods were compared, which were back propagation network (BPN), extend neuron networks (ENN), fuzzy neural network (FNN), analysis adjustment synthesis network (AASN), and genetic algorithm neural network (GANN) (Yeh 1999a, b, 2002, 2003, 2005; Leung et al. 2003; Tu and Lu 2004). Different meteorological variables are analyzed to find out their relation to economic losses and then important variables are selected as model inputs for predicting the event-based economic losses (Chen 2010b).

23.3 Experiment Design

23.3.1 Study Area and Data Collection

The research area was in Taiwan, as shown in Fig. 23.1. The predictors were selected based on the report from CWB.

The data before 2000 were collected as the reference group, and the records after 2000 were used as the verification group. These selected cases have an economic loss more than 5,000,000 NTD. The original data of typhoon loss in Taiwan was shown in Table 23.1. In this table, α is the minimum speed (kt) near typhoon center when this typhoon is in its strongest condition. β is the minimum atmospheric pressure (hPa)

Fig. 23.1 Map of the study area (Taiwan Island)

Table 23.1 The original data of typhoon in Taiwan in two groups

Year	Month	Date	Name	α	β	γ	δ	Total loss
1986	8	21–25	WAYNE	100	940	100	940	12,239,280
1986	9	16–20	ABBY	95	945	90	950	7,501,473
1994	7	11	TIM	120	930	100	942	56,853,348
1996	7	8	HERB	140	910	110	935	37,890,050
1994	8	8	DOUG	130	910	100	942	8,907,992
1998	10	14–16	ZEB	155	880	90	950	8,099,891
2000	10	30–11.01	XANGSANE	90	950	70	970	5,418,116
2000	8	21–23	BILIS	140	900	115	925	7,644,368
2001	9	16–19	NARI	80	960	80	960	5,691,986
2001	7	29–31	TORAJI	75	962	70	968	14,723,151
2004	6	29–07.02	MINDULLE	90	942	60	975	6,516,458
2005	7	16–20	HAITANG	110	915	95	935	9,831,405
2008	9	9–17	SINLAKU	100	930	90	940	5,643,770
2008	9	26–29	JANGMI	105	930	100	935	7,547,679

near typhoon when typhoon is in its strongest condition. γ is the maximum wind speed (kt) near typhoon center when this typhoon is close to Taiwan within one latitude. δ is the minimum atmospheric pressure (hPa) near typhoon center when the center of this typhoon is close to Taiwan within one latitude.

23.3.2 Scenario I Experiment: Statistical Pattern Recognition by Multivariable Analysis

Typhoon risk can be represented by a linear equation rewritten as matrix equation as

$$\Psi = \Lambda X + \mathrm{E} \tag{23.1}$$

where Ψ represents the total response pattern in the risk management system, ΛX is the individual contribution of each parameter in the risk management pattern.

23.3.3 Scenario II Experiment: Pattern Recognition by CBR with Four Meteorological Parameters

The similarity level was used to choose the reference case and use the results of the reference case to predict its results. The similarity index, η, was calculated by the following equation

$$\eta_e = \frac{P_{ref}}{p_{true(4\text{variables})}} \tag{23.2}$$

where, P_{true} is the true value of economic loss, P_{ref} is the economic loss of the reference loss. The lower value means the similarity of the case is higher.

23.3.4 Scenario III Experiment: Pattern Recognition by CBR with Five Parameters

In order to increase the successful rate of prediction, the rainfall was chosen to be the new parameter. The new similarity index, ζ, is the total value of the five parameters as shown in the following equation.

$$\xi_e = \frac{P_{ref}}{p_{true(5\text{variables})}} \tag{23.3}$$

23.3.5 Scenario IV Experiment: Pattern Recognition by CBR with Remote Sensing Parameters

Typhoon is not a stable event. It has a very strong energy and could damage the environmental and economic system. Now it is possible to use the remote sensing data to estimate the economic loss of typhoon, however, how to improve the accuracy is the key point to be concerned.

To apply the data from remote sensing, the following equation was used

$$\Omega_e = \frac{P_{ref}}{p_{true-RM}} \tag{23.4}$$

where $P_{true-RM}$ is the information from the remote sensing results.

23.3.6 Scenario V Experiment: Pattern Recognition by FT with Five Meteorological Parameters

In fuzzy prediction, we have to establish the membership function for prediction. The method to construct the membership function includes the economic loss in all interval, the lowest level and upper level of economic loss. The maximum value of the economic loss is 28,078,825, and the fuzzy interval value of membership function is in [0, 1].

The economic loss is represented by T as the total economic loss. L_L is the lowest level of economic loss. L_U is the upper level of economic loss, and κ is the total evaluation coefficient. The lowest level and upper level of the membership function in the interval is

$$\Phi = \frac{A - Y}{((X - A) + (A - Y))} \tag{23.5}$$

$$\Psi = \frac{B - Y}{((X - B) + (B - Y))} \tag{23.6}$$

where Φ is the lowest level and Ψ is the upper level of membership function. A is the lowest level of economic loss; B is the upper level of economic loss. X is the maximum value of economic loss, and Y is the minimum value of economic loss. The calculated value of Φ and Ψ in the interval of [2.0, 2.09] is [0.142, 0.310]. The total evaluation coefficient can be calculated as the following

$$\kappa = \frac{\alpha}{\beta} + \frac{\gamma}{\delta} \tag{23.7}$$

23.3.7 Scenario VI Experiment: Pattern Recognition by Neural Network Model

The first stage of this experiment is the training and test mode. In this mode, the original data was used to establish the model. The input was the unit for training, the number of hidden layer, learning cycle, and speed of learning cycle, etc. The second stage of this experiment is authentication model. The authentication mode is to use the non-training data to evaluate the reliability of the model. The results can be the information of the scattering plot, matrix, or error analysis. The output variable will also be estimated to compare their error. The third stage of this experiment is inference model. The inference model is used to predict certain condition which we want to know. The experienced value was derived from the trained mode.

23.4 Results

23.4.1 Scenario I Experiment: Pattern Recognition by Multivariable Statistical Analysis

The case study consisted of the typhoon data from the year 1985 to 2008 in Taiwan. There are 89 records in the data set with 70 effective data. The variables are shown as Table 23.2.

Table 23.2 Variables selected in multivariable statistical analysis

Name	Description
VAR1	The highest speed in the strongest wind condition of this typhoon
VAR2	The minimum atmosphere pressure in typhoon center in the strongest wind condition of this typhoon
VAR3	The highest speed when the typhoon approaches Taiwan within latitude
VAR4	The minimum atmosphere pressure in typhoon center when it approaches Taiwan within longitude
VAR5	The number of house totally collapses
VAR6	The number of houses partially collapsed
VAR7	The number of dead and missing
VAR8	The number of injuries
VAR9	The amount of agricultural losses
VAR10	The amount of fishery losses
VAR11	The amount of hydraulic losses
VAR12	The amount of railway losses
VAR13	The amount of highway loss
VAR14	The amount of port losses
VAR15	The amount of other losses
VAR16	The total amount of losses

$$\begin{pmatrix} \text{VAR5} \\ \text{VAR6} \\ \text{VAR7} \\ \text{VAR8} \end{pmatrix} = \begin{pmatrix} 5.301 & 6.913 & 3.264 & 9.777 \\ 10.957 & 21.153 & 4.964 & 35.259 \\ .284 & .010 & .028 & .135 \\ .313 & .047 & .002 & .377 \end{pmatrix} \begin{pmatrix} \text{VAR1} \\ \text{VAR2} \\ \text{VAR3} \\ \text{VAR4} \end{pmatrix} + \begin{pmatrix} 2722.860 \\ 13643.805 \\ 130.161 \\ 313.261 \end{pmatrix} \tag{23.8}$$

The above equation can explain the relationship between the environmental events, typhoon, to the environmental change. It is a clearly description of the emergency response system for environmental risk management system. More relationship can be deducted from the same procedure described above.

23.4.2 Scenario II Experiment: Pattern Recognition by CBR with Four Meteorological Parameters

Table 23.2 is the predicting results of typhoon by CBR. The typhoon records after 2000 were used to obtain the outcome as shown in Table 23.3. The number to represent the typhoon is 1. TIM; 2. ZEB; 3. WAYNE; 4. ABBY; 5. HERB, DOUG.

The successful rated of Scenario I, CBR with four parameters, is only 12.5 %, as shown in Table 23.3. This means it is still necessary to improve the prediction method. The possible reason for the high unsuccessful rate may be from the fact that

Table 23.3 Predicting of typhoon loss by CBR with four parameters (%)

α	JANGMI	SINLAKU	HAITANG	MINDULLE
ABBY	9.52	5	13.64	5.56
WAYNE	4.76	0	9.09	11.11
TIM	14.29	20	9.09	33.33
DOUG	23.81	30	18.18	44.44
HERB	33.33	40	27.27	55.56
ZEB	47.62	55	40.91	72.22
β	JANGMI	SINLAKU	HAITANG	MINDULLE
ABBY	1.61	1.61	3.28	0.32
WAYNE	1.08	1.08	2.73	0.21
TIM	0	0	1.64	1.27
DOUG	2.15	2.15	0.55	3.4
HERB	2.15	2.15	0.55	3.4
ZEB	5.38	5.38	3.83	6.58
γ	JANGMI	SINLAKU	HAITANG	MINDULLE
ABBY	10	0	5.26	50
WAYNE	0	11.11	5.26	66.67
TIM	0	11.11	5.26	66.67
DOUG	0	11.11	5.26	66.67
HERB	10	22.22	15.79	83.33
ZEB	10	0	5.26	50
δ	JANGMI	SINLAKU	HAITANG	MINDULLE
ABBY	1	1.06	1.6	2.56
WAYNE	0.53	0	0.53	3.59
TIM	0.75	0.21	0.75	3.38
DOUG	0.75	0.21	0.75	3.38
HERB	0	0.53	0	4.1
ZEB	1.6	1.06	1.6	2.56
Predicted case	WAYNE TIM	WAYNE	WAYNE TIM	ABBY
Reference case	ABBY	TIM	DOUG	TIM

the parameter is not enough to describe the whole phenomenon. Therefore we have to find out other appropriate parameters to improve the prediction efficiency.

23.4.3 Scenario III Experiment: Pattern Recognition by CBR with Five Meteorological Parameters

Since typhoon brings a lot of rain, it may cause landslide and floods which are the main influence of economic loss, so we have to take into account the effect of rainfall amount. The results of prediction results with the four parameters with the new variable, rainfall, was presented in this section (Table 23.4).

Table 23.4 Predicting the economic loss of typhoon with five parameters

Year	Month	Date	Name	α	β	γ	δ	Total loss	ζ
1986	8	21–25	WAYNE	100	940	100	940	12,239,280	2,842.4
1986	9	16–20	ABBY	95	945	90	950	7,501,473	1,906.7
1994	7	11	TIM	120	930	100	942	56,853,348	2,379
1996	7	8	HERB	140	910	110	935	37,890,050	4,082
1994	8	8	DOUG	130	910	100	942	8,907,992	3,048
1998	10	14–16	ZEB	155	880	90	950	8,099,891	2,996
2000	10	30–11.01	XANGSANE	90	950	70	970	5,418,116	3,043
2000	8	21–23	BILIS	140	900	115	925	7,644,368	2,377
2001	9	16–19	NARI	80	960	80	960	5,691,986	3,356
2001	7	29–31	TORAJI	75	962	70	968	14,723,151	2,833
2004	6	29–07.02	MINDULLE	90	942	60	975	6,516,458	3,249
2005	7	16–20	HAITANG	110	915	95	935	9,831,405	3,270.5
2008	9	8–21	SINLAKU	100	930	90	940	5,643,770	3,517.7
2008	9	26–29	JANGMI	105	930	100	935	7,547,679	2,882

The successful rate after improvement with five parameters is 3/8 = 37.5 %, as shown in Table 23.5. Although the results are better than the previous one, it is still not good enough, so we use the fuzzy theory to predict the same case in the next section.

23.4.4 Scenario IV Experiment: Pattern Recognition by CBR with Three Remote Sensing Parameters

Now we try to use the technique in the image interpretation. We try to simulate the decision process of every viewer who was invited to distinguish the satellite image. The remote sensing data of typhoon and the case base reasoning method was used in this experiment. Data of typhoon from 2006 to 2009 with the economic loss over twenty billion NT dollar such as KROSA, JANGMI, MORAKOT was used as the study object. The total population, total road length, total area of green land, and total area of river were used as the prediction factor.

The similarity of the three typhoons was calculated to know the relationship of the three typhoons and the prediction factors. The results reveal that the factor has a strong correlation with typhoon damage. Therefore, the method is able to predict the typhoon loss before the event.

Table 23.5 Similarity level for CBR with five parameters

	JANGMI	SINLAKU	HAITANG	MINDULLE	TORAJI	NARI	BILIS	XANGSANE
ABBY	0.86	17.37	11.12	10.54	2.6	13.39	22.28	4.48
WAYNE	1.37	19.2	13.09	12.51	0.33	15.3	19.58	6.59
TIM	17.45	32.37	27.26	26.78	16.03	29.11	0.08	21.82
DOUG	5.76	13.35	6.8	5.76	7.59	9.18	28.23	0.16
HERB	41.64	16.04	24.81	25.64	44.09	21.63	71.73	34.14
ZEB	3.96	14.83	8.39	7.79	5.75	10.73	26.04	1.54
Predicted case	ABBY	DOUG	DOUG	DOUG	WAYNE	DOUG	TIM	DOUG
Reference case	ABBY	TIM	DOUG	TIM	WAYNE	TIM	ABBY	TIM
Y/N	Success	Fail	Success	Fail	Success	Fail	Fail	Fail

Table 23.6 The total evaluation coefficient of the typhoon after 2000

Name	κ	Module output code
XANGSANE	2.265096	4
BILIS	2.190364	4
NARI	2	4
TORAJI	2.06523	3
MINDULLE	2.466154	*
HAITANG	2.136504	4
SINLAKU	2.100473	1
JANGMI	2.044652	3

23.4.5 Scenario V Experiment: Pattern Recognition by FT with Five Parameters

The fuzzy interval of total evaluation index, κ, was shown in Table 23.6. From Tables 23.7 and 23.8, the module output code by fuzzy theory of typhoon Sangsane, Bilis, Nari is 4, which is the number of typhoon Herb, and Doug. However, the prediction results were Tim(1) and Abby(3). Therefore, the prediction results were not correct.

The total evaluation coefficient of typhoon Mindulle can not find a corresponding typhoon because there is no data in the interval [2.4 $\sim$ 2.49], it was marked by a symbol,*. Because there are two records in the interval of [2.0, 2.09] and [2.2, 2.29], therefore, the values were adjusted. For example, the new value of total economic loss and the upper and lowest level of economic loss (Tables 23.10, 23.11 and 23.12). The successful rate of prediction is 4/7 $=$ 57 %, as shown in Table 23.9.

Table 23.7 Summary statistics of typhoon before 2000

Year	Month	Date	Name	α	β	γ	δ
1986	8	21–25	WAYNE	100	940	100	940
1986	9	16–20	ABBY	95	945	90	950
1994	7	11	TIM	120	930	100	942
1996	7	8	HERB	140	910	110	935
1994	8	8	DOUG	130	910	100	942
1998	10	14–16	ZEB	155	880	90	950

Table 23.8 Summary statistics of typhoon before 2000

Year	T	L_L	L_U	κ
1986	12,239,280	9,791,424	14,687,136	2
1986	7,501,473	6,001,178.4	9,001,767.6	2.050292398
1994	5,685,348	4,548,278.4	6,822,417.6	2.187261146
1996	37,890,050	30,312,040	45,468,060	2.245989305
1994	8,907,992	7,126,393.6	10,689,590.4	2.266029724
1998	8,099,891	6,479,912.8	9,719,869.2	2.648538012

Table 23.9 The interval of total evaluation index, κ

Interval	L_L (20 %)	Economic loss	L_U (20 %)
2.0 ~ 2.09	9,791,424	12,239,280	14,687,136
	6,001,178	7,501,473	9,001,768
2.1 ~ 2.19	4,548,278	5,685,348	6,822,418
2.2 ~ 2.29	30,312,040	37,890,050	45,468,060
	7,126,394	8,907,992	10,689,590
2.3 ~ 2.39	0	0	0
2.4 ~ 2.49	0	0	0
2.5 ~ 2.59	0	0	0
2.6 ~ 2.69	6,479,913	8,099,891	9,719,869

Table 23.10 Interval of membership function and the module output code

Interval	L_L (20 %)	Economic loss	L_U (20 %)
2.0 ~ 2.09	7,896,301.2	9,870,376.5	11,844,452
2.1 ~ 2.19	4,548,278	5,685,348	6,822,417.6
2.2 ~ 2.29	18,719,217	23,399,021	28,078,825
2.3 ~ 2.39	0	0	0
2.4 ~ 2.49	0	0	0
2.5 ~ 2.59	0	0	0
2.6 ~ 2.69	6,479,912.8	8,099,891	9,719,869.2

Table 23.11 Interval of membership function and the module output code

Interval	Lowest level in the interval of membership function	Highest level in the interval of membership function	Module output code
2.0 ~ 2.09	0.142	0.31	3
2.1 ~ 2.19	0	0.097	1
2.2 ~ 2.29	0.602	1	4
2.3 ~ 2.39			
2.4 ~ 2.49			
2.5 ~ 2.59			
2.6 ~ 2.69	0.082	0.22	2

Table 23.12 The prediction results by fuzzy theory

Typhoon to be predicted	JANGMI	SINLAKU	HAITANG	MINDULLE
Corresponding typhoon	ABBY	TIM	DOUG	TIM
Typhoon to be predicted	TORAJI	NARI	BILIS	XANGSANE
Corresponding typhoon	WAYNE	TIM	ABBY	TIM

23.4.6 *Scenario VI Experiment: Pattern Recognition by Neural Network Model*

The objective of this experiment is to use the neural network method to estimate the economic loss. The input variable is the five meteorological parameters. The output variable is total loss. Figure 23.2 is the results of neural network simulation for different training times. And Table 23.13 is the estimated results for different neural network method.

The prediction results of these five neural network schemes were shown in Table 23.14.

The error of BPN to the real value is 0.998 %. The error of ENN to real value is 0.932 %. For FNN, the error is 1.607 %. For AASN and GANN, the error is 1.190 and 1.595 % (Tables 23.15 and 23.16).

23.5 Discussions

23.5.1 *Results Evaluation*

The results for the successful rate for the three scenarios, CBR with four parameters, CBR with five parameters, and FT with five parameters, are 12.54, 37.5, and 57 %, respectively. The result of CBR with five parameters is better than CBR

Fig. 23.2 Training times and influence

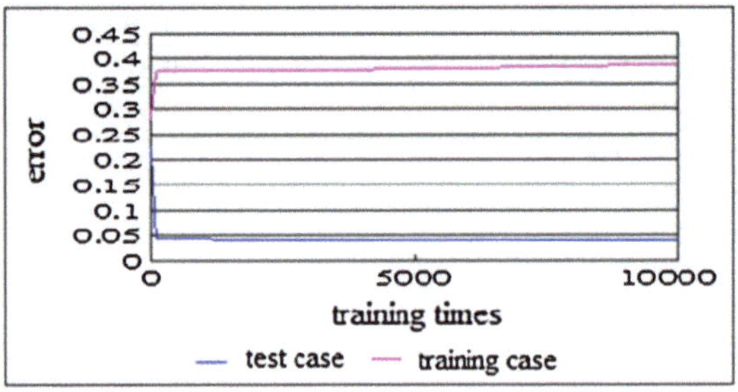

BPN

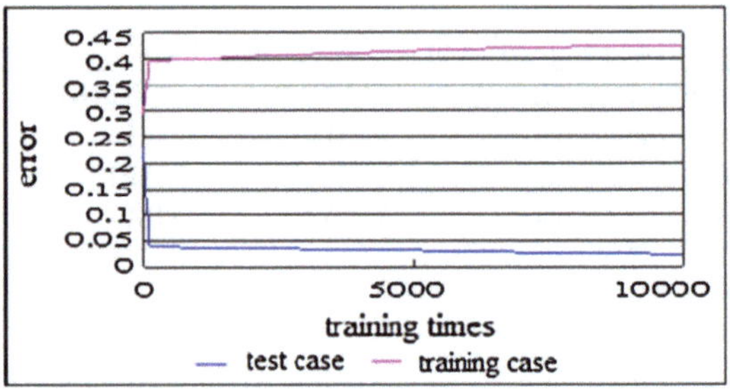

ENN

Fig. 23.2 continued

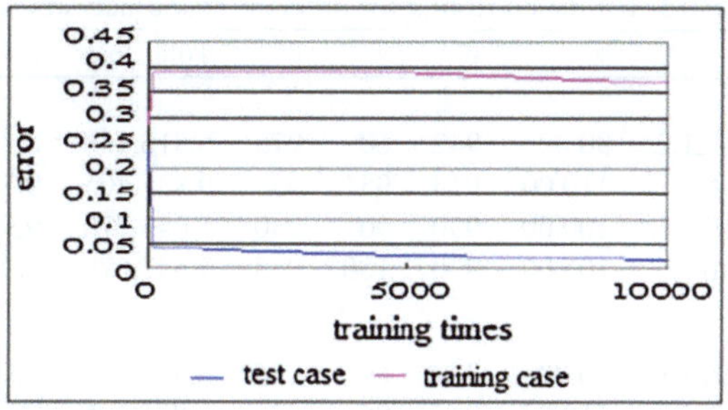

FNN

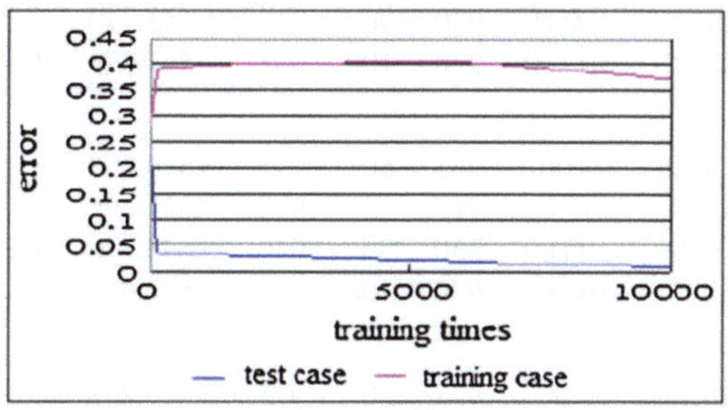

AASN

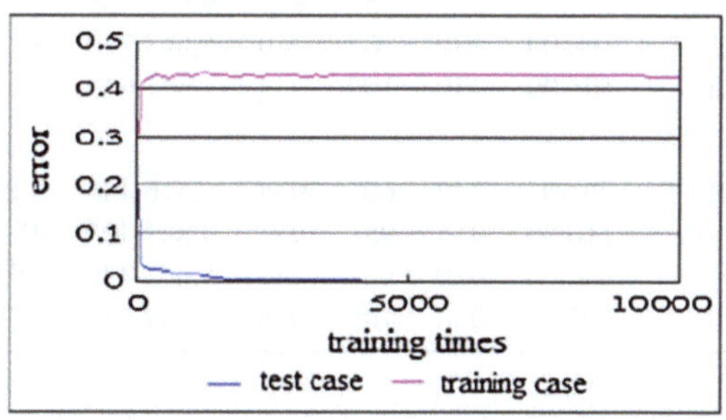

GANN

with four parameters. It shows that the CBR can be improved by the addition of appropriate parameters. The selected parameter also provides the reasonable physical meaning of the economic loss and the natural characteristic of typhoon. The newly added parameter can complete the lack of CBR for the statistically relevant data for backing and implicit generalization.

The CBR method is a method which depends on the amount in the data base. In this study, there are only six cases in the database. Therefore, it is very difficult to obtain a close estimation of the economic loss. If more data were included, the results could be increased. The scenario of FT with five parameters is the best among the three scenarios. It reveals that the fuzzy theory is better than the CBR in predicting the typhoon loss. The reason may comes from the fact that the typhoon is a very vague phenomenon, in such condition fuzzy theory has a better capability for identify the result than other calculation method. The prediction results can be improved if the fuzzy theory combines with the remote sensing information.

Table 23.13 Estimated results for different neural network method

Name	α	β	γ	δ	Total loss	Estimate loss	Error percentage (%)
BPN							
MINDULLE	90.00	942	60	975	6,516,458	8,809,955	26.03
HAITANG	110.00	915	95	935	9,831,405	9,773,094	−0.60
SINLAKU	100.00	930	90	940	5,643,770	9,967,548	43.38
JANGMI	105.00	930	100	935	7,547,679	9,994,069	24.48
ENN							
MINDULLE	90.00	942	60	975	6,516,458	13,589,920	52.05
HAITANG	110.00	915	95	935	9,831,405	13,029,570	24.55
SINLAKU	100.00	930	90	940	5,643,770	13,827,090	59.18
JANGMI	105.00	930	100	935	7,547,679	11,704,490	35.51
FNN							
MINDULLE	90.00	942	60	975	6,516,458	9,812,354	33.59
HAITANG	110.00	915	95	935	9,831,405	10,640,030	7.60
SINLAKU	100.00	930	90	940	5,643,770	10,886,470	48.16
JANGMI	105.00	930	100	935	7,547,679	11,215,900	32.71
AASN							
MINDULLE	90.00	942	60	975	6,516,458	8,095,661	19.51
HAITANG	110.00	915	95	935	9,831,405	13,056,430	24.70
SINLAKU	100.00	930	90	940	5,643,770	13,147,530	57.07
JANGMI	105.00	930	100	935	7,547,679	14,112,640	46.52
GANN							
MINDULLE	90.00	942	60	975	6,516,458	7,008,443	7.02
HAITANG	110.00	915	95	935	9,831,405	15,342,950	35.92
SINLAKU	100.00	930	90	940	5,643,770	15,095,470	62.61
JANGMI	105.00	930	100	935	7,547,679	14,594,810	48.29

Five neural network schemes were used to compare the predicting results which were BPN, ENN, FNN, AASN, and GANN. The result reveals that the BPN is the best choice, because the error is the lowest among the five schemes in this study, although the convergence rate of learning cycle is not the lowest. Therefore, we will use this model as the basic module for further research.

23.5.2 Future Research Work and Application

This study offers an ideal way for environmental risk management in the early stage or for the post examination. However, it is still in the beginning phase, many fields could be expanded as described in the following. First, it is suggested having more precise data about the patterns. Second, try to develop a new algorithm to

Table 23.14 Results of the five neural network methods

No	Level (%)	BPN (%)	ENN (%)	FNN (%)	AASN (%)	GANN (%)
23	16.59	12.88	16.85	15.94	14.65	3.90
24	0.34	1.86	3.29	0.65	2.70	2.53
25	1.65	1.85	3.91	1.04	3.64	2.51
26	2.80	2.37	4.37	1.70	4.30	3.30
27	12.02	9.57	13.35	13.01	12.41	5.11
28	6.82	4.82	8.15	6.15	8.11	4.24
29	0.07	1.98	2.93	0.63	2.32	3.32
30	3.11	−0.18	0.25	3.25	3.27	2.15
31	11.62	9.76	12.19	12.28	11.68	6.45
32	4.49	3.15	5.78	3.14	5.87	3.59
33	6.01	5.58	5.36	5.16	5.42	7.76
34	3.86	3.01	2.75	1.66	1.69	5.26
35	8.69	7.32	7.01	7.03	6.19	7.50
36	1.37	2.17	2.70	0.62	1.79	3.39
37	2.03	2.03	4.00	1.21	3.81	2.85
38	1.75	2.26	3.43	1.21	3.16	3.98
39	4.90	2.72	6.76	3.35	6.97	2.51
40	2.93	2.49	4.36	1.83	4.30	3.58
41	1.47	2.50	2.62	1.43	2.38	5.84
42	2.68	2.53	2.56	1.01	1.53	4.42
43	12.96	9.12	15.55	12.90	13.79	2.85
44	2.04	2.32	2.08	0.98	1.32	5.71
45	11.90	9.15	13.69	12.76	12.59	4.37
46	7.98	6.32	8.67	8.19	8.52	5.88
47	5.42	4.09	3.73	2.86	2.53	5.28
48	9.72	5.63	12.82	9.18	12.09	2.41
49	5.16	2.90	7.02	3.61	7.23	2.57
50	2.53	2.41	4.01	1.60	3.88	3.74
51	6.06	3.81	7.78	4.74	7.88	3.14
52	2.38	2.83	3.05	1.97	2.93	5.98
53	3.48	2.73	2.11	1.68	1.39	6.75
54	0.34	1.90	3.19	0.65	2.61	2.79
55	13.44	11.77	13.11	12.98	12.72	7.37
Mean	0.05412	0.04414	0.06346	0.04739	0.05929	0.04334

estimate the relationship between each pattern. Third, find a quantitative way to calculate the risk. Finally, have more case study with remote sensing data to improve this model.

Table 23.15 The training test data

Name	α	β	γ	δ	Total loss
Training test data					
WAYNE	100	940	100	940	12,239,280.00
ABBY	95	945	90	950	7,501,473.00
TIM	120	930	100	942	5,685,348.00
HERB	140	910	110	935	37,890,050.00
DOUG	130	910	100	942	8,907,992.00
ZEB	155	880	90	950	8,099,891.00
Validating data					
XANGSANE	90.00	950	70	970	5,418,116
BILIS	140.00	900	115	925	7,644,368
NARI	80.00	960	80	960	5,691,986
TORAJI	75.00	962	70	968	14,723,151
Inference data					
MINDULLE	90.00	942	60	975	6,516,458
HAITANG	110.00	915	95	935	9,831,405
SINLAKU	100.00	930	90	940	5,643,770
JANGMI	105.00	930	100	935	7,547,679

Table 23.16 Comparison of the error percentage

Name	BPN (%)	ENN (%)	FNN (%)	AASN (%)	GANN (%)
MINDULLE	26.0300	52.0500	33.5900	19.5100	7.0200
HAITANG	−0.6000	24.5500	7.6000	24.7000	35.9200
SINLAKU	43.3800	59.1800	48.1600	57.0700	62.6100
JANGMI	24.4800	35.5100	32.7100	46.5200	48.2900
	23.3225	42.8225	30.5150	36.9500	38.4600

23.6 Conclusion

The economic loss of typhoon was studied in this chapter. Six scenarios of experiments were conducted in this study using different pattern recognition methods such as multivariable statistic analysis (MS), case base reasoning (CBR), fuzzy theory (FT), and neural network (NN). These theories are based on the combination of typhoon process parameters and joint economic loss analysis of historical typhoon records. Different meteorological parameters from the observational records and geographical information system were used as the predictors. The results reveal that only CBR cannot precisely forecast the economic loss of typhoon disaster. However, the appropriate predictor can effectively increase the accuracy of prediction. The best method of prediction is the neural network prediction after certain times of training. The percentage of error can be reduced to thirty to forty percents.

The results in this chapter are hypothetically based, relevant model for the prevention of typhoon induced disasters. Our preliminary results show that the neural network method is capable of constructing reliable economic loss prediction. The results are also expected to assess the level of damage to economic and to provide the information for decision support. It can be widely used not only in Taiwan, but would also be useful to similar area such as Japan, Philippines, or Korean etc., as a scientific based measure in decision-making of natural disaster management and risk prevention.

Acknowledgments The present research was supported by research project awatded to DL Tang: (1) Guangdong National Science Foundation, China (2010B031900041, 8351030101000002), (2) Natural National Science Foundation of China (31061160190, 40976091, NSFC-RFBR Project-41211120181), (3) Innovation Group Program of State Key Laboratory of Tropical Oceanography (LTOZZ1201).

References

Aamodt, A., Plaza, E.: Case-based reasoning: Foundational issues, methodological variations, and system approaches. Artif. Intell. Commun. **7**(1), 39–52 (1994)

Cheetham, W., Goebel, K.: Appliance call center: A successful mixed-initiative case study. Artif. Intell. Mag. **28**(2), 89–100 (2007)

Chen, W.K.: A framework for nuclear power plant emergency response system. Nucl. Sci. Technol. **21**, 375–378 (2010a)

Chen, W.K.: An approach to pattern recognition by fuzzy category and neural network simulation. Paper presented at proceeding of the 8th international conference on machine learning and cybernetics (ICMLC), Institute of Electrical and Electronics Engineers (IEEE), Systems, Man, and Cybernetics Society (IEEE SMC Society), Qingdao (2010b)

Chen, W.K.: Environmental applications of granular computing and intelligent systems. In: Pedrycz, W., Chen, S.M. (eds.) Granular Computing and Intelligent Systems: Design with Information Granules of Higher Order and Higher Type, vol. 13, pp. 275–301. Springer, Berlin (2011)

Chen, W.K.: Fuzzy forecasting with fractal analysis for the time series of environmental pollution. In: Pedrycz, W., Chen, S.M. (eds.) Time Series Analysis, Modeling and Applications Intelligent Systems, vol. 47, pp. 199–213. Springer, Berlin (2013)

Chen, W.K., Sui, G.J., Tang, D.L.: A fuzzy intelligent decision support system for typhoon disaster management. Paper presented at proceeding of 2011 IEEE international conference on fuzzy systems, Institute of Electrical and Electronics Engineers (IEEE), Taipei (2010a)

Chen, W.K., Sui, G.J., Tang, D.L., Cai, D.L., Wang, J.S.: Study of environmental risk management by multivariable analysis of pattern structure. Paper presented at proceeding of the 4th international conference on management science and engineering management, Chungli (2010b)

Chen, W.K., Sui, G.J., Tang, D.L.: Predicting the economic loss of typhoon by case base reasoning and fuzzy theory. Paper presented at proceeding of 9th international conference on machine learning and cybernetics (ICMLC), Institute of Electrical and Electronics Engineers (IEEE), Systems, Man, and Cybernetics Society (IEEE SMC Society), Guilin (2011)

Jain, A.K., Duin, R.P.W., Mao, J.C.: Statistical Pattern Recognition: A Review. Department of Computer Science and Engineering, Michigan State University, East Lansing (2000)

Lebowitz, M.: Memory-based parsing. Artif. Intell. **21**, 363–404 (1983)

Leung, F.H.F., Lam, H.K., Ling, S.H., Tam, P.K.S.: Tuning of the structure and parameters of a neural network using an improved genetic algorithm. IEEE Trans. Neural Networks **14**(1), 79–88 (2003)

Schalkoff, R.: Pattern Recognition: Statistical, Structural, and Neural Approaches. Wiley, New York (1992)

Schuermann, J.: Pattern Recognition. Wiley, New York (1996)

Tu, Z.G., Lu, Y.: A robust stochastic genetic algorithm (StGA) for global numerical optimization. IEEE Trans. Evol. Comput. **8**(5), 456–470 (2004)

Watson, I., Gardingen, D.: A case-based reasoning system for HVAC sales support on the web. Knowl. Based Syst. J. **12**(5–6), 207–214 (1999)

Yeh, I.C.: Modeling chaotic dynamical systems using extended-neuron networks. Neural Parallel Sci. Comput. **5**(4), 429–438 (1999a)

Yeh, I.C.: Modeling chaotic two-dimensional mapping with fuzzy-neuron networks. Fuzzy Sets Syst. **105**(3), 421–427 (1999b)

Yeh, I.C.: Classification using extended-neuron networks. J. Comput. **14**(2), 6–22 (2002)

Yeh, I.C.: Modeling chaotic dynamic systems using analysis-adjustment-synthesis networks. J. Technol. **18**(3), 369–376 (2003)

Yeh, I.C.: Classification and function mapping with fuzzy-neuron networks. J. Technol. **14**(2), 153–159 (2005)

Zadeh, L.A.: Fuzzy Sets. Inf. Control **8**(3), 338–353 (1965)

Chapter 24
Disaster Management and Risk Reduction: Impacts of Sea Level Rise and Other Hazards Related to Tsunamis on Syrian Coastal Zone

Hussain Aziz Saleh and Georges Allaert

Abstract The rapid development of economic construction and urbanization, highly dense population, infrastructure and traffic, all have caused a lot of troubles to the main cities in the Syrian coastal region. In addition, this region which is located on the eastern coast of the Mediterranean Sea, and among Arabian, African and European Asian plates is suffering from increasing the number of natural and man-made disasters such as earthquakes, climate change, flash flooding, and mainly the expected sea level rise. This rise effect often depends on many elements, such as seismic hazard, vulnerability, exposure and emergency response and recovery capability. It is not possible to completely avoid this rise, but the sufferings can be minimized by creating proper awareness of this hazard and its impacts through developing an integrated system of the geographical and environmental data collection and management tools with simulation and decision tools for risk reduction and assessment. Great change becomes to integrated management and more to eco-environmental safety construction, especially to the prevention for disasters destroyed structure as sea level rise. Therefore, the purpose of this chapter is to address the need for an integrated disaster risk management in Syrian coastal zone. This will help to manage the risk of these disasters and hazards in a more effective manner through linking up disaster management more closely and consistently with urban planning and management.

Keywords Disaster management · Risk assessment · Environmental planning · Climate changes · Sea level rise · Syrian coastal zone · Mediterranean sea

---

H. A. Saleh (✉)
Higher Commission for Scientific Research, 30151 Damascus, Syria
e-mail: hussain.saleh@ugent.be; office@hcsr.gov.sy; hussainazizsaleh@gmail.com

G. Allaert
Institute for Sustainable Mobility, Ghent University, Vrijdagmarkt 10/301, 9000 Ghent, Belgium
e-mail: georges.allaert@ugent.be

D. L. Tang and G. J. Sui (eds.), *Typhoon Impact and Crisis Management*, Advances in Natural and Technological Hazards Research 40,
DOI: 10.1007/978-3-642-40695-9_24,

24.1 Introduction

World is experiencing an increasing number of disasters by a combination of changes in its physical, technological and social systems. These disasters kill thousands of people and destroy billions of dollars of habitat and property each year. Climate change and unstable land forms, coupled with deforestation, unplanned growth proliferation, tardy communication, bad Disaster Management (DM) make the disaster-prone areas mere vulnerable and suffer more or less by frequent natural disasters. Syria is not isolated from the rest of the world when it comes to the effects of these disasters. To achieve an efficient solution to Disaster Risk Reduction (DRR), this chapter addresses the need for an integrated DM in Syrian Coastal Zone (SCZ) to manage the risks of these disasters [mainly the Sea Level Rise (SLR)] in a more effective manner through linking up DM more closely and consistently with urban planning and management. Section 24.2 presents brief overview of disasters and its effects in Syria during the period 2000–2010, followed by a description of the SCZ. Section 24.3 clarifies the major concepts that revolve around dangers, damage, their interrelations and elements, while Sect. 24.4 presents the SLR and other related hazards such as tsunamis in the Mediterranean Sea in general and in SCZ in particular. In terms of the types of hazards to be classified and spatially selected, Sect. 24.5 explains the physical parameters and factors that can be associated with hazards, and it describes the current and future drivers of coastal changes in SCZ. Section 24.6 discusses characteristics, common effects and impacts of disasters in order to explain the reasoning behind the components of Risk Assessment (RA) methodology, while Sect. 24.7 describes the main effects and impacts of local SLR scenarios with focusing on the Lattakia city and its surrounding areas. Section 24.8 outlines the DM cycle and presents all the practical activities that must be carried out during all the phases of this cycle to minimize the DRR. Section 24.9 outlines the recent processes that have been made through advances in RA and its components. It explains the importance of the strategic tools for RA and their impacts on planning activities to ensure successful urban policies. Section 24.10 presents the maritime activities in SCZ and identifies the vulnerabilities to SLR, gaps, points of strength and weakness, and options to assess adaptation measures. It also outlines the suggested strategic framework for responding to SLR to protect the marine area of Syria. It insists on the importance of the capacity building in achieving successful use of the most advanced technology for DRR. This chapter ends with some recommendations, conclusions and future work.

24.2 Risk of Hazards and Disasters in Syria

Syria is a disaster-prone country with climate and geographical location make it vulnerable and suffering from increasing the number of natural, technological, and man-made disasters. It situated in southwest Asia at the eastern end of the Med.

Sea, and is exposed to significant seismic activity due to its location in the most seismic-tectonic active group in the region where the Arabian, African, and Eurasian continental plates converge as shown in Fig. 24.1. Those areas are tectonically active and cause time to time a lot of seismically events. In addition, Syria experiences related weather anomalies associated with droughts and is prone to floods, landslides and erosion resulting from the combination of heavy rain, steep topography and widespread deforestation. Climate change, industrial development, high population growth rates, rapid economic growth and urbanisation, all has increased the risk of pollution of natural environmental resources, and consequently, amplify the region's vulnerability to environmental challenges and constrain its ability to manage them. Among the major challenges that faces Syria are water scarcity, land degradation, inadequate capacities for waste management, coastal and marine environment degradation, air pollution and global warming, river and shore erosion, etc. Table 24.1 shows the top 5 national disasters with human exposure reported in the last 10 years that have caused widespread damages and losses in Syria. In this chapter, the concentration on the state of hazards in the SCZ mainly the impacts and factors of SLR and other hazards related to tsunamis.

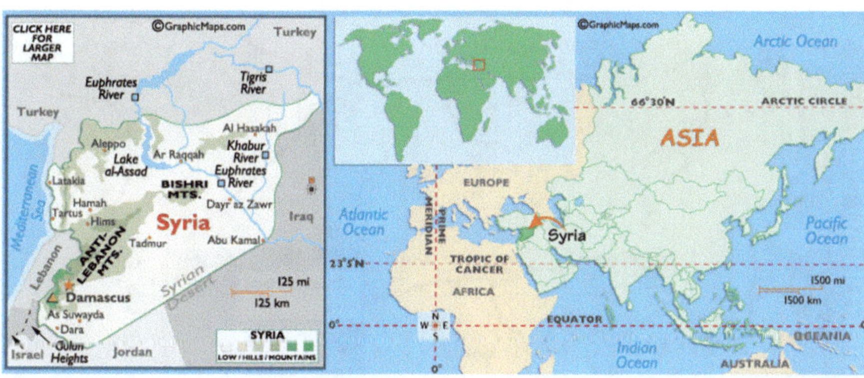

Fig. 24.1 The strategic location of Syria

Table 24.1 The profile risk in Syria in the last ten years

Risk profile		National statistics		
Human exposure		Top 5 national disaster reported		
Hazard type	Population exposed	Disaster	Date	Affected people
Drought	2,027,540	Drought	2008	1,300,000
Flood	25,572	Drought	1999	329,000
Landslide	456	Storm	2004	180
Earthquake	5,370	Storm	2001	172
Tsunami	3,759	Mass mov. wet	2002	23

Source of Data 2009 Global Assessment Report, OFDA/CRED International Disaster Database

24.2.1 The Syrian Coastal Zone

The country is divided into fourteen governorates, of which two are located along the coast, namely Lattakia and Tartous governorates as shown in Fig. 24.2. The SCZ, which is of critical importance to the country and a strategic access to the world, provides important economic, transport, residential and recreational functions, all of which depend on its physical characteristics (e.g., appealing landscape, cultural heritage, natural resources, and rich marine and terrestrial biodiversity, etc.). The region is composed of three markedly different areas: the coastal plain abundant with water and fertile soil, the hilly zone with limited water resources and lower quality agricultural land, and the mountains. The climate in SCZ is typically Mediterranean with dry and wet summers, windy winters and springs. The coastal mountain chain separate Syria's interior from the Med. coast with slopes originally covered in forests of oak and pine. The region accounts for 35 % of the national energy production, 38 % of cement production, 50 % of petroleum refining, the predominant part of the national export is shipped through the port of Lattakia. Due to abundance of freshwater and fertile soil, the area is distinguished

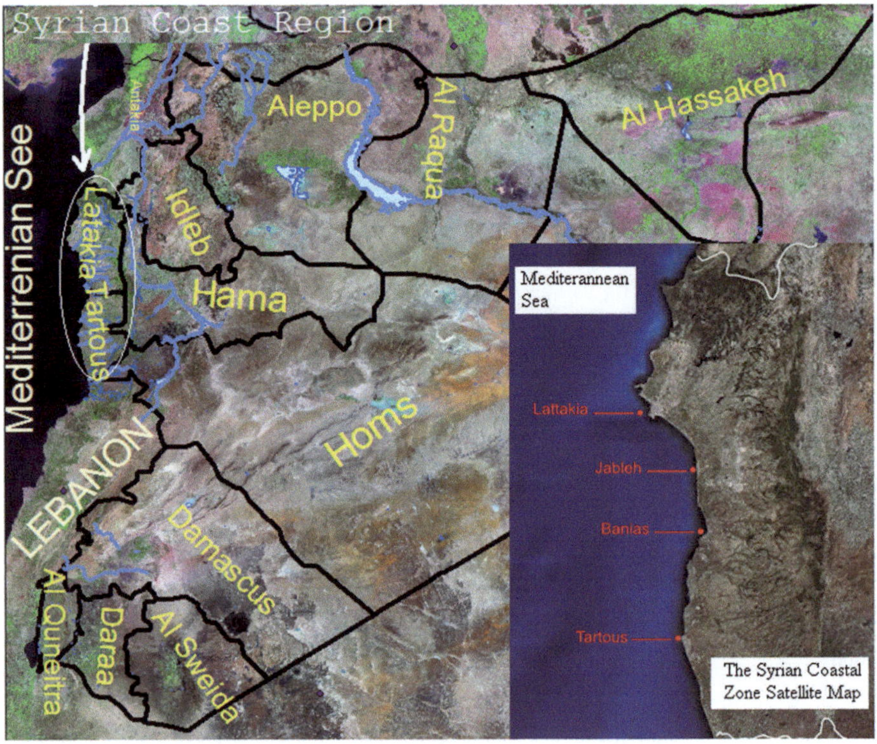

Fig. 24.2 The Syrian coastal zone

by a rich and productive agriculture of the Mediterranean type, with 16 % of cultivated land under irrigation. Syria's coast hosts many archaeological sites and ancient ruins that documented and protected, but some are still poorly excavated. Amongst the most important sites are the ruins and old cities of Lattakia, Jableh and Tartous, Ugarit north of Lattakia, El-Marquab castle near Banias, and Amrit south of Tartous. Arwad, the sole inhabited island in Syria just 3 km off Tartous, was the base of a great sea trading kingdom in the Canaanite era. Many of Syrian's coastal towns and cities have a culture and way of life stretching back over centuries.

Syria has maritime boundaries with Turkey, Cyprus and Lebanon with a 185 km long coastline which includes diverse coastal environments, from cliffs to low-lying areas. This coastline, which is bounded by mountains, represents a narrow plain that is indented by some 350 estuaries, harbours, inlets, bays or fiords. The terrain along the coastline varies from sandy shores (golden sand at Lattakia and black volcanic sand at Ras El Bassit) to rugged, rocky promontories and cliffs as shown in Fig. 24.3. It has to be considered as one of the scarce natural resources of the country, providing a narrow window to the sea for such a relatively large country. The width of the plain varies according to the reach of the nearby mountains; the plain is widest in the north near the port city of Lattakia and in the south near the Lebanese border. According to the Syrian State Planning Commission, SCZ is densely populated with 405 in Lattakia and 370 inhabitants/ km^2 in Tartous, while in the narrow coastal plain the density is almost 20 times

Fig. 24.3 The Syrian shoreline varies from sandy shores to rugged, rocky promontories and cliffs

greater than the national average, and 6 times greater than the rest of the coastal region, the hinterland (SPC 2006). This density, which is combined with its strategic and economic importance, places disproportionate pressures on SCZ.

Although the short coastline of Syria, it has a commercial fleet composed of 137 ships and 4 major ports and harbors in Banias, Jablah, Lattakia, and Tartous. The largest 2 ports are in Lattakia and Tartous, and both operate 24 h a day. Lattakia Port comprises 23 quays with a total length of 4,280 m, total area of the port is 1,500,000 m^2, of which 200,000 m^2 is the segregated container yard, and the average monthly activity is approximately 420,000 Mt. Tartous Port comprises 3 piers with 24 berths (total length of the berths is 6,366 m), and average monthly activity is approximately 350,000 Mt. These ports are connected to the railways network (2,342 km) to transport goods to and from other governorates, and neighbouring countries. The road network is back-bone for the country and is vital to the regional international transit network that connects European countries with other Arab countries. The 500 km East–West highway joins these ports on the coastal area in the west of Syria and the western border with Lebanon to the eastern border with Iraq. The Bassel Al Assad International Airport, which is located 25 km from Lattakia, is the only airport on SCZ and supports these networks.

24.2.2 Major Problems, and Challenges in Managing in Syrian Coastal Zone

Coastal margins are the transition between the ocean and the land, and the place where seawater mixes with freshwater and interacts occasionally with the fringing low-lying land during storms or extreme tides. At vulnerable coastal margins, coastal development and global warming are on an eventual collision course (if they have not collided already), which will result in further 'coastal squeeze' between the land and the sea. The developed areas around the SCZ are usually nestled in or near low-lying coastal margins (such as beaches, estuaries and harbours), and will therefore become increasingly vulnerable to the effects of global warming. It is a great national challenge on how Syria will maximize the benefit from its narrow window to the med. Sea, whilst protecting the coastal environment and natural scene. There is huge pressure to develop and occupy the SCZ (for subdivisions, marinas, roads and drainage, etc.). The dynamics of natural coastal processes, including weather systems, sediment transport mechanisms, the hydrological links between the catchments and the coast, are factors that influence the ability of the coast to sustain human activities.

Human actions have exacerbated these problems through the inappropriate location of development and the overexploitation of coastal resources. Human pressures threaten habitats and natural resources of the coastal zone, and with them, the ability of this zone to perform many of its essential functions. Increasing population, both resident and transient, is leading to increased conflict between the competing uses in the SCZ. Low impact uses are frequently being replaced by intensive uses that are more profitable in the short-term but undermine the long-term potential of the coast by reducing its resiliency. Unfortunately, there is no sign that inappropriate uses of the coastal zone are becoming less frequent. In fact, with increasing population, visitors and economic activities, the pressures are increasing. In addition, the coastline is threatened in some parts by coastal erosion resulting from development projects and engineering works. Such works have accelerated erosion of the adjacent shoreline because they did not adequately account for coastal dynamics and processes. SLR, whether as a consequence of climate change or erosion, will present increased threats and costs to sustaining infrastructure and human settlement. SLR resulting from climate change may aggravate this erosion in future and this will be discussed in Sect. 24.5.

Within the framework of this chapter, the ‘case’ to be studied is the main coastal city of Lattakia, which is a highly hazard-prone area that integrate (to a certain extent) several fields of activity, settlement development planning, industrial projects, and disaster risk management, etc. This combination of the hazards and the city’s highly vulnerable social and economic setting produces a dynamic context of risk with the permanent threat of disaster. Therefore, the selection of this case study was based mainly on: the content (i.e., the existence of a certain level of integration development planning and disaster risk management); and the context (i.e., their implementation in an urban environment). In addition, the rapid urban and industrial development in this city, and absence of pollution abatement and treatment facilities resulted in: uncontrolled ribbon development along the shoreline, sprawl of uncontrolled low-density housing development, high pollution of the coastal and marine environment, chemical pollution and bacteriological contamination of freshwater sources due to the uncontrolled disposal of untreated urban solid and liquid waste, waste from villages and farms, agricultural practices, industry and transports, and destruction of wetlands and dunes (due to sand extraction). Section 24.7 will tackles the impacts and effects of the SLR on lattakia.

24.3 Disaster, Hazard, Vulnerability and Their Elements

According to the International Strategy for Disaster Reduction, disasters, hazards, risks, and vulnerability are different definitions, but similar concepts (ISDR 2005). A natural disaster is the effect of a natural hazard (e.g., flood) that will leads to financial, environmental or human losses. The resulting loss depends on the vulnerability of the affected population to resist the hazard. This understanding is concentrated in the formulation: “disasters occur when hazards meet vulnerability”. http://en.wikipedia.org/wiki/Natural_disaster-cite_note-1. Therefore, the disaster risk is made up of hazard and vulnerability (i.e., disaster risk = hazard × vulnerability). Hence, it is clear that a risk exists only if there is vulnerability to the

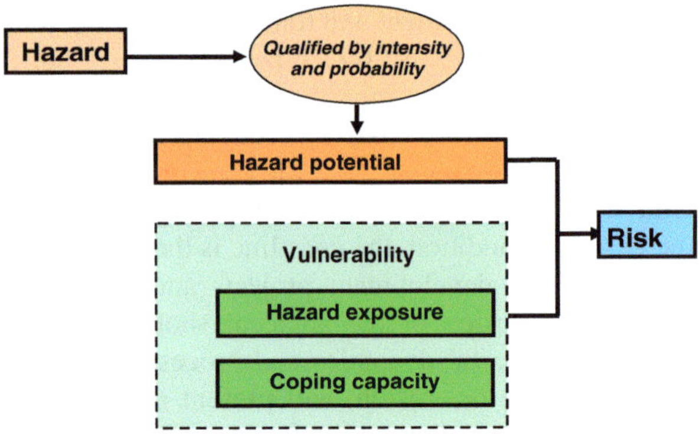

Fig. 24.4 The understanding of hazard and risk components. (*Source* Schmidt-Thomé and Kallio 2006)

hazard posed by a natural event. The following section introduces these elements in details, while Fig. 24.4 graphically depicts the relationship between them.

24.3.1 Disasters

Natural disaster can be defined as "A Disaster is a serious disruption of the functioning of a society, causing widespread human, material, or environmental losses, caused by hazards, which exceed the ability of affected society to cope using only its own resources" (Cuny 1983). In order to understand the term disaster, the term 'hazard' and the human sensitiveness towards it have to be also understood. Generally, disasters normally occur when hazards meet vulnerability, and the potential for a hazard to become a disaster mainly depends on a society's capacity to address the underlying risk factors, to reduce the vulnerability of a community, and then to be ready to respond in case of emergency (Wisner et al. 2004). It is important to note, that there are no internationally agreed minimum criteria for an event to be classified as a disaster. This is due to the variable manner in which hazards impact on population, economies and ecosystems. In this chapter, these events will be connected to the planning activities.

24.3.1.1 Classification of Disasters

It is difficult to classify disasters and draw a distinction between them, but it is useful to define disaster risk management measures. This will clarify and support the linkage among disaster, environment, spatial planning, and development. Only a few disasters, earthquakes for example, occur as purely natural phenomena, while most others (such as forest fires, floods and landslides, etc.) can come about with and without human intervention (Burton et al. 1998). Each hazard has a spatial dimension and the spatial character of a hazard can either be defined by *spatial effects* or by the possibility for an *appropriate spatial planning response*. Generally, disasters can be defined and classified according to their nature as follows:

- *Natural disasters* are of geophysical origin (e.g., earthquakes) and can result from those elements of the physical environment harmful to people and caused by forces extraneous to them. The term 'natural disaster' is not entirely correct, and they are human-made disasters exposed by natural hazards. For example, earthquake disasters show that many people were killed in poorly designed and constructed man-made structures, and not in open fields (Blaikie et al. 1994).
- *Human-induced natural disasters* are of climatic origin (e.g., floods) and caused by the human modification of the environment (e.g., cutting down forests that buffer rainfall). Then, when the flood come the blame on a natural disaster, not on these modifications and that's a human-induced disaster or, it's just poor planning on the part of short-sighted humans (Burby 2006).
- *Technological disasters* are accidental failures of design or management that could have a great perimeter of influence and affecting a relatively larger part of a country (e.g., air traffic accidents and chemical plants) (Smith 2000).
- *Man-made disasters* are resulting from man-made hazards as opposed to natural disasters resulting from natural hazards (e.g., threats having an element of human intent and negligence) (Gardoni and Murphy 2008).

24.3.2 Hazards

The United Nations International Strategy for Disaster Reduction (UNISDR) defines hazard as "A Dangerous phenomenon, substance, human activity or condition that may cause loss of life, property damage, loss of livelihoods and services, social and economic disruption, or environmental damage" (Van Westen, Terlien 1995). Hazardous events vary in magnitude, frequency, duration, area of extent, speed of onset, spatial dispersion and temporal spacing. Their origin can be purely natural (e.g., earthquake) or technological (e.g., accident in chemical plants), as well as a mixture of both (e.g. Sinking of an oil tanker in sea and subsequent coastal pollution).

Within this concept, risk can be defined as the probability of an event happening in a given time span and the magnitude of its impact when it does occur. Risk (*i.e., probable loss*) identification starts with identifying a hazard and then assesses the corresponding vulnerability *(i.e., The possible consequences)*. For natural hazards we can only attempt to reduce the risk not the hazard, either by controlling exposure to hazards or their vulnerability. The dependency of risk is on the three components of hazard, exposure, and vulnerability (Crichton 1999). This chapter deals with coastal hazards that arise from the interaction of natural processes with human use of infrastructure.

24.3.3 Vulnerability

Vulnerability is an essential part of a hazard, and it refers to the susceptibility of people, communities or regions to this hazard. It is a set of conditions resulting from physical, social, economic and environmental factors that increase this susceptibility to the impact of hazards (Oliver-Smith 2004). Vulnerability to climate change is considered to be high in developing countries due to social, economic, and environmental conditions that amplify susceptibility to negative impacts and contribute to low capacity to cope with and adapt to climate hazards. In Syria, there is an urgent need to understand the threats from climate change and to formulate policies that will lessen the risks and to take actions to cope with them (Meslmani 2010). The main factors that have to be taken into account to determine the vulnerability are:

- *Physical factors* are usually materially oriented, and come from the field of land-use and planning, engineering, and building environment (e.g., population density level, the remoteness of a settlement and the site, design and materials used for critical infrastructure and for housing, location and standards of infrastructure, etc.).
- *Social* factors are normally linked to the level of well-being of individuals, communities and societies (e.g., education levels, lack information on disasters, rapid population growth, the existence of peace and security, good governance, degree of respect for human rights, religious or political groupings, etc.).
- *Economic factors* include the economic status of individuals and communities (e.g., lacking in diversity and competition for scarce resources, rapid urbanization, inadequate access to basic socio-economic infrastructure, etc.).
- *Environmental factors* include the extent of natural resource depletion and data on resource degradation (e.g., reduced access to clean air and water, inappropriate waste management, soil degradation, deforestation, etc.).

24.4 Tsunamis and Other Hazards Related to Sea Level Rise

Tsunamis are gravity long waves generated by impulsive geophysical events of seafloor, volcanoes, asteroid impacts and landslides. They can be loosely grouped into those that are generated beyond the continental margins (distant or teletsunami), and those generated on or within the continental margins (local tsunami). Local tsunamis are generated by seismic activity (earthquakes) and volcanic activity and are likely to be in the order of 2.5–3 m above Mean Sea Level (MSL), while distant tsunamis are estimated to have a maximum height of 3.6 m above MSL (Palmer 2008). Figure 24.5 shows the global and relative plate motion in the eastern Med. Sea which can be regarded as one of the main sources of seismic hazards that cause tsunamis and SLR in this region.

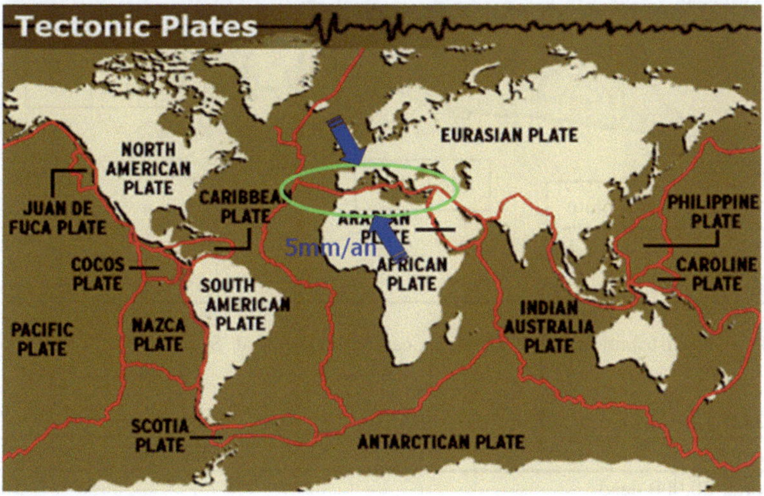

Fig. 24.5 The global seismic hazards and relative plate motion in the Eastern Mediterranean

With regards to the SLR, it results in more frequent coastal flooding that includes inundation of low-lying coastal areas caused by extreme tides and storm surge. It varies as a function of location and time at all spatial and temporal scales for many reasons. In particular, a global SLR has been evident at a rate of about 2 mm/year for the last century as determined from tide gauge records (Oliver-Smith 2009). *Global SLR* refers to the average vertical rise across the world's oceans, while *relative SLR* is the net rise relative to the landmass in a region. The relative SLR is the sum of the local subsidence (or uplift) of the coastal margin and the absolute sea-level contribution in that region. Regional relative SLR, which can be measured using tide gauges around the open coast, is what will affect us locally rather than the average global SLR.

The last assessment IPCC report has given new estimates for SLR that range 18–59 cm until the end of the next century. These estimates exclude the contribution of ice melting to the SLR, and include only the steric component of the SLR due to the heating of the ocean waters and their consequent expansion. Therefore, the numbers given by IPCC are only a lower limit of the SLR that one should expect, and it concluded that there is a consensus that global average sea level has risen by about 1.7 mm/year during the 20th century (IPCC 2007). There is also evidence from coastal tide gauges and satellite radar altimetry that the rate of increase of coastal and global sea level has accelerated from the early 1990s to 3 mm/year. On the other hand as shown in Fig. 24.6, it is predicted that with global warming, global average sea levels may rise 7–36 cm by the 2050s, 9–69 cm by the 2080s, and 30–80 cm by 2100 (Boko et al. 2007). The majority of this change will occur due to the expansion of the warmer ocean water (Nicholls and Tol 2006).

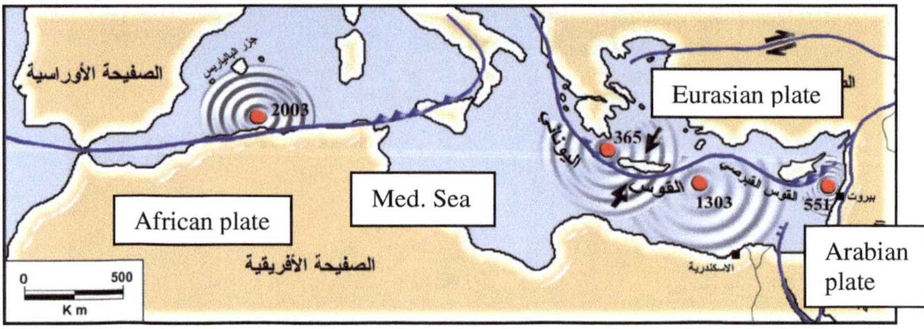

Fig. 24.6 The distribution of the Tsunamis centres in the Med. Sea

Table 24.2 The historic records of the some Tsunamis in the Mediterranean Sea

The damaged city	Magnitude	Date
Alexandria: 50,000 dead	7.0	21/07/365
Beirut: Huge number of people and boats merged	7.2	09/07/551
Alexandria: Huge number of people and boats merged	–	8/1/0303

24.4.1 Tsunamis in the Mediterranean sea

Historical documentary sources of the Med. Sea region contain much information about earthquakes and tsunamis as shown in Table 24.2, while, Fig. 24.6 depicts the distribution of the Tsunamis centres that affected the Arabian coasts in the past. Because of the active lithospheric plate convergence, this region is geodynamically characterized by high seismicity and significant volcanism, and this can be related to the surrounding tectonic sources is the Dead Sea fault system as shown in Fig. 24.5. Furthermore, coastal and submarine landslides are quite frequent, partly in response to the steep terrain that characterizes much of the basin. Some studies have suggested that massive earthquakes which are greater than magnitude 8 may strike this region roughly every 800 years (Manca et al. 2003). However, other studies outlined that not enough is known about these faults to predict how often such quakes might strike (Fukumori et al. 2003).

The first known tsunami in the Med. Sea occurred in the Syrian region around 2000 BC, while the large tsunami that hit Alexandria and killing 50,000 people was in 365 AD Weaker tsunamis have been observed more recently, notably the one generated off the coast of Algeria in 2003. The tsunamis that hit the coasts in the Mediterranean were faster despite travelling at lower speeds because the sea is not as deep as those in the Pacific Ocean (Gerassimos et al. 2005). They affect not only near-field localities, but also remote places in North Africa and in the Mediterranean side of Middle East as shown in Fig. 24.8. Three critical aspects of these tsunamis are: repeat times, maximum size, and zones of effects. However, by far the most common cause is submarine earthquake (Synolakis 2003) which

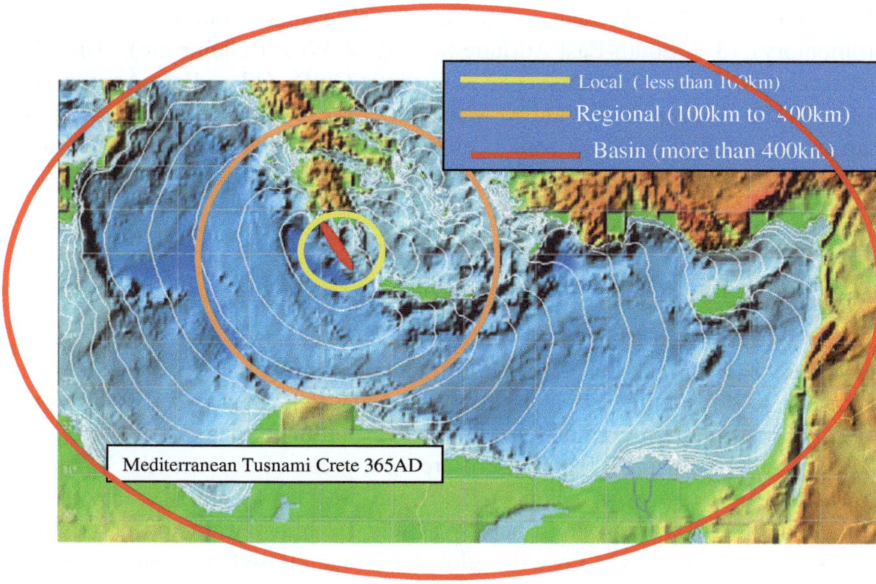

Fig. 24.7 The strong affect of the Tsunami in the Med. Sea (Papadopoulos 2003)

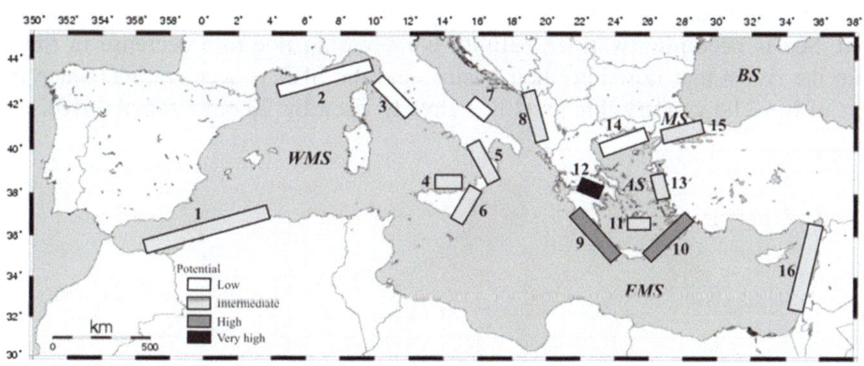

Fig. 24.8 The tsunamigenic zones of the Mediterranean Sea

owning to its occurrence the earth's crust experiences vertical deformations in the form of uplift and subsidence.

Figure 24.7 demonstrates the strong affects and impacts of 365 AD Crete tsunami waves that arrived at most of SCZ, while Fig. 24.8 illustrates a map of the known tsunamigenic sources in the Mediterranean region and a relative scale of their potential for tsunami generation calculated as a convolution of the frequency of occurrence and the intensity of tsunami events. In this figure, West Med. Sea (WMS), East Med. Sea (EMS), Aegean Sea (AS), Marmara Sea (MS), Black Sea (BS), 1 = Alboran Sea), (2 = Liguria and Cote d'Azur), (3 = Tuscany),

(4 = Calabria), (5 = Aeolian islands), (6 = Messina straits), (7 = Gargano promontory), (8 = South-East Adriatic Sea), (9 = West Hellenic arc), (10 = East Hellenic arc), (11 = Cyclades), (12 = Corinth Gulf), (13 = East Aegean Sea), (14 = North Aegean Sea), (15 = Marmara Sea), (16 = Levantine Sea). The tsunami potential of each one zone is classified in a relative scale according to the frequency of occurrence and intensity of tsunamis (Papadopoulos 2003).

24.4.2 Sea Level Rise in the Mediterranean Sea

The SLR in the Med. Sea shows a strong variability in the last century, and according to the satellite readings taken over just the last decade by the Topex-Poseidon mission corroborate observations that sea level trends in this sea are not uniform. As shown in Fig. 24.9, there are both areas of descent (the Tyrrhenian Sea and the body of water south of Italy) and areas of ascent (the eastern Med. Sea). In any case, with a rate of approximately 1.2 mm/year the observed rate of rise is significantly lower than the global average. Based on measurements of available tide-gauges the level in the Med. Sea has risen 1–1.5 mm/y since 1943 till the 1960s, and dropped few centimetres during the period 1960–1993. Then, a quick SLR of 4–5 cm took place during 1993–2000, after this there was no change. Climate figures during 1943–2008 (using marine observations) confirms that the Med. Sea is becoming warmer, salinity is increasing due to a decrease in runoff from the rivers that flow into their basins, and the SLR is accelerating. Since the start of the 21st century this level has already risen by 20 cm (Yáñez 2010).

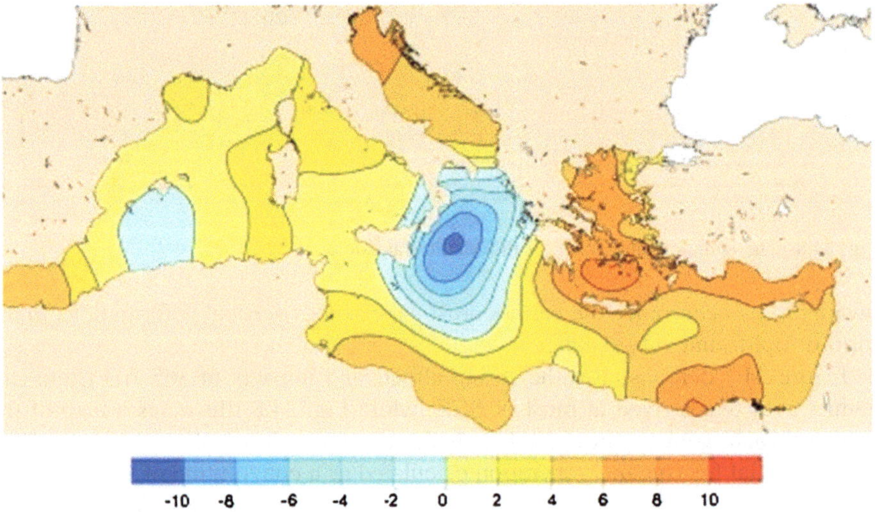

Fig. 24.9 Evolution in Med. Sea levels in mm/y from Jan. 1993 to Oct. 2004, according to the Topex-Poseidon satellite mission

This future rise during the 21st century will be mainly due to thermal expansion (30 cm), and only to a lesser degree the result of the thawing of non-polar glaciers (20 cm), and the Greenland ice sheet (10 cm) (IPCC 2001a). In addition, Med. Coast is considered vulnerable to SLR induced-flooding due to its environmental and socio-economic characteristics. It hosts valuable and sensitive habitats, such as coastal wetlands, as well as densely populated and urban developed areas and highly important economic sectors like tourism. However, SLR has been widely neglected in coastal management and planning along most of this coast (Christopher et al. 2006).

24.5 Factors of Hazards and Disasters

Disasters and hazards are tragic events to development process as they cause losing lives, disrupting social networks, and destroying economics activities. They can be narrowly confined to a locality or threaten entire regions, their intensity and probability can differ by place and this has a considerable influence on the levels of possible damage. They cut across many organizational, political, geographic, professional, topical and sociological boundaries. Therefore, there is a necessary need to integrate information and knowledge about them across many disciplines, organizations, and geographical regions (Schmidt-Thomé and Kallio 2006).

24.5.1 The Common Factors of Hazards and Disasters

These factors play a large role in determining the severity and magnitude of a disaster. They may be more or less applicable to any given society and contributes to determining the vulnerability of this society to disasters. A brief description of them can be outlined as follows: (1) *Poverty* which is the most important factor that increases the vulnerability of people to disaster. (2) *Uncontrolled population growth* that can lead to settlements in hazardous areas susceptibility to disease. (3) *Rapid urbanization and migration* which has an inevitable consequence of competition for scarce resources that can lead to man-made disasters. (4) *Environmental degradation* that can cause or exacerbate many disasters (e.g., deforestation leads to rapid rain runoff which contributes to flooding). (5) *Lack of awareness and information* which is a crucial factor in disasters that can also occur when people (who are vulnerable) have not been educated on how to get out of harm or take protective measures at the inset of a disaster.

In this chapter, the drivers and factors of coastal change and hazards in general and in Syria in particular will be discussed and analysed. They include: winds (e.g., extreme storms), waves (e.g., wave climate), sea-level variability (e.g., seasonal, interannual ENSO and interdecadel IPO cycles), river flow (e.g., extreme storms and base flows), storms and cyclones (e.g., incidence, intensity, tracks, storm surge), ocean and coastal currents, and the sediment supply to the coast (Fish et al. 2005).

24.5.2 The Common Factors of Coastal Hazards and Disasters

Coast is dynamic geographical feature and constantly changing as the land and sea interact in a variety of ways. The future changes in land use in the coastal zone will be dominated by the effects of climate change and global warming (Jeftic et al. 1992). These major effects will be due to increasing sea level in combination with possible increases in the frequency and intensity of storms, change in patterns of erosion and sedimentation, increased risk of flooding, and change in the distribution and types of coastal habitats, etc. Within this context, the relative SLR can mainly be caused by the *physical and human factors*. Physical factors can be consisted of a combination of sea level change and vertical land movement, geomorphology, storminess, waves, near shore currents, and tides, etc. On the other hands, human factors are mostly related to the coastal engineering (e.g., coastal protection structures), land reclamation, river regulation works (e.g., dam construction), marinas and commercial port development, and unregulated dredging, etc. The main factors of the coastal zone changes can be represented by:

(a) *Changes in SLR* which is very likely to have a profound impact on the shape and nature of the coastline. In terms of coastal hazards, it is the trend in relative SLR that is important (i.e. the change in sea-level relative to the local landmass), and the probable changes will include the coastal erosion and flooding (IPCC 2001b).
(b) *Coastal flooding* (the inundation of land by seawater) occurs frequently in low lying areas of the coast as a result of significant storm events coinciding with periods of high tides, storm surge and high wave energy. In addition, rainfall across catchments that raises river levels will add to the flood risk and particularly in the vicinity of river mouths with the coast (Komar 1998). On the other hands, this flooding is the most devastating natural disaster because of its rapid occurrence, little lead time for warning, and tremendous amount of water flowing with high energy.
(c) *Erosion and accretion* along the coastline is common and have large effect on its shape with time and interaction of the physical and human factors (EUROSION 2004). The rates of erosion are variable due to the range of controlling factors (rock properties, angle of wave approach, groundwater). The predicted SLR (expected to be 0.14–0.18 m by 2050, and 0.31–0.49 m by 2100) could lead to a landwards retreat of the coastline of 15–20 m at beaches along the coastline over the next century (Pilkey and Hume 2001).

(d) *Storm surges* occur when low atmospheric pressure combined with strong winds and are the greatest threat to life and property from a hurricane (Gulev and Hasse 1999). Storm surge is an abnormal rise of water generated by a storm over and above the predicted astronomical tides. The Hurricane Katrina (2005) is a good example of the damage and devastation that can be caused by surge (at least 1,500 persons lost their lives and many of those deaths occurred as a result of storm surge). Typically coastal erosion and coastal flooding of low lying land are associated with storm-surges.
(e) *Extreme weather events* are primarily ex-tropical cyclones and subtropical storms that generate storm surges and this can raise the sea level by 0.5–0.7 m (Berz 2005).
(f) *Temperature rise* which its effects can be seen in shallow estuarine waters and salt marshes or wetlands (Shaw et al. 1998; Ellis et al. 2000).
(g) *Global Warming and Climate Change* are likely to affect most of the physical processes that drive changes along coastal margins. 'Climate change' is defined as any significant change or trend in climate-natural or human-induced, and includes global warming. "There is new and stronger evidence that most of the global warming observed over the last 50 years is attributable to human activities", according to the (IPCC 2001c). Climate change will eventually affect all the drivers, either directly, or through their interaction with other drivers. For example, global ocean tides will be unaffected directly, but tidal propagation characteristics in shallow estuaries and rivers may be altered by deeper or shallower water depths (caused by changes in sediment supply) and/or a higher sea level.
(h) *Coastal habitats* are highly important on all the levels, and not only because of the scarce species of plants and birds that it supports, but also because they can act as a natural dissipater of tidal currents and waves and hence protect sea defences (Möller and Spencer 2002). They are under pressure from SLR, land reclamation for development and coastal squeezes. These squeeze result in steeper and narrower coastal zones when the ability of intertidal habitats to migrate inland in response to SLR is impeded by fixed coastal defence structures (Taylor et al. 2004).

24.5.3 Current and Future Drivers of Coastal Changes in Syria

The Syrian coastline (even it is short) has an array of geomorphological features: sandy shores, cliffs and rocky shore, hilly and flat coastal plains, narrow and wide coastal shelves, and wide variety of wetlands as shown in Fig. 24.3. Therefore, it is important to distinguish between the different environmental drivers of physical and ecological changes in coastal margins as follows:

(a) *Global Warming and Climate Change* Climate change impacts will vary locally as a result of local and regional differences in both the physical forcing functions (e.g., waves, winds, currents, sea level, etc.) and coastal types. Syria is not a major contributor in the emission of greenhouse gases, but like other countries, it is affected by the impact of probable global climate change that characterized by modifications in global precipitations and increased sea levels. Syria has recognized the importance and threats related to climate change and hence joined international efforts to combat them, ratifying the United Nations Framework Convention on Climate Change (UNFCCC) in 10 December 1995, and signed the Kyoto Protocol on 4 September 2005. Moreover, Syria has been openly realizing the importance of raising awareness on climate change, which would help the implementation of proper measures in order to reduce the possible negative impacts. The UNFCCC of 1992 is one of the recent series of Conventions which most countries have joined to combat this global challenge. The enabling activities for the preparation of Syria's initial national communication (INC project) are being implemented by the Ministry of Environmental, in collaboration with the Global Environmental Facility (GEF) and UNDP (Meslmani 2010).

(b) *Sediment Supply* Sediment sources and pathways sediments sinks can be affected by several factors including: catchment geology and rainfall, river flows, frequency and magnitude of storms river controls (e.g., dams, abstraction for irrigation), sand and gravel extraction for aggregate, ocean wave climate, prevailing winds, alongshore currents, the type of foreshore and its sedimentary composition. Sediment scarcity enhances the effects of storms

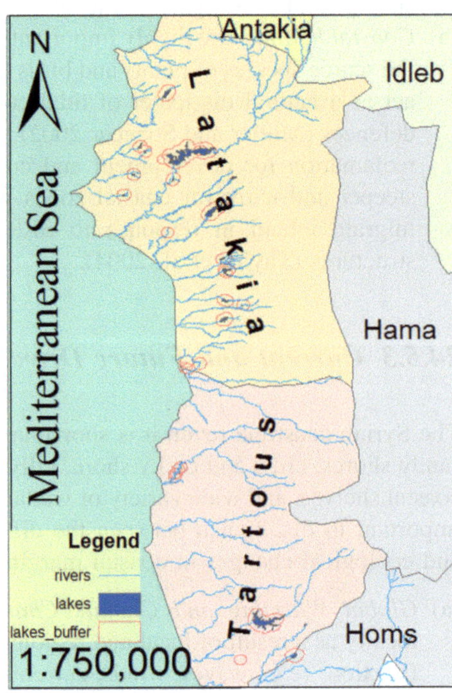

Fig. 24.10 The Surface water system in Syrian coast "Tartous and Lattakia". (Main Rivers and Lakes are surrounded by *Red circle* for clarity)

and SLR and may be more significant than both for erosion of dunes and beach. In the Syrian Coastal Basin as shown in Fig. 24.10, the surface water is the main source of the sediment supply.

(c) *Flash Flood* This disaster is a local problem and needs to be defined in its local and regional context based on: timely observation of rainfall events; and more demand on numerical weather prediction centres to produce more accurate data. The Syrian coastal region prone to flash-flood caused by torrential rain, and during the period 2009–2010, different parts in this region were subjected to heavy rain storms leaded to flash floods.

(d) *Seismic Deformation and Tectonic Changes* can over a short time frame have a much larger local impact on relative SLR than the gradual SLR expected from global warming. Syria has a long history of seismic activity, and over 2000 years of recorded history reveal more than two dozen magnitude 7 earthquakes in the vicinity of these Eastern Mediterranean countries (Ambraseys and Barazangi 1989). A major earthquake hit Syria on May 31, 526, killing approximately 250,000 people. The earthquake was followed by many aftershocks and a great fire that destroyed most of the buildings left standing by the earthquake. On 11 October, 1,138, another big earthquake

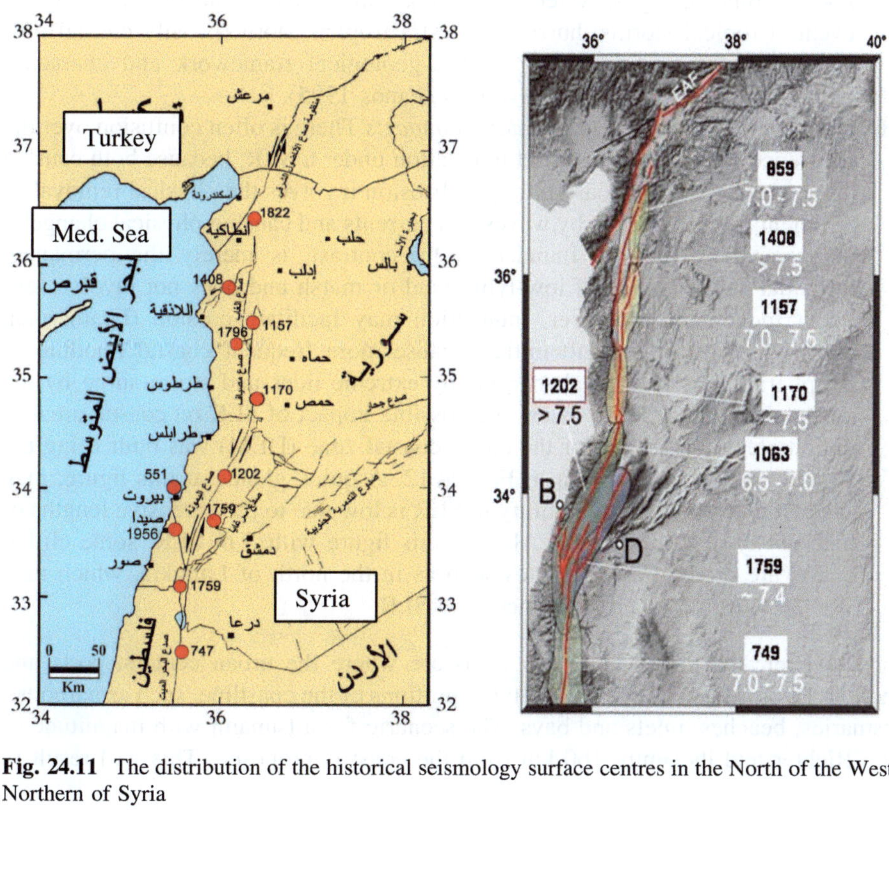

Fig. 24.11 The distribution of the historical seismology surface centres in the North of the West Northern of Syria

occurred in Aleppo and it killed 230,000 people in one of the deadliest seismic events in world history. It was felt over a part of the Eastern Med. Sea, and was accompanied by a tsunami. In 1759, a massive earthquake (estimated at more than 7.0 on the Richter scale) destroyed Damascus and the Lebanese city of Beirut. Today, the western region of the country continues to be the most susceptible to seismic activity (Gomez et al. 2001).

The bulk of the seismicity—and most of Syria's population—is concentrated in western Syria, and is related to the Syrian-African fault system (Dead Sea Fault System DSFS) between Antioch, Turkey, and the Gulf of Aquaba as shown in Fig. 24.11. DSFS formed as a result of the breakup of the Arabian plate from the African plate since the mid-Cenozoic, and is one of the largest continental strike-slip faults in the World. It represents a key tectonic feature in the Eastern Med. region. This call for critical evaluation of seismic hazard involving major cities in Syria so that proper mitigation measures (both structural and non-structural) may be undertaken before it is too late.

(e) *Human activities and other factors* These activities can be seen in coastal developing and infrastructure, coastal engineering structures, beach and dune nourishment, dredging (channels, inlets, canals), river modification (dams, levees), fluid (oil, gas, water) extraction. Other factors, can be seen in storm events (tropical storms, hurricanes, extra tropical storms), daily coastal processes (waves, currents, and winds), geological framework and character, geomorphology (slope and elevation) (Fanos 1995).
(f) *SLR, coastal flooding and inundation impacts* There is often confusion over the difference between erosion and inundation under a SLR, because both cause a loss of land from the coastal margin. Erosion involves the physical removal of sediment from the beach by waves and currents and causing physical change in the coastline structure. Inundation, by contrast, is merely the permanent intertidal submergence of low-lying land or marsh and does not involve sediment movement. However, inundation may facilitate erosion (Leatherman 2001). A rising MSL will initially cause more frequent coastal flooding of peripheral areas of coastal margins by extreme tides and storm surge by the mechanisms. In order to assess the possible impact of SLR on coastal areas, a digital elevation model for the entire coastal zone (DEM) was built using the ArcGIS software as shown in Fig. 24.12. A quick glance at this figure, may suggest that Syria's vulnerability to SLR is low due to the extensive lengths of high rocky or cliffed coast. Also, in this figure with 5 m SLR, some cliffed coasts are eroding at high rates such as in the north of Lattakia, which may worsen with increased storminess and SLR.

However, the most vulnerable areas are where the urban centers, ports and holiday resorts cluster around low-lying portions of the coastline, such as harbours, estuaries, beaches, inlets and bays. The scenario for a tsunami with magnitude of 7.5 Richter and its centre 160 km from the coast is proposed. This will result in

Fig. 24.12 The SCZ vulnerability to SLRs for different heights

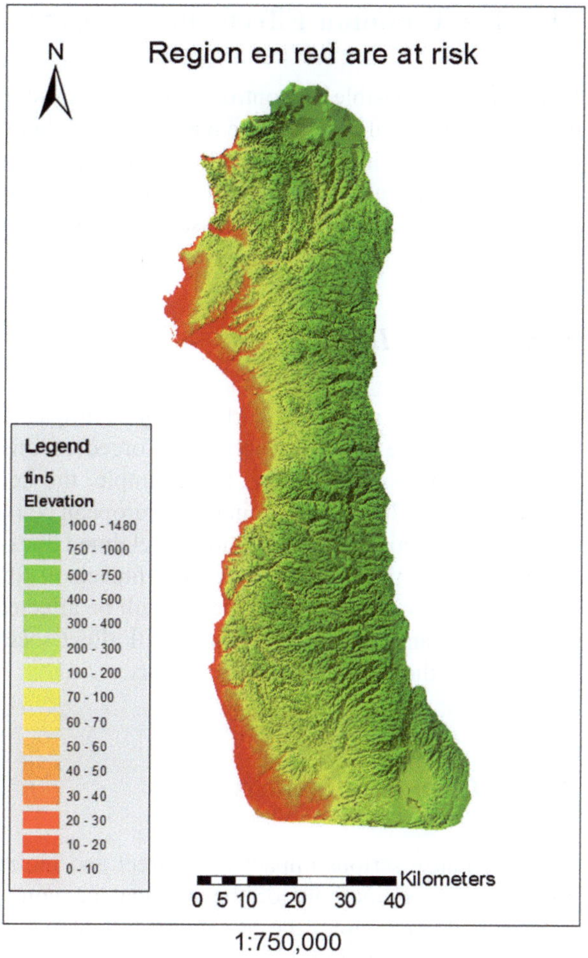

SLR caused by a wave that its high will reach till 15 m at the shore and its impacts will expand till 5 km from the coast inside the region within 20 min. With just a 1 m rise in the Med. Sea, more than one million people are predicted to be displaced and 1,000 km^2 of agricultural land will be lost. These high risk areas include parts of Lattakia, Tartoush, Banias and Jablieh. In addition, several other areas, such as those near the Turkish and Lebanese borders have also been identified as risked zones. However, the impacts may cover many aspects including impacts on water resources, agricultural and health resources, and the concentration will be on the city of Lattakia as will be discussed in Sect. 24.7.

24.6 The Common Effects and Impacts of Disasters

It may not be feasible to control nature and to stop the development of natural disasters, but the efforts could be made to avoid them and minimise their effects on human lives and infrastructure. For example, the effects of SLR will vary by location and depend on a range of physical, biological characteristics, and socio–economic factors as will be discussed in the following sections.

24.6.1 Social Impacts

Disasters destroy social network and other impacts can be related to the unemployment: people lost their jobs and forced to migrate to major cities causing problems and loss of security. For example, the heat wave of the summer 2003 claimed lives of 70,000 deaths in most European countries (EEA-JRC-WHO 2008), and the 1999 Izmit (Turkey) earthquake claimed lives of more than 17,000 fatalities. With regards to the social and cultural impacts of SLR, they can be discussed and analysed in terms of how resource use will be affected leading to displacement and resettlement, and how these changes will affect social relations regarding orientations toward the customary environments.

24.6.2 Economic Impacts

Disasters cause serious impact with direct and indirect losses on the economy by reducing the productivity of the national economy (infrastructures, commerce, industry). For example, the economic toll of natural hazards in Europe during 1998–2009 amounted to approximately 150€ billion (EM-DAT 2010).

24.6.3 Environmental, Physical, Ecological, and Geographic Impacts

These impacts can be seen on water, land/soil, crops, lake, forests, livestock, wildlife, atmosphere, energy, pollution, etc. The primary ecological impacts will stem from a rise in temperature, inundation by a higher sea level, and a loss of habitat from increased siltation in some estuaries and harbours, while in others there will be a loss of intertidal areas caused by constraining shoreline protection structures or embankments. The most serious physical impacts of SLR on coastal margins can be: (1) coastal inundation causing landward movement of estuaries, wetlands and marshes, (2) coastal erosion and shoreline change through sediment movement, (3) increased vulnerability to coastal storm damage and flooding, (4) increasing difficulty draining coastal and river lowlands, (5) the possibility of increased sediment loads to estuaries, with projected increases in rainfall intensity and run-off, (6) surface water, river water and groundwater in lowlands increasingly becoming saltier from seawater intrusion, etc. Vast areas of scarce land into the rivers and the ocean were lost due to erosions (Bell et al. 2001).

24.7 The Main Effects and Impacts of Local Sea Level Rise Scenarios: Focus on Lattakia City

SCZ constitutes particularly important regions economically, socially and culturally. In addition to increased tourism activities, tremendous move towards building new industrial complexes such cement and textile factories, power stations, petrol refineries, etc. This zone is therefore particularly vulnerable to the impact of SLR due to its relatively low elevation, salt water intrusion, soil salinization, excessive erosion rates, the deterioration of coastal tourism and the impact of extreme storms and flash floods. This in turn will directly affect the agricultural productivity and human settlements, management and access to archaeological sites, reduce tourism, and result in socio–economic impacts on the inhabitants of these areas. In addition, this coastal zone suffers from a number of problems, including a high rate of population growth, unplanned urbanization, land subsidence, land use interference, ecosystem pollution and degradation and lack of appropriate institutional management systems.

Lattakia has the main harbour and hosts about 30 % of the country's industrial capacity. This city is located in the northern western part of Syria with 229,690 ha of hilly terrenes, and total population about 1,350,000 (2010); 9.1 % working on the agricultural, 0.4 % fishing, and the rest working on industry, commerce, and professional skill, etc. In addition to the main port, there are 5 fishing ports with 677 fishing boats and 1,500–1,800 fishing men. Other vulnerable cities are Tartous, Banias, and Jablieh. Figures 24.13 and 24.14 illustrate the two scenarios for a SLR for 1 m and 5 m respectively, while Table 24.3 shows six scenarios that were developed for a SLR ranging from very low to extreme risk. The likely inundated sea shore area varies between 17.56 km^2 in a very low risk scenario to 118.90 km^2 in an extreme risk scenario. The results of these scenarios indicate that different segments of the coastline are vulnerable to a SLR as a consequence of expected climate change. This rise would have an impact on beaches, urban settings, and agricultural zones. Moreover, additional problems may arise due to salt water intrusion and increase in water and soil salinity. The socioeconomic impact of a SLR on coastal lowlands would vary depending on the flood levels, the degree of land use and development activities. Applying the extreme risk scenario shows that nearly 3.8 % of coastal populations will be affected by a SLR.

Fig. 24.13 The Lattakia's coast line vulnerability to SLR with 1 m high

24.7.1 Population Impacts

Population increase was accompanied by a remarkable increase in population density in most Syrian governorates between 1970 and 2007, and the average population density was 34 inhabitants/km^2 in 1970 and increased to 49 inhabitants/km^2 in 1981. As shown in Fig. 24.16 there is a clear disparity in the levels of population density from one governorate to another, since the highest density was noted in the Damascus governorate, with 7,090 inhabitants/km^2 in 1970 and with the very high figure of 13,152 inhabitants/km^2 in Damascus in 2006. The governorates of Lattakia and Tartous also registered high densities reaching 383 inhabitants/km^2 in Lattakia, and 376 inhabitants/km^2 in Tartous. It was also noted that Lattakia's population would be most severely impacted by SLR within the SCZ. With a 1 m SLR, approximately 8 % of Syria's population would be impacted, and most of this impact takes place in the Lattakia which will have 15 % of it affected with a 5 m SLR. The proposed scenario of SLR in Fig. 24.15 will make Lattakia a disastrous city as most of the population density is concentrated on narrow area of the cost with width of 4.5 km. Table 24.4 shows the different heights of some areas which receive various effects according to their heights. Syrian coastal population is exposed to the effects of SLR as the population is expected to double before the year 2050.

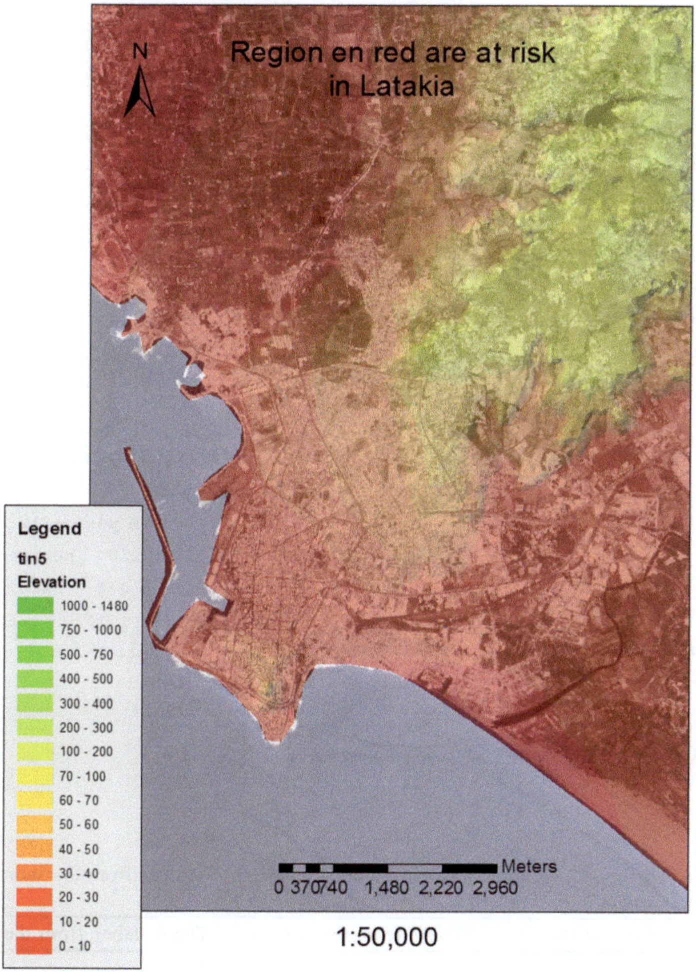

Fig. 24.14 The Lattakia's coast vulnerability to SLRs with different heights

Table 24.3 Inundated areas in 2100 according to various scenarios of a SLR

Scenario	Trend (cm/year)	Variation 2000–2100 (cm)	Inundated area (km^2)
Very low	0.6	63	17.56
Low risk	0.9	90	20.27
Moderate risk	1.3	130	23.89
Intermediate risk	1.9	1.9	27.57
High risk	2.5	250	30.35
Extreme risk	>5	500 up to 750	118.90

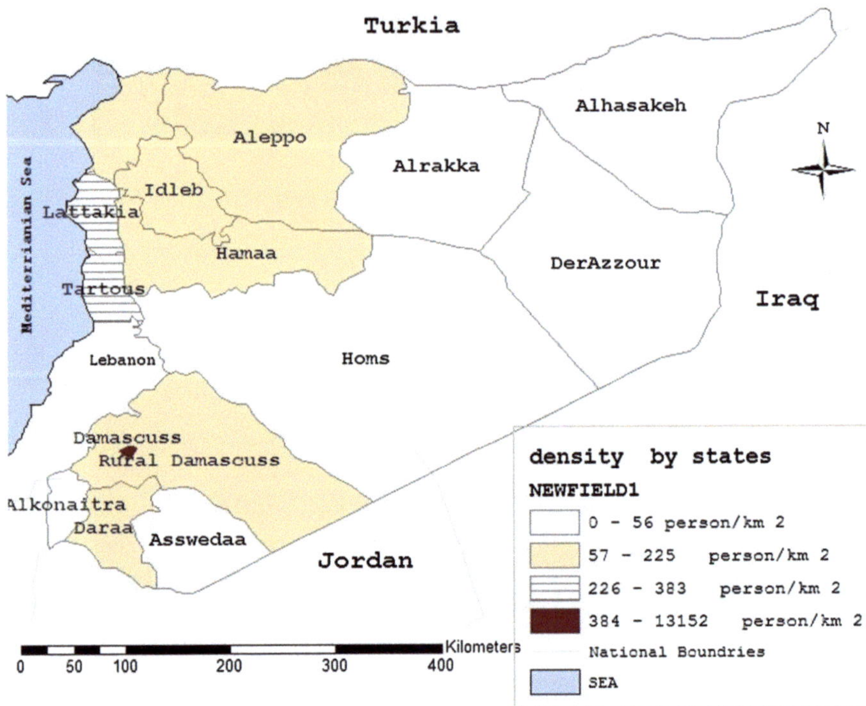

Fig. 24.15 The main Governorates and population density in Syria

Table 24.4 The heights of some areas in Lattakia and their distances from the shore that might subjected to various impacts of SLR

The main locations in the city	Height from the sea (m)	The distance from the shore (km)
Industrial City	7	4.5
The Sporting City	5	0
The Castle	75	1.5
Tishreen University	18	1
The Tobacco Factory	2	0.5
The Southern Sand Area	12	0.65
The Blue Coast Area	3	0
Al-Hussaynee Street	43	1
Saed Zaghloul Street	23	0.5
Abid Nasser Street	10	0
The Besnada Area	85	4.5
The Damsarkhou Area	8	1.5
The Southern Kournesh	30	0
The Main Train Station	15	1.8
The Bahlolieh town	177	15
The Qanjerah town	64	5.25

24.7.2 Environmental, Health and Socio–Economic Impacts

The Med. coast is considered particularly vulnerable to SLR-induced flooding due to its environmental and socio–economic characteristics. It hosts valuable and sensitive habitats, such as coastal wetlands, as well as densely populated and urban developed areas and highly important economic sectors like tourism. However, continuous SLR is expected to enhance rates of erosion of north Lattakia city's coastal zone, and the contingency plans that suggested by the GoS which aim to protect the tourism industry in the first place, but are not directly related to the impact of climate change and SLR. Additional adaptation measures are needed to target climate change and that this will be less expensive for the tourism industry than losing the beach completely. Loss of beaches will reduce the number of tourists in coastal areas, forcing tourism dependent individuals and communities to abandon their settlements and look for jobs elsewhere. The risks may be particularly severe in poor neighbourhoods and slums, where infrastructure is poorly designed. Generally, fundamental and low-lying installations in Lattakia are threatened by SLR and the recreational tourism beach facilities are endangered of partial and even full loss. Moreover, increased water logging and salinity may catalyse insect and pest problems causing health problems (Muslemani 2010). It may lead to group migration of farmers looking for jobs somewhere else. Nearly 2,000 agricultural families (and another 4,000 in the case of the extreme scenario) will be in danger of losing their economic subsistence.

Direct and indirect implications on the socioeconomic systems are important factors that have to be taken into consideration. The loss of land productivity will force a large number of farmers and/or fishermen to move away from the non-fertile land or fishing grounds and go searching for jobs. Table 24.5 shows possible economic losses due to a rise in sea level of 2.5–3 m. Reducing vulnerability to such threats is a major challenge to sustainable development and land use strategies. Coastal defence engineering is costly, while managed coastal retreat implies sacrificing private property and usable natural resources. In this study, the available land-use data, topographic and socio-economic data were used to calculate the

Table 24.5 Possible economic losses due to a SLR of 2.5–3 m along the SCZ

Scenario	Total economic loss (in millions of S.P.)
Citrus plantations	13,205
Olives	432
Greenhouses	8,303
Crops	15,023
Forest	191
Sandy soil	1,800
Urban areas	10,900
Total	49,854 ≈ 1 Millions USD

approximate numbers of people expected to be affected by SLR. It is estimated that with a SLR of 1 m in Lattakia will cause a displacement of almost 350,000 people and the loss of about 10,000 jobs. Also, changes in the ecological system of lakes will reduce fish catches and drive away a large portion of fishermen and their dependants.

24.7.3 The Impacts on the Water Resources

Syria is one of the Mediterranean countries that has proved vulnerable to water stress caused by climate change and SLR. A combination of salt water intrusion due to SLR and increased soil salinity due to increased evaporation are expected to have direct impact on quality of shallow groundwater supplies and drainage conditions, agricultural productivity and socioeconomic and health implications. SLR is expected to cause a landward shift of the salt wedge and to increase the rate of saline seepage to the topsoil of the coastal region. The main source of water in the coastal region is the surface water and due to the high rate of rain full that feed the springs rivers. The main coastal rivers are: the Northern Big Rive with length

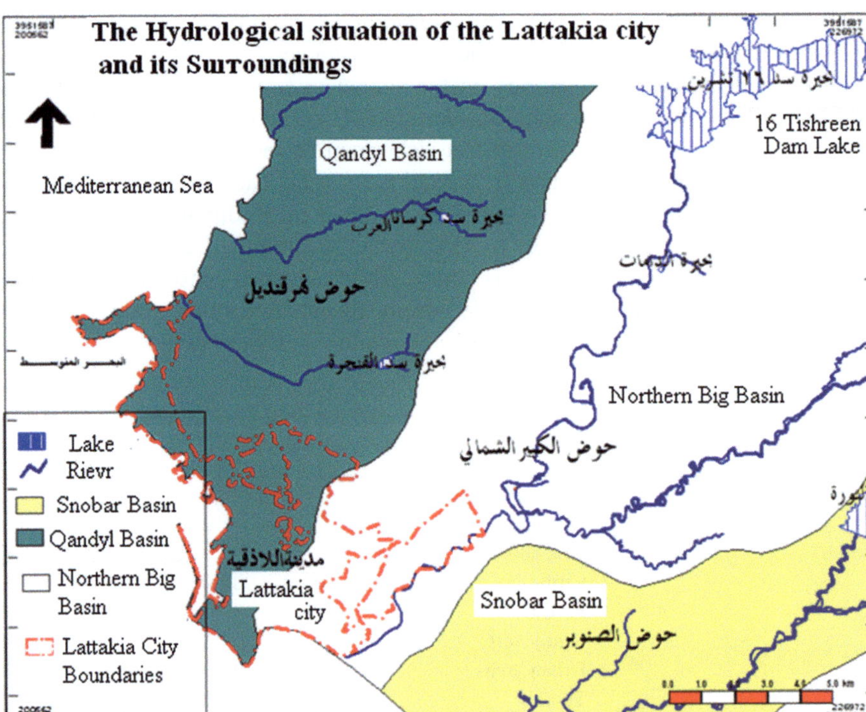

Fig. 24.16 The hydrological basin of the Lattakia city and its surroundings

of 96 km, the Southern Big River with length of 56 km, the Sin River with length of 6 km. Also, there are many lakes and surface dams with storage capacity of 357 MM^3. The accelerated SLR and the stronger influence of tidal flows penetrating these lakes will enhance the changes in the salinity conditions of these lakes which may affect their ecology and fisheries as shown in Fig. 24.16.

During the last decades, after the construction of the several dams, sediment input in some parts of the coast has been strongly reduced, and this resulted in serious shore erosion and salt water intrusion. Moreover, the construction of human-made waterways for irrigation and transportation has trapped an already depleted sediment supply to the some large coastal zones. The protective sand belt of some parts of the SCZ is facing rapid erosion, which has been a serious problem since the construction of the several earth dams on these rivers. SLR is expected to destroy weak parts of this belt, which is essential for the protection of lakes and the low-lying reclaimed lands. The impacts will be very serious as an important part of Syria's fish catches are made in these lakes.

24.7.4 The Impacts on the Agricultural and Food Resources

The agricultural sector plays a significant role in the Syrian national economy contributing about 30 % of the GDP. It supplies the overall food needs of the country and provides the domestic industry with agricultural row materials. It is expected that with a 1 m SLR, approximately 8.5 % of Syria's agricultural extent would be impacted, and this percentage reaches 25 % with a 5 m SLR. Livestock and fisheries are also vulnerable to the impacts of climatic changes and SLR. Moreover, climate change will probably affect water resources and that might pose another problem for agricultural and food production. It was deduced that the large incremental impact of SLR on agricultural areas in SCZ arises in Lattakia and its surroundings as shown in Fig. 24.18.

It is very difficult to precisely evaluate the socioeconomic impact of SLR on

Table 24.6 Impacts of SLR on Land-use/Cover

Scenario	Very low risk: 0.6 m	Extreme risk: 5–7.5 m
Land cover	Area in sq km	Area in sq km
Citrus and other plantations	3.14	37.51
Forest	0.41	2.21
Olives	0.19	1.51
Greenhouses and field crops	2.05	33.63
Sandy soil	0.99	7.98
Urban areas	5.12	16.73

local communities in affected areas. Based on land use categories in these areas, a rough examination of economic losses as a result of SLR alone reaches

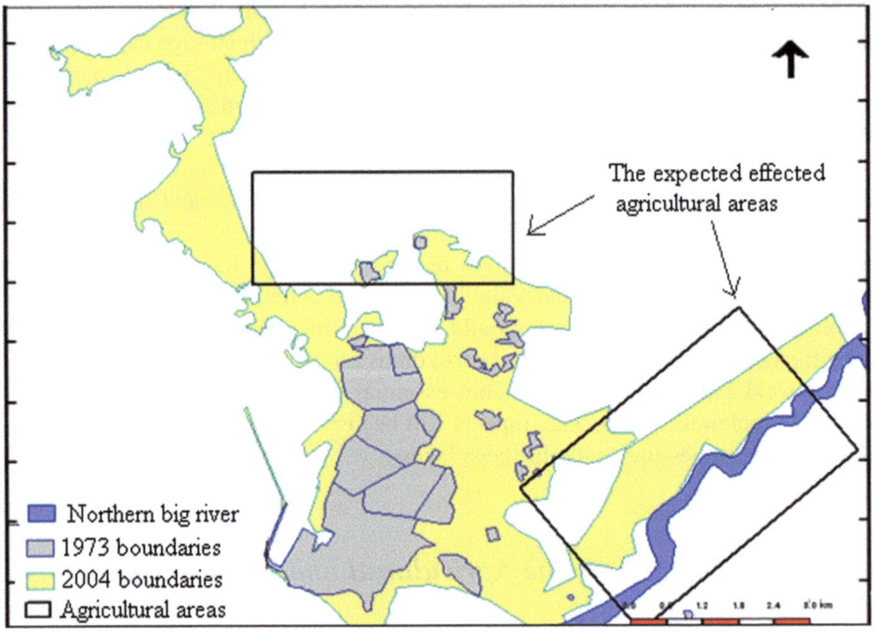

Fig. 24.17 The informal expansion of the Lattakia city and the expected effected agricultural areas by the SLR

50 billion Syrian Pounds as shown in Table 24.6. This figure represents direct average economic losses resulting from permanent disappearance of 4,108 ha of agricultural and forested areas, 450 ha of beach and 1,090 ha of urban area. However, these losses may go down to 10 billion S.P. in case of a SLR of 0.6 m and may reach 84 billion S.P. if the extreme SLR scenario is justified (FAO 2007) (Fig. 24.17).

24.8 Disaster Management Cycle and Risk Reduction Measures

Disaster Management (DM) is a multidisciplinary area that involves monitoring, preparing, warning, forecasting, supporting, evacuation, search and rescue and then re-building society when disasters occur. It requires response, incident mapping, establishing priorities, developing and implementing action plans to protect lives, property, and environment. Several and exact interconnecting steps are typically required to generate the type of action that needed by the DM community (Alexander 2002). With regards to the Risk Reduction (RR) measures, can be defined as any system, procedure, process or device that are intended to

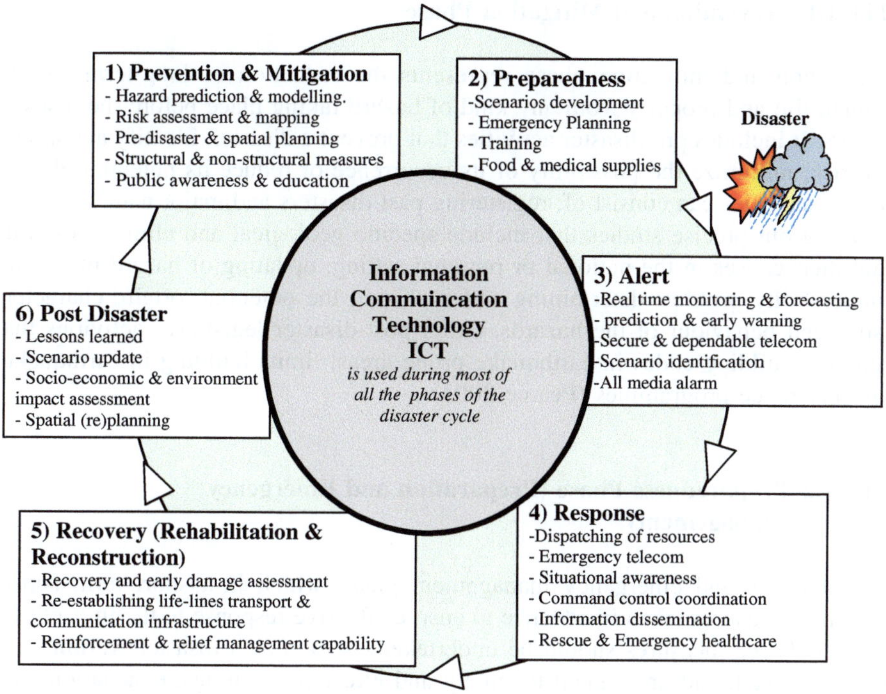

Fig. 24.18 The disaster management cycle

eliminate hazards, prevent hazardous incidents from occurring or reduce the severity of consequences of any incident that does occur. These measures can be recognised while identifying hazards and employers should be able to identify a range of control measures immediately, both the existing measures and possible alternatives. The following sections present the general framework for the DM cycle and all its phases that differ according to the type of a disaster.

24.8.1 Disaster Management Phases

DM activities, generally, can be grouped into six main phases that are often illustrated in the so-called risk or DM cycle. These phases are related by time and functions to each other, and to all types of emergencies and disasters. Each phase involves different types of skills and data from a variety of sources, and the appropriate data has to be gathered, organized, and displayed logically to determine the size and scope of DM programs. Figure 24.18 depicts the framework for DM cycle with its six phases as follows:

24.8.1.1 Prevention and Mitigation Phase

Prevention and mitigation phase represents the reduction of long-term risk to human life and property from any kind of hazard taking place before the disaster occurs. It includes pre-disaster activities that prevent a disaster, identify and assess the risk, minimize the possibility of its occurrence or reduce its potential effects. These activities can consist of: registering past disasters and major natural events, carrying out precise studies that include specific geological and climatic hazards and their causes in the national or regional setting, updating of hazard maps and vulnerability profiles, determining and analysing the potential, origin, characteristics and behaviour of the hazards. Other post-disaster legislative activities that requires building codes in earthquake prone areas, limits building in earthquake, and insurance programmes (Pearce 2003).

24.8.1.2 Preparedness Phase (Preparation and Emergency Management)

Preparedness and emergency management phase which is a short term action includes measures taken in advance to ensure effective response to the impact of a hazard. These measures should be undertaken as soon as a disaster warning has been received, and are related to timely and effective warnings, evacuation, and temporary property protection. The preparations on how to respond in case of an emergency to save lives and minimize disaster damage have to be made and developed by several complementing bodies such as governments, organizations, and individuals. Some non-exhaustive examples of these preparations are; building up and strengthening local and national disaster preparedness capabilities, implementing developmental plans that provide resilience to disasters, carrying out and mounting training disaster preparedness exercises, and installing EWS.

24.8.1.3 Alert Phase

Alert phase supports EW process such as real time monitoring and forecasting, secure and dependable telecom, and all media alarm. An effective EWS can be achieved by the provision of timely and accurate information through identified institutions that allow groups and individuals exposed to a hazard to cooperate and take serious actions that avoid or reduce the risk and prepare for effective response. Some practical studies has been carried out in the domain of flood control and management using decision support system based dynamic optimisation and spatial planning (Saleh and Allaert 2009b).

24.8.1.4 Response Phase

Response phase is the implementation of set of actions that provide short-term emergency aid assistance for causalities and save lives during and immediately following the occurrence of the disaster. These actions can include: search and rescue operations, evacuation, emergency medical services, emergency shelter, medical care, and mass feeding, etc. Response mechanisms refer to the actions to be taken and pre-planned for disasters between the community and the responding agencies. These mechanisms allow for a structured response to different disasters and ensure that response actions and resources are not duplicated (Murphy and Gardoni 2007).

24.8.1.5 Recovery Phase (Rehabilitation and Reconstruction)

Recovery phase represents the final step of post-disaster actions within the DM cycle and starts when the disaster is over. This phase includes actions that assist a community to recover and restore the living conditions of the affected population and then return to a sense of normal after a disaster. These actions are divided into main two sets:

- *Short-term recovery actions* that restore services and systems and return their vital life support to minimum operating standards. These actions can include cleanup, assuring injured people have medical care, temporary housing or shelter to citizens who have lost homes in the disaster, access to food and water, etc.
- *Long-term recovery actions* that may continue for a number of years until the entire disastrous area is either completely restored or redeveloped for entirely new purposes (that are less disaster-prone). These actions can include community planning, replacement of homes, water systems, bridges, developing measures for future prevention (e.g. watershed management, resource conservation, etc.).

24.8.1.6 Post Disaster Phase

Post Disaster phase includes lessons learned, scenario update, socio-economic and environment impact assessment, Strategic Environment Assessment (SEA) and spatial re-planning (Oliver-Smith 1992). With regards to the this phase and to the obtained results related to the SLR, it has been stated that the Government of Syria (GoS) had been working for the past 30 years on sea erosion reduction and shore protection measures particularly by constructing dams on the main rivers in the coastal areas. Furthermore, water institutions and centres are working on all the aspects to reduce or eliminate some of the negative impacts of SLR on water resources. It has planned to improve water sanitation coverage for urban and rural areas, and optimise the use of water resources by improving irrigation efficiency and agriculture drainage-water reuse. There is an urgent need to construct concrete sea walls to protect beaches from SLR. In addition, several measures could be utilised to deal with the impact on the coastal zone corridor, including beach nourishment (deposition of sand onto the beach), construction of breakwaters, setting regulations to restrict development in vulnerable areas, changes in land use and Integrated Coastal Zone Management (ICZM) principals.

With regards to the SEA, which is a decision-aiding tool, can effectively complement the process of strategic planning of the post disaster phase. It can be considered a dynamic process in which the environmental considerations are assessed and reassessed as they are changed or adapted, then investigating all the significant environmental effects and considering relevant alternatives for making decisions. The main focus is "*How SEA can effectively be linked to planning in order to contribute to DRR in SCZ?*" This can be achieved by answering the following question: How can implementation of SEA strengthen the efforts to integrate environmental considerations in planning the current and future master plan of the city of Lattakia? With regards to the environmental pollution problems in this city are solved by introducing environmental management techniques such as control of pollution at source and providing of sewage treatment facilities, etc. However, environmental risks are not being controlled completely by such solutions. Presently, in some cities, the environmental aspects are not usually considered while preparing master plans or regional plans and the process is skewed towards developmental needs. The environmental aspects are to be induced into each of the developmental activities at the planning stage itself and are to be well co-ordinated and balanced. The present work details the need for usage of environmental planning as a tool for environmental protection and the priority actions needed to be taken in Syria.

Problem analysis and necessary measures can be determined after planning alternatives have been assessed. The more complex the alternatives are, the more likely formalised assessment methods like cost-benefit analysis or value-benefit analysis have to be taken into consideration. After the discussion of all alternatives has been completed, the third phase of the planning process can begin. The relationship between comprehensive spatial planning and sectoral planning divisions is a crucial factor for mitigating spatial risks. In the following section, risk assessment and management will be understood as the systematic application of management policies, procedures and practices to the task of identifying, analysing, assessing, treating and monitoring risk. Figure 24.19 depicts the proposed geographical locations of the expanding and re-planning the city of Lattakia considering some control measures of the impacts of SLR and other related hazards such as building some dams to control the flash flooding, and where to build the settlements far from the effect of industrial city and the effect of SLR.

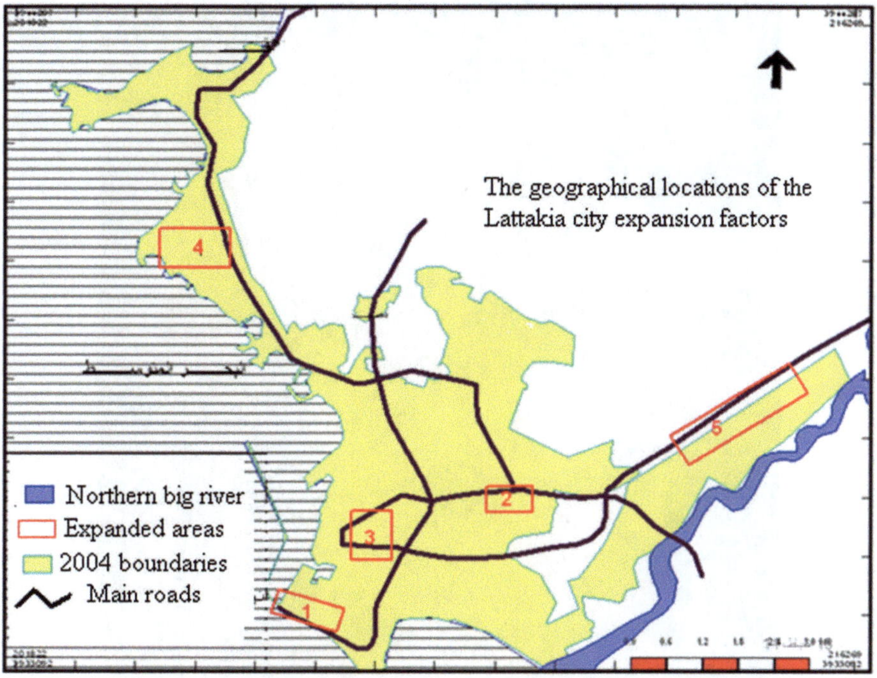

Fig. 24.19 The proposed geographical locations of the expanding Lattakia city considering some control measures of the impacts of SLR and other related hazards

24.8.2 Disaster Management and Disaster Risk Reduction Planning Activities

The square of the Lattakia city was doubled 4 times in less than of 40 years, and this indicate that the city is great centre, but in the absence of the regional planning and DM. For the proposed master plan for the city of Lattakia as shown in Fig. 24.20, the main steps of DM planning process will be described and complemented by a description of how the steps of risk assessment and management can be integrated into the spatial planning process for the city to reduce the risk of SLR and other related hazards.

During the DM planning process, several steps have to be considered to bring DRR and development planning concerns closer as follows: (1) The collection step of basic data on disaster risk and the development of planning tools for tracking the relationship between development policy and disaster risk. (2) The dissemination step of best practice in development planning and policy for reducing disaster risk. (3) The assessment step for providing a climate change scenario, and consisted of: (a) hazard assessment (historical profile of disasters, predications of trends in natural disasters related to climate change), (b) vulnerability assessment (geographical locations, transportation networks, communication networks, shelters in the event of

Fig. 24.20 The master plan of the Lattakia city

disasters, water and sanitation, health, livelihood), (c) capacity assessment (DM plans in project areas, coping strategies of communities, role of people in activities of mitigating disaster impacts, proposed adaptation measures). (4) The planning step is to find safer master plans through integration development plan into DRR. This step can include finding scenarios for training (climate change, codes) and identification of issues and properties. (5) The implementation step of the proposed plan through several projects (with cooperation of co-financing with local governments). These projects can include construction of shelters, roads, sanity, distribution of rescue and warning equipment, etc.

24.8.3 Advances in Disaster Risk Reduction and Management

In recent years, policies for disaster risk reduction and management have shifted from defence against hazards (mostly by structural measures) to a more comprehensive and integrated risk approach (as shown in Fig. 24.20). Within Integrated Risk Management (IRM), the full DM cycle should be taken into consideration when dealing with any type of disaster. The implementation of IRM is currently taking place at both international and national levels and is promoted by several initiatives (IDNDR 1999). In the 2005 World Conference on Disaster Reduction, Hyogo Framework for Action (HFA) was launched as a global plan for building the resilience of nations and communities to disasters during the period 2005–2015. HFA has been adopted by 168 governments (including Syria) with aim to reduce losses from disasters substantially by 2015 in terms of lives, social, economic and environmental assets of communities and countries. More specifically, the *three goals* of HFA are: the integration of disaster risk reduction into sustainable development policies and planning, the development and strengthening of institutions, mechanisms and capacities to build resilience to hazards, and the systematic incorporation of risk reduction approaches into the implementation of emergency preparedness, response and recovery programmes.

According to the national report on the status of progress in the implementation of the HFA in Syria (2009–2011) (SYR-HFA 2011), http://www.preventionweb.net/english/hyogo/progress/reports/v.php?id=17404&pid:223//substantial work has been done on this subject at national, regional and international levels, and can be classified as follows:

1. The development of legislation governing the structure and operation of DM, incorporating the subjects of preparedness and DRR in government development plans and linking it with the sustainable development process.
2. The database for hazards is being analyzed to be used in the future regional planning, also work is undergoing on improving coordination and cooperation between various stakeholders to unify and develop this database.
3. There are EWSs being developed for different types of risks such as wild fires through installing new equipments for predicting the occurrence of fires, and EWSs for oil spills and oil pollution on beaches. With regards to EWS towards drought, work is under going to use remote sensing, and automated monitoring system for meteorology which can help in predicting dust storms. In addition to the above, there is an EWS for predicting earthquakes placed near seismic faults.
4. With regards to the professional DRR education programmes, an MSc degree in the field of disaster and risk management in cooperation with Damascus University has been established. It aims to prepare specialized team in disaster prevention and response in which the author of this chapter has been participating in establishing the course and teaching activities.

5. The cooperation and strengthening partnerships between the public and private sector are reflected in the modern industrial cities through the identification of risks and prevention procedures. In addition to that the private sector puts under the disposal of governments agencies concerned with disaster mitigation all the available capabilities such as materials, equipments and human resources that can enhance the ability of society to response and reduce the risks.
6. Laws and legislations have been put under implementation to ensure the safety of hospitals, schools and other governmental constructions. Also, the executive instructions have been issued to ensure correct construction and taking into account public safety conditions in all buildings that are newly created. In addition, effective interests have been carried out in schools through the system of preventive maintenance and periodic inspections.
7. The 11th Five Year Plan (FYP) (2011–2015) had added a clear strategy for safety through prevention, EW, preparation, response, recovery and rehabilitation. It includes DRR in its strategies, but the essential point is in the implementation of these strategies and the availability of necessary funds.

The overall challenges that face the GoS in proper implementation of HFA are: there is still lack of knowledge towards DRR on different levels, and the emergency management plans are still lacking the prevention phase of the DM lifecycle.

24.8.4 Complementing Disasters Risk Management

This section focuses on capacity building in relation to disaster risk management, and includes principles for training and tools that can be useful to build up skills at all levels. It presents the most recent processes that have been made through advances in EW and observing systems, communications, and how this is helping to understand the physics of hazards and promote integrated observation and modelling of the disaster. In addition, it outlines the use of the field tools and data cross-checking in identifying hazards and vulnerability. Then, it discusses the role of good governance, decision support and advocacy in providing proper process of implementation of the whole DM cycle.

24.8.4.1 Capacity Building

Within the context of the DM, the capacity building can be defined as "the efforts aimed to develop human skills or societal infrastructures within a community or organization needed to reduce the level of risk" (ISDR 2005). It is essential that human and institutional resources are adequately trained to reduce disaster risks. Building national capacity for natural disaster mitigation and management can take the form of advisory services, training, workshop, field projects and

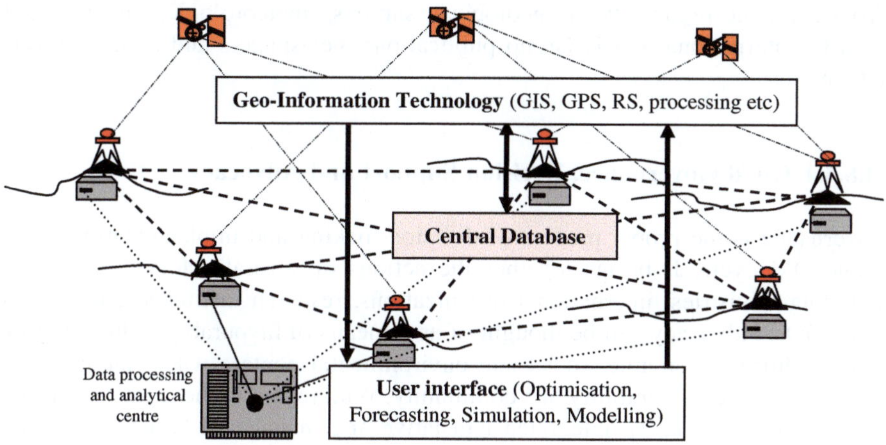

Fig. 24.21 The real-time warning network and its database structure

publications, and accesses to technology or other forms of technical assistance intended to improve institutional efficiency, etc.

24.8.4.2 Geo-Information Technology

In recent years, the focus of DM community is increasingly moving on to the more effective utilization of advanced geo-information technologies that enable communities at risk to prepare for, and to mitigate the potential damages caused due to disasters. Using these advanced technologies, a real-time EW network has been designed to consider the real life applications of DM as shown in Fig. 24.21. This network utilizes the strengths of the most advanced geo-information technologies and centralized databases, dynamic optimisation and geospatial models, data collection, internet, information communication technology, and expert systems, etc. EWSs can help to monitor natural hazards, plan response activities, identify affected populations and their needs, assess the flexibility of existing instruments or the functioning of markets or facilitate targeting of beneficiaries. This will have potential to provide valuable support to decision making through providing and representing spatial data, and dynamic models in analysing and representing temporal processes that control the disaster. More information about the scientific research based dynamic optimisation and geo-information technologies for DM&RR can be seen in (Saleh and Allaert 2011).

24.8.4.3 Field Tools and Data Cross-Checking (DRR in Practice)

It is assumed that any information gathered in the field will be cross-checked where possible with other existing data sources (e.g., government, NGOs, Red

Cross/Crescent organisations, geological surveys, meteorological data, health records, International Crisis Group publications, newspapers and academic journals, etc.).

24.8.4.4 Good Governance, Decision Support and Advocacy

Governance is the proper process of decision-making and implementation of the whole DM cycle. It brings together the actions of several actors at all levels, including ministries, international organizations, research institutes, universities, and NGOs. Advocacy can be thought of as a means of favourably influencing the wider political, economic, social and environmental context where these factors contribute to the vulnerability of a community. With regard to the legal aspects and key players in Syria, an integrated legislative and managerial base covering the entire SCZ does not exist. The development of coastal governance based on the principles of ICZM is an ongoing process in Syria as will be seen in the Sect. 24.10.

24.9 Risk Assessment (Hazard and Vulnerability Analysis)

Within the concept of DRR, risk assessment is carried out to identify which hazards are more likely to occur and to have the biggest impact on a community's or individual's assets. It is a systematic tool to integrate science with state-of-the-art geo-information technology to better understand the complex interaction of hazards, community and infrastructures, and then to help managing individuals safety risks.

24.9.1 Components of Risk Assessment

The main components of risk assessment are *hazard analysis* and *vulnerability analysis* that allow assessing the risk facing communities. This can be done by identifying the hazards which are most likely to occur within a given time-frame and to determine which of them will have the greatest magnitude of impact on the assets and livelihood options of a community. Over time, some changes can occur in terms of vulnerability of a community, types, causes, and intensity of the hazards that are face. Therefore, hazard analysis is concerned with identifying the underlying causes that influence the occurrence of hazards and provides more details about their frequency, seasonality, geographical area of the hazards' occurrence, etc. The importance of undertaking a hazard analysis can be illustrated by looking at the specific hazard of a SLR occurring in the East Med. Sea. There is high probability that the SCZ will experience a SLR sometime in the future, but as they are very infrequent, the likelihood in any one year of experiencing a SLR is very low. However, when the underlying causes that create SLR are studied, a more informed picture begins to emerge as shown in Sect. 24.5.3.

The *hazard analysis* usually includes: (a) *hazard identification*: to recognize particular types of natural disasters that have the potential of occurring within a region. (b) *profiles of hazard events*: to identify past incidences of natural disasters within each region. (c) *community profile*: to compare overall county property statistics to those within the pertinent hazard area. (d) *Estimated losses and vulnerability analysis*: will be determined using the hazard analysis, individual parcels and property asset data.

RA and risk management can be understood as corresponding instruments for achieving disaster resiliency that has to be seen as important objective for planning policy. Based on the above, risk management can be defined as adjustment policies that intensify efforts to lower the potential for loss from future extreme events. This shows that risk management is characterised by decision making of stakeholders which is a normative, politically influenced strategy about tolerating or altering risks. Therefore, within the DM planning and development, the task of urban planners relates to gathering, processing and presenting data to allow a series of questions to be answered so that decision-makers can formulate successful strategies. The following questions are of concern in this context: (1) What is the level of risk that the society is willing to accept? (2) What are the protection goals for the different protection objects that are threatened by specific hazards? or (3) What are the foreseeable environmental effects from a planned object in case of an occurred hazard? After selecting of the spatially relevant hazards, the first question in this sequence is: *what is the risk?* In other words, what would be the expected losses in human life and property? Therefore, deriving risk components have to be identified after selection the type of hazards.

24.9.2 Selection of Spatially Relevantly Hazards

Every hazard has a spatial dimension (i.e., disasters take place somewhere), hence the spatial character is defined by spatial effects that might occur if a hazard turns into a disaster. However, the occurrence of spatially relevant hazards is limited to a certain disaster area, which is regularly or irregularly prone to hazards (e.g. river flooding, SLR) (EPSON 2006). Spatially non-relevant hazards occur more or less anywhere (e.g. flash floods). The main question is: *Which of the existing hazards are of relevance in the context of the spatial planning?* The selection of hazards can be carried out following these steps: *Risk type* in which a list of possible hazards in the country is compiled firstly. Then, *spatial relevance* of the hazards is assessed secondly. Mainly, the selected hazards are classified according to the effect of climate change which is currently can be regarded as the major impact (Gornitz et al. 1994). In this research, the selection of SLR as relevantly hazard is based on its spatial effects that might occur if it is turns into a disaster that affect the whole SCZ as shown in Fig. 24.20. This figure depicts the proposed

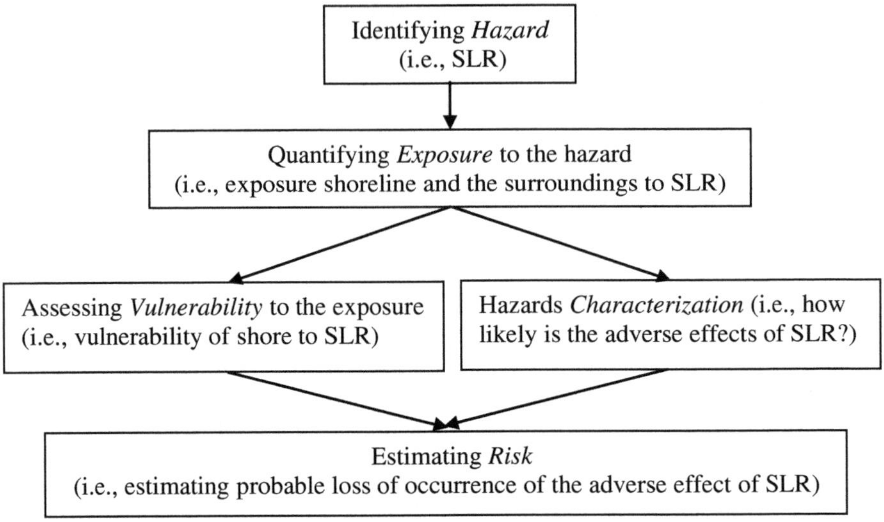

Fig. 24.22 The conceptual flowchart of risk assessment components

geographical locations of the expanding Lattakia city considering some control measures of the impacts of SLR and other related hazards

24.9.3 Analytical and Planning Tools for Deriving Risk

Risks are dynamic by nature, and therefore, a successful RA should be able to provide reliable information on where, when, how and why a hazard and disaster are likely to occur. Analytical and planning tools can help to systematically look at hazards in terms of risk they pose, their causes, characteristics and potential controllability. They help to analyse vulnerabilities and subsequent impacts in terms of their negative influence on livelihoods in order a sensible risk assessment can be made. These tools can also be used in fieldwork with communities to assist them in understanding their risks and suggesting DRR measures. The flowchart of Fig. 24.22 illustrates the methodology for risk estimation by explaining the main steps of deriving risk as follows:

24.9.3.1 Identifying the Hazard

The hazard identification step is based on scientific and technical findings, and involves the estimation of the probability of occurrence a damaging hazard (e.g., What is the probability of a SLR to occur within a specific period of time in a given area?). Usually, the sectoral planning divisions are responsible for this stage due to their specific competencies. Therefore, to achieve an effective planning process, a thorough coordination and cooperation have to be carried out between the spatial planning office who is in charge of the preparation of zoning instruments and other relevant authorities.

24.9.3.2 Quantifying Exposure to the Hazard

The step of quantifying exposure to the hazard (i.e., at risk) includes the determination of many important components within a given area that is threatened by a certain hazard (e.g., SLR). These components can be population, buildings, civil engineering works, economic activities, public services, utilities and infrastructure, etc. In ideal situation, existing hazard and risk maps (with an appropriate spatial scale) provide an important source for the necessary information that support obtaining high quality results as shown in Fig. 24.20 which presents the proposed geographical locations of the expanding the city of Lattakia under consideration of some impacts of SLR and other related hazards. This will allow various planning authorities to make the comparability between the risk assessments that were carried out to this given area that exposed to the SLR. Similar to hazard identification in first step, the spatial planning authority requires the support of the sectoral planning divisions during this second step.

24.9.3.3 Assessing Vulnerability to the Exposure

The assessing vulnerability to the exposure step (i.e., vulnerability of a building near the coast with respect to the SLR) is based on risk analysis using human geography and construction engineering, etc. It involves the estimation of the degree of loss for a given suffered situation for the different scenarios based on the impacts of the SLR on the city of Lattakia as shown in Table 24.6 [e.g., resulting from the occurrence of a SLR of a given magnitude and expressed on a scale from (0 no damage) to (1 total damage)]. This scientific and deterministic step characterizes the risk analysis as a mathematical calculation that includes the analysis of a hazard and its consequences.

24.9.3.4 Hazard Characteristics

The characteristics of a hazard can assist in ensuring the most important information that identifies this hazard. This information has to be recorded and could include the following items: (1) *Causes of the hazard* which are often a combination of the hazard itself, human practice and governance issues (e.g., the causes of drought can be related to poor land-use policies, etc.). (2) *Intensity of the hazard* which includes strength, extent and duration that are related to seasonality, frequency and location (i.e., how the SLR is severe?). (3) *Frequency of the hazard* which is the period between the occurrences of hazards (e.g., SLR every two years). (4) *Location or the boundaries* of a place that affected by a hazard (e.g., extensively alongside shore up to 5 m contour line above river height in case of the SLR). (5) *History and trends of the hazard* which are particularly important to capture the effects of environmental degradation attributed to global weather changes. (6) *Controllability* which is the degree to control the impacts of hazards as many of them are outside of the human control or partially controllable (e.g., the rainfall is outside the human control). In some cases where the community cannot influence the frequency, intensity or scale of a hazard, this community needs to focus on strengthening its capacities to respond and recover from its impacts.

24.9.3.5 Estimating the Risk

The risk estimation step (i.e., probable loss) is based on several factors such as urban planning and human geography, economy, relevant statistics, individual parcels and property asset data, etc. Therefore, risk perceptions can be incorporated in norms, practices and probability calculations. With respect to an individual's perception of risk, there are many factors this perception such as familiarity with a risk, control over the risk or its consequences, proximity in space and time, scale of the risk or general fear of the unknown ("dread factor"). An important and interesting aspect of risk perception is the variation in different cultural regional and national contexts. When the risk has been determined, planners need to decide whether it is *within tolerable limits*. The following strategic tools for RA are not meant to be all-inclusive, but rather to present and explain a few of the techniques used to develop and assess risk at a community level.

24.9.4 Strategic Tools for Risk Assessment

This section discusses the importance of the strategic tools for RA especially dealing with the general aspects of gaps of infrastructure and knowledge necessary to identify and assess present day conditions. The shortage of long-term data and information on various aspects of disasters and hazards (e.g., climatic variations and its impacts) makes it very difficult to make decisions at early times. For example, missing data with respect to the SLR can includes: time series data concerning climatic parameters, data on tide gauges at a number of strategic positions indicating land subsidence, socioeconomic and health data in highly vulnerable areas, and accurate topographic data of the vulnerable low land areas. The following tools for risk assessment will support bridging these gaps.

Table 24.7 Relative risk ranking of a hazard

Relative Risk Ranking		*Likelihood* (How many individuals get effect from the SLR)		
		Unlikely (no effect)	Likely (some effects)	Very likely (many effects)
Severity (How effects do individuals get?)	Moderate	Lower	Lower	Medium
	Serious	Lower	Medium	Higher
	Severe	Medium	Higher	Higher

24.9.4.1 Hazard Ranking by Risk

Once all potential hazards have been identified, the priority should be to work on the highest ranked hazard first, and then the risks have to be controlled by the use of appropriate procedures or devices. In ranking the hazards, attention must be given to control methods which are already in place to mitigate the hazard (e.g., elimination, substitution, engineering controls and administrative controls, etc.). However, some tasks may have specific hazards that are beyond the scope or experience of local management (e.g., chemicals and radioactive materials, etc.). In these cases managers must seek appropriate expertise to help with the assessment and development of hazard control. It is possible to rank hazards on a simple graph that plots magnitude of impact with respect to an individual or community against the probability and frequency of a specific hazard occurring. Also, hazards can be ranked using tables as shown in Table 24.7.

24.9.4.2 Strength, Weakness, Opportunity and Threat Analysis

Strength, Weakness, Opportunity and Threat (SWOT) analysis of policies, measures and programs can be a useful tool in the initial phase of a RA to guide the community to capture and identify its overall areas of development. The benefits of this analysis are the identification of the links between each of the perceived "threats" which relates to the community's "weaknesses", the "weaknesses" to related "opportunities", and the "opportunities" to related "strengths". The items at which the most links converge indicate the priority threats to be mitigated, weaknesses to be corrected, opportunities to be seized, and strengths to be reinforced.

Identification of the SWOT analysis is essential because the subsequent steps in the planning processes might come as a result from this analysis. With regards to the impacts of SLR and other related hazards on the SCZ, a critical SWOT analysis has been presented as shown in Table 24.8, and concluded that: (A) most of the coastal cities are vulnerable to the impacts of SLR not only through direct inundation, but also due to salt water intrusion. (B) shortage of institutional system for climate change and SLR has limited proactive planning and development of policy to adapt to potential impacts which cover all sectors of development. (C) shortage

Table 24.8 An example of results from a SWOT analysis for a SLR disaster

Item	Strengths	Weakness	Opportunities	Threats
Flash flood and flood management	Local knowledge of water resources, flash and seasonal floods	Lack of people centered EWSs, and other information about DM	Well established water constructions, and flood management systems	Information about floods are not reaching the population in time
Institutional structure	GoS has developed some institutional structures for adaptation including integration among vulnerable sectors, A national committee for climate change has been formulated	GoS is still in the process of developing and implementing some regulations that have proved to be efficient such as EIA, ICZM	Integrating capacities and coordinating activities easier now since all sectors and research centres are encountering problems of climate changes and SLR	Time delays means loss of opportunities for adaptation and economic loss
Disaster management	Autonomy for the local governmental administration	Lack of integration of environmental conservation in local development plans and policies	Ability to carry out development plans and policies without macro management from national level	Local government show little interest in investing in environment and disaster risk reduction
Awareness of decision makers	Most of decision makers in Syria are well aware of the SLR and other hazards related Tsunamis	Awareness does not mean action of any kind and priorities of immediate needs are preferred	The early the GoS start working on managing these hazards the more they save lives and protect properties	More programs will be implemented with no consideration to climate change, and more damages over all sectors on the regional scale
Awareness of the community	Vulnerable communities are aware of the SLR and other hazards related Tsunamis in Syria	weak enforcement of regulations, and still missing of many practices in the community	Now increasing with community feeling heat waves and flash floods	Continuation of over consumption, unplanned urban development and interference of land use

systematic observation of coastal changes, (D) lack of integrated geographic data basis of indicators (e.g., land subsidence in particular), and (E) vulnerability to SLR and its implications on water resources, food security, tourism and public health for all coastal Syrian cities cannot be overlooked.

24.9.4.3 Environmental Impact Assessment

The SWOT analysis can be complemented by Environmental Impact Assessment (EIA) which is a policymaking tool that provides information on the environmental impacts of hazards. EIA could be used to assess risk in coastal areas facing SLR, or to assess the impact on food security. With regards to the current situation of EIA in Syria, the main shortcomings can be identified: the absence of binding EIA related legislation, weak environmental institutions and authorities, lack of awareness and experience, lack of environmental data, lack of coordination and cooperation, the weak role of private and public sectors in EIA.

24.9.4.4 Hazard and Risk Maps

Following the SWOT analysis, the community can benefit from developing hazard/risk maps that display the detailed geographical SCZ that might be negatively affected by SLR as shown in Figs. 24.13 and 24.14. These maps, that created using geo-information technology, can help to locate this hazard and provide its historical record, and to identify its risk that support decision makers to find solutions or take precautions. Mapping can be carried out of a single hazard (flood map), or it can take form as a multiple hazard map which combines all the present hazards in one map to give a composite picture of the situation for a given area. Multiple maps have the possibility of providing common recommendations for mitigation techniques, outlining sub-areas that require more information, specifying the needed hazard-reduction techniques, and land-use decisions benefiting all hazard considerations simultaneously (Saleh and Allaert 2011).

24.9.4.5 Cost-Benefit Analysis of Prevention and Mitigation Measures

One of the main questions when preparing the mitigation measures to control hazards is: *If the risk is not within "tolerable limits", what are costs of the various prevention and mitigation measures?* Cost-Benefit Analysis (CBA) is a necessity tool in establishing the feasibility of prevention and reduction measures. This section illustrates how CBA and RA can be incorporated into the urban planning process from the point of view of both prevention and reduction procedures. Scenarios of these procedures can be used to illustrate how RA is used as an input in a CBA, as well as how the results of this analysis are likely to be interpreted by decision-makers. For example, the World Bank and USA Geological Survey calculated that economic losses worldwide from natural disasters in the 1990s could be reduced by US$280 billion if US$40 billion were invested in preparedness, reduction and prevention strategies. For Lattakia city, it will be wise to establish high rock barriers against SLR as a coastal hazard and this requires GoS to plan for it, and to manage development planning activities to minimize loss of property due to SLR as shown in Fig. 24.24. In this situation, the estimation cost for developing these protective structures has to be fully considered as a wise investment because the potential loss of the important infrastructure and populated areas through a SLR would be so disastrous.

24.10 Maritime Activities and National Responses to Protect Syrian Marine Area

In the last five decades, the SCZ witnessed the development of large maritime facilities including the Banias oil terminal, the commercial sea port of Lattakia and the oil terminal of Tartous, etc. Hence, the problem of climate change is being taken seriously, and low lying land in the SCZ is considered to be at risk especially from the effects of SLR. In particular, cities of Lattakia, Banias, and Tartous, which are major industrial and economic centres, are expected to experience serious environmental impacts, if no action is taken. The GoS considers several actions in cooperation with global communities to protect the risked areas and to decrease the effects of the climate change by serious research work and setting new environmental regulations (Meslmani 2010).

24.10.1 Maritime Activities

With regards to the coastal governance, protection of marine areas, and maritime activities, the establishment of effective marine governance is a challenging prospect in Syria. A wide range of activities that compete for coastal and marine space are governed by a complex and uncoordinated array of legal and administrative arrangements. In order to try and achieve a more integrated approach to coastal management issues in Syria, governance models based on the concepts of ICZM have been developed (PAP/RAC 2008), which include: (a) policy should be developed in an integrated manner, rather than on a sector by sector basis. (b) marine and coastal areas should be treated as a single zone rather than as differently administered and managed separate units. (c) a long-term approach to coastal and marine governance is required because of the long timescales over which coastal and marine systems operate. (d) a governance system that allows all stakeholders to contribute to policy formulation will be more effective. It has also been recognised that achievement of these concepts

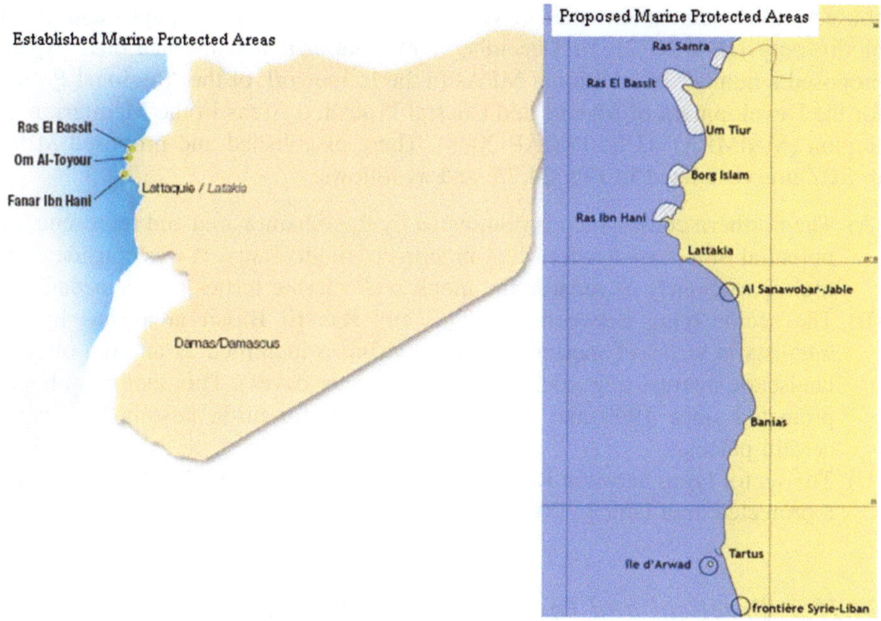

Fig. 24.23 The established and proposed marine protected areas—Syria. *Source* Policy Research Corporation based on the Network of Managers of Marine Protected Areas in the Mediterranean, www.medpan.org and RAC/SPA (2004), Regional Project for the Development of Marine and Coastal Protected Areas in the Mediterranean Region (MedMPA)—Recommendations for the elaboration of a national plan to develop marine protected areas in Syria

will need to involve the incorporation of spatial planning into the administrative process.

The GoS has recently approved new regulations to include ICZM into developmental plans needed for better management of coastal resources and protection. This makes it necessary to have a strong institutional monitoring capability in addition to a decision support capability for adoption of options for adaptation. With regards to the marine environment, Syria's continental shelf is naturally characterised by low bio-productivity due to the high salinity of the coastal water, relatively low freshwater inputs, a slow sea current and a low tidal level. Furthermore, a high degree of pollution of the coastal and marine environment puts pressure on the future of fisheries in the SCZ. The strongest impact of human activities on the marine environment is water pollution mainly caused by untreated wastewater discharges originating from the cities.

According to the network of managers of Marine Protected Areas (MPAs) in the Mediterranean (RAC/SPA 2004), three MPAs have been defined in Syria's territorial sea with a total size of about 50 km^2, and are: Fanar Ibn Hani, Om al Toyour, and Ras El Bassit as shown in Fig. 24.24. The Om al Toyour area is recognised as an Important Bird Area (IBA) which is a global partnership of conservation organisations that strives to conserve birds, their habitats and global biodiversity (IBA 2012). In December 2004, national and international experts proposed a number of additional MPAs in the framework of the 'Regional Project for the Development of Marine and Coastal Protected Areas in the Mediterranean Region (MedMPA)' (UNEP/MAP 2008). These established and proposed MPAs in SCZ are visualized in Fig. 24.23, and as follows:

(A) The northern coast which is dominated by Ras Shamra area and represents the potential of a cross-border MPA in terms of biodiversity. It was considered as high, particularly as regards the monk seal, marine turtles and cetaceans.
(B) The sector lying between Um Tiur and Ras El Bassit areas has limited interests in terms of marine biodiversity, but it identified as an area of great landscape interest (e.g. rocky cliffs, underwater caves). This sector is already protected since 1999 and there is a potential for turtle nesting beaches at certain places.
(C) The sector lying between Ras Ibn Hani and Borg Islam areas which is already a protected area (since 2000, 10 km^2) rich in marine biodiversity and turtles.

24.10.2 Practical and Institutional Adaptation Measures

For Syria to mitigate and adapt to the effects of SLR and other related hazards, the GoS will have to respond effectively to some urgent needs such as establishing a strong coastal monitoring, assessment and law enforcement system hence identifying and protecting vulnerable areas. In addition, there is an urgent need to activate ICZM committee and to incorporate climate change in the EIA, to promote awareness and community resilience, and to create new opportunities at safe areas. Another action taken by GoS is the 5 Years Development Plan prepared for coastal governorates considering several crucial suggestions for mitigating and adapting to the expected SLR. This plan encourages greater coordination between governorate bodies responsible for enforcing environmental policies, urban and developmental planning and regulations to improve the quality of coastal protection measures. It also introduces measures to reduce coastal zone erosion such as raising environmental awareness amongst governorate staff and private developers, and developing an action plan to prevent erosion and to protect Syrian coastline from further erosion (UNEP/MAP 2008).

A national adaptation strategy is in progress with consideration of the following aspects: (1) upgrading adaptive capacity through establishment of institutional system for monitoring, building data basis, modelling and upgrading awareness. (2) adopting a proactive policy in planning and enforcing regulations for follow up. (3) carrying out research on renewable energy, salt tolerant plants, and desalination. (4) considering geo-engineering activities for protection against SLR. Based on these aspects, it is therefore recommended to carry out: (a) establishing institutional capacity which includes monitoring systems and human capabilities for climate change in general and SLR impacts in particular. (b) enhancing

Fig. 24.24 The proposed high rock barriers against SLR on the coast of Lattakia

adaptive capacity through encouraging establishment of rainwater storage systems, upgrading water management and development of integrated coastal zone planning. (c) establishing monitoring and EWSs for coastal subsidence, flash flood, and SLR, etc. (d) establishing high rock barriers opposite to the density population areas, the sporting city, and other important tourism structures to protect them from SLR and reduce its impacts as shown in Fig. 24.24. These barriers can be built in an attractive shape to attract tourists or can be planted by trees. (e) planting a green built with 200 m width full of palm trees parallel to the coast, and this will separate this coast from agricultural area. This will reduce the initial effect of the tsunami and at the same it can be lung for fresh air that minimise the pollution. (f) it is suggested to establish a SLR monitoring network of 11 sea level gauges that will be positioned around the open coast and spanning the entire coast.

24.10.3 The Suggested Strategic Framework for Responding to Sea Level Rise in Syria

The urgent need for integrated adaptation infrastructure and institutional capability for monitoring, building data base, and periodic assessment and risk reduction in Syria is an important prerequisite for proactive planning and sustainable development. Therefore, a general strategic framework for responding to SLR and other hazards will be planned taking into accounts various national activities and recent experience of preparedness in the frame of DRR (UNEP/MAP 2008). The strategy which is based on the need for a tsunami and SLR warning system in the SCZ, should respond to pressure on and from coastal hazards by including: (1) building infrastructure and institutional capabilities for monitoring, modelling, vulnerability assessment and development of policies, measures and enforcement of regulations. (2) carrying out research on water availability and management, food security and salt tolerant plants, coastal extreme events and water conservation programs. (3) creating development plan and enforcement of SEA and EIA in it, and taking into account climate changes and SLR implications.

One of the main reasons to monitor the Lattakia coast and other coastal cities using advanced EWSs is the potential of earthquakes resultant from the Afro-Syrian fault as explained above. Once the earthquake is identified, the warning centres use sea level data to confirm that a tsunami was generated or, if there are no changes in sea level, to cancel the alert messages. There are three main sources/causes of tsunamis: submarine earthquakes, landslides, and volcanic eruptions (Annunziato et al. 2009) and (Dawson et al. 2004). The main elements of EWS can include the following: (1) tsunami hazard and RA, (2) seismic monitoring and earthquake detection, (3) sea level monitoring and tsunami detection, (4) dissemination of warnings and mitigation programs and public awareness. In this system, tsunami hazard assessment is a key element and requires knowledge of past tsunami occurrences and possible sources, their likelihood of occurrence and their effects along the threatened coasts. In addition, the compilation of tsunami catalogs and inundation mapping constitute the main components of tsunami hazard assessment (Saleh and Allaert 2009a; ISDR 2004).

24.11 Conclusion

It is almost impossible to prevent the occurrence of SLR and other related hazards. However, it is possible to reduce their impacts and damages by adopting suitable disaster mitigation strategies. RA and management should be incorporated within the planning activities process to achieve greater sustainability and at least resiliency of society's development. DM is a dynamic process and consists of different kinds of knowledge: technical (e.g., modelling of phenomena, simulation, data management and telecommunication), human aspect (e.g., modelling of behaviour, training and learning), and organizational and managerial aspect (e.g., planning, cooperation of stakeholders, information and communication, etc.). However, some conclusions about these aspects can be outlined as follows: technology-oriented decision support has drawbacks (e.g., modelling and simulation are essential, but there are limits, telecommunications often fail in crisis situations, GIS usefulness depends on updating, and sensors provide field data, but their reliability is limited, etc.). On the other hands, the human and organizational aspects are essential and can play important role: to face the growing complexity of situations, to cope with unplanned situations and crisis, to establish a synergy among the many stakeholders, and to learn from experience and share knowledge. In the close future, the most profound changes affecting SCZ will be driven by climate change, particularly by rising sea level. Hundreds of thousands of people are likely to be displaced by SLR, accompanying economic and ecological damage will be severe due to recurrent droughts, inequitable land distribution, and over dependence on rain-fed agriculture. Many major challenges will have to be successfully and equitably confronted by proper management and adaptation to the consequences of climate change and the competing human demands on coastal land. Adaptation plans are mainly focusing on increasing the adaptive capacity of the different systems, by changes in processes, practices, or structures to reduce risk of these disasters. In Syria, the priority of these plans is the high vulnerable systems to climate change from the perspectives of food production, rural population stabilization, and distribution of water resources. Therefore, the high vulnerability of the agricultural sector put it on top of the priority list of adaptation plans. In addition, the Syrian adaptation capacity is challenged as it comes in conjunction with high development pressure, increasing populations, water management that is already regulating most of available water resources, and agricultural systems that are often not adapted to local conditions.

This chapter summarises that most of the Syrian coastal cities are particularly vulnerable to potential impacts of SLR. It ends with a set of analytical conclusions and adaptation recommendations: (1) almost none of these cities have established an effective institutional capability for adaptation with particular emphasis on monitoring capabilities. (2) they should carry out massive programs for detailed vulnerability assessment, investigate open options for adaptation and develop strategies, policies and measures in all sectors of development. (3) they need to build up resilience for vulnerable communities and carry out proactive planning for ICZM and development in non-vulnerable coastal areas. (4) Regardless of the expected magnitude of the SLR, taking action for adaptation is necessary whether there is a SLR or not. (5) A multidisciplinary research project based on a national proposal for maximum size and disaster scenarios for extreme SLR in the eastern coast of Med. Sea has to be carried out as soon as possible. Finally, since DM is only one of many issues that decision-makers face, efforts must be made to raise awareness of the benefits of disaster mitigation.

References

Alexander, D.: Principles of Emergency Planning and Management. Oxford University Press, New York (2002)

Ambraseys, N., Barazangi, M.: The 1759 earthquake in the Beka Valley: Implications for earthquake hazard assessment in the Eastern Mediterranean region. J. Geophys. Res. **94**, 4007–4103 (1989)

Annunziato, A., Carrilho, F., Matias, L.M., Baptista, M.A., Omira, R.: Progress in the establishment of the Portuguese Tsunami Warning System. EMSC (European Mediterranean Seismological Centre) Newsletter. April 2009

Bell, G., Goring, G., Walters, A.: Advances in understanding sea-level variability around New Zealand. In: coastal engineering, Billy edge (ed.), 2, 1490–1500, Proceedings of 27th International conference on coastal engineering, Sydney, American Society of Civil Engineers, NY. http://www.mfe.govt.nz/publications/climate/effect-coastal-sep01/index.html (2001)

Berz, G.: Windstorm and storm surges in Europe: loss trends and possible count-actions from the viewpoint of an international reinsurer. Phil. Trans. R. Soc. A **363**, 1431–1440 (2005)

Blaikie, P., Cannon, T., Davis, I., Wisner, B.: Natural Hazards, people's vulnerability and disasters, 1st edn. Routledge, London (1994)

Boko, M., Niang, I., Nyong, A., Vogel, C., Githeko, A., Medany, M., Osman-Elasha, B., Tabo, R., Yanda, P.: Africa Climate Change 2007: Impacts, Adaptation and Vulnerability. Contribution of Working Group II to the 4th Assessment Report of the Intergovernmental Panel on Climate Change (2007)

Burby, R.J.: Hurricane Katrina and the paradoxes of government disaster policy: Bringing about wise governmental decisions for hazardous areas. Ann. Am. Acad. Polit. Soc. Sci. **604**(1), 171–191 (2006)

Burton, I., Smith, J.B., Lenhart, S.: Adaptation to climate change: theory and assessment. In: Handbook on Methods for Climate Change Impact Assessment and Adaptation Strategies [Feenstra, J.F., I. Burton, J.B. Smith, and R.S.J. Tol (eds.)]. UNEP and Institute for Environmental Studies, Free University of Amsterdam, pp. 5.1–5.20 (1998)

Christopher, D.G., Harley, A., Hughes, R., Hultgren, K.M., Miner, B.G., Sorte, C.J.B., Thornber, C.S., Rodriguez, L.F., Tomanek, L., Williams, S.L.: The impacts of climate change in coastal marine systems. Ecol. Lett. **9**, 228–241 (2006)

Crichton, D.: The risk triangle. In: Ingleton, Jon (ed.) Natural disaster management. Tudor Rose, London (1999)

Cuny, F.: Disaster and development. Oxford University Press, Oxford (1983)

Dawson, G., Lockett, P., Shi, S.: Tsunami hazards in Europe. Environ. Int. **30**, 577–585 (2004)

EEA-JRC-WHO: Impacts of Europe's changing climate—2008 indicator-based assessment, JRC Reference Report No. JRC47756, Luxembourg: Office for Official Publications of the European Communities (2008)

Ellis, J.I., Norkko, A., Thrush, S.F.: Broad-scale disturbance of intertidal and shallow sublittoral soft-sediment habitats: Effects on the benthic macrofauna. J. Aquat. Ecosyst. Health Recovery **7**, 57–74 (2000)

EM-DAT: The International Disaster Database, Annual Disaster Statistical Review 2009. Center of Research on the Epidemiology of Disasters (CRED). http://www.emdat.be/database (2010)

EPSON: The Spatial Effects and Management of Natural and Technological Hazards in Europe. Report to Directorate General Environment, European Commission (2006)

EUROSION: Living with coastal erosion in Europe: sediment and space for sustainability. Report to Directorate General Environment, European Commission (2004)

Fanos, A.M.: The impact of human activities on the erosion and accretion of the Nile Delta coast. J. Coastal Res. **11**, 821–833 (1995)

Fish, M.R., Cote, I.M., Gill, J.A., Jones, A.P., Renshoff, S., Watkinson, A.R.: Predicting the impact of sea-level rise on Caribbean sea turtle nesting habitat. Conserv. Biol. **19**, 482–491 (2005)

Food and Agriculture Organisation of the United Nations (FAO): Fishery Country Profile—Syrian Arab Republic; www.fao.org (2007)

Fukumori, I., Lee, T., Tang, B., and Menemenlis, D.: A basin-oscillation of the Mediterranean Sea. Ocean Sci. Meet. Suppl., EOS Trans. **84**, 52 (2003)

Gardoni, P., Murphy, C.: Recovery from natural and man-made disasters as capabilities restoration and enhancement. Int. J. Sus. Dev. Plann. **3**(4), 317–333 (2008)

Gerassimos, A., Papadopoulos, G.A., Anna, F.: Strong Tsunamis in the Mediterranean sea: A Re-evaluation. ISET J. Earthquake Technol. **42**(4), 159–170 (2005) (Paper No. 463)

Gomez, F., Meghrauoi, M., Darkal, A.N., Sbeinati, R. Darawcheh, R., Tabet, C., Khawlie, M., Charabe, M., Khair, K., Barazangi, M.: Coseismic displacements along the Serghaya fault: An active branch of the Dead Sea fault system. J. Geol. Soc. London **158,** 405–408 (2001)

Gornitz, V.M., Daniels, R.C., White, T.W., Birdwell, K.R.: The development of a coastal risk assessment database: Vulnerability to sea-level rise in the U.S. southeast. J. Coastal Res. **12**, 327–338 (1994). (Special Issue)

Gulev, S.K., Hasse, L.: Changes of wind waves in the North Atlantic over the last 30 years. Int. J. Climatol. **19**, 1091–1117 (1999)

IBA: Important Bird Area. Report of the Syrian Arab Republic on the implementation of the Agreement with IBA in the period 2006–2007. Directorate of Biodiversity and Protected Areas, The State Ministry for Environmental Affairs, Damascus, Syria. http://www.birdlife.org/ (2012)

IDNDR: The International Day was to be observed annually during the International Decade for Natural Disaster Reduction, 1990–99 (resolution 56/195) (1999)

IPCC: Climate change 2001: The scientific basis. Summary for Policymakers and Technical Summary of the Working Group I Report contribution to the 3th Assessment Report of the Intergovernmental Panel on Climate Change (IPCC). Cambridge University Press, UK. (http://www.ipcc.ch) (2001a)

IPCC: Climate change 2001: Impacts, adaptation, and vulnerability, contribution of working group II to 3th assessment report of the Intergovernmental Panel on Climate Change (IPCC). Cambridge University Press, UK. (http://www.ipcc.ch) (2001b)

IPCC: Climate Change 2001: Mitigationm, contribution of working group III to 3th assessment report of the Intergovernmental Panel on Climate Change (IPCC). Cambridge University Press, UK. (http://www.ipcc.ch) (2001c)

IPCC: Climate Change 2007: The physical science basis. Contribution of working group I to 4th assessment report of the intergovernmental panel on climate change. Cambridge University Press, Cambridge; New York (2007)

ISDR: National platforms and country profiles for "disaster risk reduction" 1994/2004. http://www.unisdr.org/eng/country-inform/introduction.htm (2004)

ISDR: Regional workshop: Program for a Tsunami early warning system in Central America. http://www.eird.org/eng/revista/n11_2005/art18.htm (2005)

Jeftic, L., J.D., Milliman, G., Sestini, (eds): Climatic Change and the Mediterranean: environmental and societal impacts of climate change and sea-level rise in the Mediterranean region. Volume I (1992)

Komar, P.D.: Beach processes and sedimentation, 2nd edn. Prentice Hall, NJ (1998)

Leatherman, S.P.: Social and economic costs of sea level rise. In: Douglas, B.C., Kearney, M.S., Leatherman, S.P. (eds.) Sea level rise—history and consequences, pp. 181–223. San Diego, USA: Academic Press (2001)

Manca, B.B., Budillon, G., Scarazzato, P., Ursella, L.: Evolution of dynamics in the Eastern Mediterranean affecting water mass structures and properties in the Ionian and Adriatic seas. J. Geophys. Res. **108**(C9), 8102 (2003)

Meslmani, Y.: Review of Air Quality in Syria: Syrian's Initial National Communication to the United Nations Framework Convention on Climate Change (UNFCCC). The Ministry of State for Environment Affairs and the United Nation Development Program (Damascus-Syria). www.inc-sy.org (2010)

Möller, I., Spencer, T.: Wave dissipation over macro-tidal salt marshes: Effects of marsh edge typology and vegetation change. J. Coast. Res. SI36: 506–521 (2002)

Murphy, C., Gardoni, P.: Determining public policy and resource allocation priorities for mitigating natural hazards: a capabilities-based approach. Sci. Eng. Ethics **13**(4), 489–504 (2007)

Nicholls, R.J., Tol, R.S.J.: Impacts and responses to sea-level rise: a global analysis of the SRES scenarios over the twenty-first century. Philos. Trans. Royal Soc. A. **364**, 1073–1095 (2006)

Oliver-Smith, A.: Theorizing vulnerability in a globalized world: A political ecological perspective. In: Bankoff, G., Frerks, G., Hilhorst, D. (eds.) Mapping vulnerability: Disasters, pp. 10–24. Development and People, Earthscan (2004)

Oliver-Smith, A.: Problems in post disaster resettlement: Cross cultural perspectives, in: Y. Aysan, I. Davis (eds.) Disasters and the small dwelling: Perspectives for the UN IDNDR, James & James, London, pp. 58–66 (1992)

Oliver-Smith, A.: Sea Level Rise and the Vulnerability of Coastal Peoples: Responding to the Local Challenges of Global Climate Change in the 21st Century, 'Interdisciplinary Security Connections' Publication Series of UNU-EHS, No. 7/2009 (2009)

Palmer, A.: Rising sea levels, The World Bank report. http://www.theworldincrisis.com/artman2/publish/climate/Rising_Sea_Levels.shtml (2008). Accessed 12 July 2011

PAP/RAC: Report on the national workshop to promote Syria's ICZM policy brief; the Coastal Management Centre. www.pap-thecoastcentre.org (2008)

Papadopoulos, G.A.: Tsunami hazard in the Eastern Mediterranean: Strong earthquakes and tsunamis in the Corinth Gulf, Central Greece. Nat. Hazards **29**(3), 437–464 (2003)

Pearce, L.: Disaster management and community planning, and public participation: How to achieve sustainable hazard mitigation. Nat. Hazards **28**, 211–228 (2003)

Pilkey, O.H., Hume, T.M.: The shoreline erosion problem: lessons from the past. NIWA Water Atmos. **9**(2), 22–23 (2001)

RAC/SPA: Regional project for the development of marine and coastal protected areas in the Mediterranean Region (MedMPA)—Recommendations for the elaboration of a national plan to develop marine protected areas in Syria. www.medpan.org (2004)

Saleh, H., Allaert, G.: Scientific research based optimisation and geo-information technologies for integrating environmental planning in disaster management. In: Tang, D., Singh, R. (eds.) Remote sensing of the changing oceans: natural hazards. Springer, Berlin (2011)

Saleh, H., Allaert, G.: Mitigating urban flood disasters in Syria: A case study of the massive Zeyzoun dam collapse. In the proceedings of the workshop on safe water services in post-conflict and post-disaster contexts, World Water Week, Stockholm, Sweden 16th–22th Aug 2009a

Saleh, H., Allaert, G.: Optimal decision making in disaster management and risk reduction using early warning system, Communications, and Geo-informatics. In the International Symposium on disaster management, Riyadh, Saudi Arabia, 3–7 Oct 2009b

Schmidt-Thomé, P., Kallio, H.: Natural and technological hazard maps of Europe. In: Schmidt-Thomé, P. (ed.) Natural and technological hazards and risks in European regions. Geol. Surv. Finl. Spec. Paper **42**, 17–63 (2006)

Shaw, J., Taylor, R.B., Forbes, D.L., Ruz, M.H., Solomon, S.: Sensitivity of the Canadian coast to SLR. Geol. Surv. Canada Bull. **505**, 114 (1998)

Smith, K.: Environmental hazards-assessing risk and reducing disaster, Routledge, 3rd edn. pp. 381 (2000)

Synolakis, C.E.: Tsunami and Seiche. In Earthquake Engineering Handbook, In: Chen W. F., Scawthorn, C. (eds.). CRC Press, US, pp. 9:1–9:90 (2003)

Syrian Arab Republic: National progress report on the implementation of the Hyogo Framework for Action, SYR-HFA: http://www.preventionweb.net/english/hyogo/progress/reports/v.php?id=17404&pid:223//http://preventionweb.net/go/17404 (2009–2011). Accessed July 10 2011

Syrian Planning Commission SPC: Analysis of the macroeconomical status Quo, 2003–2006 (2006)

Taylor, J.A., Murdock, A.P., Pontee, N.I.: A macroscale analysis of coastal steepening around the coast of England and Wales. Geogr. J. **170**, 179–188 (2004)

UNEP/MAP: Towards a strategy for Syria's Coast—ICZM policy background. METAP/SMAP III report (2008)

Van Westen, C.1., Terlien, M.T.1.: An approach towards deterministic landslide hazard analysis in GIS. A case study from Manizales (Colombia). Earth Surf. Land. **21**, 853–868 (1995)

Wisner, B., Blaikie, P., Cannon, T., Davis, I.: At risk natural hazards, people's vulnerability and disasters. 2nd ed. Routledge, London, New York (2004)

Yáñez, I.: Cambio Climático en el Mediterráneo Español, (Climate Change in the Spanish Mediterranean). Spanish Oceanography Institute (IEO), tells SINC. (2010). ISBN: 978-84-95877-48-2

Chapter 25
Using Geology as a Tool for Assessing Coastal Risk in Asia

Fengling Yu and Adam D. Switzer

Abstract In many places in Asia, the written record of typhoons (tropical cyclones) and tsunamis is too short or inconsistent to accurately assess the risk of coastal disasters from both kinds of event. Here we propose that sedimentary deposits left by overwash events attributed to past typhoons or tsunamis can be used to extend the record of overwash, and generate an analysis of recurrence interval, thus, improving long-term risk assessment. The two primary factors in assessing coastal hazard risk are frequency and magnitude. These can be addressed through field and modeling studies of deposits where they are found. Recent advances in identification of overwash deposits along with progress on attempts to distinguish between storm and tsunami sedimentation, have increased the utility of using overwash deposits to improve coastal risk assessment.

Keywords Tropical cyclones · Tsunami · Coastal hazards · Nuclear plants · Industrial parks · Storm surges · Risk · Mega cities

25.1 Introduction

Assessing coastal vulnerability (risk) has emerged as a key concept for understanding the impacts of climate change and natural hazards. Such assessments are essential for developing adequate risk management strategies. In this chapter we will consider vulnerability in terms of coastal hazards, with the basic premise that the coastal vulnerability relates to the susceptibility of both the natural system and

F. Yu · A. D. Switzer
Earth Observatory of Singapore, Nanyang Technological University, Singapore 639798, Singapore

A. D. Switzer (✉)
Division of Earth Sciences, School of Mathematical and Physical Sciences, Nanyang Technological University, Singapore 639798, Singapore
e-mail: aswitzer@ntu.edu.sg

D. L. Tang and G. J. Sui (eds.), *Typhoon Impact and Crisis Management*, Advances in Natural and Technological Hazards Research 40,
DOI: 10.1007/978-3-642-40695-9_25,

coastal societies towards coastal hazards. The vulnerability of a coastal community can be regarded as the sum of its social, economic and physiographic properties, and is a function of the natural and social coping and adaptive capacity to adverse impacts, i.e. a society's resilience. Assessing coastal vulnerability is an important prerequisite to determine "where", "why" and "how" questions related to inundation risk, e.g., where to place Nuclear Power plants (Fig. 25.1). Only when armed with such knowledge, government agencies could take steps to prepare for and reduce the risk of coastal disasters.

While the cyclone and tsunami warning systems allow us to monitor and predict the impact of modern events, it would be dangerous to ignore the history of a coastal site, or the prehistoric record preserved in the geological record. Additionally, whilst the occurrence of typhoons is seasonal, and hence reasonably predictable, the question must always be asked 'what is the biggest typhoon to have ever affected on this coast?' It is also apparent that on many coasts a similar question can and should then be asked for a tsunami. At almost all locations on the planet the only way to adequately answer this question is to turn to the geological record. This is even the case in places like China where the long history that exists is unfortunately fraught with inconsistencies that make recurrence interval analysis difficult (Lau et al. 2010). In this chapter we outline the current practices used in the investigation of geological records of past typhoon events, and outline how despite some inherent limitation such studies can be useful for coastal hazard

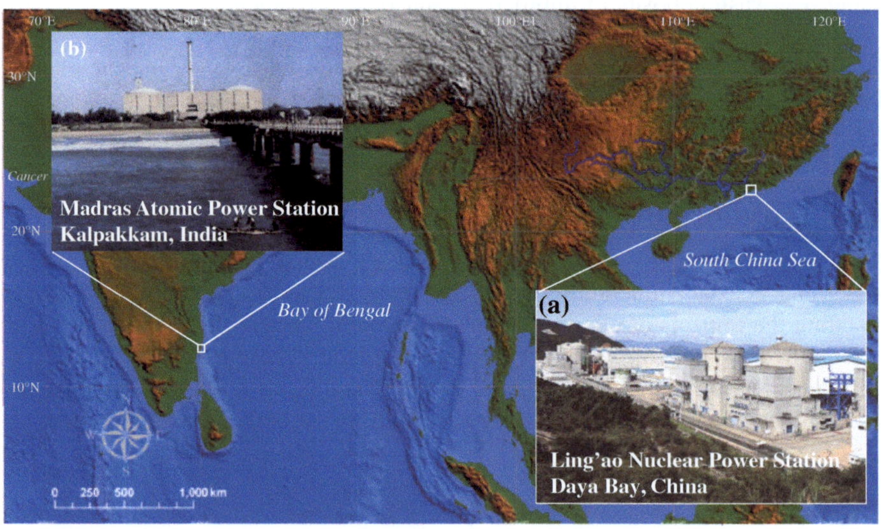

Fig. 25.1 Two of many coastal nuclear power plants in Asia. **a** The Ling'ao nuclear power plant in Daya bay, near Shenzhen, southern China; and **b** the Madras Atomic Power Station, Kalpakkam, near Chennai, India. As many Asian countries have frantically built coastal nuclear facilities in recent decades, it is apparent that in many cases they have made little use of science to determine whether these areas are safe. More than 30 plants in operation or under construction in Asia are at risk of one day being hit by a very large (Category 5) typhoon and/or tsunami

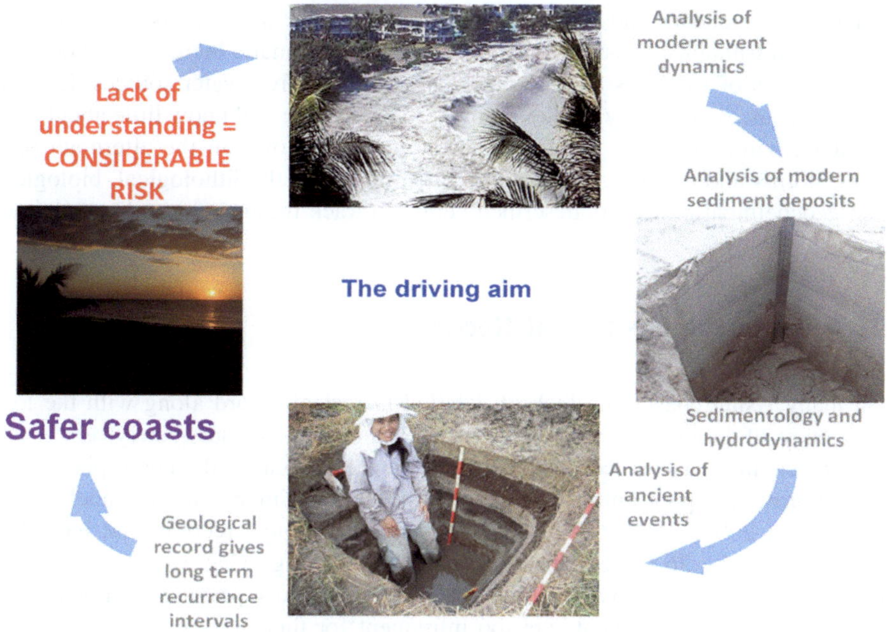

Fig. 25.2 Summary diagram of the process based approach to reconstructing coastal hazard recurrence intervals using geological record or data (in photo: Kruawun Jankaew)

assessment. Like most geological studies, the investigations of the past are driven by observations and measurements from modern events. Figure 25.2 provides a summary schematic of the basic premise of using coastal stratigraphy.

25.2 Reviewing Historical Documents

There is an immediate and obvious need to assess the risk of catastrophic inundation events (from typhoons and tsunami) along many Asian coasts. The first step in this exercise is usually to look at historical records (e.g., Liu et al. 2001; Louie et al. 2003; Lau et al. 2010; Lee et al. 2012). On some coasts this is a relatively simple exercise as the written historical record is not very long (e.g., Northern Australia). On other coasts, for example China or coasts bordering the Mediterranean Sea, the historical records are much longer and more detailed. To adequately assess the historical records of catastrophic events, a database is often compiled. The material is assessed and cross-checked to remove errors and misinterpretations, and to refine early work and ensure a more robust dataset (e.g., Liu et al. 2001, Lau et al. 2010; Lee et al. 2012). Unfortunately, as researchers step back further in time, those studying the historical record are often frustrated by inconsistencies in descriptions, inaccuracies in translation between different languages, calendars and location names. As a result of overwash events from storms and tsunamis, coastal areas may experience dramatic increases in flooding, erosion, loss of wetlands and seawater intrusion into freshwater sources. Despite recent advances and many studies on modern events, we still know little about how coastal environments are disrupted, and how they recover in the aftermath of a large storms and tsunami. It should be possible to use the lithological, biological and geochemical indicators described herein to track these changes through time.

25.3 Using the Geological Record

On many Asian coasts, the lack of detailed historical record along with the low frequency of large overwash events makes it difficult to adequately assess the recurrence intervals of coastal hazards. The geological and geomorphological record provides opportunities to assess coastal hazards more fully (Donnely et al. 2001; Nott 2004; Switzer 2008a). Analysis of geological records that indicated precursors to the 26 December 2004 Indian Ocean (e.g., Choowong et al. 2008) and 11 March 2011 Tōhoku (Minora et al. 2001) tsunamis, poignantly showed that catastrophic overwash events are too infrequent for their hazard to be characterized by historical records alone.

25.3.1 Assessing the Recurrence of Overwash Events

Research into the geological record of palaeo-washover deposits has led to the discovery of long records spanning many thousands of years in numerous locations around the world (e.g., Liu et al. 2000; Woodruff et al. 2009). One notable example is the study of Nanayama et al. (2003) on the coast of Hokkaido, Japan. Here, tsunami deposited sand sheets (some extending several kilometers inland), show that large tsunamis inundated the coast on average every 500 years, between 2,000 and 7,000 years ago. Such records provided clear evidence of large-scale palaeo-tsunamis in the region, and information from these palaeo-events could have been used to help guide management and planning procedures for the unfortunate events that were to occur on 11 March 2011.

The sand and boulder deposits left by storms and tsunamis overwash can help to assess a variety of environmental parameters about a particular storm or tsunami event (Fig. 25.3) (Paris et al. 2007, 2011; Switzer and Burston 2010; Terry and Etienne 2011). In low energy coastal environments such as coastal marshes or lakes, overwash by typhoons and tsunami can leave behind sandy deposits (Fig. 25.4). Such deposits can yield information that allows the investigation of water depths and velocities of past inundations; estimate source locations (of tsunamigenic events); and the mapping of the likely inundation distance of future events, to name but a few (Jaffe and Gelfenbuam 2007; Okal et al. 2010).

Fig. 25.3 Using modern process studies, geomorphologists (boulders) and geologists (sand deposits) are working toward using geology to differentiate storms and tsunamis. Only recently has this information started to be integrated into coastal planning, disaster management including the establishment of hazard zones, and evacuation procedures

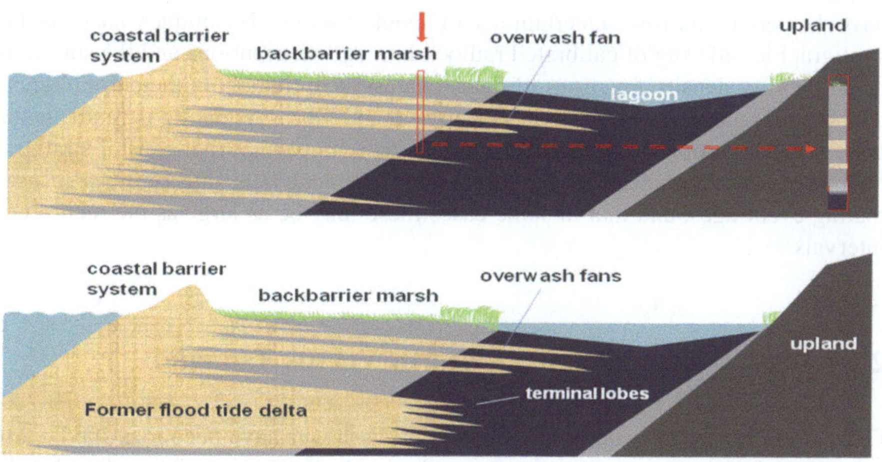

Fig. 25.4 Sand layers found in low energy settings such as coastal lagoons and marshes can yield recurrence intervals through sedimentary analysis and detailed dating. Unfortunately such work is a labourious and expensive process that requires extensive fieldwork, laboratory study and financial support. Such limitations are particularly relevant in low GDP nations (after Jeff Donnelly, personal communication, 2010)

To date, the majority of work on palaeo-washover deposits has been conducted in sub-tropical and temperate latitudes where there are extensive salt marshes and coastal lagoons. The Indian Ocean tsunami (IOT) of 2004 and Cyclone Nargis in 2008 (e.g., Fritz 2009) indicated a real need for research on palaeo-washover in tropical environments. Unfortunately such environments come with a myriad of challenges including: poor deposit preservation, intensive bioturbation, seasonal flooding and limited or difficult access. Many low GDP countries also have few resources for such studies (i.e., Bangladesh, Myanmar, and Indonesia) and considerable socio-economic issues.

25.3.2 Dating

Despite the challenges of working in coastal landforms, if long-term washover records are found, they can provide an opportunity to evaluate the recurrence of storms and tsunamis large enough to leave lasting sedimentary signatures (Woodruff et al. 2008). Determining such recurrence intervals is critical for evaluating the long-term behaviour of coastal settings and assist in determining the probability of future events. Dating deposits using techniques such as radiocarbon and optically stimulated luminescence (OSL) can develop detailed chronological histories that allow the analysis of the site specific or regional recurrence intervals of past events. Recently new analytical approaches to the treatment of age data have helped to narrow uncertainties of event timing. Techniques such as the stratigraphic ordering of calibrated radiocarbon age distributions and the summing of probability density functions of dates can assist in tightening the chronological controls on palaeowashover studies (Kelsey et al. 2005). As dating is fundamental to the determination of recurrence intervals, it is important that coastal stratigraphers face the limitations of the geological and analytical uncertainties in estimating event ages, and that in some cases these may be as large as the recurrence intervals.

25.4 Distinguishing Between Storm Versus Tsunami

Once stratigraphic or geomorphic records of washover have been identified, earth scientists often have to try to determine the likely mechanism. This often breaks down to the question of "is the deposit a result of a storm(s) or tsunami(s)?'' Although the mere presence of washover sand sheets provides a primary source of information for coastal planning it is often necessary to attempt to distinguish between tsunami and storm deposits in the geological record.

There is no single analytical technique that can unambiguously distinguish between storms and tsunamis as the depositional mechanism of palaeo-washover deposits (from other processes) (Switzer 2005; Switzer and Jones 2008b). In all palaeo-washover studies, local geomorphic and stratigraphic field criteria must be applied first. What can be compiled is a list of commonalities where combining field observations and the analysis of geomorphic, sedimentological, geochemical, and palaeontological signatures may enable positive identification. In boulder studies the results are often inconclusive (e.g., Switzer and Burston 2010; Nandasena et al. 2011). In contrast, comparative stratigraphic studies of both types of deposit reveal some differences in sedimentology, stratigraphy, faunal composition and inland height and extent (Fig. 25.5) (Goff et al. 2001; Phantuwongraj and Choowong 2011).

Generally, overwash events that deposit sand cause a decrease in the total organic matter within coastal plain sediments. The internal sedimentology (e.g., bedding), the regional physiography (e.g., height above sea level) and the three dimensional distribution (e.g., landward extent and taper and regional continuity), can also assist in determining the origin of a washover sandsheet. The existence of marine fossils or exotic (deep water) heavy mineral assemblages within a terrestrial environment may indicate transport via a tsunami as opposed to a storm (Fig. 25.6). Additionally, salinity changes caused by short-lived marine inundation events can cause notable changes in the assemblages of ostracods, diatoms, foraminifera, pollen and aquatic plants. Although several researchers have proposed criteria to distinguish these two types of deposits (Fig. 25.5), it is apparent that each deposit must be carefully considered in the context of its regional setting (Switzer and Jones 2008b).

Despite these limitations, storm or tsunami characteristics in some locations have been successfully reconstructed from geological evidence. For example, Woodruff et al. (2009) successfully reconstructed the mid-Holocene periodicity of typhoons in southern Japan providing a commentary on mid-Holocene storminess in the Western Pacific. In northwest Australia, Nott 2011 reconstructed the landfall of large tropical cyclones at Shark Bay in northern Australia. Using beach ridges, he reconstructed a 6,000 year record of large cyclone events in the area. In tsunami studies, Bondevik et al. (2005) successfully compared extensive data on the distribution and elevation of palaeo-tsunami deposits in Norway, Scotland, and the Shetland Islands with numerical simulations of the 8,000-year-old Storegga submarine slide and tsunami in the North Sea. Likewise, similar work by Atwater et al. (2005) used tsunami deposits from Japan and North America along with modelled tsunami heights that were inferred from written historical records in Japan to suggest that a M9 megathrust earthquake off northwest North America occurred on the 26 of January 1700.

25.5 Integrating Hazard Studies with Coastal Planning

While palaeo-washover deposits provide a mechanism for extending the historical record of storms and tsunamis, and for studying overwash events for which there are no written or oral records, transferring this information to well-designed

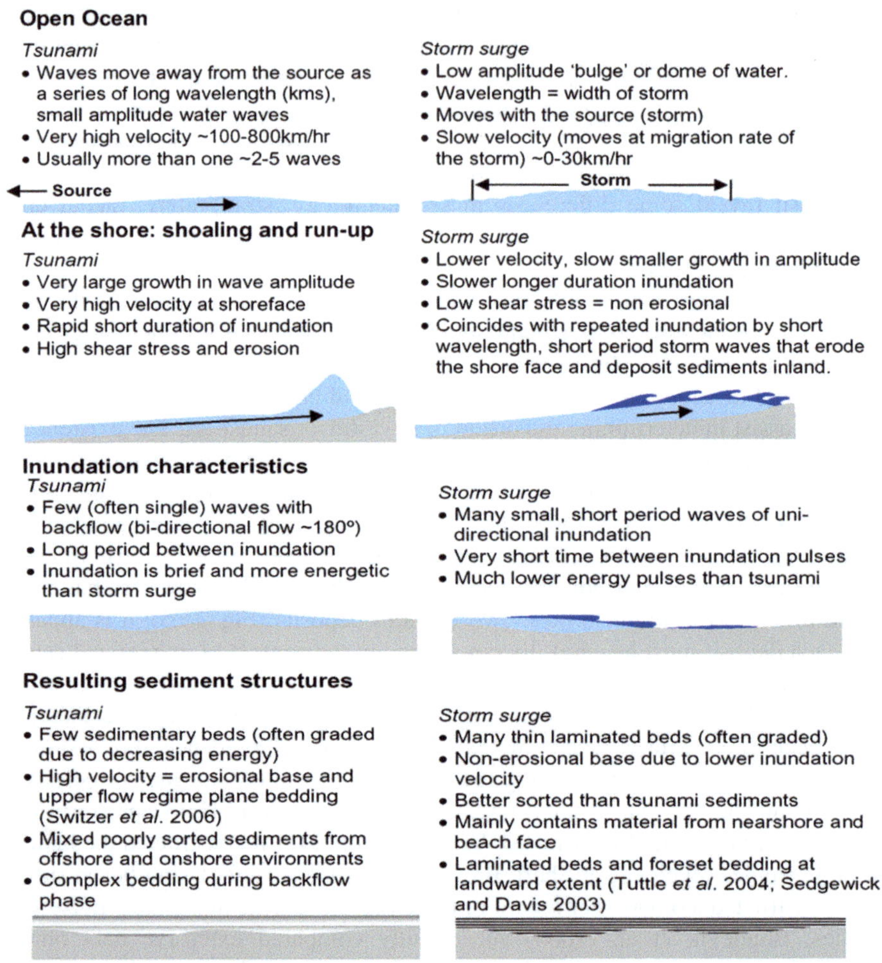

Fig. 25.5 Summary diagram of tsunami and storm surge characteristics: in the ocean, at the coast, during run-up and inundation, along with a basic description of the resulting sediments. The generation, propagation and run up of tsunami and storm surge are very different. In the open ocean the wavelength and amplitude of both events are similar. At the coast the differences in velocity, wave set-up and steepness, period and water volume mean tsunami and storm surge are differentiated in terms of velocity, period and repetition of inundation, likely bed shear stress and back flow characteristics. An additional special feature that differentiates tsunami from storm surge is the bidirectional flow of tsunami waves. Note that all washover deposits require adequate sediment supply and any deposit will reflect the supply and sedimentology of its source sediments (after Switzer and Jones 2008b)

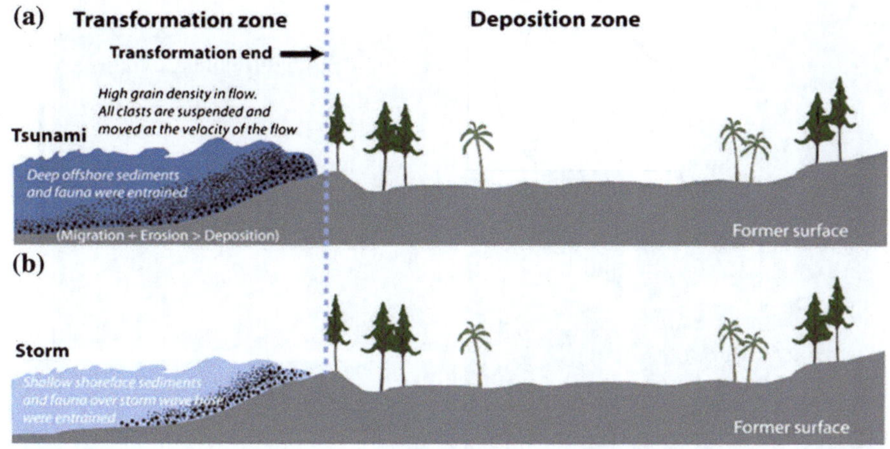

Fig. 25.6 Tsunamis (**a**) generally erode sediment from further offshore than storms. (**b**) Careful analysis of faunal and mineral content in overwash deposits may allow the differentiation between storm and tsunami deposition (after Phantuwongraj and Choowong 2011)

coastal planning is often problematic. In addition to providing information on the occurrence and frequency of coastal hazards in the stratigraphic record, detailed analysis of the facies, thickness, grain size changes, faunal content and chronological framework within the deposits has the potential to provide information on the dynamics and recurrence interval of inundation (a valuable tool for hazard mapping and evacuation planning). The investigation of washover deposits is not necessarily straightforward and there remains no 'recipe' for distinguishing between storms and tsunamis in the geological record (Switzer et al. 2011). The site-by-site approach to these studies and the fact that such work requires more interpretation than does the historic record, often means the true 'hazard' message may be lost in the scientific arguments of whether a deposit is a tsunami deposit or a storm deposit (Fig. 25.7).

The careful examination of washover deposits can yield valuable information to coastal planners and disaster management agencies. For example, it is tragically apparent that although geological evidence for previous large tsunami near Tōhoku had been identified in the past, the science didn't convey well to the coastal management of what was once a beautiful city (Fig. 25.8). Similarly if the coastal overwash stratigraphy of various locations around the northeast Indian Ocean (Thailand, Sri Lanka, India) had been carefully studied before 26 December 2004, signs of giant prehistoric tsunamis similar to the 2004 event could have been discovered (e.g., Jankaew et al. 2008).

Unfortunately, in many instances the geological and geomorphological records of large overwash events are deemed interesting scientifically, but fail to transfer into meaningful action by local and regional governments. It is only when this

Fig. 25.7 Excavated face (**a**) showing a high-energy marine deposit (Unit 2) confined between a sandy beach deposit (Unit 1) and an overlying soil [Unit 3 (**b**)]. The resence of marine fauna, articulated bivalves (**c**) and marine sand indicate a marine origin for the deposit. Although the deposit is found 2 km inland and up to 2.4 m above sea-level, a lack of definitive evidence for tsunami or storm has limited the usefulness of the implications (after Switzer et al. 2011)

information is used as a basis for signage, evacuation maps, and emergency planning, that lives may be saved in future events of a similar magnitude.

If one was to step away from the scientific arguments of storm versus tsunami and look at the basic premise of coastal hazard and planning assessment then a deposit's genesis becomes less relevant. The pure fact that the 'marine' deposit is found in an overwash setting yields basic information on the recurrence of 'large marine overwash events'. That in itself is of key importance for future coastal and hazard mapping and planning to coastlines that have short historical records.

Fig. 25.8 Tsunami deposits from below the Sendei Plain in central Japan. The sandy deposits (highlighted in *yellow boxes*) at this site (location of the Tohoku earthquake and tsunami) clearly showed that the Sendei Plain had experienced very large tsunamis in the past. With the last being the Jogan tsunami of year AD869. Coastal planning that acted on this knowledge may have saved many lives during the event. Some industries (e.g. the nuclear industry) chose to downplay the importance of such work (photos by Jody Bourgeois)

25.6 Summary

The Earth's coastal environments are dynamic, diverse and under constant change. This physical setting coupled with increasing coastal population growth, infrastructure development and the resulting escalating value of coastal property has lead to considerable divergence of opinion around how we value the coast as a place of community space. The desire to protect the natural character and cultural values of the coast must be weighed against the desire for development. In many parts of the world we must also address the potential for coastal hazards including the impacts of storms and tsunami. Such debates will be influenced by public risk perception framed around regional government planning and the demand for development. Technical experts such as engineers, geomorphologists, modelers and risk managers can contribute to the task, but often this contribution can confound the problem through differences in approach, and conflicts in contextual paradigms. In our management of coastal hazards we currently work in a paradigm of building on experiences learnt from past 'mistakes'. Such a paradigm is no longer valid and integrated coastal hazard assessment is no longer bound by historical experiences. The integration of geological and historical records with computer modeling and foresight analysis means that coastal communities are increasingly better equipped to actively manage coastal hazards. Recent coastal disasters and a global population with increasing technological savvy and access to information means that the general public will likely be participating in future coastal planning discussions and or lobbying for coastal hazard mitigation more actively. Only then will coastal hazards be adequately managed.

Acknowledgments This research is partially supported by the National Research Foundation Singapore under its Singapore NRF Fellowship scheme (National Research Fellow Award No. NRF-RF2010-04) and administered by Earth Observatory of Singapore, we appreciate suggestions from the reviewer for improving the manuscript, and we also thank SAGE, Springer and Wiley for permission of reuse of figures.

References

Atwater, B.F., Musumi-Rokkaku, S., Satake, K., Tsuji, Y., Ueda, K., Yamaguchi, D.: The Orphan Tsunami of 1700: 2005 Japanese clues to a parent earthquake in North America. Prof. Pap. U.S. Geol. Surv. **1707**, 3–123 (2005)

Bondevik, S., Løvholt, F., Harbitz, C., Mangerud, J., Dawson, A., Inge Svendsen, J.: The storegga slide tsunami—comparing field observations with numerical simulations. Mar. Pet. Geol. **22**(1–2), 195–208 (2005). doi:10.1016/j.marpetgeo.2004.10.003

Choowong, M., et al.: Flow conditions of the 2004 Indian ocean tsunami in Thailand, inferred from capping bedforms and sedimentary structures. Terra Nova **20**(2), 141–149 (2008). doi:10.1111/j.1365-3121.2008.00799.x

Donnelly, J.P., Roll, S., Wengren, M., Butler, J., Lederer, R., Webb, T.: Sedimentary evidence of intense hurricane strikes from New Jersey. Geology **29**(7), 615–618 (2001). doi: 10.1130/0091-7613(2001)029<0615:seoihs>2.0.co;2

Fritz, H.M., Blount, C.D., Thwin, S., Thu, M.K., Chan, N.: Cyclone Nargis storm surge in Myanmar. Nat. Geosci. **2**(7), 448–449 (2009). doi: http://www.nature.com/ngeo/journal/v2/n7/suppinfo/ngeo558_S1.html

Goff, J., Chagué-Goff, C., Nichol, S.: Palaeotsunami deposits: a New Zealand perspective. Sediment. Geol. **143**(1–2), 1–6 (2001). doi:10.1016/s0037-0738(01)00121-x

Jaffe, B.E., Gelfenbuam, G.: A simple model for calculating tsunami flow speed from tsunami deposits. Sed. Geol. **200**(3–4), 347–361 (2007). doi:10.1016/j.sedgeo.2007.01.013

Jankaew, K., Atwater, B.F., Sawai, Y., Choowong, M., Charoentitirat, T., Martin, M.E., Prendergast, A.: Medieval forewarning of the 2004 Indian Ocean tsunami in Thailand. Nature **455**(7217), 1228–1231 (2008). doi: http://www.nature.com/nature/journal/v455/n7217/suppinfo/nature07373_S1.html

Kelsey, H.M., Nelson, A.R., Hemphill-Haley, E., Witter, R.C.: Tsunami history of an Oregon coastal lake reveals a 4600 yr record of great earthquakes on the Cascadia subduction zone. Geol. Soc. Am. Bull. **117**(7–8), 1009–1032 (2005). doi:10.1130/b25452.1

Lau, A.Y.A., Switzer, A.D., Dominey-Howes, D., Aitchison, J.C., Zong, Y.: Written records of historical tsunamis in the northeastern South China sea—challenges associated with developing a new integrated database. Nat. Hazards Earth Syst. Sci. **10**(9), 1793–1806 (2010). doi:10.5194/nhess-10-1793-2010

Lee Y, Yu, F., Switzer, A.D., Lau, A.Y., Terry, J., Gouramanis, C.: Developing a historical typhoon database for the southeastern Chinese coastal provinces, 1951–2010. In: Proceedings of the Annual International Conference on Geological & Earth Sciences, pp. 8–12 (2012). doi: 10.5176/2251-3361_GEOS12.10

Liu, K.-b., Fearn, M.L.: Reconstruction of prehistoric landfall frequencies of Catastrophic Hurricanes in Northwestern Florida from lake sediment records. Quat. Res. **54**(2), 238–245 (2000). doi: 10.1006/qres.2000.2166

Liu, K.-b., Shen, C., Louie, K.-s.: A 1,000-year history of Typhoon landfalls in Guangdong, Southern China, reconstructed from Chinese historical documentary records. Ann. Ass. Am. Geogr. **91**(3), 453–464 (2001)

Louie, K.-s., Liu, K.-b.: Earliest historical records of typhoons in China. J. Hist. Geogr. **29**(3), 299–316 (2003). doi: 10.1006/jhge.2002.0453

Minora, K., Imamura, F., Sugawara, D., Kono, Y., Iwashita, T.: The 869 Jogan tsunami deposit and recurrence interval of large-scale tsunami on the Pacific coast of northeast Japan. J. Nat. Disaster Sci. **23**, 83–88 (2001)

Nanayama, F., Satake, K., Furukawa, R., Shimokawa, K., Atwater, B.F., Shigeno, K., Yamaki, S.: Unusually large earthquakes inferred from tsunami deposits along the Kuril trench. Nature **424**(6949), 660–663 (2003). doi: http://www.nature.com/nature/journal/v424/n6949/suppinfo/nature01864_S1.html

Nandasena, N.A.K., Paris, R., Tanaka, N.: Reassessment of hydrodynamic equations: minimum flow velocity to initiate boulder transport by high energy events (storms, tsunamis). Mar. Geol. **281**(1–4), 70–84 (2011). doi:10.1016/j.margeo.2011.02.005

Nott, J.: Palaeotempestology: the study of prehistoric tropical cyclones—a review and implications for hazard assessment. Environ. Int. **30**(3), 433–447 (2004). doi:10.1016/j.envint.2003.09.010

Nott, J.: A 6000 year tropical cyclone record from Western Australia. Quat. Sci. Rev. **30**(5–6), 713–722 (2011). doi:10.1016/j.quascirev.2010.12.004

Okal, E.A., et al.: Field survey of the Samoa Tsunami of 29 September 2009. Seismol. Res. Lett. **81**(4), 577–591 (2010). doi:10.1785/gssrl.81.4.577

Paris, R., Lavigne, F., Wassmer, P., Sartohadi, J.: Coastal sedimentation associated with the December 26, 2004 tsunami in Lhok Nga, west Banda Aceh (Sumatra, Indonesia). Mar. Geol. **238**(1–4), 93–106 (2007). doi:10.1016/j.margeo.2006.12.009

Paris, R., Naylor, L.A., Stephenson, W.J.: Boulders as a signature of storms on rock coasts. Mar. Geol. **283**(1–4), 1–11 (2011). doi:10.1016/j.margeo.2011.03.016

Phantuwongraj, S., Choowong, M.: Tsunamis versus storm deposits from Thailand. Nat. Hazards. 1–20 (2011). doi: 10.1007/s11069-011-9717-8

Switzer, A.D., Jones, B.G.: Setup, deposition, and sedimentary characteristics of two storm overwash deposits, Abrahams Bosom Beach Southeastern Australia. J. Coast. Res. **24**, 189–200 (2008a). doi:10.2112/05-0487.1

Switzer, A.D., Jones, B.G.: Large-scale washover sedimentation in a freshwater lagoon from the southeast Australian coast: sea-level change, tsunami or exceptionally large storm? The Holocene **18**(5), 787–803 (2008b). doi:10.1177/0959683608089214

Switzer, A.D., Burston, J.M.: Competing mechanisms for boulder deposition on the southeast Australian coast. Geomorphology **114**(1–2), 42–54 (2010). doi:10.1016/j.geomorph.2009.02.009

Switzer, A.D., Pucillo, K., Haredy, R.A., Jones, B.G., Bryant, E.A.: Sea level, storm, or tsunami: enigmatic sand sheet deposits in a sheltered coastal embayment from Southeastern New South Wales, Australia. J. Coast. Res. 655–663 (2005). doi:10.2112/04-0177.1

Switzer, A.D., Mamo, B.L., Dominey-Howes, D., Strotz, L.C., Courtney, C., Jones, B.G., Haslett, S.K., Everett, D.M.: On the possible origins of an unusual (mid to late Holocene) coastal deposit, Old Punt Bay, South-East Australia. Geogr. Res. **49**(4), 408–430 (2011). doi:10.1111/j.1745-5871.2011.00700.x

Terry, J., Etienne, S.: "Stones from the dangerous winds": reef platform mega-clasts in the tropical Pacific Islands. Nat. Hazards **56**(3), 567–569 (2011). doi:10.1007/s11069-010-9697-0
Woodruff, J.D., Donnelly, J.P., Mohrig, D., Geyer, W.R.: Reconstructing relative flooding intensities responsible for hurricane-induced deposits from Laguna Playa Grande Vieques, Puerto Rico. Geology **36**(5), 391–394 (2008). doi:10.1130/g24731a.1
Woodruff, J.D., Donnelly, J.P., Okusu, A.: Exploring typhoon variability over the mid-to-late Holocene: evidence of extreme coastal flooding from Kamikoshiki, Japan. Quat. Sci. Rev. **28**(17–18), 1774–1785 (2009). doi:10.1016/j.quascirev.2009.02.005

Chapter 26
A Typhoon Disaster Loss Evaluation System Based on Multi-models

AiMin Yang, GuangJun Sui, DanLing Tang, He Chen and JiangHao Lin

Abstract Typhoon disasters cause serious economic losses and human casualties every year. An accurate evaluation system for typhoon disaster losses is the prerequisite for preventing and reducing the consequences of the disasters. In this chapter, a multi-model system is introduced. Based on the databases of typhoon information, this system establishes a comprehensive evaluation model which is capable of evaluating the possible losses before, during, and after typhoon disasters and automatically generating disaster ranks. Accordingly, a report consisting of disaster losses and coping strategies is proposed by this system. This chapter illustrates the functional framework, the database design, the model construction, the evaluation methods and the generating report of the system. Besides, partial demonstrations and experimental results are also provided.

Keywords Typhoon disaster loss evaluation · Multi-model system · Typhoon databases · Disaster ranks · Strategies and measures · Coping strategy report

26.1 Introduction

Typhoon disaster is one of serious natural disasters which may easily bring heavy economic losses and casualties (Emanuel 1987; Knutson et al. 1998). By evaluating typhoon disaster losses, people can effectively launch the typhoon disaster management so as to prevent and reduce disaster losses. Therefore, an in-depth

A. M. Yang (✉) · G. J. Sui · H. Chen · J. H. Lin
Guangdong University of Foreign Studies, Guangzhou, China
e-mail: amyang18@163.com

D. L. Tang
South China Sea Institute of Oceanology, Chinese Academy of Sciences, Guangzhou, China

D. L. Tang and G. J. Sui (eds.), *Typhoon Impact and Crisis Management*, Advances in Natural and Technological Hazards Research 40,
DOI: 10.1007/978-3-642-40695-9_26, © Springer-Verlag Berlin Heidelberg 2014

study of typhoon disaster loss evaluation is of great significance (Okazaki et al. 2005; Shi et al. 2010). Methods of typhoon disaster loss evaluation include the material analytic method, the experimental analogy method, the mathematics modeling method, and the remote sensing GIS method (Sun 2001). As for the evaluation of single facet of disaster losses, the principal components analysis and the Support Vector Machine(SVM) technology could be adopted to evaluate the number of collapsed houses (Zhang and Lou 2010), while the neural network technology could be applied to evaluate the economic losses of disasters (Zhang et al. 2009). As for the comprehensive evaluation of disaster losses, scholars like Liu (2010) came up with the extension method which computes comprehensive correlation degree to judge disaster ranks and presents the evaluation results on GIS platform (Liu et al. 2010). In addition, some scholars suggested the multi-dimensional linear dependence model to evaluate the loss of China's typhoon disasters (Luĭ and Chen 1998). Others proposed a fuzzy synthetic evaluation method based on remote sensing technique, pattern recognition theory, and SVM, to identify the remote sensing images of ground conditions and evaluate the disaster losses (Hsu and Hong 2007).

Nowadays, single model evaluation is the most widely used method to estimate typhoon disaster losses. However, the evaluation results of this method may not be so precise with regard to some indicators of disaster losses. Therefore, this chapter proposes a multi-model evaluation method for typhoon disasters which integrates several trained evaluation models to a model set. This multi-model evaluation method is to comprehensively evaluate typhoon disaster losses and overcome those drawbacks of single model evaluation method.

This chapter is divided into seven parts. A brief introduction to the function and framework of the system is presented in Sect. 26.2. Section 26.3 is about typhoon database structure and design of data sheets. Section 26.4 describes the construction methodology, the algorithm, the validity as well as the overall assessment method of disaster losses of the evaluation model. Section 26.5 talks about the computational method of disaster ranks. Section 26.6 discusses the generating mechanism of the report for the disaster situation and its coping strategies, followed by the last section, the conclusion and discussion.

26.2 System Function Framework

After reviewing the major evaluation methods and models having been put forward so far, the multi-model evaluation system for typhoon disaster losses is proposed. Based on the detail records of typhoons and regional information in China as training data, this system is designed and realized via computer technology. The design of the system is modularized and its functional framework is shown in Fig. 26.1.

There are mainly five modules in this system, including a database management module, a model building module, a losses evaluation module, a coping strategies

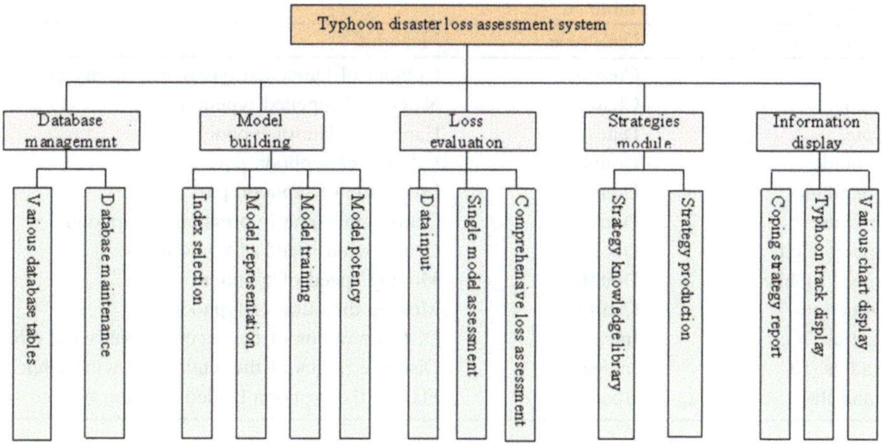

Fig. 26.1 Diagram of system function framework

module and an information display module. The database management module is responsible for the data entry, query, modification, deletion, analysis and provides sample data for the training model. The model building module focuses on the model-building factor selection, the model representation, the model training and the model test etc. The loss evaluation module is about the evaluation condition data input, the single model loss evaluation and the comprehensive assessment. The coping strategies module generates the coping strategies based on comprehensive assessment of losses and the database of strategy knowledge. Information display module relates to the strategy report, the typhoon path analysis, the map and regional information, and the display, save and print of all types of data analysis charts.

26.3 Database Design

Database is the foundation of typhoon disaster loss assessment. The data in the database is based on the typhoon records in mainland of China over the years, which mainly come from Typhoon Yearbook and the Local Economic Yearbook. This database consists of a typhoon track table, a typhoon rainfall table, a typhoon loss data sheet, a geographic information data sheet, a population information table and an economic information data sheet etc. The properties, types and meanings of various data sheets are shown in Tables 26.1, 26.2, 26.3, 26.4, 26.5, 26.6, 26.7 and 26.8. Table 26.9 is some example data of typhoon disaster losses.

Table 26.1 Typhoon path table

Attribute name	Data type	Meaning
Number	Char(6)	Number of happened typhoon (primary key)
Name	Char(20)	Name of happened typhoon
Date	Date	Happened date of typhoon
Latitude	Double	Latitude of typhoon pass
Longitude	Double	Longitude of typhoon pass
Pressure	Double	Centre atmospheric pressure of typhoon
Speed wind	Double	Centre wind speed of typhoon
Speed mobile	Double	Moving speed of typhoon
Direction	Char(10)	Moving direction of typhoon
Radius_7	Integer	Distance radius of the seventh level wind circle
Radius_10	Integer	Distance radius of the tenth level wind circle
Land flag	Boolean	Flag of the typhoon landed position firstly

Table 26.2 Typhoon rainfall table

Attribute name	Data type	Meaning
Number	Char(6)	Number of happened typhoon (primary key)
Name	Char(20)	Name of happened typhoon
Day	Date	Happened date of typhoon
Rainfall	Double	Rainfall of happened typhoon's region
Region name	Char(20)	Name for rainfall region

Table 26.3 Typhoon disaster loss table

Attribute name	Data type	Meaning
Number	Char(6)	Number of happened typhoon (primary key)
Name	Char(20)	Name of happened typhoon
Region name	Char(20)	Name of disaster region
Area	Double	Area of disaster region
Death	Integer	The number of death population
Casualty	Integer	The number of casualty population
Building damage	Integer	The number of damage building
Building collapsed	Integer	The number of collapsed building
Economy loss	Double	Financial loss

Table 26.4 Basic information table with county

Attribute name	Data type	Meaning
County code	Char(10)	District code of county (primary key)
County name	Char(20)	Name of county
Longitude	Double	Longitude of county
Latitude	Double	Latitude of county
Area	Double	Area of county
Description	Text	Description

Table 26.5 Population information table with county

Attribute name	Data type	Meaning
County code	Char(10)	District code of county
County name	Char(20)	Name of county
Total population	Integer	Total population
Male population	Integer	Male population
Female population	Integer	Female population
Non-agricultural population	Integer	Non-agricultural population
Agricultural population	Integer	Agricultural population
Staff population	Integer	Number of fully employed staff and workers
Year	Date	Year

Table 26.6 Economy information table with county

Attribute name	Data type	Meaning
County code	Char(10)	District code of county
County name	Char(20)	Name of county
Gross	Double	Gross domestic product
Primary product	Text	Primary product
Secondary product	Text	Secondary product
Tertiary	Text	Tertiary product
Gross product	Double	Per capita gross domestic product
Gross industry	Double	Gross output value of industry
Gross agriculture	Double	Gross output value of agriculture
Year	Date	Year

Table 26.7 Income information table with county

Attribute name	Data type	Meaning
County code	Char(10)	District code of county
County name	Char(20)	Name of county
Staff	Integer	Number of fully employed staff and workers
State total	Double	State-owned units total wages
State average	Double	State-owned units average wages
Collective total	Double	Urban collective-owned units total wages
Collective average	Double	Urban collective-owned units average wages
Other total	Double	Units of other types of ownership total wages
Other average	Double	Units of other types of ownership average wages
Year	Date	Year

26.4 Model Building and Loss Evaluation

In most cases, one single model or method has been applied to the typhoon disaster evaluation system and it usually fails to capture the evaluation results. This is due to the insufficient data, the complex disaster causal factors and the short disaster period etc. Our research methods adopt multiple models, in order to make a

Table 26.8 Agriculture information table with county

Attribute name	Data type	Meaning
County code	Char(10)	District code of county
County name	Char(20)	Name of county
Grain	Text	Grain description
Rice	Text	Rice description
Grain per	Double	Per capita grain
Sugarcane	Text	Sugarcane description
Peanuts	Text	Peanuts description
Vegetable	Text	Vegetable description
Hogs at	Double	Number of hogs on hand at the year-end
Hogs out	Double	Slaughtered fattened hogs
Pork	Double	Output of pork
Year	Date	Year

Table 26.9 Example data of typhoon disaster loss

Number	Name	Region name	Area	Death	Casualty	Building damage	Economy loss	Building collapsed
200010	Bilis	FuJian	277.5	48	0	36.8	356,000	12.2
200010	Bilis	JiangXi	84.9	5	1,065	1.68	23,000	1.12
200010	Bilis	GuangDong	41.01	9	0	0	60,600	0.43
200010	Bilis	ZheJiang	36.81	0	11	0	23,000	0.15
200012	Prapiroon	JiangSu	990	19	300	0	360,000	5
200012	Prapiroon	ShanDong	54	0	0	0.05	34,000	0.02
200012	Prapiroon	ShangHai	8.1	1	0	0	0	0
200013	Maria	ZheJiang	0	4	0	0	80,000	0.05
200013	Maria	HuNan	334.5	63	770	3.13	134,800	2.37
200013	Maria	GuangDong	182.715	21	1	0.15	128,000	0.71
200016	Wukong	HaiNan	531.6	7	138	6.5	139,800	0.73

comprehensive assessment for typhoon disaster and improve the precision of the assessment. Before introducing the method, some convention symbols are presented as follows for the convenience of describing.

$\mathrm{X}(x_1, x_2, \ldots, x_n)$: X indicates a set of disaster causal factors, $x_i(i = 1, 2, \ldots, n)$ represents one disaster causal factor; X is Partial Attribute and Attribute Transformation from Tables 26.1, 26.2 and 26.4–26.8 in Sect. 26.3.

$\mathrm{Y}(y_1, y_2, \ldots, y_n)$: Y indicates a set of disaster loss indicators, $y_j(j = 1, 2, \ldots, h)$ represents one indicator of disaster losses. Y is Partial Attribute and Attribute Transformation from Table 26.3 in Sect. 26.3.

$\mathrm{D}(d_1, d_2, \ldots, d_n)$: D indicates a set of descriptive factors (not participating in the training model generally), $d_l(l = 1, 2, \ldots, g)$ represents one descriptive factor, D is Partial Attribute from Tables 26.1–26.8 in Sect. 26.3.

$M(m_1, m_2, \ldots, m_p)$: M indicates a set of model building methods, $m_k(k = 1, 2, \ldots, p)$ represents one model building method, such as Partial Least

Squares (PLS), Artificial Neural Network (ANN), Support Vector Machine (SVM), Fuzzy Synthetic Evaluation Method (FSEM), Multiple Linear Interdependent Models (MLIM) and others.

M_set ($model_{j1}$, $model_{j2}$, $model_{jh}$): *M_set* indicates a model set after training, $model_{jq}$ ($q = 1, 2, h_j$) represents the qth model of the jth indicator of disasters losses.

Data_set, Tr_set, Te_set indicates total data set, training data set and test data set respectively. All of them were retrieved from the database mentioned in Sect. 26.3.

26.4.1 Model Building and Training

For gaining a precise result, we adopted multiple methods to set up the model and the training set. These methods include different causal factors, model constructions and model parameters. For a disaster loss y_j ($j = 1, 2, \ldots, h$), its model building and training is shown in Fig. 26.2.

The model construction algorithm is as follows:

Algorithm 1: *The model construction algorithm* Input: Data_set, X, Y, M, acc (Loss evaluation precision threshold), M_count (model number), $j = 1, 2, \ldots, h$

Output: M_set

Procedures:

Step 0: $j = 0$, $t = 0$.

Step 1: selecting the indicators of losses: select y_j from Y.

Step 2: generating the training and testing data sets: some necessary treatments should be done on the Data_set according to specified y_j, and generate Tr_set, Te_set randomly according to proportion.

Step 3: selecting disaster causal factors: use exhaustive method (Guo and Zhang 2010) to select some disaster causal factors.

Step 4: selecting model construction methods: select a model building method from M.

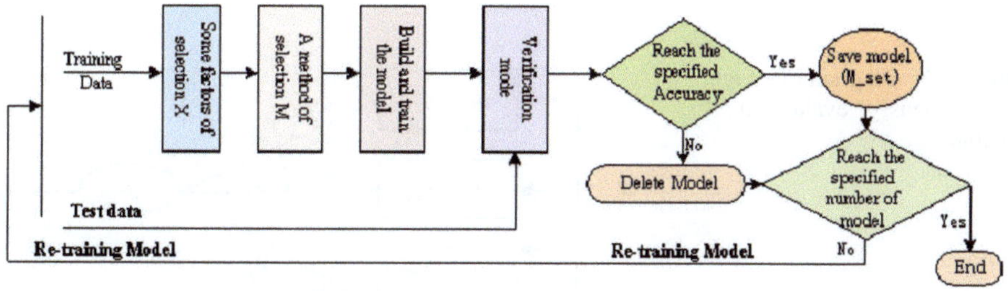

Fig. 26.2 Schemes of model constitute

Step 5: model training: do some training of the model according to Step 3 and 4.
Step 6: testing model: use Te_set data set to test the performance of the trained model and calculate average square-error and E_{jl} of testing date set. The formula is shown as follows:

$$E_{jl} = \frac{\sum_{k=1}^{num\_T} \left(y_{jlk} - \bar{y}_{jlk}\right)^2}{num\_T} \tag{26.1}$$

In this formula, num_T represents the size of testing data set Te_set, $y_{jlk}, \bar{y}_{jlk}$ ($k = 1, 2, \ldots, num\_T$) represents the real value and the model evaluation value of the testing data set.

Step 7: model saving: if the performance of the model meets the requirements (acc), then saved the model to *M_set*, and $t = t + 1$, otherwise delete it.
Step 8: model number checking: if $t >$ M_count and $j <$ h, then $j = j + 1$, $t = 0$, go to Step 1, else if $t <$ M_count, then go to Step 3.
Step 9: the end.

26.4.2 Comprehensive Evaluation Method

As described in Sect. 26.4.1, every y_j (means losses) corresponds to multiple models. When doing assessments, each model will generate an evaluation value. So how to choose a final evaluation value is a vital question to be solved. We designed an integrated evaluation method which specified on the credible weights. The principle of this method is shown in Fig. 26.3.

In Fig. 26.3, x refers to an input vector of the evaluation of the disaster causal factors. Indicator y_j has h models. Each model will generate an evaluation value, such as y_{j1}, y_{j2}, etc.; w_{j1}, $w_{j2}\ldots$ and w_{jh} refers to the credible weight of each model; f represents the integrated evaluation function, $\bar{y}_j$ indicates the final evaluation value. Results can be calculated by the following formula

$$f\left(y_{j1}, y_{j2}, \ldots, y_{yh}\right) \rightarrow \bar{y}_j = \sum_{l=1}^{h} w_{jl} \times y_{jl} \tag{26.2}$$

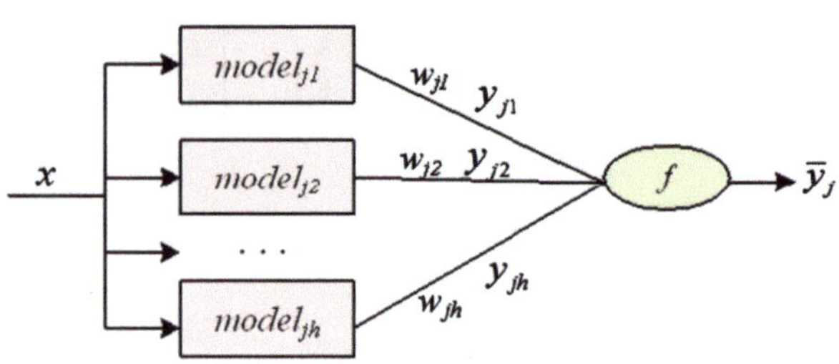

Fig. 26.3 Diagram of weight comprehensive evaluation method

The credible weight w_{jl} can be obtained through the following formula

$$w_{jl} = \frac{1/E_{jl}}{\sum_{l=1}^{h} 1/E_{jl}} \tag{26.3}$$

E_{jl} represents the mean square-error of the model evaluation ($l = 1, 2\ldots h$).

26.5 Calculation Method of Disaster Rank

We can assess the values of the indicators about the disaster losses by the methods indicated in Sects. 26.4.1 and 26.4.2, but these values could not show to the public a notion about the size of a disaster. Then, it's necessary to analyze the rank of a disaster. According to The Trial Regulation about Meteorological Disaster Investigation and Assessment, published by China Meteorological Forecasting and Disaster Mitigation Bureau in 2006, the ranks of disasters can be defined as:

- Oversize disasters: more than 100 people died, or more than 300 casualties, or more than a direct economic loss about one billion;
- Large-scale disasters: the number of the death between 100 and 300, or the economic loss between 100 million and one billion;
- Medium-sized disasters: the number of the death between 3 and 30, or the number of casualties between 30 and 100, or the economic loss between 10 million and 100 million;
- Small disasters: the number of the death between 1 and 3, or the number of casualties between 10 and 30, or the economic loss between 1 million and 10 million;
- Smaller disasters: no personnel death, or the number of casualties below 10, or direct economic loss below one million.

As we know, most of the typhoon disasters cause heavy casualties, as well as serious and wide range of damages. Many experts keep the fundamental features of typhoon disaster in mind and various studies, e.g. the one from (Zhao and Ma 1993) etc., have been conducted to assess the quality of the evaluation typhoon disaster. Then we adopt five indicators as damage area (y_1), deaths (y_2), injured population (y_3), collapsed buildings (y_4) and direct economic losses (y_5) to evaluate typhoon disasters with five ranks. Details are presented in Table 26.10.

The selected indicators of disaster losses are dimensional and different from each other. Therefore, we firstly convert each single indicator of disaster loss to ensure that the range of each transferred indicators within 0 and 1. After using the conversing function, the relation between the single indicator value of disaster level and the single indicator value of conversion function are: oversize disaster (0.8, 1), large-size disaster (0.6, 0.8), medium-size disaster (0.4, 0.6), small disaster (0.2, 0.4), smaller disaster (0, 0.2). By this method, we could reduce the

Table 26.10 Disaster grade index of standards

Indicators	Oversize disaster	Large-scale disaster	Medium-scale disaster	Small disaster	Smaller disaster
$y_1/10^4$ hm^2	[1000, +∞)	[500, 1000)	[200, 500)	[50, 200)	[0, 50)
y_2人	[100, +∞)	[20, 100)	[10, 20)	[0, 10)	0
y_3人	[2000, +∞)	[500, 2000)	[100, 500)	[10, 100)	[1, 10)
$y_4/10^4$ *houses*	[10, +∞)	[5, 10)	[1, 5)	[0.1, 1)	[0, 0.1)
$y_5/10^4$ yuan	[10^6, +∞)	[3×10^5, 10^6)	[10^5, 3×10^5)	[10^4, 10^5)	[0, 10^4)
Disaster level	5 (0.8–1.0)	4 (0.6–0.8)	3 (0.4–0.6)	2 (0.2–0.4)	1 (0–0.2)

difference between each indicator in the same dimension for different units, and normalize functions by transforming the large span of different diversions into 0 to 1. Then we could expediently do correlation analysis. It is proposed that a group value of losses is $\bar{y}_i$ $(i = 1, 2, \ldots, 5)$, then we could use linear conversion function to transfer $\bar{y}_i$ into (0, 1) in our system. The formula is as follow

$$U(\bar{y}_i) = \frac{(\bar{y}_i - min(y_{i,k})) \times 0.2}{max(y_{i,k}) - min(y_{i,k})} + G(k) \tag{26.4}$$

It may be used to define a level for a typhoon disaster. Variables in formula (26.4) are: i (i = 1, 2,…, 5) indicates the array number of loss indicator; k (k = 1, 2,…, 5) indicates the array number of disaster level; min (y_i, k), max (y_i, k) represents minimum and maximum of the kth level. G (k) refers the lower limit function of relevant level, and the oversize disaster lower limit function is G (5) = 0.8, for example. With the help of i, $\bar{y}_i$ and figures in Table 26.9, we could define one disaster's level (suppose it was k). After formula (26.4) had been applied, the indicators of the typhoon disaster were converted into an unified indicator signal that is comparable to each other. The specific standard indicators of disaster level indicator, the typhoon disaster intensity level and the relation of conversion function are shown in Table 26.9.

In a word, the transformation of the indictor of the typhoon disaster loss to dimensionless number between 0 and 1 and Grey Relational Analysis Method is applied in order to carry out comprehensive evaluation of the typhoon disaster losses (Wu et al. 2009). More specific calculation methods can be explained via the statement below.

Let reference array $U_0 = (U_{0i})$, $U_{0i} = 1$ $(i = 1, 2, \ldots, 5)$ as the standard of oversize disaster. Comparable array is $U(U_{(\bar{y}_i)})(i = 1, 2, \ldots, 5)$. To compare the disaster level via relevancy, the value array U should at first be compared with its related level between other level disasters. Comparative sequence and reference sequence correlation coefficient between various indicators are shown as 26.5 below.

$$\xi_{0i} = \frac{1}{1 + \Delta_{0i}} \tag{26.5}$$

Table 26.11 Disaster grade index of standards

Disaster level	Micro disaster	Small disaster	Medium-scale disaster	Large-scale disaster	Oversize disaster
Correlation degree r	[0.5, 0.6)	[0.6, 0.7)	[0.7, 0.8)	[0.8, 0.9)	[0.9, 1.0)

In this formula, $\Delta_{0i} = \left|U_0(U_{0i}) - U\left(U_{(y_i)}\right)\right|, (i = 1, 2, \ldots, 5)$ refers to the absolute difference between the value of the ith indicator and the absolute value between reference sequence U_0 and comparative sequence U. The larger the Δ_{0i} is, the bigger the distance would be between single indicator and the same term indicator of reference sequence, then the little the relation coefficient ξ_{0j}. On the opposite side, the little the Δ_{0i} is, the less distance between single indicator and the same term indicator of reference sequence, then the bigger the relation coefficient ξ_{0j}. If $\Delta_{0i} \in [0, 1]$, then $\xi_{0i} \in [0.5, 1]$.

We could get the coefficient value of the association between comparative sequence and reference sequence from formula (26.5). We use five indicators in our system, and in our system we adopt the average method such as processing power. Then we attain the average value r of the sum of the coefficient value of association between comparative sequence and reference sequence. The r reflects series and reference sequence of associate degree as formula (26.6) below

$$r = \frac{1}{5}\sum_{i=1}^{5} \xi_{0i} \tag{26.6}$$

The larger Correlation degree r ($r \in [0.5, 1]$) is, the more serious the disaster would be; on the opposite side, the smaller r is, the less serious the disaster would be. It can receive sequence and references in the sequence of indexes of total correlation coefficient on average and reflect series and reference sequence of associate degree as Table 26.11.

26.6 Report of the Disaster and Countermeasures

In Sect. 26.4, the system can assess the losses and evaluate disaster level according to the disaster—causing factors input. But it is necessary to report the information of the situation and investigate the strategies in research about how to play a more significant role in disaster prevention and reduction, and how to make these information quickly and conveniently understood by the relevant personnel. Our system framework shown in Fig. 26.4 produces the disaster and the measures reported.

As it is shown in the Fig. 26.4, assessment selection model, which is based on the input information, assesses and obtains disaster damage assessment by rating. It also evaluates with coping strategies from the knowledge base. The information

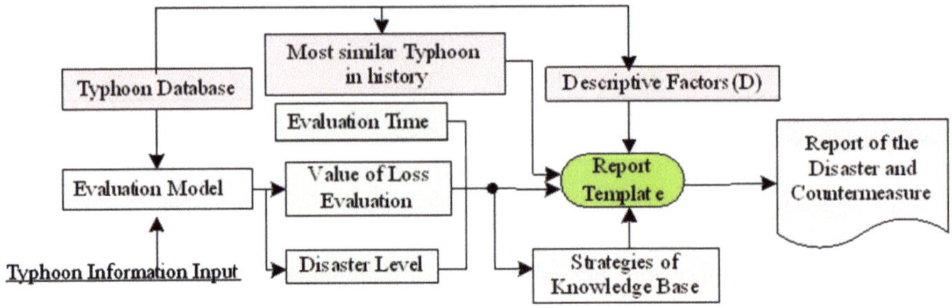

Fig. 26.4 Schematic diagram of the disaster and the measures reported forming

can be put into the report template to generate relevant information. It is necessary to research the typhoon from the database to find the most similar historical typhoon and typhoon areas of geography, economy, population and other descriptive information and send them into the report template. In addition, the typhoon path, disaster areas and other image information are also sent to the report template to generate a complete report on the disaster and countermeasures. In Fig. 26.4, the coping strategies knowledge that matched in the database are shown in Table 26.12.

Table 26.11 in the evaluation period are: (1) 12 h before landfall, (2) 6 h before landfall, (3) the process of typhoon after landfall, (4) the end of typhoon; Disaster levels: (1) Micro disaster, (2) Small disaster, (3) Medium-scale disaster, (4) Large-scale disaster, (5) Oversize disaster; strategy types described in Table 26.13.

Disaster and countermeasures report template is presented in Table 26.14.

The typhoon is a kind of natural weather activity which may bring great disasters. Thus, to investigate the occurrence and development of typhoon, the consequences of typhoon disasters and the measures to reduce disaster losses in further studies are of great significance. Our research attempts to construct databases, evaluation models and provide rapid decision making via computer technology.

The characteristics of this evaluation system possess the following three aspects: (1) Construction of multi-type databases, including not only the data of the typhoon path and disaster losses but also data from geography, industry, agriculture and population, all of which lay down a solid foundation to the evaluation of typhoon disasters. (2) Adoption of a multi-model approach to comprehensively evaluate the disaster losses, which overcomes the drawbacks of the

Table 26.12 Strategies of knowledge fields

Attribute name	Data type	Meaning
Id	Integer	Number of rule
Phase	Integer	Assessment period
Disaster level	Integer	Hazard level
Policy type	Integer	Strategy category
Policy	Text	Strategy description
Note	Text	Remarks

Table 26.13 Strategy types

Policy type	Meaning	Policy type	Meaning
1	General description of disaster	13	Information can be communicated measures
2	Measures against strong winds	14	Rescue measures
3	Measures against strong winds	15	Salvage measures
4	Measures against the tide	16	Water protection measures
5	Measures to prevent water logging	17	Emergency rescue measures
6	Measures to prevent flash floods	18	Resettlement measures for victims
7	Debris flow early warning measures	19	Measures for insurance claims
8	Personnel transfer measures	20	Property protection measures at sea
9	Personnel evacuation measures	21	Water logging measures row
10	Property protection measures	22	Organize and conduct the security measures
11	Power protection	23	Participate in department, division, and information
12	Communications security measures		–

Table 26.14 Disaster and countermeasures report template

Report of the typhoon disaster and countermeasures

Report generation time:

Operator name:

1. *Typhoon basic information*

Typhoon name:

Typhoon features:

Typhoon Time:

Typhoon location information:

2. *Typhoon disaster assessment information*

Assessment period:

Assessment of causal factors input:

Information on disaster losses may:

Disaster level:

Acquaintance with the history of the typhoon:

Damage assessment model and method of use described:

3. *Disaster prevention and mitigation measures*

(Knowledge from the coping strategies)

4. *Typhoon-related information to describe*

Description of geographic information:

Described the economic situation:

Description of population:

Description of industry:

Description of agricultural information:

5. *All kinds of pictures*

Typhoon map:

Regional plans may be affected:

single-model evaluation and improves the stability and precision of the evaluation value. (3) Provision of the evaluation value and disaster ranks as well as the automatic generation of the coping strategies, which offers important evidence to prevent and reduce the consequences of the typhoon disasters.

Nonetheless, due to the lack of historical data of typhoon disasters and the complex disaster causal factors and internal processes, the evaluation of the typhoon disaster losses became a great challenge to the scholars. It is necessary to explore new ways with regards to theoretical methods and database construction.

Acknowledgments This research was supported by: (1) Guangdong Natural Science Foundation Team Research Project, China, "Hazard Evaluation on Typhoons and Examination on Their Biological Effect in the South China Coastal Area Based on Satellite Remote Sensing" (8351030101000002). (2) Science and Technology Planning Project of Guangdong Province, China, "Development of Emergency Response System of Major Typhoon Disaster and Urban Fire Disaster" (2010B031900041). (3) Science and Technology Planning Project of Guangdong Province, China, "Research on Real Classification System of Peer-to-Peer Network Traffic based on Machine Learning Method" (2009B080701031).

References

Emanuel, K.A.: The dependence of hurricane intensity on climate. Nature **326**(6112), 483–485 (1987)

Guo, J.S., Zhang, F.: An equivalent key attack on an image cryptosystem. Acta Electronica Sin. **38**(004), 781–785 (2010)

Hsu, C.C., Hong, Z.Y.: An intelligent typhoon damage prediction system from aerial photographs. Knowledge-Based Intelligent Information and Engineering Systems Lecture Notes in Computer Science, vol. 4692, pp. 747–756 (2007)

Knutson, T.R., Tuleya, R.E., Kurihara, Y.: Simulated increase of hurricane intensities in a CO_2-warmed climate. Science **279**(5353), 1018 (1998)

Liu, S.J., Zhang, J.H., He, Z.W., Cai, D.X., Tian, G.H.: Study on assessment model of typhoon disaster losses based on GIS. J. Catastrophology **25**(002), 64–67 (2010)

Lou, W.P., Chen, H.Y., Zheng, F., Chen, R.: Economic loss assessment of typhoon based on principal component analysis and neural network. Geogr. Res. **28**(5), 1243–1254 (2009)

LuÏ, C.L., Chen, S.H.: Multiple linear interdependent models (MLIM) applied to typhoon data from China. Theor. Appl. Climatol. **61**(3), 143–149 (1998)

Okazaki, T., Watabe, H., Ishihara, T.: Development of typhoon simulation model for insurance risk estimation. Wind Eng. **12–14**, 1790–1802 (2005)

Shi, Y.G., Kou, Y.Y., Li, G.X., Wang, Y., Peng, Y., Shi, Y.: A Fuzzy Multi-Criteria Decision-Making Hybrid Approach to Evaluate the Damage Level of Typhoon: Integration of fuzzy AHP and fuzzy TOPSIS, IEEE, pp. 666–671 (2010)

Sun, S.C.: A study on the contents and methods of disaster assessment. Prog. Geogr. **20**(002), 122–130 (2001)

Wu, H., Chen, D.M., Wu, S.A., Wu, C.K.: Application of grey relation analysis for evaluation of tropical cyclone disaster. Chin. J. Trop. Crops **30**(2), 244–248 (2009)

Zhang, X.C., Lou, W.P.: Support vector machine based on principal component of the number of typhoons between the assessment of houses collapsed. J. Anhui Agric. Sci. **012**, 6339–6341 (2010)

Zhao, A.X., Ma, Z.J.: Appraising study for the loss evaluation system of natural disasters. J. Nat. Disasters **2**(3), 1–7 (1993)

Afterword

We are happy to present this book to you. It is our 2nd Springer book, and together with our 1st Springer book "Remote Sensing of the Changing Oceans", we are pleased to communicate in "the language of science", with a clear understanding of management needs when natural disasters occur.

Loving the sciences and their derived technologies has brought us together. We have studied in different countries, and we have worked in different fields. This book links us together through the natural and social sciences, and we have learned from each other.

Each chapter of this book is about typhoons: the nature of typhoons, typhoon impacts on the environment and ecosystems, and typhoon impacts on people, countries and their economies. Those diverse–but interdependent-typhoon topics have "blown" us even more closely together as a family of investigators of the oceans, the environment, and their sustainable management.

Other things, such as the writing and editing of this book together, has greatly enhanced our knowledge of, and our respect for, the forces of nature. We value this chance to share our research results with you, and dedicate this book "Typhoon Impact and Crisis Management" to all who seek to share knowledge of the forces of nature, for the well-being of future generations.

Guangzhou, China, May 2013 GuangJun Sui
DanLing Tang
Halifax, Canada Yi Sui

D. L. Tang and G. J. Sui (eds.), *Typhoon Impact and Crisis Management*, Advances in Natural and Technological Hazards Research 40,
DOI: 10.1007/978-3-642-40695-9, © Springer-Verlag Berlin Heidelberg 2014

About the Editors

Dr. DanLing (lingzis) Tang, is a professor of the South China Sea Institute of Oceanology, Chinese Academy of Sciences. She is an expert on marine ecosystems and remote sensing of marine ecology. Her major research interests include ocean dynamics of phytoplankton blooms, global environmental changes and impacts of natural hazards on marine ecosystems. She has published approximately 80 papers in international journals. Her book "Remote Sensing of the Changing Oceans" was published in 2011 by Springer. Currently, she is the President of the Pan Ocean Remote Sensing Association (PORSEC), Councilor of the American Geophysical Union (AGU), Vice Chair of Sub-Commission (A2-Ocean Dynamics and Productivity), Committee on Space Research (COASPAR), and The president (elected) of the Board of Directors of PACON International.

Dr. GuangJun Sui, a full professor, is former President of Guangdong University of Foreign Studies (GDUFS), and currently the Chairman of the University Management Committee of GDUFS, and Executive Vice-Director of the Guangdong Research Institute of International Strategy. His major research areas include industrial economy, strategy management and innovation and crisis management. He has published ten academic works including a Book Series on Case Studies of Management, Guangdong at the Turning Point, and he is the author to over 80 papers in academic journals such as Management World, and Reform.

D. L. Tang and G. J. Sui (eds.), *Typhoon Impact and Crisis Management*,
Advances in Natural and Technological Hazards Research 40,
DOI: 10.1007/978-3-642-40695-9, © Springer-Verlag Berlin Heidelberg 2014

Index

D. L. Tang and G. J. Sui (eds.), *Typhoon Impact and Crisis Management*,
Advances in Natural and Technological Hazards Research 40,
DOI: 10.1007/978-3-642-40695-9, © Springer-Verlag Berlin Heidelberg 2014

D

E

H

I

J

K

L

M

MIX
Papier aus verantwortungsvollen Quellen
Paper from responsible sources
FSC® C105338

If you have any concerns about our products,
you can contact us on
ProductSafety@springernature.com

In case Publisher is established outside the EU,
the EU authorized representative is:
Springer Nature Customer Service Center GmbH
Europaplatz 3, 69115 Heidelberg, Germany

Printed by Libri Plureos GmbH
in Hamburg, Germany